Symmetry in Graph Theory

Symmetry in Graph Theory

Special Issue Editor
Jose Manuel Rodriguez Garcia

MDPI • Basel • Beijing • Wuhan • Barcelona • Belgrade

MDPI

Special Issue Editor
Jose Manuel Rodriguez Garcia
Universidad Carlos III de Madrid
Spain

Editorial Office
MDPI
St. Alban-Anlage 66
4052 Basel, Switzerland

This is a reprint of articles from the Special Issues published online in the open access journal *Symmetry* (ISSN 2073-8994) from 2017 to 2018 (available at: https://www.mdpi.com/journal/symmetry/special issues/Graph Theory, https://www.mdpi.com/journal/symmetry/special issues/Symmetry Graph Theory)

For citation purposes, cite each article independently as indicated on the article page online and as indicated below:

LastName, A.A.; LastName, B.B.; LastName, C.C. Article Title. *Journal Name* **Year**, *Article Number, Page Range*.

ISBN 978-3-03897-658-5 (Pbk)
ISBN 978-3-03897-659-2 (PDF)

Contents

About the Special Issue Editor

Jose M. Rodriguez, Full Professor, received his M.Sc. in Mathematics in 1986 and his Ph.D. in Mathematics in 1991 from the Universidad Autonoma de Madrid, Spain. Since 2011 he has been a Full Professor in the Department of Mathematics at the Universidad Carlos III de Madrid, Spain. He acts as Director of the Programs of Master and PhD in Mathematical Engineering at the Universidad Carlos III de Madrid, Spain. He has published more than 100 papers in Mathematics and its Applications. His current research interests include Geometric Function Theory, Graph Theory, Mathematical Chemistry, Approximation Theory, and Riemannian Geometry.

symmetry

MDPI

Editorial
Graph Theory

Jose M. Rodriguez

Departamento de Matemáticas, Universidad Carlos III de Madrid, Avenida de la Universidad,
30 CP-28911 Leganés, Madrid, Spain; jomaro@math.uc3m.es

Received: 19 January 2018; Accepted: 19 January 2018; Published: 22 January 2018

This book contains the successful invited submissions [1–10] to a special issue of *Symmetry* on the subject area of 'graph theory'.

Although symmetry has always played an important role in graph theory, in recent years, this role has increased significantly in several branches of this field, including, but not limited to: Gromov hyperbolic graphs, metric dimension of graphs, domination theory, and topological indices. This Special issue invites contributions addressing new results on these topics, both from a theoretical and an applied point of view.

This special issue includes the novel techniques and tools for graph theory, such as:

- Local metric dimension of graphs [1].
- Gromov hyperbolicity on geometric graphs [2,3,5].
- Beta-differential of graphs [4].
- Path ordinal method [6].
- Neural networks on multi-centrality-index diagrams [7] and complex networks [8].
- Connectivity indices and movement directions at path segments [9].
- Independent (1, 2)-sets in cylindrical networks [10].

The response to our call had the following statistics:

- Submissions (40);
- Publications (10);
- Rejections (30);
- Article types: Research Article (10);

Our authors' geographical distribution (published papers) is:

- Spain (8)
- Japan (4)
- Mexico (4)
- Austria (2)
- Korea (2)
- Luxembourg (1)
- Poland (1)
- Egypt (1)

Published submissions are related to local metric dimension, Gromov hyperbolicity, differential, path ordinal method, neural networks, connectivity indices, and independent sets, as well as their applications.

We found the edition and selections of papers for this book very inspiring and rewarding. We also thank the editorial staff and reviewers for their efforts and help during the process.

Conflicts of Interest: The author declares no conflict of interest.

Symmetry **2018**, *10*, 32

References

1. Barragán-Ramírez, G.; Estrada-Moreno, A.; Ramírez-Cruz, Y.; Rodríguez-Velázquez, J. The Simultaneous Local Metric Dimension of Graph Families. *Symmetry* **2017**, *9*, 132. [CrossRef]
2. Granados, A.; Pestana, D.; Portilla, A.; Rodríguez, J. Gromov Hyperbolicity in Mycielskian Graphs. *Symmetry* **2017**, *9*, 131. [CrossRef]
3. Martínez-Pérez, Á. Generalized Chordality, Vertex Separators and Hyperbolicity on Graphs. *Symmetry* **2017**, *9*, 199. [CrossRef]
4. Basilio, L.; Bermudo, S.; Leaños, J.; Sigarreta, J. β-Differential of a Graph. *Symmetry* **2017**, *9*, 205. [CrossRef]
5. Hernández-Gómez, J.; Reyes, R.; Rodríguez, J.; Sigarreta, J. Mathematical Properties on the Hyperbolicity of Interval Graphs. *Symmetry* **2017**, *9*, 255. [CrossRef]
6. Kamal, H.; Larena, A.; Bernabeu, E. Analytical Treatment of Higher-Order Graphs: A Path Ordinal Method for Solving Graphs. *Symmetry* **2017**, *9*, 288. [CrossRef]
7. Mizui, Y.; Kojima, T.; Miyagi, S.; Sakai, O. Graphical Classification in Multi-Centrality-Index Diagrams for Complex Chemical Networks. *Symmetry* **2017**, *9*, 309. [CrossRef]
8. Lee, Y.; Sohn, I. Reconstructing Damaged Complex Networks Based on Neural Networks. *Symmetry* **2017**, *9*, 310. [CrossRef]
9. Taczanowska, K.; Bielański, M.; González, L.; Garcia-Massó, X.; Toca-Herrera, J. Analyzing Spatial Behavior of Backcountry Skiers in Mountain Protected Areas Combining GPS Tracking and Graph Theory. *Symmetry* **2017**, *9*, 317. [CrossRef]
10. Carreño, J.; Martínez, J.; Puertas, M. Efficient Location of Resources in Cylindrical Networks. *Symmetry* **2018**, *10*, 24. [CrossRef]

symmetry

MDPI

Article

The Simultaneous Local Metric Dimension of Graph Families

Gabriel A. Barragán-Ramírez [1], Alejandro Estrada-Moreno [1], Yunior Ramírez-Cruz [2,*]
and Juan A. Rodríguez-Velázquez [1]

[1] Departament d'Enginyeria Informàtica i Matemàtiques, Universitat Rovira i Virgili, Av. Països Catalans 26, 43007 Tarragona, Spain; gbrbcn@gmail.com (G.A.B.-R.); alejandro.estrada@urv.cat (A.E.-M.); juanalberto.rodriguez@urv.cat (J.A.R.-V.)
[2] Interdisciplinary Centre for Security, Reliability and Trust, University of Luxembourg, 6 av. de la Fonte, L-4364 Esch-sur-Alzette, Luxembourg
* Correspondence: yunior.ramirez@uni.lu

Received: 10 May 2017; Accepted: 24 July 2017; Published: 27 July 2017

Abstract: In a graph $G = (V, E)$, a vertex $v \in V$ is said to distinguish two vertices x and y if $d_G(v, x) \neq d_G(v, y)$. A set $S \subseteq V$ is said to be a local metric generator for G if any pair of adjacent vertices of G is distinguished by some element of S. A minimum local metric generator is called a local metric basis and its cardinality the local metric dimension of G. A set $S \subseteq V$ is said to be a simultaneous local metric generator for a graph family $\mathcal{G} = \{G_1, G_2, \ldots, G_k\}$, defined on a common vertex set, if it is a local metric generator for every graph of the family. A minimum simultaneous local metric generator is called a simultaneous local metric basis and its cardinality the simultaneous local metric dimension of $\mathcal{G}$. We study the properties of simultaneous local metric generators and bases, obtain closed formulae or tight bounds for the simultaneous local metric dimension of several graph families and analyze the complexity of computing this parameter.

Keywords: local metric dimension; simultaneity; corona product; lexicographic product; complexity

1. Introduction

A generator of a metric space is a set S of points in the space with the property that every point of the space is uniquely determined by its distances from the elements of S. Given a simple and connected graph $G = (V, E)$, we consider the function $d_G : V \times V \to \mathbb{N}$, where $d_G(x, y)$ is the length of the shortest path between u and v and $\mathbb{N}$ is the set of non-negative integers. Clearly, (V, d_G) is a metric space, i.e., d_G satisfies $d_G(x, x) = 0$ for all $x \in V$, $d_G(x, y) = d_G(y, x)$ for all $x, y \in V$ and $d_G(x, y) \leq d_G(x, z) + d_G(z, y)$ for all $x, y, z \in V$. A vertex $v \in V$ is said to distinguish two vertices x and y if $d_G(v, x) \neq d_G(v, y)$. A set $S \subseteq V$ is said to be a metric generator for G if any pair of vertices of G is distinguished by some element of S.

Metric generators were introduced by Blumental [1] in the general context of metric spaces. They were later introduced in the context of graphs by Slater in [2], where metric generators were called locating sets, and, independently, by Harary and Melter in [3], where metric generators were called resolving sets. Applications of the metric dimension to the navigation of robots in networks are discussed in [4] and applications to chemistry in [5,6]. This invariant was studied further in a number of other papers including, for instance [7–20].

As pointed out by Okamoto et al. in [21], there exist applications where only neighboring vertices need to be distinguished. Such applications were the basis for the introduction of the local metric dimension. A set $S \subseteq V$ is said to be a local metric generator for G if any pair of adjacent vertices of G is distinguished by some element of S. A minimum local metric generator is called a local metric basis and its cardinality the local metric dimension of G, denoted by $\dim_l(G)$. Additionally,

Jannesari and Omoomi [16] introduced the concept of adjacency resolving sets as a result of considering the two-distance in $V(G)$, which is defined as $d_{G,2}(u,v) = \min\{d_G(u,v),2\}$ for any two vertices $u,v \in V(G)$. A set of vertices S' such that any pair of vertices of $V(G)$ is distinguished by an element s in S' considering the two-distance in $V(G)$ is called an adjacency generator for G. If we only ask S' to distinguish the pairs of adjacent vertices, we call S' a local adjacency generator. A minimum local adjacency generator is called a local adjacency basis, and the cardinality of any such basis is the local adjacency dimension of G, denoted $\mathrm{adim}_l(G)$.

The notion of simultaneous metric dimension was introduced in the framework of the navigation problem proposed in [4], where navigation was studied in a graph-structured framework in which the navigating agent (which was assumed to be a point robot) moves from node to node of a "graph space". The robot can locate itself by the presence of distinctively-labeled "landmark" nodes in the graph space. On a graph, there is neither the concept of direction, nor that of visibility. Instead, it was assumed in [4] that a robot navigating on a graph can sense the distances to a set of landmarks. Evidently, if the robot knows its distances to a sufficiently large set of landmarks, its position on the graph is uniquely determined. This suggests the following problem: given a graph G, what are the fewest number of landmarks needed and where should they be located, so that the distances to the landmarks uniquely determine the robot's position on G? Indeed, the problem consists of determining the metric dimension and a metric basis of G. Now, consider the following extension of this problem, introduced by Ramírez-Cruz, Oellermann and Rodríguez-Velázquez in [22]. Suppose that the topology of the navigation network may change within a range of possible graphs, say $G_1, G_2, ..., G_k$. This scenario may reflect several situations, for instance the simultaneous use of technologically-differentiated redundant sets of landmarks, the use of a dynamic network whose links change over time, etc. In this case, the above-mentioned problem becomes determining the minimum cardinality of a set S, which must be simultaneously a metric generator for each graph G_i, $i \in \{1, ..., k\}$. Therefore, if S is a solution for this problem, then each robot can be uniquely determined by the distance to the elements of S, regardless of the graph G_i that models the network at each moment. Such sets we called simultaneous metric generators in [22], where, by analogy, a simultaneous metric basis was defined as a simultaneous metric generator of minimum cardinality, and this cardinality was called the simultaneous metric dimension of the graph family $\mathcal{G}$, denoted by $\mathrm{Sd}(\mathcal{G})$.

In this paper, we recover Okamoto et al.'s observation that in some applications, it is only necessary to distinguish neighboring vertices. In particular, we consider the problem of distinguishing neighboring vertices in a multiple topology scenario, so we deal with the problem of finding the minimum cardinality of a set S, which must simultaneously be a local metric generator for each graph G_i, $i \in \{1, ..., k\}$.

Given a family $\mathcal{G} = \{G_1, G_2, ..., G_k\}$ of connected graphs $G_i = (V, E_i)$ on a common vertex set V, we define a simultaneous local metric generator for $\mathcal{G}$ as a set $S \subseteq V$ such that S is simultaneously a local metric generator for each G_i. We say that a minimum simultaneous local metric generator for $\mathcal{G}$ is a simultaneous local metric basis of $\mathcal{G}$ and its cardinality the simultaneous local metric dimension of $\mathcal{G}$, denoted by $\mathrm{Sd}_l(\mathcal{G})$ or explicitly by $\mathrm{Sd}_l(G_1, G_2, ..., G_k)$. An example is shown in Figure 1, where the set $\{v_3, v_4\}$ is a simultaneous local metric basis of $\{G_1, G_2, G_3\}$.

It will also be useful to define the simultaneous local adjacency dimension of a family $\mathcal{G} = \{G_1, G_2, \ldots, G_k\}$ of connected graphs $G_i = (V, E_i)$ on a common vertex set V, as the cardinality of a minimum set $S \subseteq V$ such that S is simultaneously a local adjacency generator for each G_i. We denote this parameter as $\mathrm{Sad}_l \mathcal{G}$.

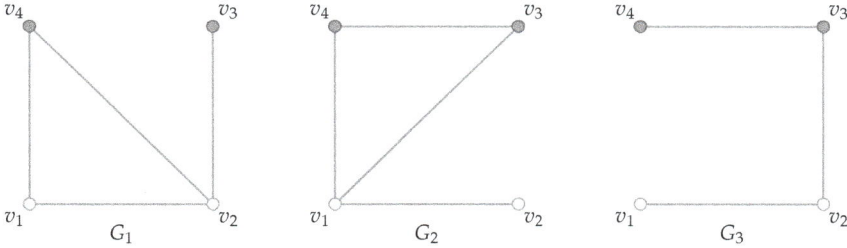

Figure 1. The set $\{v_3, v_4\}$ is a simultaneous local metric basis of $\{G_1, G_2, G_3\}$. Thus, $\mathrm{Sd}_l(G_1, G_2, G_3) = 2$.

In what follows, we will use the notation K_n, $K_{r,s}$, C_n, N_n and P_n for complete graphs, complete bipartite graphs, cycle graphs, empty graphs and path graphs of order n, respectively. Given a graph $G = (V, E)$ and a vertex $v \in V$, the set $N_G(v) = \{u \in V : u \sim v\}$ is the open neighborhood of v, and the set $N_G[v] = N_G(v) \cup \{v\}$ is the closed neighborhood of v. Two vertices $x, y \in V(G)$ are true twins in G if $N_G[x] = N_G[y]$, and they are false twins if $N_G(x) = N_G(y)$. In general, two vertices are said to be twins if they are true twins or they are false twins. As usual, a set $A \subseteq V(G)$ is a vertex cover for G if for every $uv \in E(G)$, $u \in A$ or $v \in A$. The vertex cover number of G, denoted by $\beta(G)$, is the minimum cardinality of a vertex cover of G. The remaining definitions will be given the first time that the concept appears in the text.

The rest of the article is organized as follows. In Section 2, we obtain some general results on the simultaneous local metric dimension of graph families. Section 3 is devoted to the case of graph families obtained by small changes on a graph, while in Sections 4 and 5, we study the particular cases of families of corona graphs and families of lexicographic product graphs, respectively. Finally, in Section 6, we show that the problem of computing the simultaneous local metric dimension of graph families is NP-hard, even when restricted to families of graphs that individually have a (small) fixed local metric dimension.

2. Basic Results

Remark 1. *Let $\mathcal{G} = \{G_1, \ldots, G_k\}$ be a family of connected graphs defined on a common vertex set V, and let $G' = (V, \cup E(G_i))$. The following results hold:*

1. $\mathrm{Sd}_l(\mathcal{G}) \geq \max\limits_{i \in \{1, \ldots, k\}} \{\dim_l(G_i)\}.$
2. $\mathrm{Sd}_l(\mathcal{G}) \leq \mathrm{Sd}(\mathcal{G}).$
3. $\mathrm{Sd}_l(\mathcal{G}) \leq \min \left\{ \beta(G'), \sum\limits_{i=1}^{k} \dim_l(G_i) \right\}.$

Proof. (1) is deduced directly from the definition of simultaneous local metric dimension. Let B be a simultaneous metric basis of $\mathcal{G}$, and let $u, v \in V - B$ be two vertices not in B, such that $u \sim_{G_i} v$ in some G_i. Since in G_i there exists $x \in B$ such that $d_{G_i}(u, x) \neq d_{G_i}(v, x)$, B is a simultaneous local metric generator for $\mathcal{G}$, so (2) holds. Finally, (3) is obtained from the following facts: (a) the union of local metric generators for all graphs in $\mathcal{G}$ is a simultaneous local metric generator for $\mathcal{G}$, which implies that $\mathrm{Sd}_l(\mathcal{G}) \leq \sum_{i=1}^{k} \dim_l(G_i)$; (b) any vertex cover of G' is a local metric generator of G_i, for every $G_i \in \mathcal{G}$, which implies that $\mathrm{Sd}_l(\mathcal{G}) \leq \beta(G')$. □

The inequalities above are tight. For example, the graph family $\mathcal{G}$ shown in Figure 1 satisfies $\mathrm{Sd}_l(\mathcal{G}) = \mathrm{Sd}(\mathcal{G})$, whereas $\mathrm{Sd}_l(\mathcal{G}) = 2 = \dim_l(G_1) = \dim_l(G_2) = \max\limits_{i \in \{1,2,3\}} \{\dim_l(G_i)\}$. Moreover, the family $\mathcal{G}$ shown in Figure 2 satisfies $\mathrm{Sd}_l(\mathcal{G}) = 3 = |V| - 1 < \sum\limits_{i=1}^{6} \dim_l(G_i) = 12$, whereas the family $\mathcal{G} = \{G_1, G_2\}$ shown in Figure 3 satisfies $\mathrm{Sd}_l(\mathcal{G}) = 4 = \dim_l(G_1) + \dim_l(G_2) < |V| - 1 = 7$.

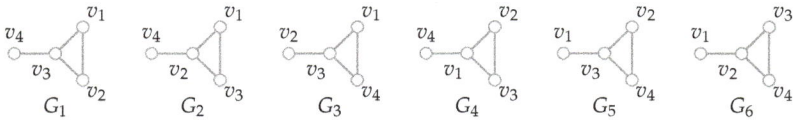

Figure 2. The family $\mathcal{G} = \{G_1, \ldots, G_6\}$ satisfies $\mathrm{Sd}_l(\mathcal{G}) = |V| - 1 = 3$.

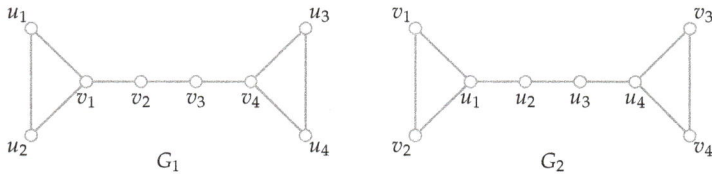

Figure 3. The family $\mathcal{G} = \{G_1, G_2\}$ satisfies $\mathrm{Sd}_l(\mathcal{G}) = \dim_l(G_1) + \dim_l(G_2) = 4$.

We now analyze the extreme cases of the bounds given in Remark 1.

Corollary 1. *Let $\mathcal{G}$ be a family of connected graphs on a common vertex set. If $K_n \in \mathcal{G}$, then:*

$$\mathrm{Sd}_l(\mathcal{G}) = n - 1.$$

As shown in Figure 2, the converse of Corollary 1 does not hold. In general, the cases for which the upper bound $\mathrm{Sd}_l(\mathcal{G}) \leq |V| - 1$ is reached are summarized in the next result.

Theorem 1. *Let $\mathcal{G}$ be a family of connected graphs on a common vertex set V. Then, $\mathrm{Sd}_l(\mathcal{G}) = |V| - 1$ if and only if for every $u, v \in V$, there exists a graph $G_{uv} \in \mathcal{G}$ such that u and v are true twins in G_{uv}.*

Proof. We first note that for any connected graph $G = (V, E)$ and any vertex $v \in V$, it holds that $V - \{v\}$ is a local metric generator for G. Therefore, if $\mathrm{Sd}_l(\mathcal{G}) = |V| - 1$, then for any $v \in V$, the set $V - \{v\}$ is a simultaneous local metric basis of $\mathcal{G}$, and as a consequence, for every $u \in V - \{v\}$, there exists a graph $G_{uv} \in \mathcal{G}$, such that the set $V - \{u, v\}$ is not a local metric generator for G_{uv}, i.e., u and v are adjacent in G_{uv} and $d_{G_{u,v}}(u, x) = d_{G_{u,v}}(v, x)$ for every $x \in V - \{u, v\}$. Therefore, u and v are true twins in $G_{u,v}$.

Conversely, if for every $u, v \in V$ there exists a graph $G_{uv} \in \mathcal{G}$ such that u and v are true twins in G_{uv}, then for any simultaneous local metric basis B of $\mathcal{G}$, it holds that $u \in B$ or $v \in B$. Hence, all but one element of V must belong to B. Therefore, $|B| \geq |V| - 1$, which implies that $\mathrm{Sd}_l \mathcal{G} = |V| - 1$. $\square$

Notice that Corollary 1 is obtained directly from the previous result. Now, the two following results concern the limit cases of Item (1) of Remark 1.

Theorem 2. *A family $\mathcal{G}$ of connected graphs on a common vertex set V satisfies $\mathrm{Sd}_l(\mathcal{G}) = 1$ if and only if every graph in $\mathcal{G}$ is bipartite.*

Proof. If every graph in the family is bipartite, then for any $v \in V$, the set $\{v\}$ is a local metric basis of every $G_i \in \mathcal{G}$, so $\mathrm{Sd}_l(\mathcal{G}) = 1$.

Let us now consider a family $\mathcal{G}$ of connected graphs on a common vertex set V such that $\mathrm{Sd}_l(\mathcal{G}) = 1$ and assume that some $G \in \mathcal{G}$ is not bipartite. It is shown in [21] that $\dim_l(G) \geq 2$, so Item (1) of Remark 1 leads to $\mathrm{Sd}_l(\mathcal{G}) \geq 2$, which is a contradiction. Thus, every $G \in \mathcal{G}$ is bipartite. $\square$

Paths, trees and even-order cycles are bipartite. The following result covers the case of families composed of odd-order cycles.

Theorem 3. *Every family $\mathcal{G}$ composed of cycle graphs on a common odd-sized vertex set V satisfies $\mathrm{Sd}_l(\mathcal{G}) = 2$, and any pair of vertices of V is a simultaneous local metric basis of $\mathcal{G}$.*

Proof. For any cycle $C_i \in \mathcal{G}$, the set $\{v\}$, $v \in V$, is not a local metric generator, as the adjacent vertices $v_{j+\lfloor \frac{|V|}{2} \rfloor}$ and $v_{j-\lfloor \frac{|V|}{2} \rfloor}$ (subscripts taken modulo $|V|$) are not distinguished by v, so Item (1) of Remark 1 leads to $\mathrm{Sd}_l(\mathcal{G}) \geq \max_{G \in \mathcal{G}}\{\dim_l(G)\} \geq 2$. Moreover, any set $\{v, v'\}$ is a local metric generator for every $C_i \in \mathcal{G}$, as the single pair of adjacent vertices not distinguished by v is distinguished by v', so that $\mathrm{Sd}_l(\mathcal{G}) \leq 2$. □

The following result allows us to study the simultaneous local metric dimension of a family $\mathcal{G}$ from the family of graphs composed by all non-bipartite graphs belonging to $\mathcal{G}$.

Theorem 4. *Let $\mathcal{G}$ be a family of graphs on a common vertex set V, not all of them bipartite. If $\mathcal{H}$ is the subfamily of $\mathcal{G}$ composed of all non-bipartite graphs belonging to $\mathcal{G}$, then:*

$$\mathrm{Sd}_l(\mathcal{G}) = \mathrm{Sd}_l(\mathcal{H}).$$

Proof. Since $\mathcal{H}$ is a non-empty subfamily of $\mathcal{G}$, we conclude that $\mathrm{Sd}_l(\mathcal{G}) \geq \mathrm{Sd}_l(\mathcal{H})$. Since any vertex of a bipartite graph G is a local metric generator for G, if $B \subseteq V$ is a simultaneous local metric basis of $\mathcal{H}$, then B is a simultaneous local metric generator for $\mathcal{G}$ and, as a result, $\mathrm{Sd}_l(\mathcal{G}) \leq |B| = \mathrm{Sd}_l(\mathcal{H})$. □

Some interesting situations may be observed regarding the simultaneous local metric dimension of some graph families versus its standard counterpart. In particular, the fact that false twin vertices need not be distinguished in the local variant leads to some cases where both parameters differ greatly. For instance, consider any family $\mathcal{G}$ composed of three or more star graphs having different centers. It was shown in [22] that any such family satisfies $\mathrm{Sd}(\mathcal{G}) = |V| - 1$, yet by Theorem 2, we have that $\mathrm{Sd}_l(\mathcal{G}) = 1$.

Given a family $\mathcal{G} = \{G_1, G_2, \dots, G_k\}$ of graphs $G_i = (V, E_i)$ on a common vertex set V, we define a simultaneous vertex cover of $\mathcal{G}$ as a set $S \subseteq V$, such that S is simultaneously a vertex cover of each G_i. The minimum cardinality among all simultaneous vertex covers of $\mathcal{G}$ is the simultaneous vertex cover number of $\mathcal{G}$, denoted by $\beta(\mathcal{G})$.

Theorem 5. *For any family $\mathcal{G}$ of connected graphs with common vertex set V,*

$$\mathrm{Sd}_l(\mathcal{G}) \leq \beta(\mathcal{G}).$$

Furthermore, if for every $uv \in \cup_{G \in \mathcal{G}} E(G)$ there exists $G' \in \mathcal{G}$ such that u and v are true twins in G', then $\mathrm{Sd}_l(\mathcal{G}) = \beta(\mathcal{G})$.

Proof. Let $B \subseteq V$ be a simultaneous vertex cover of $\mathcal{G}$. Since $V - B$ is a simultaneous independent set of $\mathcal{G}$, we conclude that $\mathrm{Sd}_l(\mathcal{G}) \leq \beta(\mathcal{G})$.

We now assume that for every $uv \in \cup_{G \in \mathcal{G}} E(G)$, there exists $G' \in \mathcal{G}$, such that u and v are true twins in G', and suppose, for the purpose of contradiction, that $\mathrm{Sd}_l(\mathcal{G}) < \beta(\mathcal{G})$. In such a case, there exists a simultaneous local metric basis $C \subseteq V$, which is not a simultaneous vertex cover of $\mathcal{G}$. Hence, there exist $u', v' \in V - C$ and $G \in \mathcal{G}$ such that $u'v' \in E(G)$, ergo $u'v' \in \cup_{G \in \mathcal{G}} E(G)$. As a consequence, u' and v' are true twins in some graph $G' \in \mathcal{G}$, which contradicts the fact that C is a simultaneous local metric basis of $\mathcal{G}$. Therefore, the strict inequality does not hold, hence $\mathrm{Sd}_l(\mathcal{G}) = \beta(\mathcal{G})$. □

3. Families Obtained by Small Changes on a Graph

Consider a graph G whose local metric dimension is known. In this section, we address two related questions:

- If a series of small changes is repeatedly performed on $E(G)$, thus producing a family $\mathcal{G}$ of consecutive versions of G, what is the behavior of $\mathrm{Sd}_l(\mathcal{G})$ with respect to $\dim_l(G)$?
- If several small changes are performed on $E(G)$ in parallel, thus producing a family $\mathcal{G}$ of alternative versions of G, what is the behavior of $\mathrm{Sd}_l(\mathcal{G})$ with respect to $\dim_l(G)$?

Addressing this issue in the general case is hard, so we will analyze a number of particular cases. First, we will specify three operators that describe some types of changes that may be performed on a graph G:

- Edge addition: We say that a graph G' is obtained from a graph G by an edge addition if there is an edge $e \in E(\overline{G})$ such that $G' = (V(G), E(G) \cup \{e\})$. We will use the notation $G' = \mathrm{add}_e(G)$.
- Edge removal: We say that a graph G' is obtained from a graph G by an edge removal if there is an edge $e \in E(G)$ such that $G' = (V(G), E(G) - \{e\})$. We will use the notation $G' = \mathrm{rmv}_e(G)$.
- Edge exchange: We say that a graph G' is obtained from a graph G by an edge exchange if there is an edge $e \in E(G)$ and an edge $f \in E(\overline{G})$ such that $G' = (V(G), (E(G) - \{e\}) \cup \{f\})$. We will use the notation $G' = \mathrm{xch}_{e,f}(G)$.

Now, consider a graph G and an ordered k-tuple of operations $O_k = (\mathrm{op}_1, \mathrm{op}_2, \ldots, \mathrm{op}_k)$, where $\mathrm{op}_i \in \{\mathrm{add}_{e_i}, \mathrm{rmv}_{e_i}, \mathrm{xch}_{e_i,f_i}\}$. We define the class $\mathcal{C}_{O_k}(G)$ containing all graph families of the form $\mathcal{G} = \{G, G_1', G_2', \ldots, G_k'\}$, composed by connected graphs on the common vertex set $V(G)$, where $G_i' = \mathrm{op}_i(G_{i-1}')$ for every $i \in \{1, \ldots, k\}$. Likewise, we define the class $\mathcal{P}_{O_k}(G)$ containing all graph families of the form $\mathcal{G} = \{G_1', G_2', \ldots, G_k'\}$, composed by connected graphs on the common vertex set $V(G)$, where $G_i' = \mathrm{op}_i(G)$ for every $i \in \{1, \ldots, k\}$. In particular, if $\mathrm{op}_i = \mathrm{add}_{e_i}$ ($\mathrm{op}_i = \mathrm{rmv}_{e_i}$, $\mathrm{op}_i = \mathrm{xch}_{e_i,f_i}$) for every $i \in \{1, \ldots, k\}$, we will write $\mathcal{C}_{A_k}(G)$ ($\mathcal{C}_{R_k}(G)$, $\mathcal{C}_{X_k}(G)$) and $\mathcal{P}_{A_k}(G)$ ($\mathcal{P}_{R_k}(G)$, $\mathcal{P}_{X_k}(G)$).

We have that performing an edge exchange on any tree T (path graphs included) either produces another tree or a disconnected graph. Thus, the following result is a direct consequence of this fact and Theorem 2.

Remark 2. *For any tree T, any $k \geq 1$ and any graph family $\mathcal{T} \in \mathcal{C}_{X_k}(T) \cup \mathcal{P}_{X_k}(T)$,*

$$\mathrm{Sd}_l(\mathcal{T}) = 1.$$

Our next result covers a large class of families composed by unicyclic graphs that can be obtained by adding edges, in parallel, to a path graph.

Remark 3. *For any path graph P_n, $n \geq 4$, any $k \geq 1$ and any graph family $\mathcal{G} \in \mathcal{P}_{A_k}(P_n)$,*

$$1 \leq \mathrm{Sd}_l(\mathcal{G}) \leq 2.$$

Proof. Every graph $G \in \mathcal{G}$ is either a cycle or a unicyclic graph. If the cycle subgraphs of every graph in the family have even order, then $\mathrm{Sd}_l(\mathcal{G}) = 1$ by Theorem 2. If $\mathcal{G}$ contains at least one non-bipartite graph, then $\mathrm{Sd}_l(\mathcal{G}) \geq 2$. We now proceed to show that in this case, $\mathrm{Sd}_l(\mathcal{G}) \leq 2$. To this end, we denote by $V = \{v_1, \ldots, v_n\}$ the vertex set of P_n, where $v_i \sim v_{i+1}$ for every $i \in \{1, \ldots, n-1\}$. We claim that $\{v_1, v_n\}$ is a simultaneous local metric generator for the subfamily $\mathcal{G}' \subset \mathcal{G}$ composed by all non-bipartite graphs of $\mathcal{G}$. In order to prove this claim, consider an arbitrary graph $G \in \mathcal{G}'$, and let $e = v_p v_q$, $1 \leq p < q \leq n$ be the edge added to $E(P_n)$ to obtain G. We differentiate the following cases:

1. $e = v_1 v_n$. In this case, G is an odd-order cycle graph, so $\{v_1, v_n\}$ is a local metric generator.

2. $1 < p < q = n$. In this case, G is a unicyclic graph where v_p has degree three, v_1 has degree one and the remaining vertices have degree two. Consider two adjacent vertices $u, v \in V - \{v_1, v_n\}$. If u or v belong to the path from v_1 to v_p, then v_1 distinguishes them. If both, u and v, belong to the cycle subgraph of G, then $d(u, v_1) = d(u, v_p) + d(v_p, v_1)$ and $d(v, v_1) = d(v, v_p) + d(v_p, v_1)$. Thus, if v_p distinguishes u and v, so does v_1, otherwise v_n does.

3. $1 = p < q < n$. This case is analogous to Case 2.

4. $1 < p < q < n$. In this case, G is a unicyclic graph where v_p and v_q have degree three, v_1 and v_n have degree one and the remaining vertices have degree two. Consider two adjacent vertices $u, v \in V - \{v_1, v_n\}$. If u or v belong to the path from v_1 to v_p (or to the path from v_q to v_n), then v_1 (or v_n) distinguishes them. If both u and v belong to the cycle, then $d(u, v_1) = d(u, v_p) + d(v_p, v_1)$, $d(v, v_1) = d(v, v_p) + d(v_p, v_1)$, $d(u, v_n) = d(u, v_q) + d(v_q, v_n)$ and $d(v, v_n) = d(v, v_q) + d(v_q, v_n)$. Thus, if v_p distinguishes u and v, so does v_1, otherwise v_q distinguishes them, which means that v_n also does.

According to the four cases above, we conclude that $\{v_1, v_n\}$ is a local metric generator for G, so it is a simultaneous local metric generator for $\mathcal{G}'$. Thus, by Theorem 4, $\mathrm{Sd}_l(\mathcal{G}) = \mathrm{Sd}_l(\mathcal{G}') \le 2$. $\square$

Remark 4. *Let C_n, $n \ge 4$, be a cycle graph, and let e be an edge of its complement. If n is odd, then*

$$\dim_l(\mathrm{add}_e(C_n)) = 2.$$

Otherwise,

$$1 \le \dim_l(\mathrm{add}_e(C_n)) \le 2.$$

Proof. Consider $e = v_i v_j$. We have that C_n is bipartite for n even. If, additionally, $d_{C_n}(v_i, v_j)$ is odd, then the graph $\mathrm{add}_e(C_n)$ is also bipartite, so $\dim_l(\mathrm{add}_e(C_n)) = 1$. For every other case, $\dim_l(\mathrm{add}_e(C_n)) \ge 2$. From now on, we assume that $n \ge 5$ and proceed to show that $\dim_l(\mathrm{add}_e(C_n)) \le 2$. Note that $\mathrm{add}_e(C_n)$ is a bicyclic graph where v_i and v_j are vertices of degree three and the remaining vertices have degree two. We denote by C_{n_1} and C_{n-n_1+2} the two graphs obtained as induced subgraphs of $\mathrm{add}_e(C_n)$, which are isomorphic to a cycle of order n_1 and a cycle of order $n - n_1 + 2$, respectively. Since $n \ge 5$, we have that $n_1 > 3$ or $n - n_1 + 2 > 3$. We assume, without loss of generality, that $n_1 > 3$. Let $a, b \in V(C_{n_1})$ are two vertices such that:

- if n_1 is even, $ab \in E(C_{n_1})$ and $d(v_i, a) = d(v_j, b)$,
- if n_1 is odd, $ax, xb \in E(C_{n_1})$, where $x \in V(C_{n_1})$ is the only vertex such that $d(x, v_i) = d(x, v_j)$.

We claim that $\{a, b\}$ is a local metric generator for $\mathrm{add}_e(C_n)$. Consider two adjacent vertices $u, v \in V(\mathrm{add}_e(C_n)) - \{a, b\}$. We differentiate the following cases, where the distances are taken in $\mathrm{add}_e(C_n)$:

1. $u, v \in V(C_{n_1})$. It is simple to verify that $\{a, b\}$ is a local metric generator for C_{n_1}, hence $d(u, a) \ne d(v, a)$ or $d(u, b) \ne d(v, b)$.

2. $u \in V(C_{n_1})$ and $v \in V(C_{n-n_1+2}) - \{v_i, v_j\}$. In this case, $u \in \{v_i, v_j\}$ and $d(u, a) < d(v, a)$ or $d(u, b) < d(v, b)$.

3. $u, v \in V(C_{n-n_1+2}) - \{v_i, v_j\}$. In this case, if $d(u, a) = d(v, a)$, then $d(u, v_i) = d(v, v_i)$, so $d(u, v_j) \ne d(v, v_j)$ and, consequently, $d(u, b) \ne d(v, b)$.

According to the three cases above, $\{a, b\}$ is a local metric generator for $\mathrm{add}_e(C_n)$, and as a result, the proof is complete. $\square$

The next result is a direct consequence of Remarks 1 and 4.

Remark 5. *Let C_n, $n \ge 4$, be a cycle graph. If e, e' are two different edges of the complement of C_n, then:*

$$1 \le \mathrm{Sd}_l(\mathrm{add}_e(C_n), \mathrm{add}_{e'}(C_n)) = \mathrm{Sd}_l(C_n, \mathrm{add}_e(C_n), \mathrm{add}_{e'}(C_n)) \le 4.$$

4. Families of Corona Product Graphs

Let G and H be two graphs of order n and n', respectively. The corona product $G \odot H$ is defined as the graph obtained from G and H by taking one copy of G and n copies of H and joining by an edge each vertex from the i-th copy of H with the i-th vertex of G. Notice that the corona graph $K_1 \odot H$ is isomorphic to the join graph $K_1 + H$. Given a graph family $\mathcal{G} = \{G_1, \ldots, G_k\}$ on a common vertex set and a graph H, we define the graph family:

$$\mathcal{G} \odot H = \{G_1 \odot H, \ldots, G_k \odot H\}.$$

Several results presented in [23,24] describe the behavior of the local metric dimension on corona product graphs. We now analyze how this behavior extends to the simultaneous local metric dimension of families composed by corona product graphs.

Theorem 6. *In references* [23,25], *Let G be a connected graph of order $n \geq 2$. For any non-empty graph H,*

$$\dim_l(G \odot H) = n \cdot \mathrm{adim}_l(H).$$

As we can expect, if we review the proof of the result above, we check that if A is a local metric basis of $G \odot H$, then A does not contain elements in $V(G)$. Therefore, any local metric basis of $G \odot H$ is a simultaneous local metric basis of $\mathcal{G} \odot H$. This fact and the result above allow us to state the following theorem.

Theorem 7. *Let $\mathcal{G}$ be a family of connected non-trivial graphs on a common vertex set V. For any non-empty graph H,*

$$\mathrm{Sd}_l(\mathcal{G} \odot H) = |V| \cdot \mathrm{adim}_l(H).$$

Given a graph family $\mathcal{G}$ on a common vertex set and a graph family $\mathcal{H}$ on a common vertex set, we define the graph family:

$$\mathcal{G} \odot \mathcal{H} = \{G \odot H : G \in \mathcal{G} \text{ and } H \in \mathcal{H}\}.$$

The following result generalizes Theorem 7. In what follows, we will use the notation $\langle v \rangle$ for the graph $G = (V, E)$ where $V = \{v\}$ and $E = \emptyset$.

Theorem 8. *For any family $\mathcal{G}$ of connected non-trivial graphs on a common vertex set V and any family $\mathcal{H}$ of non-empty graphs on a common vertex set,*

$$\mathrm{Sd}_l(\mathcal{G} \odot \mathcal{H}) = |V| \cdot \mathrm{Sad}_l(\mathcal{H}).$$

Proof. Let $n = |V|$, and let V' be the vertex set of the graphs in $\mathcal{H}$, V_i' the copy of V' corresponding to $v_i \in V$, $\mathcal{H}_i$ the i-th copy of $\mathcal{H}$ and $H_i \in \mathcal{H}_i$ the i-th copy of $H \in \mathcal{H}$.

We first need to prove that any $G \in \mathcal{G}$ satisfies $\mathrm{Sd}_l(G \odot \mathcal{H}) = n \cdot \mathrm{Sad}_l(\mathcal{H})$. For any $i \in \{1, \ldots, n\}$, let S_i be a simultaneous local adjacency basis of $\mathcal{H}_i$. In order to show that $X = \bigcup_{i=1}^n S_i$ is a simultaneous local metric generator for $\mathcal{G} \odot \mathcal{H}$, we will show that X is a local metric generator for $G \odot H$, for any $G \in \mathcal{G}$ and $H \in \mathcal{H}$. To this end, we differentiate the following four cases for two adjacent vertices $x, y \in V(G \odot H) - X$.

1. $x, y \in V_i'$. Since S_i is an adjacency generator of H_i, there exists a vertex $u \in S_i$ such that $|N_{H_i}(u) \cap \{x, y\}| = 1$. Hence,

$$d_{G \odot H}(x, u) = d_{\langle v_i \rangle + H_i}(x, u) \neq d_{\langle v_i \rangle + H_i}(y, u) = d_{G \odot H}(y, u).$$

2. $x \in V_i'$ and $y \in V$. If $y = v_i$, then for $u \in S_j$, $j \neq i$, we have:

$$d_{G \odot H}(x, u) = d_{G \odot H}(x, y) + d_{G \odot H}(y, u) > d_{G \odot H}(y, u).$$

Now, if $y = v_j$, $j \neq i$, then we also take $u \in S_j$, and we proceed as above.

3. $x = v_i$ and $y = v_j$. For $u \in S_j$, we find that:

$$d_{G \odot H}(x, u) = d_{G \odot H}(x, y) + d_{G \odot H}(y, u) > d_{G \odot H}(y, u).$$

4. $x \in V_i'$ and $y \in V_j'$, $j \neq i$. In this case, for $u \in S_i$, we have:

$$d_{G \odot H}(x, u) \leq 2 < 3 \leq d_{G \odot H}(u, y).$$

Hence, X is a local metric generator for $G \odot H$, and since $G \in \mathcal{G}$ and $H \in \mathcal{H}$ are arbitrary graphs, X is a simultaneous local metric generator for $\mathcal{G} \odot \mathcal{H}$, which implies that:

$$\mathrm{Sd}_l(\mathcal{G} \odot \mathcal{H}) \leq \sum_{i=1}^{n} |S_i| = n \cdot \mathrm{Sad}_l(\mathcal{H}).$$

It remains to prove that $\mathrm{Sd}_l(\mathcal{G} \odot \mathcal{H}) \geq n \cdot \mathrm{Sad}_l(\mathcal{H})$. To do this, let W be a simultaneous local metric basis of $\mathcal{G} \odot \mathcal{H}$, and for any $i \in \{1, \ldots, n\}$, let $W_i = V_i' \cap W$. Let us show that W_i is a simultaneous adjacency generator for $\mathcal{H}_i$. To do this, consider two different vertices $x, y \in V_i' - W_i$, which are adjacent in $G \odot H$, for some $H \in \mathcal{H}$. Since no vertex $a \in V(G \odot H) - V_i'$ distinguishes the pair x, y, there exists some $u \in W_i$, such that $d_{G \odot H}(x, u) \neq d_{G \odot H}(y, u)$. Now, since $d_{G \odot H}(x, u) \in \{1, 2\}$ and $d_{G \odot H}(y, u) \in \{1, 2\}$, we conclude that $|N_{H_i}(u) \cap \{x, y\}| = 1$, and consequently, W_i must be an adjacency generator for H_i; and since $H \in \mathcal{H}$ is arbitrary, W_i is a simultaneous local adjacency generator for $\mathcal{H}_i$. Hence, for any $i \in \{1, \ldots, n\}$, $|W_i| \geq \mathrm{Sad}_l(H_i)$. Therefore,

$$\mathrm{Sd}_l(\mathcal{G} \odot \mathcal{H}) = |W| \geq \sum_{i=1}^{n} |W_i| \geq \sum_{i=1}^{n} \mathrm{Sad}_l(\mathcal{H}_i) = n \cdot \mathrm{Sad}_l(\mathcal{H}).$$

This completes the proof. □

The following result is a direct consequence of Theorem 8.

Corollary 2. *For any family $\mathcal{G}$ of connected non-trivial graphs on a common vertex set V and any family $\mathcal{H}$ of non-empty graphs on a common vertex set,*

$$\mathrm{Sd}_l(\mathcal{G} \odot \mathcal{H}) \geq |V| \cdot \mathrm{Sd}_l(\mathcal{H}).$$

Furthermore, if every graph in $\mathcal{H}$ has diameter two, then:

$$\mathrm{Sd}_l(\mathcal{G} \odot \mathcal{H}) = |V| \cdot \mathrm{Sd}_l(\mathcal{H}).$$

Now, we give another result, which is a direct consequence of Theorem 8 and shows the general bounds of $\mathrm{Sd}_l(\mathcal{G} \odot \mathcal{H})$.

Corollary 3. *For any family $\mathcal{G}$ of connected graphs on a common vertex set V, $|V| \geq 2$ and any family $\mathcal{H}$ of non-empty graphs on a common vertex set V',*

$$|V| \leq \mathrm{Sd}_l(\mathcal{G} \odot \mathcal{H}) \leq |V|(|V'| - 1).$$

We now consider the case in which the graph H is empty.

Theorem 9. *In reference* [24]*, Let G be a connected non-trivial graph. For any empty graph H,*

$$\dim_l(G \odot H) = \dim_l(G).$$

The result above may be extended to the simultaneous scenario.

Theorem 10. *Let $\mathcal{G}$ be a family of connected non-trivial graphs on a common vertex set. For any empty graph H,*

$$\mathrm{Sd}_l(\mathcal{G} \odot H) = \mathrm{Sd}_l(\mathcal{G}).$$

Proof. Let B be a simultaneous local metric basis of $\mathcal{G} = \{G_1, G_2, \ldots, G_k\}$. Since H is empty, any local metric generator $B' \subseteq B$ of G_i is a local metric generator for $G_i \odot H$, so B is a simultaneous local metric generator for $\mathcal{G} \odot H$. As a consequence, $\mathrm{Sd}_l(\mathcal{G} \odot H) \leq \mathrm{Sd}_l(\mathcal{G})$.

Suppose that A is a simultaneous local metric basis of $\mathcal{G} \odot H$ and $|A| < |B|$. If there exists $x \in A \cap V_{ij}$ for the j-th copy of H in any graph $G_i \odot H$, then the pairs of vertices of $G_i \odot H$ that are distinguished by x can also be distinguished by v_j. As a consequence, the set A' obtained from A by replacing by v_j each vertex $x \in A \cap V_{ij}, i \in \{1, \ldots, k\}, j \in \{1, \ldots, n\}$ is a simultaneous local metric generator for $\mathcal{G}$ such that $|A'| \leq |A| < \mathrm{Sd}_l(\mathcal{G})$, which is a contradiction, so $\mathrm{Sd}_l(\mathcal{G} \odot H) \geq \mathrm{Sd}_l(\mathcal{G})$. $\square$

Theorem 11. *In reference* [24]*, Let H be a non-empty graph. The following assertions hold.*

1. *If the vertex of K_1 does not belong to any local metric basis of $K_1 + H$, then for any connected graph G of order n,*

$$\dim_l(G \odot H) = n \cdot \dim_l(K_1 + H).$$

2. *If the vertex of K_1 belongs to a local metric basis of $K_1 + H$, then for any connected graph G of order $n \geq 2$,*

$$\dim_l(G \odot H) = n \cdot (\dim_l(K_1 + H) - 1).$$

As for the previous case, the result above is extensible to the simultaneous setting.

Theorem 12. *Let $\mathcal{G}$ be a family of connected non-trivial graphs on a common vertex set V, and let $\mathcal{H}$ be a family of non-empty graphs on a common vertex set. The following assertions hold.*

1. *If the vertex of K_1 does not belong to any simultaneous local metric basis of $K_1 + \mathcal{H}$, then:*

$$\mathrm{Sd}_l(\mathcal{G} \odot \mathcal{H}) = |V| \cdot \mathrm{Sd}_l(K_1 + \mathcal{H}).$$

2. *If the vertex of K_1 belongs to a simultaneous local metric basis of $K_1 + \mathcal{H}$, then:*

$$\mathrm{Sd}_l(\mathcal{G} \odot \mathcal{H}) = |V| \cdot (\mathrm{Sd}_l(K_1 + \mathcal{H}) - 1).$$

Proof. As above, let $n = |V|$, and let V' be the vertex set of the graphs in $\mathcal{H}$, V_i' the copy of V' corresponding to $v_i \in V$, $\mathcal{H}_i$ the i-th copy of $\mathcal{H}$ and $H_i \in \mathcal{H}_i$ the i-th copy of $H \in \mathcal{H}$.

We will apply a reasoning analogous to the one used for the proof of Theorem 11 in [24]. If $n = 1$, then $\mathcal{G} \odot \mathcal{H} \cong K_1 + \mathcal{H}$, so the result holds. Assume that $n \geq 2$, Let S_i be a simultaneous local metric basis of $\langle v_i \rangle + \mathcal{H}_i$, and let $S_i' = S_i - \{v_i\}$. Note that $S_i' \neq \emptyset$ because $\mathcal{H}_i$ is the family of non-empty graphs and v_i does not distinguish any pair of adjacent vertices belonging to V_i'. In order to show that $X = \cup_{i=1}^n S_i'$ is a simultaneous local metric generator for $\mathcal{G} \odot \mathcal{H}$, we differentiate the following cases for two vertices x, y, which are adjacent in an arbitrary graph $G \odot H$:

1. $x, y \in V_i'$. Since v_i does not distinguish x, y, there exists $u \in S_i'$ such that $d_{G \odot H}(x, u) = d_{\langle v_i \rangle + H_i}(x, u) \neq d_{\langle v_i \rangle + H_i}(y, u) = d_{G \odot H}(y, u)$.
2. $x \in V_i'$ and $y = v_j$. For $u \in S_j', j \neq i$, we have $d_{G \odot H}(x, u) = 1 + d_{G \odot H}(y, u) > d_{G \odot H}(y, u)$.
3. $x = v_i$ and $y = v_j$. For $u \in S_j'$, we have $d_{G \odot H}(x, u) = 2 = d_{G \odot H}(x, y) + 1 > 1 = d_{G \odot H}(y, u)$.

Hence, X is a local metric generator for $G \odot H$, and since $G \in \mathcal{G}$ and $H \in \mathcal{H}$ are arbitrary graphs, X is a simultaneous local metric generator for $\mathcal{G} \odot \mathcal{H}$.

Now, we shall prove (1). If the vertex of K_1 does not belong to any simultaneous local metric basis of $K_1 + \mathcal{H}$, then $v_i \notin S_i$ for every $i \in \{1, ..., n\}$, and as a consequence,

$$\mathrm{Sd}_l(\mathcal{G} \odot \mathcal{H}) \leq |X| = \sum_{i=1}^{n} |S_i'| = \sum_{i=1}^{n} \mathrm{Sd}_l(\langle v_i \rangle + \mathcal{H}_i) = n \cdot \mathrm{Sd}_l(K_1 + \mathcal{H}).$$

Now, we need to prove that $\mathrm{Sd}_l(\mathcal{G} \odot \mathcal{H}) \geq n \cdot \mathrm{Sd}_l(K_1 + \mathcal{H})$. In order to do this, let W be a simultaneous local metric basis of $\mathcal{G} \odot \mathcal{H}$, and let $W_i = V_i' \cap W$. Consider two adjacent vertices $x, y \in V_i' - W_i$ in $G \odot H$. Since no vertex $a \in W - W_i$ distinguishes the pair x, y, there exists $u \in W_i$ such that $d_{\langle v_i \rangle + H_i}(x, u) = d_{G \odot H}(x, u) \neq d_{G \odot H}(y, u) = d_{\langle v_i \rangle + H_i}(y, u)$. Therefore, we conclude that $W_i \cup \{v_i\}$ is a simultaneous local metric generator for $\langle v_i \rangle + \mathcal{H}_i$. Now, since v_i does not belong to any simultaneous local metric basis of $\langle v_i \rangle + \mathcal{H}_i$, we have that $|W_i| + 1 = |W_i \cup \{v_i\}| > \mathrm{Sd}_l(\langle v_i \rangle + \mathcal{H}_i)$ and, as a consequence, $|W_i| \geq \mathrm{Sd}_l(\langle v_i \rangle + \mathcal{H}_i)$. Therefore,

$$\mathrm{Sd}_l(\mathcal{G} \odot \mathcal{H}) = |W| \geq \sum_{i=1}^{n} |W_i| \geq \sum_{i=1}^{n} \mathrm{Sd}_l(\langle v_i \rangle + \mathcal{H}_i) = n \cdot \mathrm{Sd}_l(K_1 + \mathcal{H}),$$

and the proof of (1) is complete.

Finally, we shall prove (2). If the vertex of K_1 belongs to a simultaneous local metric basis of $K_1 + \mathcal{H}$, then we assume that $v_i \in S_i$ for every $i \in \{1, ..., n\}$. Suppose that there exists B such that B is a simultaneous local metric basis of $\mathcal{G} \odot \mathcal{H}$ and $|B| < |X|$. In such a case, there exists $i \in \{1, ..., n\}$ such that the set $B_i = B \cap V_i'$ satisfies $|B_i| < |S_i'|$. Now, since no vertex of $B - B_i$ distinguishes the pairs of adjacent vertices belonging to V_i', the set $B_i \cup \{v_i\}$ must be a simultaneous local metric generator for $\langle v_i \rangle + \mathcal{H}_i$. Therefore, $\mathrm{Sd}_l(\langle v_i \rangle + \mathcal{H}_i) \leq |B_i| + 1 < |S_i'| + 1 = |S_i| = \mathrm{Sd}_l(\langle v_i \rangle + \mathcal{H}_i)$, which is a contradiction. Hence, X is a simultaneous local metric basis of $\mathcal{G} \odot \mathcal{H}$, and as a consequence,

$$\mathrm{Sd}_l(\mathcal{G} \odot \mathcal{H}) = |X| = \sum_{i=1}^{n} |S_i'| = \sum_{i=1}^{n} (\mathrm{Sd}_l(\langle v_i \rangle + \mathcal{H}_i) - 1) = n(\mathrm{Sd}_l(K_1 + \mathcal{H}) - 1).$$

The proof of (2) is now complete. $\square$

Corollary 4. *Let G be a connected graph of order $n \geq 2$, and let $\mathcal{H} = \{K_{r_1, n' - r_1}, K_{r_2, n' - r_2}, \ldots, K_{r_k, n' - r_k}\}$, $1 \leq r_i \leq n' - 1$, be a family composed by complete bipartite graphs on a common vertex set V'. Then,*

$$\mathrm{Sd}_l(G \odot \mathcal{H}) = n.$$

Proof. For every $x \in V'$, the set $\{v, x\}$ is a simultaneous local metric basis of $\langle v \rangle + \mathcal{H}$, so $\mathrm{Sd}(G \odot \mathcal{H}) = n \cdot (\mathrm{Sd}(K_1 + \mathcal{H}) - 1) = n$. $\square$

Lemma 1. *In reference [24], Let H be a graph of radius $r(H)$. If $r(H) \geq 4$, then the vertex of K_1 does not belong to any local metric basis of $K_1 + H$.*

Note that an analogous result holds for the simultaneous scenario.

Lemma 2. *Let $\mathcal{H}$ be a graph family on a common vertex set V, such that $r(H) \geq 4$ for every $H \in \mathcal{H}$. Then, the vertex of K_1 does not belong to any simultaneous local metric basis of $K_1 + \mathcal{H}$.*

Proof. Let B be a simultaneous local metric basis of $\{K_1 + H_1, \ldots, K_1 + H_k\}$. We suppose that the vertex v of K_1 belongs to B. Note that $v \in B$ if and only if there exists $u \in V - B$, such that $B \subseteq N_{K_1 + H_i}(u)$ for some $H_i \in \mathcal{H}$. If $r(H_i) \geq 4$, proceeding in a manner analogous to that of the proof of Lemma 1 as given in [24], we take $u' \in V$ such that $d_{H_i}(u, u') = 4$ and a shortest path $uu_1u_2u_3u'$. In such a case, for every

$b \in B - \{v\}$, we will have that $d_{K_1+H_i}(b, u_3) = d_{K_1+H_i}(b, u') = 2$, which is a contradiction. Hence, v does not belong to any simultaneous local metric basis of $\{K_1 + H_1, K_1 + H_2, \ldots, K_1 + H_k\}$. $\square$

As a direct consequence of item (1) of Theorem 12 and Lemma 2, we obtain the following result.

Proposition 1. *For any family $\mathcal{G}$ of connected graphs on a common vertex set V and any graph family $\mathcal{H}$ on a common vertex set V' such that $r(H) \geq 4$ for every $H \in \mathcal{H}$,*

$$\mathrm{Sd}_l(\mathcal{G} \odot \mathcal{H}) = |V| \cdot \mathrm{Sd}_l(K_1 + \mathcal{H}).$$

5. Families of Lexicographic Product Graphs

Let $\mathcal{G} = \{G_1, \ldots, G_r\}$ be a family of connected graphs with common vertex set $V = \{u_1, \ldots, u_n\}$. For each $u_i \in V$, let $\mathcal{H}^i = \{H_{i1}, \ldots H_{is_i}\}$ be a family of graphs with common vertex set V_i. For each $i = 1, \ldots, n$, choose $H_{ij} \in \mathcal{H}^i$ and consider the family $\mathcal{H}_j = \{H_{1j}, H_{2j}, \ldots, H_{nj}\}$. Notice that the families $\mathcal{H}^i$ can be represented in the following scheme where the columns correspond to the families $\mathcal{H}_j$.

$$
\begin{array}{ccccccl}
\mathcal{H}^1 = & \{H_{11}, & \ldots & H_{1j}, & \ldots & H_{1s_1}\} & \text{defined on } V_1 \\
\vdots & \vdots & & \vdots & & \vdots & \\
\mathcal{H}^i = & \{H_{i1}, & \ldots & H_{ij}, & \ldots & H_{is_i}\} & \text{defined on } V_i \\
\vdots & \vdots & & \vdots & & \vdots & \\
\mathcal{H}^n = & \{H_{n1}, & \ldots & H_{nj}, & \ldots & H_{ns_n}\} & \text{defined on } V_n
\end{array}
$$

For a graph $G_k \in \mathcal{G}$ and the family $\mathcal{H}_j$, we define the lexicographic product of G_k and $\mathcal{H}_j$ as the graph $G_k \circ \mathcal{H}_j$ such that $V(G_k \circ \mathcal{H}_j) = \bigcup_{u_i \in V}(\{u_i\} \times V_i)$ and $(u_{i_1}, v)(u_{i_2}, w) \in E(G_k \circ \mathcal{H}_j)$ if and only if $u_{i_1} u_{i_2} \in E(G_k)$ or $i_1 = i_2$ and $vw \in E(H_{i_1j})$. Let $\mathcal{H} = \{\mathcal{H}_1, \mathcal{H}_2, \ldots \mathcal{H}_s\}$. We are interested in the simultaneous local metric dimension of the family:

$$\mathcal{G} \circ \mathcal{H} = \{G_k \circ \mathcal{H}_j : G_k \in \mathcal{G}, \mathcal{H}_j \in \mathcal{H}\}.$$

The relation between distances in a lexicographic product graph and those in its factors is presented in the following remark.

Remark 6. *If (u, v) and (u', v') are vertices of $G \circ H$, then:*

$$
d_{G \circ H}((u, v), (u', v')) = \begin{cases} d_G(u, u'), & \text{if } u \neq u', \\ \\ \min\{d_H(v, v'), 2\}, & \text{if } u = u'. \end{cases}
$$

We point out that the remark above was stated in [26,27] for the case where $H_{ij} \cong H$ for all $H_{ij} \in \mathcal{H}_j$. By Remark 6, we deduce that if $u \in V - \{u_i\}$, then two adjacent vertices $(u_i, w), (u_i, y)$ are not distinguished by $(u, v) \in V(\mathcal{G} \circ \mathcal{H})$. Therefore, we can state the following remark.

Remark 7. *If B is a simultaneous local metric generator for the family of lexicographic product graphs $\mathcal{G} \circ \mathcal{H}$, then $B_i = \{v : (u_i, v) \in B\}$ is a simultaneous local adjacency generator for $\mathcal{H}^i$.*

In order to state our main result (Theorem 13), we need to introduce some additional notation. Let B be a simultaneous local adjacency generator for a family of non-trivial connected graphs $\mathcal{H}^i = \{H_{i1}, \ldots, H_{is}\}$ on a common vertex set V_i, and let $\mathcal{G} \circ \mathcal{H}$ be family of lexicographic product graphs defined as above.

- $D[\mathcal{H}^i, B] = \{v \in V_i : B \subseteq N_{H_{ij}}(v) \text{ for some } H_{ij} \in \mathcal{H}^i\}$.

- If $D[\mathcal{H}^i, B] \neq \emptyset$, then we define the graph $\mathcal{D}[\mathcal{H}^i, B]$ in the following way. The vertex set of $\mathcal{D}[\mathcal{H}^i, B]$ is $D[\mathcal{H}^i, B]$, and two vertices v, w are adjacent in $\mathcal{D}[\mathcal{H}^i, B]$ if and only if for for every $H_{ij} \in \mathcal{H}^i$, $vw \notin E(H_{ij})$.

- If $D[\mathcal{H}^i, B] = \emptyset$, then define $\Psi(B) = |B|$, otherwise $\Psi(B) = \gamma(\mathcal{D}[\mathcal{H}^i, B]) + |B|$, where $\gamma(\mathcal{D}[\mathcal{H}^i, B])$ represents the domination number of $\mathcal{D}[\mathcal{H}^i, B]$.

- $\Gamma(\mathcal{H}^i) = \{C \subseteq V_i : C \text{ is a simultaneous local adjacency generator for } \mathcal{H}^i\}$

- $\Psi(\mathcal{H}^i) = \min\{\Psi(B) : B \in \Gamma(\mathcal{H}^i)\}$.

- $\mathcal{S}_0$ is a family composed by empty graphs.

- $\Phi(V, \mathcal{H}) = \{u_i \in V : \mathcal{H}^i \subseteq \mathcal{S}_0\}$

- $I(V, \mathcal{H}) = \{u_i \in V : \Psi(\mathcal{H}^i) > \mathrm{Sad}_l(\mathcal{H}^i)\}$. Notice that $\Phi(V, \mathcal{H}) \subseteq I(V, \mathcal{H})$.

- $Y(V, \mathcal{H})$ is the family of subsets of $I(V, \mathcal{H})$ as follows. We say that $A \in Y(V, \mathcal{H})$ if for every $u', u'' \in I(V, \mathcal{H}) - A$ such that $u'u'' \in E(G_k)$, for some $G_k \in \mathcal{G}$, there exists $u \in (A \cup (V - \Phi(V, \mathcal{H}))) - \{u', u''\}$ such that $d_{G_k}(u, u') \neq d_{G_k}(u, u'')$.

- $\mathbf{G}(\mathcal{G}, I(V, \mathcal{H}))$ is the graph with vertex set $I(V, \mathcal{H})$ and edge set $\mathbf{E}$ such that $u_i u_j \in \mathbf{E}$ if and only if there exists $G_k \in \mathcal{G}$ such that $u_i u_j \in E(G_k)$.

Remark 8. $\Psi(\mathcal{H}^i) = 1$ *if and only if* $H_{i,j} \cong N_{|V_i|}$ *for every* $H_{i,j} \in \mathcal{H}^i$.

Proof. If $H_{i,j} \cong N_{|V_i|}$ for every $H_{i,j} \in \mathcal{H}^i$, then $B = \emptyset$ is the only simultaneous local adjacency basis of $\mathcal{H}^i$, $\mathcal{D}[\mathcal{H}^i, \emptyset] \cong K_{|V_i|}$, and then, $\Psi(\mathcal{H}^i) = \gamma(K_{|V_i|}) = 1$. On the other hand, suppose that $H_{i,j} \not\cong N_{|V_i|}$ for some $H_{i,j} \in \mathcal{H}^i$. In this case, $\mathrm{Sad}_l(\mathcal{H}^i) \geq 1$. If $\mathrm{Sad}_l(\mathcal{H}^i) > 1$, then we are done. Suppose that $\mathrm{Sad}_l(\mathcal{H}^i) = 1$. For any simultaneous local adjacency basis $B = \{v_1\}$ of $\mathcal{H}^i$, there exists $v_2 \in N_{H_{ij}}(v_1)$ for some H_{ij}, which implies that $D[\mathcal{H}^i, \{v_2\}] \neq \emptyset$ and so $|\gamma(\mathcal{D}[\mathcal{H}^i, \{v_2\}])| \geq 1$. Therefore, $\Psi(\mathcal{H}^i) \geq 2$, and the result follows. $\square$

As we will show in the next example, in order to get the value of $\Psi(\mathcal{H}^i)$, it is interesting to remark about the necessity of considering the family $\Gamma(\mathcal{H}^i)$ of all simultaneous local adjacency generators and not just the family of simultaneous local adjacency bases of $\mathcal{H}^i$.

Example 1. *Let* $H_1 \cong H_2 \cong P_5$ *be two copies of the path graph on five vertices such that* $V(H_1) = V(H_2) = \{v_1, v_2, \ldots, v_5\}$*, whereas* $E(H_1) = \{v_1v_2, v_2v_3, v_3v_4, v_4v_5\}$ *and* $E(H_2) = \{v_2v_1, v_1v_3, v_3v_5, v_5v_4\}$*. Consider the family* $\mathcal{H} = \{H_1, H_2\}$*. We have that* $B_1 = \{v_3\}$ *is a simultaneous local adjacency basis of* $\mathcal{H}$ *and* $B_2 = \{v_1, v_4\}$ *is a simultaneous local adjacency generator for* $\mathcal{H}$*. Then,* $D[\mathcal{H}, B_1] = \{v_1, v_2, v_4, v_5\}$*,* $E(\mathcal{D}[\mathcal{H}, B_1]) = \{v_1v_4, v_4v_2, v_2v_5, v_5v_1\}$*,* $\gamma(\mathcal{D}[\mathcal{G}, B_1]) = 2$*,* $\Psi(B_1) = 2 + 1 = 3$*. However,* $D[\mathcal{H}, B_2] = \emptyset$ *and* $\Psi(B_2) = 2$.

We define the following graph families.

- $\mathcal{S}_1$ is the family of graphs having at least two non-trivial components.
- $\mathcal{S}_2$ is the family of graphs having at least one component of radius at least four.
- $\mathcal{S}_3$ is the family of graphs having at least one component of girth at least seven.
- $\mathcal{S}_4$ is the family of graphs having at least two non-singleton true twin equivalence classes U_1, U_2 such that $d(U_1, U_2) \geq 3$.

Lemma 3. *Let* $\mathcal{H} \not\subseteq \mathcal{S}_0$ *be a family of graphs on a common vertex set* V*. If* $\mathcal{H} \subseteq \bigcup_{i=0}^{4} \mathcal{S}_i$*, then:*

$$\Psi(\mathcal{H}) = \mathrm{Sad}_l(\mathcal{H}).$$

Proof. Let B be a simultaneous local adjacency generator for $\mathcal{H}$ and $v \in V$. We claim that $B \not\subseteq N_H(v)$. To see this, we differentiate the following cases for $H \in \mathcal{H}$.

- H has two non-trivial connected components J_1, J_2. In this case, $B \cap J_1 \neq \emptyset$ and $B \cap J_2 \neq \emptyset$, which implies that $B \not\subseteq N_H(v)$.

- H has one non-trivial component J such that $r(J) \geq 4$. If H has two non-trivial components, then we are in the first case. Therefore, we can assume that J is the only non-trivial component of H. Suppose that $B \subseteq N_H(v)$, and get $v' \in V$ such that $d_H(v, v') = 4$. If $vv_1v_2v_3v'$ is a shortest path from v to v', then v_3 and v' are adjacent, and they are not distinguished by the elements in B, which is a contradiction.

- H has one non-trivial component J of girth $g(J) \geq 7$. In this case, if H has two non-trivial components, then we are in the first case. Therefore, we can assume that H has just one non-trivial component of girth $g(J) \geq 7$. Suppose that $B \subseteq N_H(v)$. For each cycle $v_1v_2\ldots v_nv_1$, there exists $v_iv_{i+1} \in E(J)$ such that $d_H(v, v_i) \geq 3$ and $d_H(v, v_{i+1}) \geq 3$; therefore, for each $b \in B$, we have $d_H(b, v_i) \geq 2$ and $d_H(b, v_{i+1}) \geq 2$, which is a contradiction.

- H has two non-singleton true twin equivalence classes U_1, U_2 such that $d_H(U_1, U_2) \geq 3$. Since $B \cap U_1 \neq \emptyset$ and $B \cap U_2 \neq \emptyset$, we can conclude that $B \nsubseteq N_H(v)$.

- $H \cong N_{|V|}$. Notice that $B \neq \emptyset$, as $\mathcal{H} \nsubseteq S_0$, so that $B \nsubseteq \emptyset = N_H(v)$.

According to the five cases above, $\mathcal{H} \subseteq \cup_{i=0}^{4}\mathcal{S}_i$ leads to $D[\mathcal{H}, B] = \emptyset$, for any simultaneous local adjacency generator, which implies that $\Psi(\mathcal{H}) = \mathrm{Sad}_l(\mathcal{H})$. $\square$

Remark 9. *If $A \in Y(V, \mathcal{H})$, then $A \cup (V - \Phi(V, \mathcal{H}))$ is a simultaneous local metric generator for $\mathcal{G}$. However, the converse is not true, as we can see in the following example.*

Example 2. *Consider the family of connected graphs $\mathcal{G} = \{G_1, G_2, G_3\}$ on a common vertex set $V = \{u_1, \ldots, u_8\}$ with $E(G_i) = \{u_1u_2, u_1u_{2i+1}, u_2u_{2i+2}, u_ju_{2i+1}, u_ju_{2i+2}, \text{for } j \notin \{1, 2, 2i+1, 2i+2\}\}$. Let $\mathcal{H}^i$ be the family consisting of only one graph H_i, as follows: $H_1 \cong H_2 \cong K_2$, $H_3 \cong H_4 \cong \cdots \cong H_8 \cong N_2$. We have that $\mathcal{G} \circ \mathcal{H} = \{G_i \circ \{H_1, \ldots, H_8\}, i = 1, 2, 3\}$ and $I(V, \mathcal{H}) = V$. If we take $A = \emptyset$, then $A \cup (V - \Phi(V, \mathcal{H})) = \{u_1, u_2\} \subseteq I(V, \mathcal{H})$ is a simultaneous local metric basis of $\mathcal{G}$. However, $\emptyset \notin Y(V, \mathcal{H})$ because u_1 is adjacent to u_2 in G_i, $i \in \{1, 2, 3\}$, and $(V - \Phi(V, \mathcal{H})) - \{u_1, u_2\} = \emptyset$.*

Lemma 4. *Let $\mathcal{G} \circ \mathcal{H}$ be a family of lexicographic product graphs. Let $B \subseteq V$ be a simultaneous local metric generator for $\mathcal{G}$. Then, $B \cap I(V, \mathcal{H}) \in Y(V, \mathcal{H})$.*

Proof. Let $A = B \cap I(V, \mathcal{H})$ and $u_i, u_j \in I(V, \mathcal{H}) - A = I(V, \mathcal{H}) - B$. Since $B \subseteq V$ is a simultaneous local metric generator for $\mathcal{G}$, for each $G_k \in \mathcal{G}$, there exists $b \in B$ such that $d_{G_k}(b, u_i) \neq d_{G_k}(b, u_j)$. If $b \notin I(V, \mathcal{H})$, then necessarily $b \in (V - I(V, \mathcal{H})) \subseteq ((V - \Phi(V, \mathcal{H})) - \{u_i, u_j\})$, and if $b \in I(V, \mathcal{H})$, then $b \in A - \{u_i, u_j\}$; and we are done. $\square$

Corollary 5. *If there exists a simultaneous local metric generator B for $\mathcal{G}$ such that $B \subseteq V - I(V, \mathcal{H})$ or the graph $G(\mathcal{G}, I(V, \mathcal{H}))$ is empty, then $\emptyset \in Y(V, \mathcal{H})$.*

Remark 10. *If B is a vertex cover of $G(\mathcal{G}, I(V, \mathcal{H}))$, then $B \in Y(V, \mathcal{H})$.*

Lemma 5. *Let $\mathcal{G} \circ \mathcal{H}$ be a family of lexicographic product graphs. For each $u_i \in V$, let $B_i \subseteq V_i$ be a simultaneous local adjacency generator for $\mathcal{H}^i$, and let $C_i \subseteq V_i$ be a dominating set of $D[\mathcal{H}^i, B_i]$. Then, for any $A \in Y(V, \mathcal{H})$, the set $B = (\cup_{u_i \in A}\{u_i\} \times (B_i \cup C_i)) \cup (\cup_{u_i \notin A}\{u_i\} \times B_i)$ is a local metric generator for $\mathcal{G} \circ \mathcal{H}$.*

Proof. In order to prove the lemma, let $G_k \in \mathcal{G}$, $H_j \in \mathcal{H}$, and let $(u_{i_1}, v_1), (u_{i_2}, v_2)$ be a pair of adjacent vertices of $G_k \circ H_j$. If $i_1 = i_2$, then there exists $v \in B_{i_1}$ such that (u_{i_1}, v) distinguishes the pair. Otherwise, $i_1 \neq i_2$, and we consider the following cases:

1. $|\{u_{i_1}, u_{i_2}\} \cap I(V, \mathcal{H})| \leq 1$, say $u_{i_1} \notin I(V, \mathcal{H})$. In this case, there exists $v \in B_{i_1}$ such that $vv_1 \notin E(H_{i_1j})$, and then, (u_{i_1}, v) distinguishes the pair.

2. $u_{i_1}, u_{i_2} \in I(V, \mathcal{H})$ and $\{u_{i_1}, u_{i_2}\} \cap A = \varnothing$. In this case, by definition of A, there exists $u_{i_3} \in (A \cup (V - \Phi(V, \mathcal{H}))) - \{u_{i_1}, u_{i_2}\}$ such that $d_{G_k}(u_{i_3}, u_{i_1}) \neq d_{G_k}(u_{i_3}, u_{i_2})$. For any $v \in B_{i_3} \cup C_{i_3}$,

$$d_{G_k \circ \mathcal{H}_j}((u_{i_3}, v), (u_{i_1}, v_1)) = d_{G_k}(u_{i_3}, u_{i_1}) \neq$$
$$d_{G_k}(u_{i_3}, u_{i_2}) = d_{G_k \circ \mathcal{H}_j}((u_{i_3}, v), (u_{i_2}, v_2)).$$

3. $u_{i_1}, u_{i_2} \in I(V, \mathcal{H})$ and $|\{u_{i_1}, u_{i_2}\} \cap A| \geq 1$, say $u_{i_1} \in A$. In this case, if there exists $v \in B_{i_1}$ such that $vv_1 \notin E(H_{i_1 j})$, then (u_{i_1}, v) distinguishes the pair. Otherwise, v_1 is a vertex of $\mathcal{D}[\mathcal{H}^{i_1}, B_{i_1}]$, and either $v_1 \in C_{i_1}$ and $(u_{i_1}, v_1) \in B$ distinguishes the pair or there exists $v \in C_{i_1}$, such that $vv_1 \in E(\mathcal{D}[\mathcal{H}^{i_1}, B_{i_1}])$, which means $vv_1 \notin E(H_{i_1 j})$; then, (u_{i_1}, v) distinguishes the pair. $\square$

Corollary 6. *Let $\mathcal{G} \circ \mathcal{H}$ be a family of lexicographic product graphs. Then:*

$$\mathrm{Sd}_l(\mathcal{G} \circ \mathcal{H}) \leq \min_{A \in Y(V, \mathcal{H})} \left\{ \sum_{u_i \in A} \Psi(\mathcal{H}^i) + \sum_{u_i \notin A} \mathrm{Sad}_l(\mathcal{H}^i) \right\}.$$

Proof. Let $A \in Y(V, \mathcal{H})$. For each $u_i \notin A$, let $B_i \subseteq V_i$ be a simultaneous local adjacency basis of $\mathcal{H}^i$. For each $u_i \in A$, let B_i be a local adjacency generator for $\mathcal{H}^i$ and $C_i \subseteq V_i$ a dominating set of $\mathcal{D}(\mathcal{H}^i, B_i)$ such that $|B_i \cup C_i| = \Psi(\mathcal{H}^i)$. Let:

$$B = (\cup_{u_j \in A} \{u_j\} \times (B_j \cup C_j)) \bigcup (\cup_{u_i \notin A} \{u_i\} \times B_i)$$

then, by Lemma 5, B is a simultaneous local metric generator for $\mathcal{G} \circ \mathcal{H}$, and:

$$\mathrm{Sd}_l(\mathcal{G} \circ \mathcal{H}) \leq |B| = \sum_{u_i \in A} \Psi(\mathcal{H}^i) + \sum_{u_i \notin A} \mathrm{Sad}_l(\mathcal{H}^i)$$

As $A \in Y(V, \mathcal{H})$ is arbitrary:

$$\mathrm{Sd}_l(\mathcal{G} \circ \mathcal{H}) \leq \min_{A \in Y(V, \mathcal{H})} \left\{ \sum_{u_i \in A} \Psi(\mathcal{H}^i) + \sum_{u_i \notin A} \mathrm{Sad}_l(\mathcal{H}^i) \right\}$$

and the result follows. $\square$

Lemma 6. *Let F be a simultaneous local metric basis of $\mathcal{G} \circ \mathcal{H}$. Let $F_i = \{v \in V_i : (u_i, v) \in F\}$ and $X_F = \{u_i \in I(V, \mathcal{H}) : |F_i| \geq \Psi(\mathcal{H}^i)\}$. Then, $X_F \in Y(V, \mathcal{H})$.*

Proof. Suppose, for the purpose of contradiction, that $X_F \notin Y(V, \mathcal{H})$. That means that there exist $u_{i_1}, u_{i_2} \in I(V, \mathcal{H}) - X_F$ and $G_k \in \mathcal{G}$ such that $u_{i_1} u_{i_2} \in E(G_k)$, and $d_{G_k}(u, u_{i_1}) = d_{G_k}(u, u_{i_2})$ for every $u \in (X_F \cup (V - \Phi(V, \mathcal{H}))) - \{u_{i_1}, u_{i_2}\}$. As $u_{i_1}, u_{i_2} \in I(V, \mathcal{H}) - X_F$, $|F_{i_1}| < \Psi(\mathcal{H}^{i_1})$ and $|F_{i_2}| < \Psi(\mathcal{H}^{i_2})$, so that there exist $H_{i_1 j_1} \in \mathcal{H}^{i_1}$ and $H_{i_2 j_2} \in \mathcal{H}^{i_2}$ such that for some $v_1 \in V_{i_1}, v_2 \in V_{i_2}, F_{i_1} \subseteq N_{H_{i_1 j_1}}(v_1)$ and $F_{i_2} \subseteq N_{H_{i_2 j_2}}(v_2)$. Let $\mathcal{H}_j$ be such that $H_{i_1 j_1}, H_{i_2 j_2} \in \mathcal{H}_j$. Consider the pair of vertices $(u_{i_1}, v_1), (u_{i_2}, v_2)$ adjacent in $G_k \circ \mathcal{H}_j$. As F is a simultaneous local metric generator, there exists $(u_{i_3}, v) \in F$ that resolves the pair, which implies that $F_{i_3} \neq \varnothing$. By hypothesis $u_{i_3} \in (\Phi(V, \mathcal{H}) - X_F) \cup \{u_{i_1}, u_{i_2}\}$, and so, $u_{i_3} \in \{u_{i_1}, u_{i_2}\}$. Without loss of generality, we assume that $u_{i_3} = u_{i_1}$ and, in this case,

$$d_{G_k \circ \mathcal{H}_j}((u_{i_3}, v), (u_{i_1}, v_1)) = d_{H_{i_1 j_1}, 2}(v, v_1)$$
$$= d_{G_k}(u_{i_3}, u_{i_2})$$
$$= d_{G_k \circ \mathcal{H}_j}((u_{i_3}, v), (u_{i_2}, v_2)),$$

which is a contradiction. Therefore, $X_F \in Y(V, \mathcal{H})$. $\square$

Theorem 13. *Let $\mathcal{G} \circ \mathcal{H}$ be a family of lexicographic product graphs.*

$$\mathrm{Sd}_l(\mathcal{G} \circ \mathcal{H}) = \min_{A \in Y(V, \mathcal{H})} \left\{ \sum_{u_i \in A} \Psi(\mathcal{H}^i) + \sum_{u_i \notin A} \mathrm{Sad}_l(\mathcal{H}^i) \right\}$$

Proof. Let B be a simultaneous local metric basis of $\mathcal{G} \circ \mathcal{H}$. Let $B_i = \{v \in V_i : (u_i, v) \in B\}$ and $X_B = \{u_i \in I(V, \mathcal{H}) : |B_i| \geq \Psi(\mathcal{H}^i)\}$. By Remark 7, $|B_i| \geq \mathrm{Sad}_l(\mathcal{H}^i)$ for every $u_i \in V$, so that Lemma 6 leads to:

$$\min_{A \in Y(V, \mathcal{H})} \left\{ \sum_{u_i \in A} \Psi(\mathcal{H}^i) + \sum_{u_i \notin A} \mathrm{Sad}_l(\mathcal{H}^i) \right\} \leq \sum_{u_i \in X_B} \Psi(\mathcal{H}^i) + \sum_{u_i \notin X_B} \mathrm{Sad}_l(\mathcal{H}^i) \leq |B|$$

and the result follows by Corollary 6. $\square$

Now, we will show some cases where the calculation of $\mathrm{Sd}_l(\mathcal{G} \circ \mathcal{H})$ is easy. At first glance, we have two main types of simplification: first, to simplify the calculation of $\Psi(\mathcal{H}^i)$ and, second, the calculation of the $A \in Y(V, \mathcal{H})$ that makes the sum achieves its minimum.

For the first type of simplification, we can apply Lemma 3 to deduce the following corollary.

Corollary 7. *If for each i, $\mathcal{H}^i \not\subseteq S_0$ and $\mathcal{H}^i \subseteq \bigcup_{j=0}^{4} S_j$, then:*

$$\mathrm{Sd}_l(\mathcal{G} \circ \mathcal{H}) = \sum \mathrm{Sad}_l(\mathcal{H}^i).$$

Given a family $\mathcal{G}$ of graphs on a common vertex set V and a graph H, we define the family of lexicographic product graphs:

$$\mathcal{G} \circ H = \{G \circ H : G \in \mathcal{G}\}.$$

By Theorem 13, we deduce the following result.

Corollary 8. *Let $\mathcal{G}$ be a family of graphs on a common vertex set V, and let H be a graph. If for every local adjacency basis B of H, $B \not\subseteq N_H(v)$ for every $v \in V(H) - B$, then:*

$$\mathrm{Sd}_l(\mathcal{G} \circ H) = |V| \, \mathrm{adim}_l(H).$$

By Corollary 5 and Theorem 13, we have the following result.

Proposition 2. *If $V - I(V, \mathcal{H})$ is a simultaneous local metric generator for $\mathcal{G}$ or the graph $G(\mathcal{G}, I(V, \mathcal{H}))$ is empty, then:*

$$\mathrm{Sd}_l(\mathcal{G} \circ \mathcal{H}) = \sum \mathrm{Sad}_l(\mathcal{H}^i)$$

For the second type of simplification, we have the following remark.

Remark 11. *As $\mathrm{Sad}_l(\mathcal{H}^i) \leq \Psi(\mathcal{H}^i)$, if $A \subseteq B \subseteq V$, then:*

$$\sum_{u_i \in A} \Psi(\mathcal{H}^i) + \sum_{u_i \notin A} \mathrm{Sad}_l(\mathcal{H}^i) \leq \sum_{u_i \in B} \Psi(\mathcal{H}^i) + \sum_{u_i \notin B} \mathrm{Sad}_l(\mathcal{H}^i)$$

From Remark 11, we can get some consequences of Theorem 13.

Proposition 3. *Let $\mathcal{G} \circ \mathcal{H}$ be a family of lexicographic product graphs. For any vertex cover B of $G(\mathcal{G}, I(V, \mathcal{H}))$,*

$$\mathrm{Sd}_l(\mathcal{G} \circ \mathcal{H}) \leq \sum_{u_i \in B} \Psi(\mathcal{H}^i) + \sum_{u_i \notin B} \mathrm{Sad}_l(\mathcal{H}^i)$$

Proposition 4. *Let $\mathcal{G}$ be a family of connected graphs with common vertex set V, and let $\mathcal{G} \circ \mathcal{H}$ be a family of lexicographic product graphs. The following statements hold.*

1. *If the subgraph of G_j induced by $I(V, \mathcal{H})$ is empty for every $G_j \in \mathcal{G}$, then:*

$$\mathrm{Sd}_l(\mathcal{G} \circ \mathcal{H}) = \sum_{u_i \in V} \mathrm{Sad}_l(\mathcal{H}^i).$$

2. *Let $u_{i_0} \in I(V, \mathcal{H})$ be such that $\Psi(\mathcal{H}^{i_0}) = \max\{\Psi(u_i) : u_i \in I(V, \mathcal{H})\}$. If $\mathrm{Sd}_l(\mathcal{G}) = |V| - 1$ and $|I(V, \mathcal{H})| \geq 2$, then:*

$$\mathrm{Sd}_l(\mathcal{G} \circ \mathcal{H}) = \sum_{u_i \notin I(V, \mathcal{H})} \mathrm{Sad}_l(\mathcal{H}^i) + \sum_{u_i \in I(V, \mathcal{H}) - \{u_{i_0}\}} \Psi(\mathcal{H}^i) + \mathrm{Sad}_l(\mathcal{H}^{i_0})$$

Proof. It is clear that if the subgraph of G_j induced by $I(V, \mathcal{H})$ is empty for every $G_j \in \mathcal{G}$, then $\varnothing \in Y(V, \mathcal{H})$, so that Theorem 13 leads to (1). On the other hand, let $\mathcal{G}$ be a family of connected graphs with common vertex set V such that $\mathrm{Sd}_l(\mathcal{G}) = |V| - 1$ and $|I(V, \mathcal{H})| \geq 2$. By Lemma 1, for every $u_i, u_j \in I(V, \mathcal{H})$, there exists $G_{ij} \in \mathcal{G}$ such that u_i, u_j are true twins in G_{ij}. Hence, no vertex $u \notin \{u_i, u_j\}$ resolves u_i and u_j. Therefore, $A \in Y(V, \mathcal{H})$ implies $|A| = |I(V, \mathcal{H})| - 1$, and (2) follows from Theorem 13 and Remark 11. $\square$

Proposition 5. *Let $\mathcal{G}$ be a family of non-trivial connected graphs with common vertex set V. For any family of lexicographic product graphs $\mathcal{G} \circ \mathcal{H}$,*

$$\mathrm{Sd}_l(\mathcal{G} \circ \mathcal{H}) \geq \mathrm{Sd}_l(\mathcal{G}).$$

Furthermore, if $\mathcal{H} = \{N_{|V_1|}, \dots, N_{|V_n|}\}$, then:

$$\mathrm{Sd}_l(\mathcal{G} \circ \mathcal{H}) = \mathrm{Sd}_l(\mathcal{G}).$$

Proof. Let W be a simultaneous local metric basis of $\mathcal{G} \circ \mathcal{H}$ and $W_V = \{u \in V : (u, v) \in W\}$. We suppose that W_V is not a simultaneous local metric generator for $\mathcal{G}$. Let $u_i, u_j \notin W_V$ and $G \in \mathcal{G}$ such that $u_i u_j \in E(G)$ and $d_G(u_i, u) = d_G(u_j, u)$ for every $u \in W_V$. Thus, for any $v \in V_i$, $v' \in V_j$ and $(x, y) \in W$, we have:

$$d_{G \circ H_i}((x, y), (u_i, v)) = d_G(x, u_i) = d_G(x, u_j) = d_{G \circ H_j}((x, y), (u_j, v')),$$

which is a contradiction. Therefore, W_V is a simultaneous local metric generator for $\mathcal{G}$ and, as a result, $\mathrm{Sd}_l(\mathcal{G}) \leq |W_V| \leq |W| = \mathrm{Sd}_l(\mathcal{G} \circ \mathcal{H})$.

On the other hand, if $\mathcal{H} = \{N_{|V_1|}, \dots, N_{|V_n|}\}$, then $V = I(V, \mathcal{H}) = \Phi(V, \mathcal{H})$. Let $B \subseteq V$ be a simultaneous local metric basis of $\mathcal{G}$. Now, for each $u_i \in B$, we choose $v_i \in V_i$, and by Remark 9, we claim that $B' = \{(u_i, v_i) : u_i \in B\}$ is a simultaneous local metric generator for $\mathcal{G} \circ \mathcal{H}$. Thus, $\mathrm{Sd}_l(\mathcal{G} \circ \mathcal{H}) \leq |B'| = |B| = \mathrm{Sd}_l(\mathcal{G})$. $\square$

Proposition 6. *Let $\mathcal{G} \neq \{K_2\}$ be a family of non-trivial connected bipartite graphs with common vertex set V and $\mathcal{H} \neq \{\mathcal{H}_1, \dots, \mathcal{H}_n\}$ such that $\mathcal{H}_j \not\subseteq S_0$, for some j. If $V = I(V, \mathcal{H})$ and there exist $u_1, u_2 \in V$ and $G_k \in \mathcal{G}$ such that $V - \Phi(V, \mathcal{H}) = \{u_1, u_2\}$ and $u_1 u_2 \in E(G_k)$, then:*

$$\mathrm{Sd}_l(\mathcal{G} \circ \mathcal{H}) = \sum \mathrm{Sad}_l(\mathcal{H}^i) + 1,$$

otherwise,

$$\mathrm{Sd}_l(\mathcal{G} \circ \mathcal{H}) = \sum \mathrm{Sad}_l(\mathcal{H}^i).$$

Proof. If $V = I(V, \mathcal{H})$ and there exist $u_1, u_2 \in V$ and $G_k \in \mathcal{G}$ such that $V - \Phi(V, \mathcal{H}) = \{u_1, u_2\}$ and $u_1 u_2 \in E(G_k)$, then $\emptyset \notin Y(V, \mathcal{H})$ because no vertex in $(V - \Phi(V, \mathcal{H})) - \{u_1, u_2\} = \emptyset$ distinguishes u_1 and u_2. Let $x, y \in I(V, \mathcal{H})$ such that $xy \in \cup_{G \in \mathcal{G}} E(G)$. Since any $u_i \in \Phi(V, \mathcal{H})$ distinguishes x and y, we can conclude that $\{u_i\} \in Y(V, \mathcal{H})$, and by Remark 8, $\Psi(\mathcal{H}^i) = 1$. Therefore, Theorem 13 leads to $\mathrm{Sd}_l(\mathcal{G} \circ \mathcal{H}) = \sum \mathrm{Sad}_l(\mathcal{H}^i) + 1$.

Assume that there exists $u_i \in V - I(V, \mathcal{H})$, or $V - \Phi(V, \mathcal{H}) = \{u_i\}$, or $V - \Phi(V, \mathcal{H}) = \{u_i, u_j\}$ and, for every $G_k \in \mathcal{G}$, $u_i u_j \notin E(G_k)$ or $\{u_i, u_j, u_k\} \subseteq V - \Phi(V, \mathcal{H})$. In any one of these cases $\{u_i\}$ is a simultaneous local metric basis of $\mathcal{G}$ and, for every pair u_1, u_2 of adjacent vertices in some $G_k \in \mathcal{G}$ such that $u_i \notin \{u_1, u_2\}$, u_i distinguishes the pair. Since $u_i \in V - \Phi(V, \mathcal{H})$, we can claim that $\emptyset \in Y(V, \mathcal{H})$, and by Theorem 13, $\mathrm{Sd}_l(\mathcal{G} \circ \mathcal{H}) = \sum \mathrm{Sad}_l(\mathcal{H}^i)$. $\square$

5.1. Families of Join Graphs

For two graph families $\mathcal{G} = \{G_1, \dots, G_{k_1}\}$ and $\mathcal{H} = \{H_1, \dots, H_{k_2}\}$, defined on common vertex sets V_1 and V_2, respectively, such that $V_1 \cap V_2 = \emptyset$, we define the family:

$$\mathcal{G} + \mathcal{H} = \{G_i + H_j : 1 \leq i \leq k_1, 1 \leq j \leq k_2\}.$$

Notice that, since for any $G_i \in \mathcal{G}$ and $H_j \in \mathcal{H}$ the graph $G_i + H_j$ has diameter two,

$$\mathrm{Sd}_l(\mathcal{G} + \mathcal{H}) = \mathrm{Sad}_l(\mathcal{G} + \mathcal{H}).$$

The following result is a direct consequence of Theorem 13.

Corollary 9. *For any pair of families $\mathcal{G}$ and $\mathcal{H}$ of non-trivial graphs on common vertex sets V_1 and V_2, respectively,*

$$\mathrm{Sd}_l(\mathcal{G} + \mathcal{H}) = \min\{\mathrm{Sd}_{A,l}(\mathcal{G}) + \Psi(\mathcal{H}), \mathrm{Sd}_{A,l}(\mathcal{H}) + \Psi(\mathcal{G})\}$$

Remark 12. *Let $\mathcal{G}$ be a family of graphs defined on a common vertex set V_1. If there exists B a simultaneous local adjacency basis of $\mathcal{G}$ such that $D[\mathcal{G}, B] = \emptyset$, then for every $\mathcal{H}$ family of graphs defined on a common vertex set V_2, we have:*

$$\mathrm{Sd}_l(\mathcal{G} + \mathcal{H}) = \mathrm{Sad}_l(\mathcal{G}) + \mathrm{Sad}_l(\mathcal{H})$$

By Lemma 3 and Remark 12, we deduce the following result.

Proposition 7. *Let $\mathcal{G}$ and $\mathcal{H}$ be two families of non-trivial connected graphs on a common vertex set V_1 and V_2, respectively. If $\mathcal{G} \subseteq \cup_{i=1}^4 S_i$, then:*

$$\mathrm{Sd}_l(\mathcal{G} + \mathcal{H}) = \mathrm{Sad}_l(\mathcal{G}) + \mathrm{Sad}_l(\mathcal{H}).$$

6. Computability of the Simultaneous Local Metric Dimension

In previous sections, we have seen that there is a large number of classes of graph families for which the simultaneous local metric dimension is well determined. This includes some cases of graph families whose simultaneous metric dimension is hard to compute, e.g., families composed by trees [22], yet the simultaneous local metric dimension is constant. However, as proven in [23], the problem of finding the local metric dimension of a graph is NP-hard in the general case, which trivially leads to the fact that finding the simultaneous local metric dimension of a graph family is also NP-hard in the general case.

Here, we will focus on a different aspect, namely that of showing that the requirement of simultaneity adds to the computational difficulty of the original problem. To that end, we will show that there exist families composed by graphs whose individual local metric dimensions are constant, yet it is hard to compute their simultaneous local metric dimension.

To begin with, we will formally define the decision problems associated with the computation of the local metric dimension of one graph and the simultaneous local metric dimension of a graph family.

Local metric Dimension (LDIM)
Instance: A graph $G = (V, E)$ and an integer p, $1 \le p \le |V(G)| - 1$.
Question: Is $\dim_l(G) \le p$?

Simultaneous Local metric Dimension (SLD)
Instance: A graph family $\mathcal{G} = \{G_1, G_2, \ldots, G_k\}$ on a common vertex set V and an integer p, $1 \le p \le |V| - 1$.
Question: Is $\mathrm{Sd}_l(\mathcal{G}) \le p$?

As we mentioned above, LDIM was proven to be NP-complete in [23]. Moreover, it is simple to see that determining whether a vertex set $S \subseteq V$, $|S| \le p$, is a simultaneous local metric generator can be done in polynomial time, so SLD is in NP. In fact, SLD can be easily shown to be NP-complete, since for any graph $G = (V, E)$ and any integer $1 \le p \le |V(G)| - 1$, the corresponding instance of LDIM can be trivially transformed into an instance of SLD by making $\mathcal{G} = \{G\}$.

For the remainder of this section, we will address the issue of the complexity added by the requirement of simultaneity. To this end, we will consider families composed by the so-called tadpole graphs [28]. An (h, t)-tadpole graph (or (h, t)-tadpole for short) is the graph obtained from a cycle graph C_h and a path graph P_t by joining with an edge a leaf of P_t to an arbitrary vertex of C_h. We will use the notation $T_{h,t}$ for (h, t)-tadpoles. Since $(2q, t)$-tadpoles are bipartite, we have that $\dim_l(T_{2q,t}) = 1$. In the case of $(2q + 1, t)$-tadpoles, we have that $\dim_l(T_{2q+1,t}) = 2$, as they are not bipartite (so, $\dim_l(T_{2q+1,t}) \ge 2$), and any set composed by two vertices of the subgraph C_{2q+1} is a local metric generator (so, $\dim_l(T_{2q+1,t}) \le 2$). Additionally, consider the sole vertex v of degree three in $T_{2q+1,t}$ and a local metric generator for $T_{2q+1,t}$ of the form $\{v, x\}$, $x \in V(C_{2q+1}) - \{v\}$. It is simple to verify that for any vertex $y \in V(P_t)$, the set $\{y, x\}$ is also a local metric generator for $T_{2q+1,t}$.

Consider a family $\mathcal{T} = \{T_{h_1,t_1}, T_{h_2,t_2}, \ldots, T_{h_k,t_k}\}$ composed by tadpole graphs on a common vertex set V. By Theorem 4, we have that $\mathrm{Sd}_l(\mathcal{T}) = \mathrm{Sd}_l(\mathcal{T}')$, where $\mathcal{T}'$ is composed by $(2q + 1, t)$-tadpoles. As we discussed previously, $\dim_l(T_{2q+1,t}) = 2$. However, by Remark 1 and Theorem 1, we have that $2 \le \mathrm{Sd}_l(\mathcal{T}') \le |V| - 1$. In fact, both bounds are tight, since the lower bound is trivially satisfied by unitary families, whereas the upper bound is reached, for instance, by any family composed by all different labeled graphs isomorphic to an arbitrary $(3, t)$-tadpole, as it satisfies the premises of Theorem 1. Moreover, as we will show, the problem of computing $\mathrm{Sd}_l(\mathcal{T}')$ is NP-hard, as its associated decision problem is NP-complete. We will do so by showing a transformation from the hitting set problem, which was shown to be NP-complete by Karp [29]. The hitting set problem is defined as follows:

Hitting Set Problem (HSP)
Instance: A collection $\mathcal{C} = \{C_1, C_2, \ldots, C_k\}$ of non-empty subsets of a finite set S and a positive integer $p \le |S|$.
Question: Is there a subset $S' \subseteq S$ with $|S'| \le p$ such that S' contains at least one element from each subset in $\mathcal{C}$?

Theorem 14. *The Simultaneous Local metric Dimension problem (SLD) is NP-complete for families of $(2q + 1, t)$-tadpoles.*

Proof. As we discussed previously, determining whether a vertex set $S \subseteq V$, $|S| \le p$, is a simultaneous local metric generator for a graph family $\mathcal{G}$ can be done in polynomial time, so SLD is in NP.

Now, we will show a polynomial time transformation of HSP into SLD. Let $S = \{v_1, v_2, \ldots, v_n\}$ be a finite set, and let $\mathcal{C} = \{C_1, C_2, \ldots, C_k\}$, where every $C_i \in \mathcal{C}$ satisfies $C_i \subseteq S$. Let p be a positive integer such that $p \le |S|$. Let $A = \{w_1, w_2, \ldots, w_k\}$ such that $A \cap S = \emptyset$. We construct the family

$\mathcal{T} = \{T_{2q_1+1,t_1}, T_{2q_2+1,t_2}, \ldots, T_{2q_k+1,t_k}\}$ composed by $(2q + 1, t)$-tadpoles on the common vertex set $V = S \cup A \cup \{u\}$, $u \notin S \cup A$, by performing one of the two following actions, as appropriate, for every $r \in \{1, \ldots, k\}$:

- If $|C_r|$ is even, let C_{2q_r+1} be a cycle graph on the vertices of $C_r \cup \{u\}$; let P_{t_r} be a path graph on the vertices of $(S - C_r) \cup A$; and let T_{2q_r+1,t_r} be the tadpole graph obtained from C_{2q_r+1} and P_{t_r} by joining with an edge a leaf of P_{t_r} to a vertex of C_{2q_r+1} different from u.
- If $|C_r|$ is odd, let C_{2q_r+1} be a cycle graph on the vertices of $C_r \cup \{u, w_r\}$; let P_{t_r} be a path graph on the vertices of $(S - C_r) \cup (A - \{w_r\})$; and let T_{2q_r+1,t_r} be the tadpole graph obtained from C_{2q_r+1} and P_{t_r} by joining with an edge the vertex w_r to a leaf of P_{t_r}.

Figure 4 shows an example of this construction.

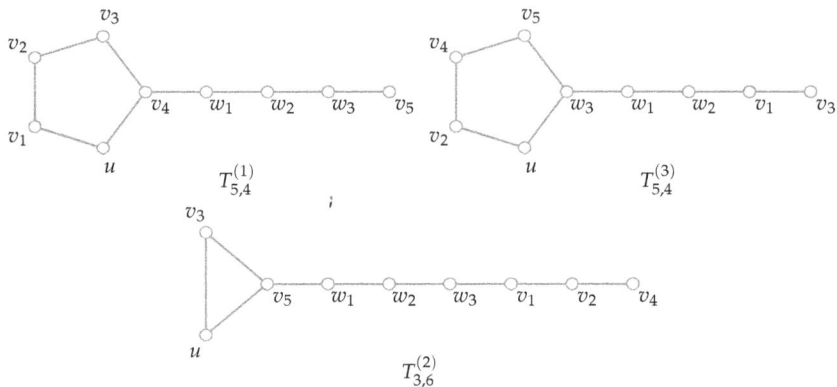

Figure 4. The family $\mathcal{T} = \{T_{5,4}^{(1)}, T_{3,6}^{(2)}, T_{5,4}^{(3)}\}$ is constructed for transforming an instance of the Hitting Set Problem (HSP), where $S = \{v_1, v_2, v_3, v_4, v_5\}$ and $\mathcal{C} = \{\{v_1, v_2, v_3, v_4\}, \{v_3, v_5\}, \{v_2, v_4, v_5\}\}$, into an instance of Simultaneous Local metric Dimension (SLD) for families of $(2q + 1, t)$-tadpoles.

In order to prove the validity of this transformation, we claim that there exists a subset $S'' \subseteq S$ of cardinality $|S''| \le p$ that contains at least one element from each $C_r \in \mathcal{C}$ if and only if $\mathrm{Sd}_l(\mathcal{T}) \le p+1$.

To prove this claim, first assume that there exists a set $S'' \subseteq S$, which contains at least one element from each $C_r \in \mathcal{C}$ and satisfies $|S''| \le p$. Recall that any set composed by two vertices of C_{2q_r+1} is a local metric generator for T_{2q_r+1,t_r}, so $S'' \cup \{u\}$ is a simultaneous local metric generator for $\mathcal{T}$. Thus, $\mathrm{Sd}_l(\mathcal{T}) \le p+1$.

Now, assume that $\mathrm{Sd}_l(\mathcal{T}) \le p+1$, and let W be a simultaneous local metric generator for $\mathcal{T}$ such that $|W| = p+1$. For every $T_{2q_r+1,t_r} \in \mathcal{T}$, we have that $u \in V(C_{2q_r+1})$ and $\delta_{T_{2q_r+1,t_r}}(u) = 2$, so $|((W - \{x\}) \cup \{u\}) \cap V(C_{2q_r+1})| \ge |W \cap V(C_{2q_r+1})|$ for any $x \in W$. As a consequence, if $u \notin W$, any set $(W - \{x\}) \cup \{u\}$, $x \in W$, is also a simultaneous local metric generator for $\mathcal{T}$, so we can assume that $u \in W$. Moreover, applying an analogous reasoning for every set $C_r \in \mathcal{C}$ such that $W \cap C_r = \emptyset$, we have that, firstly, there is at least one vertex $v_{r_i} \in C_r$ such that $v_{r_i} \in V(C_{2q_r+1}) - \{u\}$ and $\delta_{T_{2q_r+1,t_r}}(v_{r_i}) = 2$, and secondly, there is at least one vertex $x_r \in W \cap (\{w_r\} \cup V(P_{t_r}))$, which can be replaced by v_{r_i}. Then, the set:

$$W' = \bigcup_{W \cap C_r = \emptyset} ((W - \{x_r\}) \cup \{v_{r_i}\})$$

is also a simultaneous local metric generator for $\mathcal{T}$ of cardinality $|W'| = p+1$ such that $u \in W'$ and $(W' - \{u\}) \cap C_r \ne \emptyset$ for every $C_r \in \mathcal{C}$. Thus, the set $S'' = W' - \{u\}$ satisfies $|S''| \le p$ and contains at least one element from each $C_r \in \mathcal{C}$.

To conclude our proof, it is simple to verify that the transformation of HSP into SLD described above can be done in polynomial time. □

7. Conclusions

In this paper we introduced the notion of simultaneous local dimension of graph families. We studied the properties of this new parameter in order to obtain its exact value, or sharp bounds, on several graph families. In particular, we focused on families obtained as the result of small changes in an initial graph and families composed by graphs obtained through well-known operations such as the corona and lexicographic products, as well as the join operation (viewed as a particular case of the lexicographic product). Finally, we analysed the computational complexity of the new problem, and showed that computing the simultaneous local metric dimension is computationally difficult even for families composed by graphs whose (individual) local metric dimensions are constant and well known.

Author Contributions: The results presented in this paper were obtained as a result of collective work sessions involving all authors. The process was organized and led by Juan A. Rodríguez-Velázquez and the writing of the final version of the paper was conducted by Juan A. Rodríguez-Velázquez and Yunior Ramírez-Cruz.

Conflicts of Interest: The authors declare no conflict of interest.

References

1. Blumenthal, L.M. *Theory and Applications of Distance Geometry*; Oxford University Press: Oxford, UK, 1953.
2. Slater, P.J. Leaves of trees. *Congr. Numer.* **1975**, *14*, 549–559.
3. Harary, F.; Melter, R.A. On the metric dimension of a graph. *Ars Comb.* **1976**, *2*, 191–195.
4. Khuller, S.; Raghavachari, B.; Rosenfeld, A. Landmarks in graphs. *Discret. Appl. Math.* **1996**, *70*, 217–229.
5. Johnson, M. Structure-activity maps for visualizing the graph variables arising in drug design. *J. Biopharm. Stat.* **1993**, *3*, 203–236.
6. Johnson, M. Browsable structure-activity datasets. In *Advances in Molecular Similarity*; Carbó-Dorca, R., Mezey, P., Eds.; JAI Press Inc.: Stamford, CT, USA, 1998; Chapter 8, pp. 153–170.
7. Melter, R.A.; Tomescu, I. Metric bases in digital geometry. *Comput. Vis. Graph. Image Process.* **1984**, *25*, 113–121.
8. Chartrand, G.; Eroh, L.; Johnson, M.A.; Oellermann, O.R. Resolvability in graphs and the metric dimension of a graph. *Discret. Appl. Math.* **2000**, *105*, 99–113.
9. Sebő, A.; Tannier, E. On metric generators of graphs. *Math. Oper. Res.* **2004**, *29*, 383–393.
10. Saenpholphat, V.; Zhang, P. Conditional resolvability in graphs: A survey. *Int. J. Math. Math. Sci.* **2004**, *2004*, 1997–2017.
11. Hernando, C.; Mora, M.; Pelayo, I.M.; Seara, C.; Cáceres, J.; Puertas, M.L. On the metric dimension of some families of graphs. *Electron. Notes Discret. Math.* **2005**, *22*, 129–133.
12. Haynes, T.W.; Henning, M.A.; Howard, J. Locating and total dominating sets in trees. *Discret. Appl. Math.* **2006**, *154*, 1293–1300.
13. Cáceres, J.; Hernando, C.; Mora, M.; Pelayo, I.M.; Puertas, M.L.; Seara, C.; Wood, D.R. On the metric dimension of Cartesian product of graphs. *SIAM J. Discret. Math.* **2007**, *21*, 423–441.
14. Yero, I.G.; Kuziak, D.; Rodríguez-Velázquez, J.A. On the metric dimension of corona product graphs. *Comput. Math. Appl.* **2011**, *61*, 2793–2798.
15. Bailey, R.F.; Meagher, K. On the metric dimension of grassmann graphs. *Discret. Math. Theor. Comput. Sci.* **2011**, *13*, 97–104.
16. Jannesari, M.; Omoomi, B. The metric dimension of the lexicographic product of graphs. *Discret. Math.* **2012**, *312*, 3349–3356.
17. Feng, M.; Wang, K. On the metric dimension of bilinear forms graphs. *Discret. Math.* **2012**, *312*, 1266–1268.
18. Guo, J.; Wang, K.; Li, F. Metric dimension of some distance-regular graphs. *J. Comb. Optim.* **2013**, *26*, 190–197.
19. Imran, M.; Bokhary, S.A.U.H.; Ahmad, A.; Semaničová-Feňovčíková, A. On classes of regular graphs with constant metric dimension. *Acta Math. Sci.* **2013**, *33*, 187–206.

20. Saputro, S.; Simanjuntak, R.; Uttunggadewa, S.; Assiyatun, H.; Baskoro, E.; Salman, A.; Bača, M. The metric dimension of the lexicographic product of graphs. *Discret. Math.* **2013**, *313*, 1045–1051.

21. Okamoto, F.; Phinezy, B.; Zhang, P. The local metric dimension of a graph. *Math. Bohem.* **2010**, *135*, 239–255.

22. Ramírez-Cruz, Y.; Oellermann, O.R.; Rodríguez-Velázquez, J.A. The simultaneous metric dimension of graph families. *Discret. Appl. Math.* **2016**, *198*, 241–250.

23. Fernau, H.; Rodríguez-Velázquez, J.A. On the (Adjacency) Metric Dimension of Corona and Strong Product Graphs and Their Local Variants: Combinatorial and Computational Results (arXiv:1309.2275). Available online: https://arxiv.org/abs/1309.2275 (accessed on 26 July 2017).

24. Rodríguez-Velázquez, J.A.; Barragán-Ramírez, G.A.; García Gómez, C. On the local metric dimension of corona product graphs. *Bull. Malays. Math. Sci. Soc.* **2016**, *39* (Suppl. 1), S157–S173.

25. Fernau, H.; Rodríguez-Velázquez, J.A. *Notions of Metric Dimension of Corona Products: Combinatorial and Computational Results*; Springer International Publishing: Cham, Switzerland, 2014; pp. 153–166.

26. Hammack, R.; Imrich, W.; Klavžar, S. *Handbook of Product Graphs*, 2nd ed.; CRC Press: Boca Raton, FL, USA, 2011.

27. Imrich, W.; Klavžar, S. *Product Graphs, Structure and Recognition*; Wiley: New York, NY, USA, 2000.

28. Koh, K.M.; Rogers, D.G.; Teo, H.K.; Yap, K.Y. Graceful graphs: Some further results and problems. *Congr. Numer.* **1980**, *29*, 559–571.

29. Karp, R. Reducibility among combinatorial problems. In *Complexity of Computer Computations*; Miller, R., Thatcher, J., Eds.; Plenum Press: New York, NY, USA, 1972; pp. 85–103.

symmetry

MDPI

Article

Gromov Hyperbolicity in Mycielskian Graphs

Ana Granados [1], Domingo Pestana [2,*], Ana Portilla [1] and José M. Rodríguez [2]

[1] Department of Mathematics and Computer Science, Saint Louis University, Avenida del Valle 34, 28003 Madrid, Spain; agranado@slu.edu (A.G.); aportil2@slu.edu (A.P.)
[2] Department of Mathematics, Universidad Carlos III de Madrid, Avenida de la Universidad 30, 28911 Leganés, Spain; jomaro@math.uc3m.es
* Correspondence: domingo.pestana@uc3m.es; Tel.: +34-91-624-9098

Academic Editor: Angel Garrido
Received: 21 June 2017; Accepted: 21 July 2017; Published: 27 July 2017

Abstract: Since the characterization of Gromov hyperbolic graphs seems a too ambitious task, there are many papers studying the hyperbolicity of several classes of graphs. In this paper, it is proven that every Mycielskian graph G^M is hyperbolic and that $\delta(G^M)$ is comparable to $\mathrm{diam}(G^M)$. Furthermore, we study the extremal problems of finding the smallest and largest hyperbolicity constants of such graphs; in fact, it is shown that $5/4 \le \delta(G^M) \le 5/2$. Graphs G whose Mycielskian have hyperbolicity constant $5/4$ or $5/2$ are characterized. The hyperbolicity constants of the Mycielskian of path, cycle, complete and complete bipartite graphs are calculated explicitly. Finally, information on $\delta(G)$ just in terms of $\delta(G^M)$ is obtained.

Keywords: extremal problems on graphs; Mycielskian graphs; geodesics; Gromov hyperbolicity

MSC: Primary 05C75; Secondary 05C12

1. Introduction

Hyperbolic spaces (see Section 2 for definitions) were introduced by Mikhail Gromov in the 1980s in the context of geometric group theory (see [1–4]). Classical Riemannian geometry states that negatively-curved spaces possess an interesting property known as geodesic stability. Namely, near-optimal paths (quasi-geodesics) remain in a neighborhood of the optimal path (geodesic). When Mario Bonk proved in 1996 that geodesic stability was, in fact, equivalent to Gromov hyperbolicity (see [5]), the theory of hyperbolic spaces became a way to grasp the essence of negatively-curved spaces and to translate it to the simpler and more general setting of metric spaces. In this way, a simple concept led to a very rich general theory (see [1–4]) and, in particular, made hyperbolic spaces applicable to graphs. The theory has also been extensively used in discrete spaces like trees, the Cayley graphs of many finitely-generated groups and random graphs (see, e.g., [6–9]).

Hyperbolic spaces were initially applied to the study of automatic groups in the science of computation (see, e.g., [10]); indeed, it was proven that hyperbolic groups are strongly geodesically automatic, i.e., there is an automatic structure on the group [11]. The concept of hyperbolicity appears also in discrete mathematics, algorithms and networking [12]. For example, it has been shown empirically in [13] that the Internet topology embeds with better accuracy into a hyperbolic space than into a Euclidean space of comparable dimension (formal proofs that the distortion is related to the hyperbolicity can be found in [14]); furthermore, it is evidenced that many real networks are hyperbolic (see, e.g., [15–19]). Recently, among the practical network applications, hyperbolic spaces were used to study secure transmission of information on the Internet or the way viruses are spread through the network (see [20,21]); also to traffic flow and effective resistance of networks [22–24].

In fact, there is a new and growing interest for graph theorists in the study of the mathematical properties of Gromov hyperbolic spaces. (see, for example, [6–9,14,18–21,25–35]).

Several researchers have shown interest in proving that the metrics used in geometric function theory are Gromov hyperbolic. For instance, the Kobayashi and Klein–Hilbert metrics are Gromov hyperbolic (under certain conditions on the domain of definition; see [36–38]), the Gehring–Osgood *j*-metric is Gromov hyperbolic but the Vuorinen *j*-metric is not Gromov hyperbolic, except in the punctured space (see [39]). Furthermore, in [40], the hyperbolicity of the conformal modulus metric μ and the related so-called Ferrand metric λ^* have been studied. Gromov hyperbolicity of the Poincaré and the quasi-hyperbolic metrics is also the subject of [34,41–46].

Mycielskian graphs (see Section 2 for definitions) are a construction for embedding any undirected graph into a larger graph with higher chromatic number, but avoiding the creation of additional triangles. For example, the simple path graph with two vertices and one edge has chromatic number two, but its Mycielskian graph (which is the cycle graph with five vertices) raises that number to three. Actually, Jan Mycielski proved that there exist triangle-free graphs with an arbitrarily large chromatic number by applying this construction repeatedly to a starting triangle-free graph (see [47]). This means that, on the one hand, this construction enlarges small graphs in order to increase their chromatic number, but on the other hand (as we prove in the present paper), the resulting graph is Gromov hyperbolic; furthermore, its hyperbolicity constant is strongly constrained to a small interval. In this work, we also characterize which graphs yield Mycielskian graphs with hyperbolicity constant in the boundary cases. Note that a constraint value of the hyperbolicity constant is relevant, since it gives an idea of the tree-likeness of the space in question (see [35]).

Computing the hyperbolicity constant of a space is a difficult goal: For any arbitrary geodesic triangle T, the minimum distance from any point p of T to the union of the other two sides of the triangle to which p does not belong must be calculated. Then, the hyperbolicity constant is the supremum over all the minimum distances of possible choices for p and then over all of the possible choices for T in that space. Anyhow, notice that in general, the main obstacle is locating the geodesics in the space. In [2], the equivalence of the hyperbolicity of any geodesic metric space and the hyperbolicity of a graph associated with it are proven; similar results for Riemannian surfaces (with a very simple graph) can be found in [34,44–46]; hence, it is very useful to know hyperbolicity criteria for graphs. It is possible to compute the hyperbolicity constant of a finite graph with n vertices in time $O(n^{3.69})$ [48] (this result is improved in [17,49]). There is an algorithm that allows to decide if a Cayley graph (of a presentation with a solvable word problem) is hyperbolic [50]. However, there is no easy method to decide if a general infinite graph is hyperbolic or not.

Thus, a way to approach the problem is to study the hyperbolicity for particular types of graphs. For example, some other authors have obtained results on hyperbolicity for the complement of graphs, chordal graphs, vertex-symmetric graphs, lexicographic product graphs, corona and join of graphs, line graphs, bipartite and intersection graphs, bridged graphs, expanders and median graphs [24,27,28,32,33,35,51–56].

If G is a graph, G^M denotes its Mycielskian graph (see Section 2 for definitions), and $\delta(G^M)$ stands for its hyperbolicity constant. As usual, we denote by $V(G)$ and $E(G)$ as the set of vertices and edges of G, respectively. Let us also denote by $J(G)$ the union of the set $V(G)$ and the midpoints of the edges of G. The diameter of a graph G (the maximum distance between any two points of G) will be denoted by $\text{diam}(G)$ and the diameter of the set of vertices $V(G)$ of G by $\text{diam}\,V(G)$.

The main results of this work deal with the hyperbolicity constant of Mycielskian graphs, as said above. The first of them states that Mycielskian graphs are always hyperbolic and solves the extremal problems of finding the smallest and largest hyperbolicity constants of such graphs. The second and third ones characterize graphs with hyperbolicity constants $5/2$ and $5/4$, the maximum and the minimum values, respectively. The fourth result gives an accurate estimate of $\delta(G^M)$, and the fifth one allows us to obtain information on $\delta(G)$ just in terms of $\delta(G^M)$.

Theorem 1. *Let G be any graph. Then, G^M is hyperbolic, $2 \leq \text{diam}\,V(G^M) \leq 4$ and $5/2 \leq \text{diam}(G^M) \leq 5$. Furthermore, if G is a complete graph, then $\delta(G^M) = 5/4$; otherwise:*

$$\frac{5}{4} \leq \delta(G^M) \leq \min\left\{\frac{5}{2}, \frac{1 + \operatorname{diam} V(G)}{2}\right\}$$

and both bounds of $\delta(G^M)$ are sharp.

Theorem 2. *Let G be any graph. Then, $\delta(G^M) = 5/2$ if and only if there exists a geodesic triangle $T := \{x, y, z\}$ in G and G^M with:*

$$x, y \in J(G) \setminus V(G), z \in J(G), \tag{1}$$
$$d_G(x, y) = 5, \ d_G(x, z) \leq 5, \ d_G(y, z) \leq 5, \tag{2}$$
the midpoint p in [xy] is a vertex of G and $d_G(p, [xz] \cup [yz]) = 5/2$. $\tag{3}$

Theorem 10 characterizes the graphs G with $\delta(G^M) = 5/4$. Since this characterization is not easy to state briefly, we present here nice necessary and sufficient conditions on G for $\delta(G^M) = 5/4$.

Theorem 3. *Let G be any graph:*

$$\text{If } \operatorname{diam}(G) \leq 2, \text{ then } \delta(G^M) = 5/4, \tag{4}$$
$$\text{If } \delta(G^M) = 5/4, \text{ then } \operatorname{diam}(G) \leq 5/2. \tag{5}$$

Furthermore, the converses of (4) and (5) do not hold.

The hyperbolicity of a metric space is at most half of its diameter. The following result states an unexpected fact: $\delta(G^M)$ is not only upper bounded by $\frac{1}{2}\operatorname{diam}(G^M)$; in this case, that upper bound is close to the actual value of the hyperbolicity constant.

Theorem 4. *Let G be any graph. Then:*

$$\frac{1}{2}\operatorname{diam} V(G^M) \leq \delta(G^M) \leq \frac{1}{2}\operatorname{diam}(G^M).$$

So far, our main results have obtained information about G^M in terms of G. However, it is also interesting to consider what can be said about $\delta(G)$ in terms of $\delta(G^M)$. Our next theorem gives a partial answer to this question.

Theorem 5. *If $\delta(G^M) \leq 3/2$, then $\delta(G) \leq \delta(G^M)$.*

The outline of the paper will be as follows. In Section 2, some definitions and previous results will be stated. Section 3 contains the proof of the main parts of Theorem 1. Sections 4 and 5 will present the proofs of Theorem 2 and Theorem 3, respectively. In Section 6, the hyperbolicity constants of the Mycielskian of path, cycle, complete and complete bipartite graphs are calculated explicitly. Apart from the intrinsic interest of these results, they are also employed in the proofs of some of the main results of the paper. Furthermore, it contains the proofs of Theorems 1, 4 and 5. Finally, in Section 7, a characterization for graphs with $\delta(G^M) = 5/4$ is given.

Since the hypotheses in most theorems are simple to check, the main results in this paper can be applied to every graph. An exception is Theorem 10, but even in this case, a rough algorithm is provided, which allows one to check the hypotheses in polynomial time. Furthermore, information on $\delta(G^M)$ from G and on $\delta(G)$ in terms of G^M is found. The main inequalities obtained in this work are applied in Section 6 in order to compute explicitly the hyperbolicity constants of the Mycielskian of some classical examples, such as path, cycle, complete and complete bipartite graphs. Finally, note that Mycielskian graphs are not difficult to identify computationally.

2. Definitions and Background

If $\gamma : [a, b] \longrightarrow X$ is a continuous curve in a metric space (X, d), we can define the length of γ as:

$$l(\gamma) := \sup \left\{ \sum_{i=1}^{n} d(\gamma(t_{i-1}), \gamma(t_i)) : a = t_0 < t_1 < \cdots < t_n = b \right\}.$$

The curve γ is a geodesic if we have $l(\gamma|_{[t,s]}) = d(\gamma(t), \gamma(s)) = |t - s|$ for every $s, t \in [a, b]$ (then γ is equipped with an arc-length parametrization). The metric space X is said to be geodesic if for every couple of points in X, there exists a geodesic joining them; we denote by $[xy]$ any geodesic joining x and y; this notation is ambiguous, since in general we do not have the uniqueness of geodesics, but it is very convenient. Consequently, any geodesic metric space is connected. The graph G consists of a collection of vertices, denoted by $V(G) = \{v_i\}$ and a collection of edges joining vertices, $E(G)$; the edge joining vertices v_i and v_j will be denoted by $\{v_i, v_j\}$. Furthermore, $N_G(v_i)$ will stand for the set of neighbors of (or adjacent to) the vertex v_i, that is the set of all vertices $v \in V(G)$ for which $\{v, v_i\} \in E(G)$. All throughout this paper, the metric space X considered is a graph with every edge of length one. In order to consider a graph G as a geodesic metric space, we identify (by an isometry) any edge $\{v_i, v_j\} \in E(G)$ with the interval $[0, 1]$ in the real line. Thus, the points in G are the vertices and the points in the interiors of the edges of G. In this way, any connected graph G has a natural distance defined on its points, induced by taking the shortest paths in G, and we can see G as a metric graph. Such a distance will be denoted by d_G. Throughout this paper, G denotes a connected (finite or infinite) simple (i.e., without loops and multiple edges) graph such that every edge has length one and $E(G) \neq \varnothing$. These properties guarantee that G is a geodesic metric space and that G^M can be defined. Note that excluding multiple edges and loops is not an important loss of generality, since ([57], Theorems 8 and 10) they reduce the problem of computing the hyperbolicity constant of graphs with multiple edges and/or loops to the study of simple graphs.

A cycle in G is a simple closed curve containing adjacent vertices $v_1, \ldots, v_n$. It will be denoted by $[v_1, v_2, \ldots, v_n, v_1]$. The notation $a \sim b$ means that the vertices a and b are adjacent.

Given a graph G with $V(G) = \{v_1, \ldots, v_n\}$, the Mycielskian graph G^M of G contains G itself as a subgraph, together with $n + 1$ additional vertices $\{u_1, \ldots, u_n, w\}$, where each vertex u_i is the mirror of the vertex v_i of G and w is the supervertex. Each vertex u_i is connected by an edge to w. In addition, for each edge $\{v_i, v_j\}$ of G, the Mycielskian graph includes two edges, $\{u_i, v_j\}$ and $\{v_i, u_j\}$ (in Figure 1, the process of the construction of the Mycielskian graph for the path graph P_3 is shown). Thus, if G has n vertices and m edges, then G^M has $2n + 1$ vertices and $3m + n$ edges.

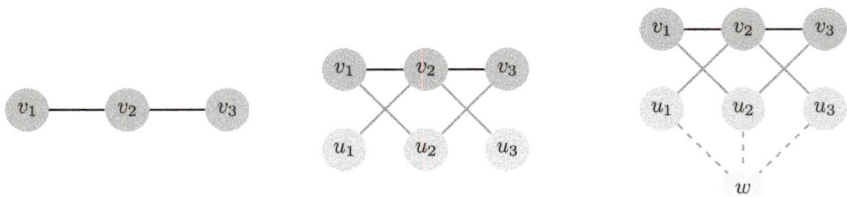

Figure 1. Construction of the Mycielskian graph of P_3.

Trivially, for all i, j, $d_{G^M}(w, v_i) = d_{G^M}(u_i, u_j) = 2$ and $d_{G^M}(w, u_i) = 1$; also $d_{G^M}(v_i, v_j) = \min\{d_G(v_i, v_j), 4\}$ and for any point $a \in G^M$, $d_{G^M}(w, a) \leq 5/2$. Note that a can be either a vertex of the graph or any other point belonging to an edge of it. Moreover, the definition of G^M makes sense also if $n = \infty$, i.e., if G is an infinite graph. Since G^M is always connected, it would be possible to consider disconnected graphs G, but Theorem 9 shows that, in order to study $\delta(G^M)$, it suffices to consider just connected graphs.

Finally, let us recall the definition of Gromov hyperbolicity that we will use (we use the notations of [3]).

If X is a geodesic metric space and $x_1, x_2, x_3 \in X$, the union of three geodesics $[x_1 x_2]$, $[x_2 x_3]$ and $[x_3 x_1]$ is a geodesic triangle that will be denoted by $T = \{x_1, x_2, x_3\}$, and we will say that x_1, x_2 and x_3 are the vertices of T; it is usual to write also $T = \{[x_1 x_2], [x_2 x_3], [x_3 x_1]\}$. We say that T is δ-thin if any side of T is contained in the δ-neighborhood of the union of the two other sides. We denote by $\delta(T)$ the sharp thin constant of T, i.e., $\delta(T) := \inf\{\delta \geq 0 : T \text{ is } \delta\text{-thin}\}$. Given a constant $\delta \geq 0$, the space X is δ-hyperbolic if every geodesic triangle in X is δ-thin. We denote by $\delta(X)$ the sharp hyperbolicity constant of X, i.e., $\delta(X) := \sup\{\delta(T) : T \text{ is a geodesic triangle in } X\} \in [0, \infty]$. We say that X is hyperbolic if X is δ-hyperbolic for some constant $\delta \geq 0$, i.e., $\delta(X) < \infty$. If we have a geodesic triangle with two identical vertices, we call it a bigon. Obviously, every bigon in a δ-hyperbolic space is δ-thin. In the classical references on this subject (see, e.g., [2,3]) appear several different definitions of Gromov hyperbolicity, which are equivalent in the sense that if X is δ-hyperbolic with respect to one definition, then it is δ'-hyperbolic with respect to another definition (for some δ' related to δ). We have chosen this definition by its deep geometric meaning [3].

Trivially, any bounded metric space X is $(\operatorname{diam} X)$-hyperbolic. A normed real linear space is hyperbolic if and only if it has dimension one. A geodesic space is zero-hyperbolic if and only if it is a metric tree. The hyperbolic plane (with curvature -1) is $\log(1 + \sqrt{2})$-hyperbolic. Every simply-connected complete Riemannian manifold with sectional curvature verifying $K \leq -k^2$, for some positive constant k, is hyperbolic. See the classical [1,3] in order to find more examples and further results.

Trees are one of the main examples of hyperbolic graphs. Metric trees are precisely those spaces X with $\delta(X) = 0$. Therefore, the hyperbolicity constant of a geodesic metric space can be seen as a measure of how "tree-like" that space is. This alternative view of the hyperbolicity constant is an interesting subject since the tractability of a problem in many applications is related to the tree-like degree of the space under investigation. (see, e.g., [58]). Furthermore, it is well known that any Gromov hyperbolic space with n points embeds into a tree metric with distortion $O(\delta \log n)$ (see, e.g., [3], p. 33).

If G_1 and G_2 are isomorphic, we write $G_1 \simeq G_2$. It is clear that if $G_1 \simeq G_2$, then $\delta(G_1) = \delta(G_2)$.

The following well-known result will be used throughout the paper (see, e.g., ([59], Theorem 8) for a proof).

Lemma 1. *Let G be any graph. Then, $\delta(G) \leq \operatorname{diam}(G)/2$.*

Consider the set $\mathbb{T}_1$ of geodesic triangles T in G that are cycles and such that the three vertices of the triangle T belong to $J(G)$, and denote by $\delta_1(G)$ the infimum of the constants λ such that every triangle in $\mathbb{T}_1$ is λ-thin.

The following results, which appear in [60] (Theorems 2.5, 2.7 and 2.6), will be used throughout the paper.

Lemma 2. *For every graph G, we have $\delta_1(G) = \delta(G)$.*

Lemma 3. *For any hyperbolic graph G, there exists a geodesic triangle $T \in \mathbb{T}_1$ such that $\delta(T) = \delta(G)$.*

The next result will narrow the possible values for the hyperbolicity constant δ.

Lemma 4. *Let G be any graph. Then, $\delta(G)$ is always a multiple of $1/4$.*

The following results deal with isometric subgraphs and how the hyperbolicity constant behaves.

A subgraph G_0 of the graph G is an isometric subgraph if for all $x, y \in G_0$, we have $d_{G_0}(x, y) = d_G(x, y)$. This is equivalent to the fact that $d_{G_0}(u, v) = d_G(u, v)$ for every $u, v \in V(G_0)$. It is known that isometric subgraphs have a lesser hyperbolicity constant (see [57], Lemma 7):

Lemma 5. *Let G_0 be an isometric subgraph of G. Then, $\delta(G_0) \leq \delta(G)$.*

The coming lemma states that the Mycielskians preserve isometric subgraphs, and a nice corollary follows.

Lemma 6. *If G_0 is an isometric subgraph of G, then G_0^M is an isometric subgraph of G^M.*

Proof. Let $x, y \in V(G_0^M)$. Throughout the proof, $V(G_0) = \{v_1, \ldots, v_n\}$, $W = \{u_1, \ldots, u_n\}$ and $V(G_0^M) = V(G_0) \cup W \cup \{w\} = \{v_1, \ldots, v_n, u_1, \ldots, u_n, w\}$ with w being the supervertex.

Without loss of generality, $x, y \in V(G_0)$ or $x \in V(G_0)$ and $y \in W$. Otherwise, trivially, if $x, y \in \{u_1, \ldots, u_n\}$ or if $x = w$ and $y \in V(G_0)$, then $d_{G_0^M}(x, y) = d_{G^M}(x, y) = 2$; also trivially, if $x = w$ and $y \in W$, then $d_{G_0^M}(x, y) = d_{G^M}(x, y) = 1$.

First, let $x = v_i$ and $y = v_j$. If $d_{G_0}(v_i, v_j) = d_G(v_i, v_j) = 1$, trivially, $d_{G_0^M}(v_i, v_j) = d_{G^M}(v_i, v_j) = 1$.

If $d_{G_0}(v_i, v_j) = d_G(v_i, v_j) = 2$, there exists $v_r \in V(G_0)$ so that $\{v_i, v_r\}, \{v_r, v_j\} \in E(G_0)$. Notice also that $d_{G^M}(v_i, v_j) \leq d_{G^M}(v_i, v_r) + d_{G^M}(u_r, v_j)$. Suppose there is a path $\gamma \subset G^M$ from v_i to v_j such that $l(\gamma) = 1$; then, $\{v_i, v_j\}] \in E(G^M)$, and thus, $\{v_i, v_j\} \in E(G)$, giving $d_G(v_i, v_j) = 1$, which contradicts $d_G(v_i, v_j) = 2$.

Additionally, if $d_{G_0}(v_i, v_j) = d_G(v_i, v_j) > 2$, then there is a geodesic path γ in G_0 and $v_k \neq v_l \in V(G_0)$ with $\{v_i, v_k\}, \{v_l, v_j\} \in E(G_0)$ and $v_i, v_k, v_l, v_j \in \gamma$. Clearly, $u_k \neq u_l$. By the triangle inequality, $d_{G^M}(v_i, v_j) \leq d_{G^M}(v_i, u_k) + d_{G^M}(u_k, w) + d_{G^M}(u_l, w) + d_{G^M}(u_l, v_j) \leq 4$. Suppose there is a path $\tilde{\gamma} \subset G^M$ from v_i to v_j such that $l(\tilde{\gamma}) \leq 3$, then $w \notin \tilde{\gamma}$, and there exists $u_r \in V(MG_0)$ with $\{v_i, u_r\}, \{u_r, v_j\} \in E(G^M)$; thus, $\{v_i, v_r\}, \{v_r, v_j\} \in E(G)$, giving $d_G(v_i, v_j) = 2$, which is a contradiction.

Next, let $x = v_i \in V(G_0)$ and $y = u_j \in W$. Notice that $d_{G^M}(v_i, u_j) \leq d_{G^M}(v_i, w) + d_{G^M}(w, u_j) = 3$. If $\{v_i, u_j\} \in E(G_0^M)$, then the result trivially holds. If $d_{G_0^M}(v_i, u_j) = 2$, then there exists $v_r \in V(G_0)$ with $\{v_i, v_r\} \in E(G_0)$ and $\{v_r, u_j\} \in E(G_0^M)$, and thus, $d_{G^M}(v_i, u_j) = 2$. Finally, if $d_{G_0^M}(v_i, u_j) = 3$, then there is a geodesic γ both in G_0^M an G^M, which contains w. □

This result has a straightforward consequence.

Corollary 1. *If G_0 is an isometric subgraph of G, then $\delta(G^M) \geq \delta(G_0^M)$.*

Denote by G' the subgraph of G^M induced by $V(G^M) \setminus \{w\}$. The following result is elementary.

Proposition 1. *Let G be any graph. Then, G is an isometric subgraph of G'.*

Proof. Consider the vertices $v_i, v_j \in V(G)$ and a geodesic γ in G' from v_i to v_j, say, $\gamma = \cup_{k=1}^s \{x_{r_k}, x_{r_{k+1}}\}$ with $x = v$ or $x = u$, $r_1 = i$ and $r_s = j$. It suffices to find a curve γ_0 in G from v_i to v_j with $l(\gamma) = l(\gamma_0)$. Since, by construction, if $\{x_{r_k}, x_{r_{k+1}}\} \in E(G')$, then $\{v_{r_k}, v_{r_{k+1}}\} \in E(G)$, we have that $\gamma_0 = \cup_{k=1}^s \{v_{r_k}, v_{r_{k+1}}\}$ joins v_i and v_j with $l(\gamma) = l(\gamma_0)$. □

3. Proof of the Main Parts of Theorem 1

In order to prove the main parts of Theorem 1, consider first the following weaker version:

Theorem 6. *Let G be any graph. Then, G^M is hyperbolic, $2 \leq \text{diam } V(G^M) \leq 4$, $5/2 \leq \text{diam}(G^M) \leq 5$ and:*

$$\frac{5}{4} \leq \delta(G^M) \leq \frac{5}{2}.$$

Proof. For the upper bounds, notice that for any two vertices $u, v \in G^M$, $d_{G^M}(u, v) \leq 4$, therefore diam $V(G^M) \leq 4$, diam$(G^M) \leq 5$, and thus, $\delta(G^M) \leq 5/2$ by Lemma 1.

For the lower bounds, recall that $d_{GM}(u, w) = 2$ for every $u \in V(G)$, and thus, diam $V(G^M) \geq 2$. If $\{v_1, v_2\} \in E(G)$ and p is the midpoint of $\{v_1, v_2\}$, then $d_{GM}(p, w) = 5/2$ and diam$(G^M) \geq 5/2$.

For $\delta(G^M)$, observe that, since $E(G) \neq \emptyset$, there exists $\{v_1, v_2\} \in E(G)$, and thus, G^M always contains a cycle of length of five, namely $C := [v_1, u_2, w, u_1, v_2, v_1]$. Let x be the midpoint of $\{v_1, v_2\}$ and consider the geodesic bigon $T = \{x, w\}$ with geodesics $\gamma_1 := [xv_1] \cup \{v_1, u_2\} \cup \{u_2, w\}$ and $\gamma_2 := [xv_2] \cup \{v_2, u_1\} \cup \{u_1, w\}$. If p is the midpoint of the geodesic γ_1, then one gets $\delta(G^M) \geq d_{GM}(p, \gamma_2) \geq 5/4$. $\square$

The following lemmas relate diam $V(G^M)$ with diam $V(G)$ for small diameters.

Lemma 7. *Let G be any graph. Then:*

(i) *If $x, y \in G$ and $d_G(x, y) \leq 9/2$, then $d_{GM}(x, y) = d_G(x, y)$.*
(ii) *If $x, y \in J(G) \setminus V(G)$ and $d_G(x, y) \leq 5$, then $d_{GM}(x, y) = d_G(x, y)$.*

Proof. Since G is a subgraph of G^M, we have $d_{GM}(x, y) \leq d_G(x, y)$.

Assume that $x, y \in G$ and $d_G(x, y) \leq 9/2$, and let γ be a geodesic $\gamma = [xy]$ in G. Define x_0 (respectively, y_0) as the closest vertex to x (respectively, y) from $\gamma \cap V(G)$ (it is possible to have $x = x_0$ and/or $y = y_0$). Since $d_G(x_0, y_0)$ is an integer number and $d_G(x, y) \leq 9/2$ by hypothesis, we have that:

(1) $d_G(x, x_0) + d_G(y, y_0) \leq 1/2$ and $d_G(x_0, y_0) \leq 4$,

or

(2) $d_G(x, x_0) + d_G(y, y_0) \leq 3/2$ and $d_G(x_0, y_0) \leq 3$.

Seeking for a contradiction, assume that $d_{GM}(x, y) < d_G(x, y)$. Thus, there exists a geodesic γ^M joining x and y in G^M with $l(\gamma^M) < l(\gamma)$, and $w \in \gamma^M$ by Proposition 1. Define x_0^M (respectively, y_0^M) as the closest vertex to x (respectively, y) from $\gamma^M \cap V(G^M)$.

Since (1) or (2) holds, we have $d_G(x, x_0^M) + d_G(y, y_0^M) \geq 1/2$. Since $w \in \gamma^M$, $d_{GM}(x_0^M, y_0^M) = 4$ and:

$$l(\gamma^M) = d_{GM}(x, y) = d_G(x, x_0^M) + d_{GM}(x_0^M, y_0^M) + d_G(y_0^M, y)$$
$$\geq \frac{1}{2} + 4 \geq d_G(x, y) = l(\gamma),$$

which is a contradiction. Then, $d_{GM}(x, y) = d_G(x, y)$ and $i)$ holds.

Assume now that $x, y \in J(G) \setminus V(G)$ and $d_G(x, y) \leq 5$. If $d_G(x, y) \leq 9/2$, then $i)$ gives the result. If $d_G(x, y) > 9/2$, then $d_G(x, y) = 5$. The argument in the proof of $i)$ gives:

$$d_G(x, x_0) = d_G(y, y_0) = \frac{1}{2}, \qquad d_{GM}(x, x_0^M) = d_{GM}(y, y_0^M) = \frac{1}{2},$$

and:

$$d_{GM}(x, y) = d_G(x, x_0^M) + d_{GM}(x_0^M, y_0^M) + d_G(y_0^M, y) = \frac{1}{2} + 4 + \frac{1}{2} = 5 = d_G(x, y).$$

Therefore, $ii)$ holds. $\square$

Lemma 8. *Let G be any graph. Then:*

$$2 \leq \text{diam } V(G) \leq 4 \qquad \text{iff} \qquad \text{diam } V(G^M) = \text{diam } V(G).$$

Proof. Assume first that $2 \leq \text{diam } V(G) \leq 4$.

Let $k := \text{diam } V(G)$. Thus:

1. If $v_i, v_j \in V(G)$, then clearly $d_G(v_i, v_j) \leq \text{diam } V(G) \leq 4$, and by Lemma 7, we conclude $d_{GM}(v_i, v_j) = d_G(v_i, v_j)$.
2. If $u_i, u_j \in V(G') \setminus V(G)$, then $d_{GM}(u_i, u_j) = 2 \leq k$.

3. If $v_i \in V(G)$ and $u_j \in V(G') \setminus V(G)$, then $d_{G^M}(v_i, u_i) = 2 \leq k$ and $d_{G^M}(v_i, u_j) \leq d_G(v_i, v_j) \leq k$ (if $i \neq j$).
4. If $\alpha \in V(G')$, then $d_{G^M}(\alpha, w) \leq 2 \leq k$.

Therefore, diam $V(G^M) \leq$ diam $V(G)$.

For the other direction, let $v_i, v_j \in V(G)$ be so that $d_G(v_i, v_j) \leq 4$. By Lemma 7, we have $d_{G^M}(v_i, v_j) = d_G(v_i, v_j)$, and thus, diam $V(G^M) \geq$ diam $V(G)$.

Assume now that diam $V(G^M) =$ diam $V(G)$. Theorem 6 gives $2 \leq$ diam $V(G^M) \leq 4$, and then, $2 \leq$ diam $V(G) \leq 4$. □

We can prove now the main parts of Theorem 1.

Theorem 7. *Let G be any graph. Then, G^M is hyperbolic, $2 \leq$ diam $V(G^M) \leq 4$ and $5/2 \leq$ diam$(G^M) \leq 5$. Furthermore, if G is not a complete graph, then:*

$$\frac{5}{4} \leq \delta(G^M) \leq \min\left\{\frac{5}{2}, \frac{1 + \text{diam } V(G)}{2}\right\}. \tag{6}$$

Proof. Theorem 6 proves all of the statements of Theorem 7 except for the upper bound for $\delta(G^M)$ in the case where diam $V(G) < 4$.

However, if diam $V(G) < 4$, then either diam $V(G) = 1$, in which case G would be isomorphic to a complete graph, or $2 \leq$ diam $V(G) < 4$, in which case $\delta(G^M) \leq$ diam$(G^M)/2 \leq (1 +$ diam $V(G^M))/2 = (1 +$ diam $V(G))/2$, where the last equality follows from the above result, Lemma 8. □

As a consequence of Lemma 8, we obtain the following result.

Corollary 2. *If G is not a complete graph, then* diam $V(G^M) \leq$ diam $V(G)$.

The following result provides information about $\delta(G)$ in terms of $\delta(G^M)$.

Theorem 8. *Let G be any graph. If* diam $V(G) \leq 4$*, then* $\delta(G) \leq \delta(G^M)$.

Proof. By Lemma 3, there is a geodesic triangle $T = \{x_1, x_2, x_3\}$ in G (with geodesics γ_{ij} joining x_i and x_j in G) and $p \in \gamma_{12}$ with $d_G(p, \gamma_{13} \cup \gamma_{23}) = \delta(G)$ and $T \in \mathbb{T}_1$. If $l(\gamma_{ij}) \leq 9/2$, then γ_{ij} is also a geodesic in G^M by Lemma 7. If $l(\gamma_{ij}) > 9/2$, then $l(\gamma_{ij}) = 5$ and $x_i, x_j \in J(G) \setminus V(G)$; hence, Lemma 7 gives that γ_{ij} is a geodesic in G^M. Therefore, T is also a geodesic triangle in G^M. Since $d_G(p, \gamma_{13} \cup \gamma_{23}) \leq d_G(p, \{x_1, x_2\}) \leq l(\gamma_{12})/2 \leq 5/2$, then $\delta(G^M) \geq d_{G^M}(p, \gamma_{13} \cup \gamma_{23}) = d_G(p, \gamma_{13} \cup \gamma_{23}) = \delta(G)$ by Lemma 7. □

We say that a vertex v of a (connected) graph Γ is a *cut-vertex* if $\Gamma \setminus \{v\}$ is not connected. A graph is *bi-connected* if it does not contain cut-vertices. Given any edge in Γ, we consider the maximal bi-connected subgraph containing it.

Finally, we have the following result regarding disconnected graphs. Even though in order to study Gromov hyperbolicity, connected graphs are needed, since G^M is always connected, the original graph G does not need to be:

Theorem 9. *Let G be any disconnected graph with connected components $\{G_j\}_j$. Then:*

$$\delta(G^M) = \max_j \delta(G_j^M).$$

Proof. It is well known that the hyperbolicity constant of a graph is the maximum value of the hyperbolicity constant of its maximal bi-connected components. Since G is not connected, the supervertex is the unique cut-vertex in G^M, and the subgraphs $\{G_j^M\}_j$ are the maximal bi-connected components of G^M. Hence, the equality holds. □

Note that, in order to study $\delta(G^M)$, it suffices to consider just (connected) graphs G by Theorem 9.

4. Proof of Theorem 2

The next lemma will be a key tool:

Lemma 9. *Let G be any graph. If $\delta(G^M) = 5/2$, $T^M := \{x_0, y_0, z_0\}$ is a geodesic triangle in G^M, such that $\delta(T^M) = 5/2$ and $p \in [x_0 y_0]$ with $d_{GM}(p, [x_0 z_0] \cup [y_0 z_0]) = 5/2$, then $d_{GM}(x_0, y_0) = 5$, $d_{GM}(p, x_0) = d_{GM}(p, y_0) = 5/2$, $x_0, y_0 \in J(G) \setminus V(G)$ and $p \in V(G^M)$.*

Proof. Since $d_{GM}(p, \{x_0, y_0\}) \geq d_{GM}(p, [x_0 z_0] \cup [y_0 z_0]) = 5/2$, and $p \in [x_0 y_0]$, $d_{GM}(x_0, y_0) = d_{GM}(x_0, p) + d_{GM}(p, y_0) \geq 5$, then $d_{GM}(x_0, y_0) = 5$ by Theorem 6. Thus, $d_{GM}(p, x_0) = d_{GM}(p, y_0) = 5/2$. The equality $d_{GM}(x_0, y_0) = 5$ and the fact that diam $V(G^M) \leq 4$, guarantee that $x_0, y_0 \in J(G^M) \setminus V(G^M)$. In fact, $x_0 \in G$ for otherwise $d_{GM}(x_0, w) \leq 3/2$, and since $d_{GM}(y_0, w) \leq 5/2$, the triangle inequality would give $d_{GM}(x_0, y_0) \leq (3/2) + (5/2) = 4 < 5$. Therefore, $x_0, y_0 \in J(G) \setminus V(G)$. This in turn implies that $p \in V(G^M)$. □

When comparing hyperbolicity constants of G and G^M, it is useful to compare triangles in those graphs. A useful tool will be a projection, which allows one to associate triangles $T \subset G$ with given triangles $T^M \subset G^M$, which do not contain the supervertex, that is $T^M \subset G'$. Namely:

Definition 1. *The projection $\Pi : G' \to G$ is defined as follows:*
For a point $a \in G'$, if $a \in G$, then $\Pi(a) = a$. If $a = u_i \in V(G') \setminus V(G)$, then $\Pi(a) = \Pi(u_i) = v_i \in V(G)$. If $a \in G' \setminus \{G \cup V(G')\}$, then a lies on an edge, say $a \in \{v_m, u_k\}$; then, $\Pi(a) \in \{v_m, \Pi(u_k)\} \subset G$, so that $d_G(\Pi(a), v_m) = d_{GM}(a, v_m)$.

Remark 1. *Observe the following:*

1. *If $\gamma = [xy]$ is a geodesic in G^M and $\gamma \subset G'$, there exists a geodesic $[\Pi(x)\Pi(y)] \subseteq \Pi([xy])$; in general, equality does not hold, since two different edges in γ might project onto the same edge in $E(G)$ (for instance, if $\gamma = \{v_i, v_j\} \cup \{v_j, u_i\}$, then $\Pi(\gamma) = \{v_i, v_j\}$; note that $l(\gamma) = 2$); therefore, $l(\Pi(\gamma)) \leq l(\gamma)$.*
2. *If $l([xy]) \leq 1$ and $x \in V(G)$, then $[\Pi(x)\Pi(y)] = \Pi([xy])$.*
3. *If T^M is a geodesic triangle in G^M and $\Pi(T^M)$ is a geodesic triangle in G and G^M, then $\Pi(T^M)$ does not need to be a cycle even if T^M is. For instance, if $\gamma_1 = \{v_i, v_j\} \cup \{v_j, v_k\}$ and $\gamma_2 = \{v_i, u_j\} \cup \{u_j, v_k\}$, then $T^M = \{\gamma_1, \gamma_2\}$ is a geodesic bigon in G^M, $\Pi(\gamma_1) = \Pi(\gamma_2) = \gamma_1$ are geodesics in G, but $\Pi(T^M)$ is not a cycle, although T^M is; note that $l(T^M) = 4$ (see Figure 2).*

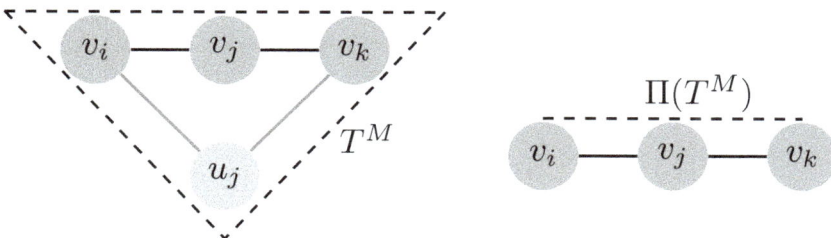

Figure 2. Projection of a cycle in G' of length four onto a path in G of length two.

However, the following holds:

Lemma 10. *Let G be any graph and Π the projection in Definition 1. If $[xy] \subset G'$ and $l([xy]) > 2$, then $\Pi([xy])$ is a geodesic in G joining $\Pi(x)$ and $\Pi(y)$. Furthermore, if $u \in (V(G') \setminus V(G)) \cap [xy]$, then $\Pi(u) \notin [xy]$.*

Proof. The statement is trivial if $[xy] \subset G$. Otherwise, it is a consequence of the following fact: If $l([xy]) > 2$, then we have either $\{u_i, v_j\} \subset [xy]$ or $\{v_i, v_j\} \subset [xy]$ for some $u_i \in V(G') \setminus V(G)$ and $v_i, v_j \in V(G)$. If $\{u_i, v_j\} \subset [xy]$, then $\{v_i, v_j\} \cap [xy] = v_j$; if $\{v_i, v_j\} \subset [xy]$, then $\{u_i, v_j\} \cap [xy] = v_j$. $\square$

Proof of Theorem 2. Assume first that $\delta(G^M) = 5/2$.

By Lemma 3, there is a geodesic triangle $T^M := \{x_0, y_0, z_0\} \subset G^M$ that is a cycle with $x_0, y_0, z_0 \in J(G^M)$ and $p \in [x_0 y_0]$, so that $d_{G^M}(p, [x_0 z_0] \cup [y_0 z_0]) = 5/2$. Then, by Lemma 9, $x_0, y_0 \in J(G) \setminus V(G)$, $p \in V(G^M)$, p is the midpoint in $[x_0 y_0]$, and $d_{G^M}(x_0, y_0) = 5$.

The goal is to produce a triangle T from this T^M, which is geodesic both in G and G^M.

To this end, it will be first shown that the supervertex $w \notin T^M$.

Clearly, $w \neq p$ for $d_{G^M}(w, [x_0 z_0] \cup [y_0 z_0]) \leq 2 < 5/2$, since $d_{G^M}(w, v) \leq 2$ for all $v \in V(G^M)$. For this same reason, $w \notin [x_0 z_0] \cup [y_0 z_0]$, for $d_{G^M}(p, [x_0 z_0] \cup [y_0 z_0]) \leq d_{G^M}(p, w) \leq 2$, since $p \in V(G^M)$. Finally, $w \notin [x_0 y_0]$, since $w, p \in V(G^M)$ and $x_0, y_0 \in J(G) \setminus V(G)$, gives $d_{G^M}(w, x_0) = d_{G^M}(w, y_0) = 5/2 = d_{G^M}(x_0, y_0)/2$, which would mean that $w = p$, which is a contradiction.

Therefore, $w \notin T^M$, $T^M \subset G'$, and $d_{G'}(x_0, y_0) = d_{G^M}(x_0, y_0) = 5$, $d_{G'}(x_0, z_0) = d_{G^M}(x_0, z_0) \leq 5$ and $d_{G'}(y_0, z_0) = d_{G^M}(y_0, z_0) \leq 5$.

If we define $x := \Pi(x_0) = x_0$, $y := \Pi(y_0) = y_0$, $z := \Pi(z_0)$, then (1) stated in Theorem 2 holds, and Lemma 7 gives $d_G(x, y) = d_{G^M}(x_0, y_0) = 5$, $d_G(x, z) = d_{G^M}(x_0, z_0) \leq 5$, $d_G(y, z) = d_{G^M}(y_0, z_0) \leq 5$. If $l([x_0 z_0]) > 2$ and $l([y_0 z_0]) > 2$, then $\Pi([x_0 y_0]) = [xy]$, $\Pi([x_0 z_0]) = [xz]$ and $\Pi([y_0 z_0]) = [yz]$ are geodesics in G by Lemma 10. Otherwise, $[x_0 z_0]$ or $[y_0 z_0]$ have length at most two. By symmetry, we can assume that $l([y_0 z_0]) \leq 2$; since $d_{G^M}(x_0, y_0) = 5$, we have $l([x_0 z_0]) \geq 3$, and Lemma 10 gives that $\Pi([x_0 y_0]) = [xy]$ and $\Pi([x_0 z_0]) = [xz]$ are geodesics in G. Furthermore, $\Pi([y_0 z_0])$ contains a geodesic $[yz]$ in G. This gives the existence of the geodesic triangle T of Theorem 2 and (2) stated in this same Theorem.

Next, let us show that $d_{G^M}(\Pi(p), [x_0 z_0] \cup [y_0 z_0]) = d_{G^M}(p, [x_0 z_0] \cup [y_0 z_0]) = 5/2$. If $p \in G$, then $\Pi(p) = p$, and the statement trivially holds. Assume that $p \notin G$ (thus, $p \in V(G') \setminus V(G)$). Since $l([x_0 y_0]) = 5 > 2$ and $p \in [x_0 y_0]$, Lemma 10 gives that $\Pi(p) \notin [x_0 y_0]$. Since $d_{G^M}(p, [x_0 z_0] \cup [y_0 z_0]) = 5/2$, we have $\Pi(p) \notin [x_0 z_0] \cup [y_0 z_0]$, and we conclude $\Pi(p) \notin T^M$.

Let η be a geodesic joining $\Pi(p)$ and $[x_0 z_0] \cup [y_0 z_0]$ in G^M with $l(\eta) = d_{G^M}(\Pi(p), [x_0 z_0] \cup [y_0 z_0])$, and let $q_1 \in V(G^M) \cap \eta$, so that q_1 is adjacent to $\Pi(p) \in V(G)$ (so $q_1 \neq w$). We have $d_{G^M}(\Pi(p), [x_0 z_0] \cup [y_0 z_0]) \leq d_{G^M}(\Pi(p), x_0) = d_{G^M}(p, x_0) = 5/2$.

If $q_1 \in V(G)$, then $\{p, q_1\} \in E(G')$, and take η' to be the curve (not necessarily geodesic) $(\eta \setminus \{\Pi(p), q_1\}) \cup \{p, q_1\}$; then $l(\eta) = l(\eta') \geq d_{G^M}(p, [x_0 z_0] \cup [y_0 z_0]) = 5/2$, and we conclude $d_{G^M}(\Pi(p), [x_0 z_0] \cup [y_0 z_0]) = d_{G^M}(p, [x_0 z_0] \cup [y_0 z_0]) = 5/2$, the desired result.

If $q_1 \in V(G') \setminus V(G)$, let $q_2 \in (V(G^M) \cap \eta) \setminus \{\Pi(p)\}$ so that q_2 is adjacent to q_1. Notice that q_2 cannot be the supervertex, for then $l(\eta) \geq 3 > 5/2$, and it would not be a geodesic; thus, $q_2 \in V(G)$. Moreover, $q_2 \notin [x_0 z_0] \cup [y_0 z_0]$ since $d_{G^M}(\Pi(p), q_2) = d_{G^M}(p, q_2) = 2$ and $d_{G^M}(p, [x_0 z_0] \cup [y_0 z_0]) = 5/2$. Since $d_{G^M}(q_2, J(G^M) \setminus \{q_2\}) = 1/2$, one gets $l(\eta) \geq 5/2$, and we conclude $l(\eta) = 5/2$.

Therefore, $d_{G^M}(\Pi(p), [x_0 z_0] \cup [y_0 z_0]) = 5/2$. Thus, in order to finish the proof of (3), one must show that $d_G(\Pi(p), [xz] \cap [yz]) = 5/2$.

Since $T = \{x, y, z\} := \{[xy], [xz], [yz]\}$ is a geodesic triangle in G and G^M, $\delta(T) \leq \delta(G^M) = 5/2$. Let us show that $\delta(T) \geq 5/2$, thus finishing the proof.

Let $D_0 := d_G(\Pi(p), [xz] \cup [yz])$. Seeking for a contradiction, assume $D_0 < 5/2$, then, in fact, $D_0 \leq 2$ and $D_0 \in \{0, 1, 2\}$.

If $D_0 = 0$, then $\Pi(p) \in [xz] \cup [yz] \subset G$. Since $p \notin [x_0z_0] \cup [y_0z_0]$ and $\Pi(p) \notin [x_0z_0] \cup [y_0z_0]$, we have $p = \Pi(p) \in G$. Let p' be the mirror vertex of p. Since $p \notin [x_0z_0] \cup [y_0z_0]$, then $p' \in [x_0z_0] \cup [y_0z_0]$. There is $v \in V(G)$, with $\{v, p\} \subset E(G)$, and so, $d_{GM}(p, [x_0z_0] \cup [y_0z_0]) \leq d_{GM}(p, p') = d_{GM}(p, v) + d_{GM}(v, p') = 2 < 5/2$, a contradiction.

If $D_0 = 1$ and $v \in [xz] \cup [yz]$ satisfies $d_G(\Pi(p), v) = 1$, then $\Pi(p)$ and v are adjacent. Let u be a vertex with $\Pi(u) = v$ and $u \in [x_0z_0] \cup [y_0z_0]$; so $5/2 = d_{GM}(p, [x_0z_0] \cup [y_0z_0]) \leq d_{GM}(p, u) \leq 2$, which is a contradiction.

Finally, if $D_0 = 2$, then the vertex $v \in [x_0z_0] \cup [y_0z_0]$, which gives the minimum distance to p, either belongs to the mirror vertices or to G, but in any case, it is adjacent to an adjacent vertex of p; thus, $d_{GM}(p, [x_0z_0] \cup [y_0z_0]) = 2$, which is a contradiction.

In summary, $d_{GM}(p, [xz] \cup [yz]) = d_{GM}(p, [x_0z_0] \cup [y_0z_0]) = 5/2$.

Let us show the other implication to conclude that $\delta(G^M) = 5/2$.

By Theorem 6, $\delta(G^M) \leq 5/2$.

Let $T := \{x, y, z\}$ be the geodesic triangle in G in the hypothesis. Properties (1) and (2) stated in Theorem 2 and Lemma 7 give that T is also a geodesic triangle in G^M. Since $d_G(p, [xz] \cup [yz]) = 5/2$, Lemma 7 gives $\delta(G^M) \geq d_{GM}(p, [xz] \cup [yz]) = d_G(p, [xz] \cup [yz]) = 5/2$, finishing the proof. $\square$

The proof of Theorem 2 has the following consequence.

Corollary 3. *If the geodesic triangle T in G satisfies (1), (2) and (3) stated in Theorem 2, then T is also a geodesic triangle in G^M.*

We also have the following corollaries of Theorem 2.

Corollary 4. *If G is a graph with* diam $V(G) = 4$ *and $\delta(G) = 5/2$, then $\delta(G^M) = 5/2$.*

Corollary 5. *If G is a graph such that it does not contain a cycle σ satisfying $10 \leq l(\sigma) \leq 15$, then:*

$$5/4 \leq \delta(G^M) \leq 9/4.$$

Remark 2. *If G contains a cycle σ with $10 \leq l(\sigma) \leq 15$, one cannot say anything more precise about $\delta(G^M)$ apart from the bounds obtained in Theorem 6, which are applicable to all Mycielskian graphs, i.e., $5/4 \leq \delta(G^M) \leq 5/2$. In the particular case when G is the cycle graph C_n, with $10 \leq n \leq 15$, then Proposition 4 gives $\delta(G^M) = 5/2$; and if G is the complete graph K_n, with $n \geq 10$, then $\delta(G^M) = 5/4$ by Proposition 5.*

Given a graph G, we define its circumference as:

$$c(G) := \sup\{l(\sigma) \mid \sigma \text{ is a cycle in } G\}. \tag{7}$$

Proposition 2. *If G is a graph with $c(G) < 10$, then:*

$$\max\{5/4, \delta(G)\} \leq \delta(G^M) \leq 9/4.$$

Proof. By Corollary 5, it suffices to show $\delta(G) \leq \delta(G^M)$. By Lemma 3, there is a geodesic triangle T in G that is a cycle with $\delta(T) = \delta(G)$. Since $l(T) \leq c(G) \leq 9$, if $a, b \in T$, then $d_G(a, b) \leq l(T)/2 \leq 9/2$ and $d_{GM}(a, b) = d_G(a, b)$ by Lemma 7. Therefore, T is a geodesic triangle in G^M and $\delta(G^M) \geq \delta(T) = \delta(G)$. $\square$

Remark 3. *If $c(G) \geq 10$, one cannot say anything more precise about $\delta(G^M)$ apart from the bounds obtained in Theorem 6, which are applicable to all Mycielskian graphs, i.e., $5/4 \leq \delta(G^M) \leq 5/2$. In particular, if $v_1 \in V(C_{10})$, $v_2 \in V(C_n)$ with $n \geq 10$, and G is the graph obtained from C_{10} and C_n by identifying the* vertices v_1 and v_2, then Corollary 1, Theorem 6 and Proposition 4 give $5/2 = \delta(C_{10}^M) \leq \delta(G^M) \leq 5/2$ and $\delta(G^M) = 5/2$; if G is the complete graph K_n, with $n \geq 10$, then $\delta(G^M) = 5/4$ by Proposition 5.

5. Proof of Theorem 3

Proof. Assume first that $\mathrm{diam}(G) \leq 2$. In order to prove $\delta(G^M) = 5/4$, it will be enough to show that $\mathrm{diam}(G^M) \leq 5/2$, by Lemma 1.

Note that since $\mathrm{diam}\, V(G) \leq 2$, then $\mathrm{diam}\, V(G^M) = 2$. Consider $x, y \in J(G^M)$. If x or y is a vertex, then $d_{G^M}(x,y) \leq 5/2$. Without loss of generality, $x, y \in J(G^M) \setminus V(G^M)$. Note that $d_G(z,v) \leq 3/2$ for every $z \in J(G) \setminus V(G)$ and $v \in V(G)$, since $\mathrm{diam}(G) \leq 2$.

Trivially, for $x, y \in G$, or $x, y \in G^M \setminus G'$, or $x \in G'$ and $y \in G^M \setminus G'$, one gets $d_{G^M}(x,y) \leq 2$.

When $x \in G$ and $y \in G' \setminus G$, since $d_G(x,v) \leq 3/2$ for all vertices v in G, we have $d_{G^M}(x,y) \leq 2$.

Assume that $x, y \in G' \setminus G$. Let $a, b \in V(G)$, $u, v \in V(G') \setminus V(G)$ be so that x and y are the midpoints of the edges $\{u,a\}$ and $\{v,b\}$, respectively. It is enough to consider the case where $d_{G^M}(a,b) = 2$ and $d_{G^M}(a,y) = 5/2$. Since $\mathrm{diam}(G) \leq 2$, we have $d_G(\{a, \Pi(u)\}, b) = 1$; since $d_G(a,b) = 2$, we deduce $d_G(\Pi(u), b) = 1$, and then, $\{u, b\} \in E(G^M)$. Thus, $d_{G^M}(x,y) \leq 2$.

Finally, consider $x \in G$ and $y \in G^M \setminus G'$. Let $a, b \in V(G)$ so that $x \in \{a, b\}$, and let $y \in \{w, u\}$ for some u vertex in the mirror of G. Since $\mathrm{diam}(G) \leq 2$, $d_G(x, \Pi(u)) \leq 3/2$, and thus, $d_{G^M}(\Pi(u), \{a,b\}) = 1$. By symmetry, one can assume that $d_{G^M}(\Pi(u), b) = 1$, and so, $d_{G^M}(u, b) = 1$ and $d_{G^M}(x,y) \leq d_{G^M}(x,b) + d_{G^M}(b,u) + d_{G^M}(u,y) = 1/2 + 1 + 1/2 = 2$. This gives the statement (4).

Next, let us show that $\mathrm{diam}(G) > 5/2$ implies $\delta(G^M) > 5/4$, proving statement (5) of the Theorem. To this end, we shall construct a geodesic bigon with sides of length three.

Let $a, b \in J(G)$ be so that $d_G(a,b) = 3$. If $a, b \in V(G)$, then $\mathrm{diam}\, V(G) \geq 3$, and Proposition 3 and Corollary 6 give $\delta(G^M) \geq 3/2$ (of course, Theorem 3 is not used in the proofs of Proposition 3 and Corollary 6). Therefore, one can assume that $a, b \in J(G) \setminus V(G)$.

Let $v_i, v_{i+1}, v_j, v_{j+1} \in V(G)$ with $a \in \{v_i, v_{i+1}\}$ and $b \in \{v_j, v_{j+1}\}$; note that $d_G(\{v_i, v_{i+1}\}, \{v_j, v_{j+1}\}) = 2$. By symmetry, we can assume that $d_G(v_{i+1}, v_j) = 2$. Let $x, y \in J(G) \setminus V(G)$ with $x \in [u_i v_{i+1}]$ and $y \in [v_j u_{j+1}]$. Consider the geodesic bigon $T := \gamma_1 \cup \gamma_2$ where $\gamma_1 := [x u_i] \cup [u_i w] \cup [w u_{j+1}] \cup [u_{j+1} y]$, and $\gamma_2 := [x v_{i+1}] \cup [v_{i+1} v_j] \cup [v_j y]$. Therefore, taking $p = w$, $d_{G^M}(p, \gamma_2) = 3/2$, and thus, $\delta(G^M) \geq \delta(T) \geq 3/2$.

This finishes the proof of both statements.

To show that the converse of (4) does not hold, consider the graph G with four vertices $\{v_i\}_{i=0}^3$ and edges $\{v_0, v_1\}, \{v_1, v_2\}, \{v_2, v_3\}, \{v_3, v_1\}$. Then, $\mathrm{diam}(G) = 5/2$ and $\mathrm{diam}(G^M) = 3$, where the only two points $x, y \in G^M$ that realize the diameter are $x \in \{v_2, v_3\}$ and $y \in \{u_0, w\}$. Consider the geodesic bigon $T = \{\gamma_1, \gamma_2\}$ given by $\gamma_1 := [x v_2] \cup [v_2 v_1] \cup [v_1 u_0] \cup [u_0 y]$ and $\gamma_2 := [y w] \cup [w u_2] \cup [u_2 v_3] \cup [v_3 x]$; let $p \in \gamma_1$ be so that $d_{G^M}(p, v_1) = 1/4$. Then, $d_{G^M}(p, \gamma_2) = d_{G^M}(p, v_1) + d_{G^M}(v_1, u_2) = 1/4 + 1$, and this bigon has $\delta(T) = 5/4$.

Since $\mathrm{diam}(G^M) = 3$, we have $\delta(G^M) \leq 3/2$. Notice that in order to have a geodesic triangle T with $\delta(T) = 3/2$, one of its sides must have length three and, therefore, must have x, y as its endpoints. As in the case of the bigon above, for any of these triangles, every vertex in $[xy]$ is at a distance at most one of the other two sides. Therefore, all of these triangles satisfy $\delta(T) = 5/4$. Therefore, $\delta(G^M) = 5/4$.

Proposition 4 gives $\delta(C_5^M) = 3/2$. Since $\mathrm{diam}(C_5) = 5/2$, the converse of (5) does not hold. $\square$

6. Hyperbolicity Constant for Some Particular Mycielskian Graphs and the Proof of Theorems 1, 4 and 5

Computing the hyperbolicity constant of some graphs turns out to be specially useful. Namely, many graphs contain path graphs as isomorphic subgraphs; the cycle graph appeared naturally as a boundary situation for hyperbolicity constant of $5/2$; and finally, graphs isomorphic to the complete one were of interest. In this section, the precise hyperbolicity constants of the Mycielskian graphs for such graphs is calculated.

Proposition 3. *Let P_n be the path graph with n vertices. Then:*

$$\delta(P_n^M) = \begin{cases} 5/4, & \text{if } n = 2,3, \\ 3/2, & \text{if } n = 4, \\ 2, & \text{if } n = 5,6,7, \\ 9/4, & \text{if } n \geq 8. \end{cases}$$

Proof. Denote by $v_1, \ldots, v_n$ the vertices of P_n with $\{v_i, v_{i+1}\} \in E(P_n)$ for $i = 1, \ldots, n-1$. One can easily check that $\text{diam}(P_3^M) = 5/2$ (see Figure 3). Since $\text{diam}(P_2^M) = \text{diam}(P_3^M) = 5/2$, Lemma 1 gives $\delta(P_n^M) \leq 5/4$ for $n = 2,3$. Then, Theorem 6 let us get the desired result for $n = 2,3$.

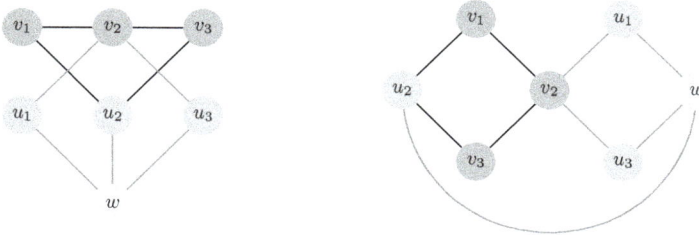

Figure 3. P_3^M has diameter $5/2$.

For P_4^M and P_5^M, a similar argument will be used. First, a triangle T with the desired δ will be constructed, giving the lower bound; then, the diameter of the Mycielskian will give the upper bound.

In P_4^M, consider the triangle $T := \{x, y, z\}$, where $x = v_1$, $y = v_4$, $z = w$, and take p as the midpoint of $[v_2 v_3]$. Here, $d_{GM}(p, [xz] \cup [yz]) = 3/2$. One can check that $\text{diam}(P_4^M) = 3$. Therefore, Lemma 1 gives $3/2 \leq \delta(T) \leq \delta(P_4^M) \leq 1/2 \text{diam}(P_4^M) = 3/2$.

A similar argument works for P_5^M.

A simple argument will give the result for $n \geq 8$.

For $n \geq 8$, Theorems 2, 6 and Lemma 4 give $\delta(P_n^M) \leq 9/4$. Consider the geodesic triangle $T := \{x, y, z\}$ with $x = v_2$, y the midpoint in $\{v_6, v_7\}$ and $z = w$, and take p the midpoint in $[xy] \subset P_n$ (see Figure 4). Then, $\delta(P_n^M) = 9/4$, since $\delta(P_n^M) \geq d_{P_n^M}(p, [xz] \cup [yz]) = 9/4$.

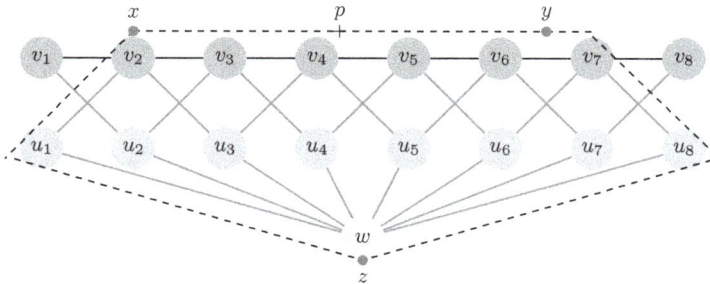

Figure 4. For $n = 8$, a geodesic triangle $T = \{x, y, z\}$ in P_n^M with $\delta(T) = 9/4$.

All that is left is to prove that $\delta(P_7^M) = 2$, which will automatically imply $\delta(P_6^M) = 2$ as well by Corollary 1, since $P_5 \subset P_6 \subset P_7$.

By Lemma 3, let $T := \{x, y, z\} \subset P_7^M$ be a geodesic triangle, so that $x, y, z \in J(P_7^M)$, and let $p \in [xy]$ be so that $d_{P_7^M}(p, [xz] \cup [yz]) = \delta(T) = \delta(P_7^M)$. Seeking for a contradiction, assume that

$\delta(P_7^M) > 2$. This would mean $\delta(P_7^M) \in \{9/4, 5/2\}$ by Lemma 4 and Theorem 6. However, by Theorem 2, $\delta(P_7^M) \neq 5/2$, since every geodesic triangle in P_7 has a hyperbolicity constant equal to zero, like any geodesic triangle in any tree. That is:

$$\delta(P_7^M) = \frac{9}{4}, \qquad d_{P_7^M}(p, \{x, y\}) = \frac{9}{4}, \tag{8}$$

$$d_{P_7^M}(x, y) \in \left\{\frac{9}{2}, 5\right\}. \tag{9}$$

Some observations: $x, y \in P_7$ for otherwise $d_{P_7^M}(x, y) \leq 4$ (via w), contradicting (9). Moreover, without loss of generality, we can assume that $x \in J(P_7) \setminus V(P_7)$ always and $y \in J(P_7) \setminus V(P_7)$ if $d_{P_7^M}(x, y) = 5$ and $y \in V(P_7)$ if $d_{P_7^M}(x, y) = 9/4$. Last, $w \in T$ for otherwise T would be a cycle of diameter two, and thus, $\delta(T) = 1$.

Suppose that $w \in [xy]$. By (8) and (9), $d_{P_7^M}(p, w) \leq 1/4$. It is easy to check that $[xz] \cup [yz] \subset P_7$. Thus, we deduce $d_{P_7^M}(w, [xz] \cup [yz]) = 2$.

Since $x \in J(P_7) \setminus V(P_7)$, we have $p \in [xw]$. Since $d_{P_7^M}(w, [xz] \cup [yz]) = 2$, (8) and (9) give $d_{P_7^M}(p, w) = 1/4$. By symmetry, we can assume that x is the midpoint of either $\{v_1, v_2\}$ or $\{v_2, v_3\}$. Assume that $x \in \{v_1, v_2\}$ (the case $x \in \{v_2, v_3\}$ is similar). Since T is a cycle and $[xz] \cup [yz] \subset P_7$, we have $v_3 \in [xz] \cup [yz] \subset P_7$, $[xv_1] \cup \{v_1, u_2\} \cup \{u_2, w\} \subset [zy]$ and $p \in \{u_2, w\}$. Then, $9/4 = d_{P_7^M}(p, [xz] \cup [yz]) \leq d_{P_7^M}(p, v_3) \leq d_{P_7^M}(p, u_2) + d_{P_7^M}(u_2, v_3) = 7/4$, a contradiction. Hence, $w \notin [xy]$.

A similar argument shows that $w \notin [xz] \cup [yz]$, which contradicts the fact that $w \in T$. One concludes that $\delta(P_7^M) \leq 2$, which together with the fact that $\delta(P_7^M) \geq \delta(P_5^M) = 2$ gives the desired result. □

Remark 4. *There are several definitions of Gromov hyperbolicity. They are all equivalent in the sense that if X is δ-hyperbolic with respect to the definition A, then it is δ'-hyperbolic with respect to the definition B, for some δ' (see, e.g., [2,3]).*

The definition that we have chosen in the present paper is known as the Rips condition. We decided to select it among others due to its deep geometric meaning (see, e.g., [3]). As an example, the simplest existing proof of the invariance of hyperbolicity by quasi-isometries uses the Rips condition (see [3]) and so does the proof for geodesic stability (see [5]). Furthermore, some results that employ a different definition (such as the four-point condition) also require the Rips condition in their proofs (see, e.g., [27]). The Rips condition also comes up in a natural way when graphs with arbitrarily large edges are considered.

Experience has shown that, although the definitions of hyperbolicity are equivalent, the values of the hyperbolicity constants of a space obtained when different definitions are considered have different behaviors actually.

As an example, the analogue of Proposition 3 for the hyperbolicity constant obtained applying the four-point condition (that we shall denote by δ_{4PC}) says that $\delta_{4PC}(P_2^M) = 1/2$ and $\delta_{4PC}(P_n^M) = 1$ for every $n \geq 3$.

The following corollary is straightforward, but it is presented here because it is used to simplify the arguments in some other proofs in the paper.

Corollary 6. *Let G be any graph. Then, $\delta(G^M) \geq \delta\left(P_{\text{diam}V(G)+1}^M\right)$.*

Observe that this follows from Corollary 1, since if $\text{diam}\, V(G) = r$, then there exists a geodesic $g_0 \subseteq G$ joining two vertices with length r. That is, P_{r+1} is isomorphic to g_0, and besides, g_0 is an isometric subgraph.

Proposition 3 and Corollaries 5 and 6 have the following consequence.

Corollary 7. *If G is a graph with* diam $V(G) \geq 7$ *such that it does not contain a cycle* σ *satisfying* $10 \leq l(\sigma) \leq 15$, *then* $\delta(G^M) = 9/4$.

By Lemmas 1 and 8, Proposition 3 and Corollary 6, we obtain the following result.

Corollary 8. *Let G be any graph. Then:*

(i) *If* diam $V(G) = 2$, *then* $5/4 \leq \delta(G^M) \leq 3/2$.
(ii) *If* diam $V(G) = 3$, *then* $3/2 \leq \delta(G^M) \leq 2$.

The next proposition deals with cycle graphs, which illustrate well the result of Theorem 2.

Proposition 4. *Let C_n be the cycle graph with n vertices. Then:*

$$\delta(C_n^M) = \begin{cases} 5/4, & \text{if } n = 3,4, \\ 3/2, & \text{if } n = 5,6, \\ 7/4, & \text{if } n = 7, \\ 2, & \text{if } n = 8, \\ 9/4, & \text{if } n = 9 \text{ or } n \geq 16, \\ 5/2, & \text{if } 10 \leq n \leq 15. \end{cases}$$

Proof. Denote by $v_1, \ldots, v_n$ the vertices of C_n with edges $\{v_n, v_1\} \cup \left(\cup_{i=1}^{n-1} \{v_i, v_{i+1}\} \right)$. With the usual notation, u_i is the mirror vertex of v_i, and w is the supervertex.

For $n = 3,4$, diam$(C_n^M) = 5/2$, and thus, $\delta(C_n^M) \leq 5/4$. On the other hand, Theorem 6 gives $\delta(C_n^M) \geq 5/4$.

For $n = 5$, diam$(C_5^M) = 3$, giving $3/2$ as an upper bound for the hyperbolicity constant. On the other hand, consider the bigon $T = \{x, y\}$ with x the midpoint of $\{v_2, v_3\}$ and y the midpoint of $\{u_1, w\}$. Denote by $[xy]$ the geodesic in T with $v_1 \notin [xy]$; thus, $\delta(C_5^M) \geq d_{C_5^M}(v_1, [xy]) = 3/2$.

For $n \in \{6,7,8,9\}$, diam$(C_n^M) = n/2$; thus, the upper bound is $n/4$. Consider the bigon T of antipodal vertices x, y in C_n, with $T \subset C_n$. Let p be the midpoint of $[xy]$; then, $\delta(T) = n/4$ and thus $\delta(C_n^M) \geq n/4$, which together with the upper bound, gives $\delta(C_n^M) = n/4$.

The range $n \in \{10, 11, 12, 13, 14, 15\}$ automatically follows from Theorem 2.

Finally, if $n \geq 16$, then diam $V(C_n) \geq 8$, and Corollary 7 gives the result. □

Remark 5. *The analogue of Proposition 4 for the hyperbolicity constant of the four-point condition says that* $\delta_{4PC}(C_n^M) = 1$ *for* $3 \leq n \leq 7$, $\delta_{4PC}(C_8^M) = 2$, $\delta_{4PC}(C_n^M) = 3/2$ *for* $9 \leq n \leq 10$ *and* $\delta_{4PC}(C_n^M) = 1$ *for every* $n \geq 11$.

The complete graph has a very constant behavior.

Proposition 5. *Let K_n be the complete graph with n vertices. Then:*

$$\text{diam}(K_n^M) = \frac{5}{2}, \qquad \delta(K_n^M) = \frac{5}{4}, \qquad \forall n \geq 2.$$

Proof. Since $K_2 = P_2$ and $K_3 = C_3$, by Propositions 3 and 4, one gets the result if $n < 4$.

For $n \geq 4$, Theorem 6 already gives $\delta(K_n^M) \geq 5/4$, so it suffices to estimate the diameter of K_n^M.

Notice diam $V(K_n^M) = 2$. Without loss of generality, take $x, y \in J(K_n^M)$; clearly, if $y = v \in V(K_n^M)$, then $d_{K_n^M}(x, v) \leq 5/2$ with equality if $x \in J(K_n) \setminus V(K_n)$ and $v = w$; if $x, y \in J(K_n^M) \setminus V(K_n^M)$, then $d_{K_n^M}(x, v) \leq 2$ with equality if $x \in \{v_i, v_j\}, y \in \{u_k, w\}$.

Summing up, $\text{diam}(K_n^M) = 5/2$; thus, $\delta(K_n^M) \leq 5/4$, which together with Theorem 6 prove the result. $\quad\square$

Remark 6. *The analogue of Proposition 5 for the hyperbolicity constant of the four-point condition says that* $\delta_{4PC}(K_2^M) = 1/2$ *and* $\delta_{4PC}(K_n^M) = 1$ *for every* $n \geq 3$.

The argument in the proof of Proposition 5 also gives the following result.

Corollary 9. *If* $\text{diam}(G^M) \leq 5/2$, *then* $\delta(G^M) = 5/4$.

We can now finish the proof of Theorem 1.

Proof of Theorem 1: The main part follows from Theorem 7 for non-complete graphs. If G is a complete graph, Proposition 5 gives that $\delta(G^M) = 5/4$

For the sharpness of $\delta(G^M) \leq 5/2$, consider the graphs P_2 and C_{10}, where P_2 stands for the path graph of vertices $\{v_0, v_1, v_2, v_3\}$ and edges $\{v_0, v_1\}$, $\{v_1, v_2\}$, $\{v_2, v_3\}$, and C_{10} is the the cyclic graph with 10 vertices. From Propositions 3 and 4, we get $\delta(P_2^M) = 5/4$ and $\delta(C_{10}^M) = 5/2$. $\quad\square$

Proof of Theorem 4. Lemma 1 gives the upper bound. If $\text{diam} \, V(G^M) \leq 2$, then Theorem 6 gives $\delta(G^M) \geq 5/4 > 1 \geq \frac{1}{2} \text{diam} \, V(G^M)$. If $\text{diam} \, V(G^M) > 2$, then Theorem 6 gives $3 \leq \text{diam} \, V(G^M) \leq 4$. Therefore, G is not a complete graph by Proposition 5. Thus, Theorem 8, Proposition 3 and Corollary 6 give $\delta(G^M) \geq \delta\left(P_{\text{diam} \, V(G)+1}^M\right) \geq \delta\left(P_{\text{diam} \, V(G^M)+1}^M\right)$. Hence, if $\text{diam} \, V(G^M) = 3$, then we have $\delta(G^M) \geq 3/2 = \frac{1}{2} \text{diam} \, V(G^M)$ by Proposition 3. If $\text{diam} \, V(G^M) = 4$, then we have $\delta(G^M) \geq 2 = \frac{1}{2} \text{diam} \, V(G^M)$ by Proposition 3. $\quad\square$

Proof of Theorem 5. By Theorem 4, we have $\text{diam} \, V(G^M) \leq 3$. One can check that $\text{diam} \, V(G) \leq 3$, and Theorem 8 gives $\delta(G) \leq \delta(G^M)$. $\quad\square$

Proposition 6. *Let* $K_{n,m}$ *be the complete bipartite graph with* $n + m$ *vertices. Then:*

$$\delta\left(K_{n,m}^M\right) = \frac{5}{4}, \qquad \forall n, m \geq 2.$$

Proof. As in the proof of Proposition 5, it suffices to show that $\text{diam}(K_{n,m}^M) = 5/2$.

One can check that $\text{diam} \, V(K_{n,m}^M) = 2$, and for every $x \in J(K_{n,m}^M) \setminus V(K_{n,m}^M)$, we have $\max_{y \in K_{n,m}^M} d(x, y) = 5/2$ (see Figure 5); then, $\text{diam}(K_{n,m}^M) = 5/2$. $\quad\square$

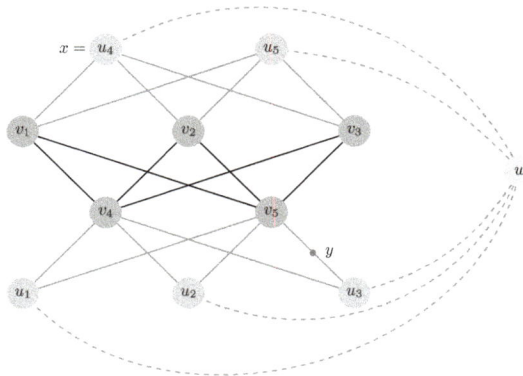

Figure 5. x and y are two points furthest apart in $K_{3,2}^M$.

7. The Case of 5/4

In order to characterize the Mycielskian graphs with hyperbolicity constant 5/4, we define some families of graphs.

Denote by C_n the cycle graph with $n \geq 3$ vertices and by $V(C_n) := \{v_1^{(n)}, \ldots, v_n^{(n)}\}$ the set of their vertices, such that $\{v_n^{(n)}, v_1^{(n)}\} \in E(C_n)$ and $\{v_i^{(n)}, v_{i+1}^{(n)}\} \in E(C_n)$ for $1 \leq i \leq n-1$. Let $\mathcal{C}_6$ be the set of graphs obtained from C_6 by adding a (proper or not) subset of the set of edges $\{\{v_2^{(6)}, v_6^{(6)}\}, \{v_4^{(6)}, v_6^{(6)}\}\}$. Let us define the set of graphs:

$$\mathcal{F}_6 := \{\text{graphs containing, as the induced subgraph, an isomorphic graph to} \\ \text{some element of } \mathcal{C}_6\}.$$

Let $\mathcal{C}_7$ be the set of graphs obtained from C_7 by adding a (proper or not) subset of the set of edges $\{\{v_2^{(7)}, v_6^{(7)}\}, \{v_2^{(7)}, v_7^{(7)}\}, \{v_4^{(7)}, v_6^{(7)}\}, \{v_4^{(7)}, v_7^{(7)}\}\}$. Define:

$$\mathcal{F}_7 := \{\text{graphs containing, as the induced subgraph, an isomorphic graph to} \\ \text{some element of } \mathcal{C}_7\}.$$

Let $\mathcal{C}_8$ be the set of graphs obtained from C_8 by adding a (proper or not) subset of the set $\{\{v_2^{(8)}, v_6^{(8)}\}, \{v_2^{(8)}, v_8^{(8)}\}, \{v_4^{(8)}, v_6^{(8)}\}, \{v_4^{(8)}, v_8^{(8)}\}\}$. Furthermore, let $\mathcal{C}_8'$ be the set of graphs obtained from C_8 by adding a (proper or not) subset of $\{\{v_2^{(8)}, v_8^{(8)}\}, \{v_4^{(8)}, v_6^{(8)}\}, \{v_4^{(8)}, v_7^{(8)}\}, \{v_4^{(8)}, v_8^{(8)}\}\}$. Define:

$$\mathcal{F}_8 := \{\text{graphs containing, as the induced subgraph, an isomorphic graph to} \\ \text{some element of } \mathcal{C}_8 \cup \mathcal{C}_8'\}.$$

Let $\mathcal{C}_9$ be the set of graphs obtained from C_9 by adding a (proper or not) subset of the set of edges $\{\{v_2^{(9)}, v_6^{(9)}\}, \{v_2^{(9)}, v_9^{(9)}\}, \{v_4^{(9)}, v_6^{(9)}\}, \{v_4^{(9)}, v_9^{(9)}\}\}$. Define:

$$\mathcal{F}_9 := \{\text{graphs containing, as the induced subgraph, an isomorphic graph to} \\ \text{some element of } \mathcal{C}_9\}.$$

Finally, we define the set $\mathcal{F}$ by:

$$\mathcal{F} := \mathcal{F}_6 \cup \mathcal{F}_7 \cup \mathcal{F}_8 \cup \mathcal{F}_9.$$

In [53] (Lemma 3.21) appears the following result.

Lemma 11. *Let G be any graph. Then, $G \in \mathcal{F}$ if and only if there is a geodesic triangle $T = \{x, y, z\}$ in G that is a cycle with $x, y, z \in J(G)$, $L([xy]), L([yz]), L([zx]) \leq 3$ and $\delta(T) = 3/2 = d(p, [yz] \cup [zx])$ for some $p \in [xy] \cap V(G)$.*

Finally, we obtain a simple characterization of the Mycielskian graphs with hyperbolicity constant 5/4.

Theorem 10. *Let G be any graph. Then, $\delta(G^M) = 5/4$ if and only if diam $V(G) \leq 2$ and $G^M \notin \mathcal{F}$.*

Proof. If $\delta(G^M) = 5/4$, then $G^M \notin \mathcal{F}$. If diam $V(G) > 2$, then Proposition 3 and Corollary 6 give $\delta(G^M) \geq 3/2$.

Assume now diam $V(G) \leq 2$ and $G^M \notin \mathcal{F}$.

If diam $V(G) = 1$, then G is a complete graph, and Proposition 5 gives $\delta(G^M) = 5/4$.

If diam $V(G) = 2$, then Lemma 8 gives diam $V(G^M) = 2$, and thus, diam$(G^M) \leq 3$.

If diam$(G^M) \leq 5/2$, then Lemma 1 gives $\delta(G^M) \leq 5/4$, and we conclude $\delta(G^M) = 5/4$ by Theorem 1. $\qquad\blacksquare$

If $\mathrm{diam}(G^M) = 3$, then $\delta(G^M) \le 3/2$ by Lemma 1. Besides, $\delta(G^M) \ge 5/4$ by Theorem 1. Hence, Lemma 4 implies $\delta(G^M) \in \{5/4, 3/2\}$. Seeking for a contradiction, assume that $\delta(G^M) = 3/2$. By Lemma 3, there exists a geodesic triangle $T = \{x, y, z\}$ that is a cycle with $x, y, z \in J(G^M)$ and $\delta(T) = 3/2 = d(p, [yz] \cup [zx])$ for some $p \in [xy]$. Then, $d_{G^M}(p, \{x, y\}) \ge d_{G^M}(p, [yz] \cup [zx]) = 3/2$ and $d_{G^M}(x, y) \ge 3$. Since $\mathrm{diam}(G^M) = 3$ and T is a cycle, we have $L([xy]) = 3$, $L([yz])$, $L([zx]) \le 3$. Since $\mathrm{diam}\, V(G^M) = 2$, $x, y \in J(G^M) \setminus V(G^M)$, p is the midpoint of $[xy]$, and it is a vertex of G^M. Thus, Lemma 11 gives $G \in \mathcal{F}$, which is the contradiction we were looking for. Hence, $\delta(G^M) \ne 3/2$, and we conclude $\delta(G^M) = 5/4$. □

We finish this work with a computational remark about Theorem 10.

Let us consider a graph Γ with m edges, a vertex with degree Δ and the other vertices with degree at most $\Delta_0 \le \Delta$. By choosing an edge $\{v_{i_1}, v_{i_2}\} \in E(\Gamma)$, an edge $\{v_{i_2}, v_{i_3}\} \in E(\Gamma)$,..., and an edge $\{v_{i_{a-1}}, v_{i_a}\} \in E(\Gamma)$, we can obtain the set of all paths of length $a - 1$ in time $O(m\Delta\Delta_0^{a-3})$; hence, we can compute all cycles with length a in time $O(m\Delta\Delta_0^{a-3})$. Therefore, it is possible to find a subgraph isomorphic to a fixed graph in $\mathcal{C}_a$ (or in $\mathcal{C}_a'$) in time $O(m\Delta\Delta_0^{a-3})$. Note that there are 4, 16, 16, 16 and 16 graphs in $\mathcal{C}_6, \mathcal{C}_7, \mathcal{C}_8, \mathcal{C}_8'$ and $\mathcal{C}_9$, respectively. Hence, we can find every subgraph of Γ isomorphic to a graph in $\mathcal{C}_6 \cup \mathcal{C}_7 \cup \mathcal{C}_8 \cup \mathcal{C}_8' \cup \mathcal{C}_9$ in time $O(m\Delta\Delta_0^6)$.

If G is a graph with n vertices, m edges and maximum degree Δ, then G^M is a graph with $3m + n$ edges, a vertex with degree n and the other vertices with degree at most 2Δ. Since $\Delta \le n - 1$, we can know if either $G^M \in \mathcal{F}$ or $G^M \notin \mathcal{F}$ in time $O((3m + n)\max\{2\Delta, n\}(2\Delta)^6) = O(nm\Delta^6)$. Hence, to check the hypothesis $G^M \notin \mathcal{F}$ is a tractable problem from a computational viewpoint, by using the algorithm sketched before.

Acknowledgments: We would like to thank the referees for their valuable comments, which have improved the paper. This work was supported in part by two grants from Ministerio de Economía y Competititvidad (MTM2013-46374-P and MTM2015-69323-REDT), Spain.

Author Contributions: Ana Granados, Domingo Pestana, Ana Portilla and José M. Rodríguez contributed equally in the results, writing the manuscript and proofreading. All authors have read and approved the final manuscript.

Conflicts of Interest: The authors declare no conflict of interest.

References

1. Alonso, J.; Brady, T.; Cooper, D.; Delzant, T.; Ferlini, V.; Lustig, M.; Mihalik, M.; Shapiro, M.; Short, H. Notes on word hyperbolic groups. In *Group Theory from a Geometrical Viewpoint*; Ghys, E., Haefliger, A., Verjovsky, A., Eds.; World Scientific: Singapore, 1992.

2. Bowditch, B.H. Notes on Gromov's hyperbolicity criterion for path-metric spaces. In *Group Theory from a Geometrical Viewpoint, Trieste, 1990*; Ghys, E., Haefliger, A., Verjovsky, A., Eds.; World Scientific: River Edge, NJ, USA, 1991; pp. 64–167.

3. Ghys, E.; de la Harpe, P. Sur les Groupes Hyperboliques d'après Mikhael Gromov. In *Progress in Mathematics 83*; Birkhäuser Boston Inc.: Boston, MA, USA, 1990.

4. Gromov, M. *Hyperbolic Groups, in "Essays in Group Theory"*; Gersten, S.M., Publ, M.S.R.I., Eds.; Springer: Berlin, Germany, 1987; Volume 8, pp. 75–263.

5. Bonk, M. Quasi-geodesics segments and Gromov hyperbolic spaces. *Geom. Dedicata* **1996**, *62*, 281–298.

6. Chen, W.; Fang, W.; Hu, G.; Mahoney, M.W. On the hyperbolicity of small-world and treelike random graphs. *Internet Math.* **2013**, *9*, 434–491.

7. Jonckheere, E.A.; Lohsoonthorn, P.; Bonahon, F. Scaled Gromov hyperbolic graphs. *J. Graph Theory* **2007**, *2*, 157–180.

8. Shang, Y. Lack of Gromov-hyperbolicity in small-world networks. *Cent. Eur. J. Math.* **2012**, *10*, 1152–1158.

9. Shang, Y. Non-hyperbolicity of random graphs with given expected degrees. *Stoch. Models* **2013**, *29*, 451–462.

10. Oshika, K. *Discrete Groups, AMS Bookstore, Iwanami Series on Modern Mathematics, Translations of Mathematical Monographs*; Amer Mathematical Society: Providence, RI, USA, 2002; Volume 207.

11. Charney, R. Artin groups of finite type are biautomatic. *Math. Ann.* **1992**, *292*, 671–683.

12. Edwards, K.; Kennedy, W.S.; Saniee, I. Fast Approximation Algorithms for p-Centres in Large δ-Hyperbolic Graphs. Available online: https://arxiv.org/pdf/1604.07359.pdf (accessed on 25 April 2016)

13. Shavitt, Y.; Tankel, T. On internet embedding in hyperbolic spaces for overlay construction and distance estimation. *IEEE/ACM Trans. Netw.* **2008**, *16*, 25–36.

14. Verbeek, K.; Suri, S. Metric embeddings, hyperbolic space and social networks. In Proceedings of the 30th Annual Symposium on Computational Geometry, Kyoto, Japan, 8–11 June 2014; pp. 501–510.

15. Abu-Ata, M.; Dragan, F.F. Metric tree-like structures in real-life networks: An empirical study. *Networks* **2016**, *67*, 49–68.

16. Adcock, A.B.; Sullivan, B.D.; Mahoney, M.W. Tree-like structure in large social and information networks. In Proceedings of the 2013 IEEE 13th International Conference Data Mining (ICDM), Dallas, TX, USA, 7–10 December 2013; pp. 1–10.

17. Cohen, N.; Coudert, D.; Lancin, A. *Exact and Approximate Algorithms for Computing the Hyperbolicity of Large-Scale Graphs*; Rapport de Recherche RR-8074; INRIA: Rocquencourt, France, September 2012.

18. Krioukov, D.; Papadopoulos, F.; Kitsak, M.; Vahdat, A.; Boguñá, M. Hyperbolic geometry of complex networks. *Phys. Rev. E* **2010**, *82*, 036106.

19. Montgolfier, F.; Soto, M.; Viennot, L. Treewidth and Hyperbolicity of the Internet. In Proceedings of the 2011 IEEE 10th International Symposium on Network Computing and Applications (NCA), Washington, DC, USA, 25–27 August 2011; pp. 25–32.

20. Jonckheere, E.A. Contrôle du trafic sur les réseaux à géométrie hyperbolique: Vers une théorie géométrique de la sécurité de l'acheminement de l'information. *J. Eur. Syst. Autom.* **2002**, *8*, 45–60.

21. Jonckheere, E.A.; Lohsoonthorn, P. Geometry of network security. In Proceedings of the 2004 American Control Conference, Boston, MA, USA, 30 June–2 July 2004; pp. 111–151.

22. Chepoi, V.; Dragan, F.F.; Vaxès, Y. Core congestion is inherent in hyperbolic networks. In Proceedings of the Twenty-Eighth Annual ACM-SIAM Symposium on Discrete Algorithms, Barcelona, Spain, 16–19 January 2017.

23. Grippo, E.; Jonckheere, E.A. Effective resistance criterion for negative curvature: application to congestion control. In Proceedings of the IEEE Multi-Conference on Systems and Control, Buenos Aires, Argentina, 19–22 September 2016.

24. Li, S.; Tucci, G.H. Traffic Congestion in Expanders, (p, δ)-Hyperbolic Spaces and Product of Trees. *Internet Math.* **2015**, *11*, 134–142, doi:10.1080/15427951.2014.884513.

25. Bandelt, H.-J.; Chepoi, V. 1-Hyperbolic Graphs. *SIAM J. Discret. Math.* **2003**, *16*, 323–334.

26. Boguñá, M.; Papadopoulos, F.; Krioukov, D. Sustaining the Internet with Hyperbolic Mapping. *Nat. Commun.* **2010**, *1*, 18.

27. Brinkmann, G.; Koolen, J.; Moulton, V. On the hyperbolicity of chordal graphs. *Ann. Comb.* **2001**, *5*, 61–69.

28. Carballosa, W.; Rodríguez, J.M.; Sigarreta, J.M.; Villeta, M. Gromov hyperbolicity of line graphs. *Electr. J. Comb.* **2011**, *18*, 210.

29. Chepoi, V.; Dragan, F.F.; Estellon, B.; Habib, M.; Vaxès, Y. Notes on diameters, centers, and approximating trees of δ-hyperbolic geodesic spaces and graphs. *Electr. Notes Discr. Math.* **2008**, *31*, 231–234.

30. Diestel, R.; Müller, M. Connected tree-width. *Combinatorica* **2017**, doi:10.1007/s00493-016-3516-5.

31. Frigerio, R.; Sisto, A. Characterizing hyperbolic spaces and real trees. *Geom. Dedicata* **2009**, *142*, 139–149.

32. Koolen, J.H.; Moulton, V. Hyperbolic Bridged Graphs. *Europ. J. Comb.* **2002**, *23*, 683–699.

33. Sigarreta, J.M. Hyperbolicity in median graphs. *Proc. Math. Sci.* **2013**, *123*, 455–467.

34. Tourís, E. Graphs and Gromov hyperbolicity of non-constant negatively curved surfaces. *J. Math. Anal. Appl.* **2011**, *380*, 865–881.

35. Wu, Y.; Zhang, C. Chordality and hyperbolicity of a graph. *Electr. J. Comb.* **2011**, *18*, 43.

36. Balogh, Z.M.; Bonk, M. Gromov hyperbolicity and the Kobayashi metric on strictly pseudoconvex domains. *Comment. Math. Helv.* **2000**, *75*, 504–533.

37. Benoist, Y. Convexes hyperboliques et fonctions quasisymétriques. *Publ. Math. Inst. Hautes Études Sci.* **2003**, *97*, 181–237.

38. Karlsson, A.; Noskov, G.A. The Hilbert metric and Gromov hyperbolicity. *Enseign. Math.* **2002**, *48*, 73–89.

39. Hästö, P. A. Gromov hyperbolicity of the j_G and $\tilde{j}_G$ metrics. *Proc. Am. Math. Soc.* **2006**, *134*, 1137–1142.

40. Lindén, H. Gromov hyperbolicity of certain conformal invariant metrics. *Ann. Acad. Sci. Fenn. Math.* **2007**, *32*, 279–288.

41. Balogh, Z.M.; Buckley, S.M. Geometric characterizations of Gromov hyperbolicity. *Invent. Math.* **2003**, *153*, 261–301.
42. Bonk, M.; Heinonen, J.; Koskela, P. *Uniformizing Gromov Hyperbolic Spaces*; Astérisque; Amer Mathematical Society: Providence, RI, USA, 2001; Volume 270.
43. Hästö, P.A.; Portilla, A.; Rodríguez, J.M.; Tourís, E. Gromov hyperbolic equivalence of the hyperbolic and quasihyperbolic metrics in Denjoy domains. *Bull. Lond. Math. Soc.* **2010**, *42*, 282–294.
44. Portilla, A.; Rodríguez, J.M.; Tourís, E. Gromov hyperbolicity through decomposition of metric spaces II. *J. Geom. Anal.* **2004**, *14*, 123–149.
45. Rodríguez, J.M.; Tourís, E. Gromov hyperbolicity through decomposition of metric spaces. *Acta Math. Hung.* **2004**, *103*, 53–84.
46. Rodríguez, J.M.; Tourís, E. Gromov hyperbolicity of Riemann surfaces. *Acta Math. Sin.* **2007**, *23*, 209–228.
47. Mycielski, J. Sur le coloriage des graphes. *Colloq. Math.* **1955**, *3*, 161–162.
48. Fournier, H.; Ismail, A.; Vigneron, A. Computing the Gromov hyperbolicity of a discrete metric space. *J. Inform. Proc. Lett.* **2015**, *115*, 576–579.
49. Coudert, D.; Ducoffe, G. Recognition of C_4-Free and 1/2-Hyperbolic Graphs. *SIAM J. Discret. Math.* **2014**, *28*, 1601–1617.
50. Papasoglu, P. An algorithm detecting hyperbolicity. In *Geometric and Computational Perspectives on Infinite Groups, DIMACS—Series in Discrete Mathematics and Theoretical Computer Science*; AMS (Amer Mathematical Society): Providence, RI, USA, 1996; Volume 25, pp. 193–200.
51. Bermudo, S.; Rodríguez, J.M.; Sigarreta, J.M.; Tourís, E. Hyperbolicity and complement of graphs. *Appl. Math. Lett.* **2011**, *24*, 1882–1887.
52. Calegari, D.; Fujiwara, K. Counting subgraphs in hyperbolic graphs with symmetry. *J. Math. Soc. Jpn.* **2015**, *67*, 1213–1226.
53. Carballosa, W.; de la Cruz, A.; Rodríguez, J.M. Gromov Hyperbolicity in Lexicographic Product Graphs. Available online: http://gama.uc3m.es/index.php/jomaro.html (accessed on 19 June 2015).
54. Carballosa, W.; Rodríguez, J.M.; Sigarreta, J.M. Hyperbolicity in the corona and join of graphs. *Aequ. Math.* **2015**, *89*, 1311–1328, doi:10.1007/s00010-014-0324-0.
55. Coudert, D.; Ducoffe, G. On the hyperbolicity of bipartite graphs and intersection graphs. *Discret. App. Math.* **2016**, *214*, 187–195
56. Martínez-Pérez, A. Chordality properties and hyperbolicity on graphs. *Electr. J. Comb.* **2016**, *23*, P3.51.
57. Bermudo, S.; Rodríguez, J.M.; Sigarreta, J.M.; Vilaire, J.-M. Gromov hyperbolic graphs. *Discret. Math.* **2013**, *313*, 1575–1585.
58. Chen, B.; Yau, S.-T.; Yeh, Y.-N. Graph homotopy and Graham homotopy. *Discret. Math.* **2001**, *241*, 153–170.
59. Rodríguez, J.M.; Sigarreta, J.M.; Vilaire, J.-M.; Villeta, M. On the hyperbolicity constant in graphs. *Discr. Math.* **2011**, *311*, 211–219.
60. Bermudo, S.; Rodríguez, J.M.; Sigarreta, J.M. Computing the hyperbolicity constant. *Comp. Math. Appl.* **2011**, *62*, 4592–4595.

symmetry

MDPI

Article

Generalized Chordality, Vertex Separators and Hyperbolicity on Graphs

Álvaro Martínez-Pérez

Department of Economic analysis and Finances, Universidad de Castilla-La Mancha,
Avda. Real Fábrica de Sedas s/n, 45600 Talavera de la Reina, Spain; alvaro.martinezperez@uclm.es;
Tel.: +34-902-204-100

Received: 21 August 2017; Accepted: 19 September 2017 ; Published: 24 September 2017

Abstract: A graph is chordal if every induced cycle has exactly three edges. A vertex separator set in a graph is a set of vertices that disconnects two vertices. A graph is δ-hyperbolic if every geodesic triangle is δ-thin. In this paper, we study the relation between vertex separator sets, certain chordality properties that generalize being chordal and the hyperbolicity of the graph. We also give a characterization of being quasi-isometric to a tree in terms of chordality and prove that this condition also characterizes being hyperbolic, when restricted to triangles, and having stable geodesics, when restricted to bigons.

Keywords: infinite graph; geodesic; Gromov hyperbolic; chordal; bottleneck property; vertex separator

---

1. Introduction

M. Gromov defined in [1] his notion of hyperbolicity for the study of finitely-generated groups. Since then, Gromov hyperbolic spaces have been studied from a geometric point of view providing a wide variety of results and making them an important subclass of metric spaces [2–6]. In particular, Gromov hyperbolicity is an important property to be studied in graphs [7–25]. Gromov hyperbolicity has found also interesting applications in phylogenetics [26,27], complex networks [28–31], virus propagation and secure transmission of information [32,33] and congestion in hyperbolic networks [34].

Given a metric space (X, d) and two points $x, y \in X$, a geodesic from x to y is an isometry, $\gamma : [0, l] \to X$, from a closed interval $[0, l]$ of the real line to X such that $\gamma(0) = x$ and $\gamma(l) = y$. We will make no distinction between the geodesic and its image. X is a geodesic metric space if for every pair of points $x, y \in X$, there is some geodesic joining x to y. Although geodesics need not be unique, for convenience, $[xy]$ will denote any such geodesic.

Herein, we consider the graphs always endowed with the usual length metric where every edge has length one. Thus, for any pair of points in G, the distance between them will be the length of the shortest path in G joining them. Notice that we are considering also the interior points of the edges as points in G. Therefore, G with the length metric is a geodesic metric space. Let us also assume that the graphs are connected.

Gromov hyperbolicity, in the context of geodesic metric spaces, can be characterized by the Rips condition as follows. If X is a geodesic metric space and $x_1, x_2, x_3 \in X$, the union of three geodesics $[x_1 x_2]$, $[x_2 x_3]$ and $[x_3 x_1]$ is called a geodesic triangle and will be denoted by $T = \{x_1, x_2, x_3\}$. If two vertices are identical then it is called a bigon. A triangle T is δ-thin if any side of T is contained in the δ-neighborhood of the union of the two other sides. A geodesic metric space X is δ-hyperbolic if every geodesic triangle is δ-thin. By $\delta(X)$, we denote the sharp hyperbolicity constant of X, this is, $\delta(X) := \inf\{\delta \,|\, \text{every triangle in } X \text{ is } \delta\text{-thin}\}$. A metric space X is hyperbolic if it is δ-hyperbolic for some $\delta \geq 0$. There exist other equivalent definitions of Gromov hyperbolicity. See [4].

A graph G is said to be chordal if every induced cycle has exactly three edges. Chordal graphs form an important subclass of perfect graphs, and as is pointed out in [35] (see the further references therein), they have applications in scheduling, Gaussian elimination on sparse matrices and relational database systems. Furthermore, chordal graphs have applications in computer science; see [36]. In [37], it is proved that chordal graphs are hyperbolic. Wu and Zhang extended this result in [38] proving that k-chordal graphs are hyperbolic where a graph is k-chordal if induced cycles have at most k edges. In [39], the authors defined some more natural generalizations of being chordal as being (k, m)-edge-chordal and $(k, \frac{k}{2})$-path-chordal proving that (k, m)-edge-chordal graphs are hyperbolic and that hyperbolic graphs are $(k, \frac{k}{2})$-path-chordal. In [40], we continue this work and define being ε-densely (k, m)-path-chordal and ε-densely k-path-chordal. In [39,40], edges were allowed to have any finite length, but in this work, we assume that all edges have length one. Therefore, the distinction between edge and path is unnecessary, and these properties are referred as (k, m)-chordal and ε-densely k-chordal. The main results in [40] (with this simplified notation) state that:

$$(k, 1)\text{-chordal} \Rightarrow \varepsilon\text{-densely } (k, m)\text{-chordal} \Rightarrow \delta\text{-hyperbolic}$$

and:

$$\delta\text{-hyperbolic} \Rightarrow \varepsilon\text{-densely } k\text{-chordal} \Rightarrow k\text{-chordal}.$$

We also proved that the converse is false for all these implications, giving counterexamples, and that a graph is hyperbolic if and only if certain chordality property is satisfied on the triangles.

Herein, we continue this study analyzing some relations between these properties and vertex separators. There are some well-known relations between chordality and vertex separators. For example, Dirac proved in [41] that a graph is chordal if and only if every minimal vertex separator is complete. Furthermore, the set of minimal vertex separators of a chordal graph allows one to decompose the graph into subgraphs that are again chordal, and the process can be continued until the subgraphs are cliques [35]. Generalized versions of chordality are also related to minimal vertex separator [42]. For further results about chordality and vertex separators, see also [36] and the references therein. For an important application of minimal vertex separators in machine learning, see [43]. Our main results are the following.

In Section 2, we prove that being $(k, 1)$-chordal implies that every minimal vertex separator has a uniformly-bounded diameter. We also obtain that, for uniform graphs, if every minimal vertex separator has a uniformly-bounded diameter, then the graph is ε-densely (k, m)-chordal and therefore hyperbolic.

Section 3 studies the relation between generalized chordality and the bottleneck property, which is an important property on hyperbolic geodesic spaces. J. Manning defined it in [44] and proved that a geodesic metric space satisfies bottleneck property, (BP), if and only if it is quasi-isometric to a tree. This characterization has proven to be very useful; see for example [45]. For some other relations with (BP), see [46,47] and the references therein.

Here, we prove that a graph satisfies (BP) if and only if it is ε-densely (k, m)-chordal, providing a characterization of being quasi-isometric to a tree in terms of chordality. Furthermore, the characterization of hyperbolicity from [40] is re-written obtaining that a graph is hyperbolic if and only if it is ε-densely (k, m)-chordal on the cycles that are geodesic triangles.

Furthermore, we prove that if G is a uniform graph and every minimal vertex separator has a uniformly-bounded diameter, then the graph satisfies (BP), and therefore, it is quasi-isometric to a tree. Finally, we prove directly that being $(k, 1)$-chordal implies (BP).

In Section 4, we generalize the concept of vertex separators defining vertex r-separators. It is proven that if, in a uniform graph, all minimal vertex r-separators have a uniformly-bounded diameter, then the graph is ε-densely (k, m)-chordal and, therefore, quasi-isometric to a tree.

Section 5 introduces neighbor separators, generalizing also vertex separators. This concept allows one to characterize (BP) in terms of having a neighbor-separator vertex.

In Section 6, we define neighbor obstructors. We use them to characterize the graphs where geodesics between vertices are stable and to prove that geodesics between vertices are stable if and only if the graph is ε-densely (k, m)-chordal on the bigons defined by two vertices. We also prove that, in general, geodesics are stable if and only if the graph is ε-densely (k, m)-chordal on the bigons.

2. Generalized Chordality and Minimal Vertex Separators

We are assuming that every path is finite and simple, that is, it has finite length and distinct vertices. By a cycle, we mean a simple closed curve, that is, a path where all the vertices are different except from the first one and the last one, which are the same.

Let γ be a path or a cycle. A shortcut in γ is a path σ joining two vertices p, q in γ such that $L(\sigma) < d_\gamma(p, q)$ where $L(\sigma)$ denotes the length of the path σ and d_γ denotes the length metric on γ. A shortcut σ in γ is strict if $\sigma \cap \gamma = \{p, q\}$. In this case, we say that p, q are shortcut vertices in γ associated with σ. A shortcut with length k is called a k-shortcut.

Remark 1. *Suppose σ is a k-shortcut in a cycle C joining two vertices, p, q. Then, σ contains a strict shortcut, and there are two shortcut vertices p', q' such that $d_C(p, p'), d_C(q, q') < k$.*

Definition 1. *A metric graph G is k-chordal if for any cycle C in G with $L(C) \geq k$, there exists a shortcut σ of C.*

Definition 2. *A metric graph G is (k, m)-chordal if for any cycle C in G with $L(C) \geq k$, there exists a shortcut σ of C such that $L(\sigma) \leq m$. Notice that being chordal is equivalent to being $(4, 1)$-chordal.*

Remark 2. *Notice that in the definitions of k-chordal and (k, m)-chordal, it makes no sense to consider $k \leq 3$ nor $k < 2m$. Therefore, let us assume always that $k \geq 4$ and $k \geq 2m$.*

Definition 3. *A subset $S \subset V(G)$ is a separator if $G \setminus S$ has at least two connected components. Two vertices a and b are separated by S if they are in different connected components of $G \setminus S$. If a and b are two vertices separated by S, then S is said to be an ab-separator.*

Let us call a path joining the vertices a, b an ab-path.

Definition 4. *S is a minimal separator if no proper subset of S is a separator. Similarly, S is a minimal ab-separator if no proper subset of S separates a and b. Finally, S is a minimal vertex separator if it is a minimal separator for some pair of vertices.*

Note that being a minimal vertex separator does not imply being a minimal separator. See Figure 1.

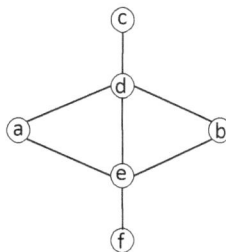

Figure 1. The set $\{d, e\}$ is a minimal ab-separator, but it is not a minimal separator.

Remark 3. *Let S be a minimal ab-separator, and let G_a, G_b be the connected components of $G \setminus S$ containing a and b, respectively. Then, notice that every vertex v in S is adjacent to both G_a and G_b. Otherwise, $S \setminus \{v\}$ is an ab-separator.*

Proposition 1. *If G is $(k, 1)$-chordal, then every minimal vertex separator has a diameter less than $\frac{k}{2}$.*

Proof. Let S be a minimal ab-separator, and suppose that $diam(S) \geq \frac{k}{2}$. Let $x, y \in S$ such that $d(x, y) \geq \frac{k}{2}$. Then, there are vertices a_1, a_n in G_a adjacent to x and y respectively, and since G_a is connected, there is a path $\gamma_1 = \{x, a_1, ..., a_n, y\}$ with $a_i \in G_a \; \forall 1 \leq i \leq n$. Similarly, there exist vertices b_1, b_m in G_b adjacent to y and x respectively and a path $\gamma_2 = \{y, b_1, ..., b_m, x\}$ with $b_i \in G_b \; \forall 1 \leq i \leq m$. Moreover, let us assume that γ_1, γ_2 have minimal length. Then, $C = \gamma_1 \cup \gamma_2$ defines a cycle in G, and since $d(x, y) \geq \frac{k}{2}$, $L(C) \geq k$. Then, since G is $(k, 1)$-chordal, there is a shortcut σ in C with $L(\sigma) = 1$. However, since S is an ab-separator, vertices in G_a and G_b cannot be adjacent, and since γ_1, γ_2 are supposed minimal, there is no possible one-shortcut on γ_i for $i = 1, 2$. Thus, x, y need to be adjacent, leading to a contradiction. $\square$

The converse is not true.

Example 1. *Consider the graph G_0 whose vertices are $V(G_0) = \{n \in \mathbb{N} \, | \, n \geq 3\}$ and edges joining consecutive numbers. Now, let us define the graph G such that for every $n \geq 3$, there is cycle C_n whose vertices are all adjacent to the vertex n in G_0. See Figure 2.*

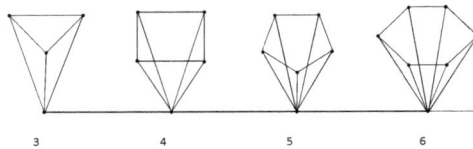

Figure 2. Every minimal vertex separator has diameter at most two, but the graph is not $(k, 1)$-chordal for any $k > 0$.

It is trivial to check that G is not $(k, 1)$-chordal for any $k > 0$ since the cycles C_n have no one-shortcut in G.

Let us see that every minimal vertex separator has diameter at most two. Consider any pair of non-adjacent vertices a, b in G.

If $a, b \in C_n$ for some n, then every vertex separator S must contain the vertex n and at least two vertices x_1, x_2 in C_n. If S is minimal, then $S = \{n, x_1, x_2\}$ and $diam(S) = 2$.

If $a, b \notin C_n$ for any n, then the geodesic $[ab]$ is contained in G_0. Therefore, any ab-separator must contain some vertex $m \in [ab]$ and m separates a and b. Thus, if S is minimal, then S is just a vertex and $diam(S) = 0$.

Remark 4. *Given two vertices a, b, a path γ joining them and a vertex $v \in \gamma$ distinct from a, b, there may not exist a minimal ab-separator containing $\{v\}$. Consider, for example four vertices x_0, x_1, x_2, x_3 with edges $x_{i-1} x_i$ for every $1 \leq i \leq 3$ and an edge $x_0 x_2$. Then, there is no minimal $x_0 x_3$-separator containing x_1.*

Given a graph G and a subgraph, $A \subset G$, let us denote $V(A)$ the vertices in A.

Definition 5. *A graph Γ is said to be μ-uniform if each vertex p of V has at most μ neighbors, i.e.,*

$$\sup \{ |N(p)| \, | \, p \in V(\Gamma) \} \leq \mu.$$

If a graph Γ is μ-uniform for some constant μ, we say that Γ is uniform.

For any vertex $v \in V(G)$ and any constant $\varepsilon > 0$, let us denote:

$$S_\varepsilon(v) := \{ w \in V(G) \, | \, d(v, w) = \varepsilon \},$$

$$B_\varepsilon(v) := \{ w \in V(G) \, | \, d(v, w) < \varepsilon \},$$

$$N_\varepsilon(v) := \{w \in V(G) \,|\, d(v,w) \le \varepsilon\}.$$

Lemma 1. *Let G be a uniform graph. Given two vertices a, b, a geodesic [ab] joining them and a vertex $v_0 \in [ab]$ distinct from a, b, then there is a minimal ab-separator containing $\{v_0\}$.*

Proof. Suppose any geodesic $[ab]$ and $v_0 \in [ab]$ with $0 < d(a, v_0) < d(a, b)$, and define $\varepsilon = d(a, v_0)$. Since G is uniform, for every vertex $v \in G$, the set $S_0 := S(v, \varepsilon)$ is finite for every $\varepsilon \in \mathbb{N}$. It is immediate to check that S_0 is an ab-separator and $[ab] \cap S_0 = \{v_0\}$. Since S_0 is finite, then there is a minimal subset $S \subset S_0$ that is also an ab-separator. Finally, since $[ab] \cap S_0 = \{v_0\}$, $v_0 \in S$. □

Let us recall that a graph Γ is countable if $|V(\Gamma)| \le \aleph_0$, i.e., if it has a countable number of vertices.

Remark 5. *In the case of countable graphs and using the axiom of choice, Lemma 1 can be slightly improved. See Lemma 2 below.*

Lemma 2. *Let G be a uniform countable graph. Given two vertices a, b, a path γ_0 joining them and a vertex $v_0 \in \gamma_0$ distinct from a, b, then either there is a one-shortcut in γ_0 or there is a minimal ab-separator containing $\{v_0\}$.*

Proof. If there is no ab-path in $G \setminus \{v_0\}$, it suffices to consider $S := \{v_0\}$. If there is an ab-path γ_1 in $G \setminus \{v_0\}$ such that $V(\gamma_1) \subset V(\gamma_0)$, then there is a one-shortcut in γ_0. Thus, let us suppose that every ab-path γ in $G \setminus \{v_0\}$ contains a vertex, which is not in γ_0, and that there is at least one of these ab-paths.

Since $|V(G)|$ is countable and G is uniform, there exist at most $\aleph_0^k$ ab-paths of length k. Then, there exists at most a countable number (a countable union of countable sets) of ab-paths, $\{\gamma_i\}_{i \in I \subset \mathbb{N}}$ in $G \setminus \{v_0\}$ where $I = \{1, \ldots, m\}$ if there exist exactly m such paths or $I = \mathbb{N}$ if the number of those paths is not finite.

For every $i \in I$, consider some vertex x_i in $V(\gamma_i) \setminus V(\gamma_0)$, and let $X = \{x_i\}_{i \in I}$. Now, let $S_0 := X$, and for every $0 < i \in I$, define:

$$S_i = \begin{cases} S_{i-1} \setminus \{x_i\} & \text{if } V(\gamma_j) \cap \left(S_{i-1} \setminus \{x_i\}\right) \ne \varnothing \text{ for every } j \le i, \\ S_{i-1} & \text{if } V(\gamma_j) \cap \left(S_{i-1} \setminus \{x_i\}\right) = \varnothing \text{ for some } j \le i. \end{cases}$$

Notice that for every i, $S_i \subset S_{i-1}$, and let $S := \cap_{i \in I} S_i$.

Claim: S is a minimal ab-separator containing v_0.

First, let us see that S is an ab-separator. Consider any ab-path, γ_j. Suppose $V(\gamma_j) \cap X = \{x_{j_1}, x_{j_2}, \ldots, x_{j_k}\}$, and assume $j_l < j_k$ for every $l < k$. Then, it is trivial to check that there exist some vertex $x_{j_r} \in S_{j_k} \cap V(\gamma_j)$ and, by construction, $x_{j_r} \in S$.

To check that S is minimal, first notice that, since $x_i \notin V(\gamma_0)$ for every $i \in I$, $V(\gamma_0) \cap (S \setminus \{v_0\}) = \varnothing$ and $S \setminus \{v_0\}$ is not an ab-separator. Now, suppose that there is some vertex $x_j \in S$ with $j \in I$ such that $S \setminus \{x_j\}$ is also an ab-separator. Since $x_j \in S \subset S_j$, there is some $k \le j$ such that $V(\gamma_k) \cap \left(S_{j-1} \setminus \{x_j\}\right) = \varnothing$ and, in particular, $V(\gamma_k) \cap \left(S \setminus \{x_j\}\right) = \varnothing$, leading to a contradiction. Thus, S is a minimal ab-separator containing v_0. □

Given a metric space (X, d) and any $\varepsilon > 0$, a subset $A \subset X$ is ε-dense if for every $x \in X$, there exists some $a \in A$ such that $d(a, x) < \varepsilon$.

Definition 6. *A metric graph (G, d) is ε-densely k-chordal if for every cycle C with length $L(C) \ge k$, there exist strict shortcuts $\sigma_1, \ldots, \sigma_r$ such that their associated shortcut vertices define an ε-dense subset in (C, d_C).*

Definition 7. *A graph (G, d) is ε-densely (k, m)-chordal if for every cycle C with length $L(C) \geq k$, there exist strict shortcuts $\sigma_1, ..., \sigma_r$ with $L(\sigma_i) \leq m \ \forall \ i$ and such that their associated shortcut vertices define an ε-dense subset in (C, d_C).*

Theorem 1. *Let G be a uniform graph. If every minimal vertex separator in G has diameter at most m, then G is $(m + \epsilon)$-densely $(4m, 2m - 1)$-chordal for any $\epsilon > \frac{1}{2}$.*

Proof. Let C be any cycle with $L(C) \geq 4m$. Let v be any vertex in C, and let a, b be the two vertices in C such that $d_C(a, v) = m = d_C(v, b)$. Let γ_0 be the ab-path in C containing v. Then, by Lemma 1, either there is a shortcut in γ_0 or there is a minimal ab-separator containing v.

If there is a shortcut in γ_0, then it has length at most $2m - 1$. Therefore, it defines a shortcut in C with an associated shortcut vertex v' such that $d_C(v, v') \leq m$. Suppose, otherwise, that S is a minimal ab-separator containing v. By hypothesis, $diam(S) \leq m$. Let γ_1 be the ab-path in C not containing v. Since S is an ab-separator, there is some vertex $w \in S \cap V(\gamma_1)$ and $d(v, w) \leq m < d_C(v, w)$. Hence, there is an m-shortcut in C joining v to w and, by Remark 1, an associated shortcut vertex v' such that $d_C(v, v') < m$.

Thus, for every vertex v, there is a shortcut vertex v' such that $d_C(v, v') \leq m$, and therefore, shortcut vertices define a $(m + \epsilon)$-dense subset in C for any $\epsilon > \frac{1}{2}$. $\square$

If the graph is countable, then we can improve quantitatively this result.

Theorem 2. *Let G be a uniform countable graph. If every minimal vertex separator in G has diameter at most m, then G is $(m + \epsilon)$-densely $(2m + 2, m)$-chordal for any $\epsilon > \frac{1}{2}$.*

Proof. Let C be any cycle with $L(C) \geq 2m + 2$. Let v be any vertex in C, and let a, b be the two vertices in C such that $d_C(a, v) = m = d_C(v, b)$. Let γ_0 be the ab-path in C containing v. Then, by Lemma 2, either there is a one-shortcut in γ_0 or there is a minimal ab-separator containing v.

If there is a one-shortcut in γ_0, then, in particular, there is an associated shortcut vertex v' such that $d_C(v, v') \leq m$. Suppose, otherwise, that S is a minimal ab-separator containing v. By hypothesis, $diam(S) \leq m$. Let γ_1 be the ab-path in C not containing v. Since S is an ab-separator, there is some vertex $w \in S \cap V(\gamma_1)$ and $d(v, w) \leq m < d_C(v, w)$. Hence, there is an m-shortcut in C joining v to w and, by Remark 1, an associated shortcut vertex v' such that $d_C(v, v') < m$.

Thus, for every vertex v, there is a shortcut vertex v' such that $d_C(v, v') \leq m$, and therefore, shortcut vertices define a $(m + \epsilon)$-dense subset in C for any $\varepsilon > \frac{1}{2}$. $\square$

Let us recall the following result:

Theorem 3. *(Theorem 4 [40]). If G is ε-densely (k, m)-chordal, then G is hyperbolic. Moreover, $\delta(G) \leq \max\{\frac{k}{4}, \varepsilon + m\}$.*

Therefore, from Theorems 1–3, we obtain:

Corollary 1. *Let G be a uniform graph. If every minimal vertex separator in G has diameter at most m, then G is hyperbolic. Moreover, $\delta(G) \leq 3m - \frac{1}{2}$.*

Corollary 2. *Let G be a uniform countable graph. If every minimal vertex separator in G has diameter at most m, then G is hyperbolic. Moreover, $\delta(G) \leq 2m + \frac{1}{2}$.*

3. Bottleneck Property

Let us recall the following definition from [44]:

Definition 8. *A geodesic metric space (X, d) satisfies the bottleneck property (BP) if there exists some constant $\Delta > 0$ so that given any two distinct points $x, y \in X$ and a midpoint z such that $d(x, z) = d(z, y) = \frac{1}{2}d(x, y)$, then every xy-path intersects $N_\Delta(z)$.*

Remark 6. *This definition, although not being exactly the same, is equivalent to Manning's. In the original definition, J. Manning asked only for the existence of such a midpoint for any pair of points x, y. However, by Theorem 4 below, (BP) implies that the space is quasi-isometric to a tree and therefore δ-hyperbolic. Hence, it is an easy exercise in hyperbolic spaces to prove that if there is always a midpoint z such that every xy-path intersects $N_\Delta(z)$, then this condition holds in general for any midpoint, possibly with a different constant depending only on Δ and δ. See, for example, Chapter 2, Proposition 25 in [5].*

Definition 9. *A graph G satisfies (BP) on the vertices if there exists some constant $\Delta' > 0$ so that given any two distinct vertices $v, w \in V(G)$ and a midpoint c such that $d(v, c) = d(c, w) = \frac{1}{2}d(v, w)$, then every vw-path intersects $N_{\Delta'}(c)$.*

Proposition 2. *A graph G satisfies (BP) if and only if it satisfies (BP) on the vertices. Moreover, if G satisfies (BP) on the vertices with constant Δ', it satisfies (BP) with $\Delta = \Delta' + \frac{3}{2}$.*

Proof. The only if condition is trivial. Let us see that it suffices to check the property on the pairs of vertices.

Consider any pair of points $x, y \in G$, and let z be a midpoint of a geodesic $[xy]$. If $d(x, y) \leq 2$, then (BP) is trivial with $\Delta = 1$. Suppose $d(x, y) > 2$. Then, the geodesic $[xy]$ is a path $xv_1 \cup v_1 v_2 \cup \cdots \cup v_k y$ with $v_1, \ldots, v_k \in V(G)$ and $k \geq 2$. Let $v = x$ if x is a vertex and $v = v_1$ otherwise, and let $w = y$ if y is a vertex and $w = v_k$ otherwise. Then, there is a geodesic $[vw] \subset [xy]$ (possibly equal), and its midpoint, c, satisfies that $d(c, z) \leq \frac{1}{2}$.

Consider any xy-path γ, and let us define a vw-path γ' as follows: First, if $v \in \gamma$, let $\gamma_0 := \gamma \setminus [xv]$ and if $v \notin \gamma$, let $\gamma_0 := [vx] \cup \gamma$.

Then, if $y \neq w$ and $w \in \gamma$, let $\gamma' := \gamma_0 \setminus [yw]$ and if $y \neq w$ and $w \notin \gamma$, let $\gamma' := [wy] \cup \gamma_0$. By hypothesis, γ' passes through $N_{\Delta'}(c)$. Since $d(a, v), d(b, w) \leq 1$ and $d(c, z) \leq \frac{1}{2}$, it is immediate to check that γ passes through $N_{\Delta' + \frac{3}{2}}(z)$. $\square$

A map between metric spaces, $f : (X, d_X) \to (Y, d_Y)$, is said to be a quasi-isometric embedding if there are constants $\lambda \geq 1$ and $C > 0$ such that $\forall x, x' \in X$,

$$\frac{1}{\lambda} d_X(x, x') - C \leq d_Y(f(x), f(x')) \leq \lambda d_X(x, x') + C.$$

If there is a constant $D > 0$ such that $\forall y \in Y, d(y, f(X)) \leq D$, then f is a quasi-isometry, and X, Y are quasi-isometric.

Theorem 4. *(Theorem 4.6 [44]). A geodesic metric space (X, d) is quasi-isometric to a tree if and only if it satisfies (BP).*

Theorem 5. *A graph G satisfies (BP) if and only if it is ε-densely (k, m)-chordal.*

Proof. Suppose that G satisfies (BP) with parameter Δ and consider any cycle C with $L(C) \geq 2\Delta + 4$. Consider any vertex $x \in C$ and the two vertices a, b such that $d_C(a, x) = d_C(x, b) = \Delta + 1$. Thus, C defines two ab-paths, γ_1, γ_2. Let us assume that $x \in \gamma_1$. If γ_1 is not geodesic, then there is a shortcut with length at most $2\Delta + 1$ and a shortcut vertex in $N_{\Delta+1}(x)$. Otherwise, since G satisfies (BP) with parameter Δ, there is a vertex y in γ_2 such that $d(x, y) \leq \Delta$. Since $d_C(x, y) > \Delta$ and by Remark 1, there is a shortcut vertex z such that $d_C(x, z) < \Delta$. Therefore, G is $(\Delta + 1 + \varepsilon)$-densely $(4\Delta + 4, 2\Delta + 1)$-chordal for any $\varepsilon > \frac{1}{2}$.

Suppose that G is ε-densely (k,m)-chordal and it does not satisfy (BP) with parameter $\Delta = \max\{\frac{k}{4}, \varepsilon + m\}$. Then, there are two points, a, b, a geodesic $[ab]$ with midpoint c and a path γ such that $\gamma \cap N_\Delta(c) = \emptyset$. Then, it is immediate to check that there exist two points $a', b' \in \gamma \cap [ab]$ such that the restriction of $[ab]$, $[a'b']$, and the restriction of γ, γ', joining a' to b' define a cycle C with $L(C) > k$. Since G is ε-densely (k,m)-chordal, there is a strict shortcut σ with $L(\sigma) \leq m$ with an associated shortcut vertex w such that $d_C(c, w) < \varepsilon < \Delta$. Therefore, $w \in [a'b']$, and since $[a'b']$ is geodesic, the shortcut must join w to a vertex z in $\gamma' \subset \gamma$. Hence, $d(z, c) < \varepsilon + m$ and $\gamma \cap N_\Delta(c) \neq \emptyset$, leading to a contradiction. $\square$

Corollary 3. *A graph G is quasi-isometric to a tree if and only if it is ε-densely (k,m)-chordal.*

Definition 10. *Given any family $\mathcal{F}$ of cycles, a metric graph (G, d) is ε-densely (k,m)-chordal on $\mathcal{F}$ if for every $C \in \mathcal{F}$ with length $L(C) \geq k$, there exist strict shortcuts $\sigma_1, \ldots, \sigma_r$ with $L(\sigma_i) \leq m \; \forall i$ and such that their associated shortcut vertices define an ε-dense subset in (C, d_C).*

Let us recall the following:

Lemma 3. *(Lemma 2.1 [48]). Let X be a geodesic metric space. If every geodesic triangle in X which is a cycle is δ-thin, then X is δ-hyperbolic.*

Let $\mathcal{T}$ be the family of cycles that are geodesic triangles. It is immediate to check that, using Lemma 3, the proof of Theorem 13 in [40] can be trivially re-written (we include it for completeness) to obtain the following:

Theorem 6. *G is δ-hyperbolic if and only if G is ε-densely (k,m)-chordal on $\mathcal{T}$.*

Proof. Suppose that G is ε-densely (k,m)-path-chordal on $\mathcal{T}$. Let us see that $\delta(G) \leq \max\{\frac{k}{4}, \varepsilon + m\}$. Consider any cycle that is a geodesic triangle $T = \{x, y, z\}$. If $L(T) < k$, it follows that every side of the triangle has length at most $\frac{k}{2}$. Therefore, the hyperbolic constant is at most $\frac{k}{4}$. Then, let $L(T) \geq k$, and let us prove that T is $(\varepsilon + m)$-thin. Consider any point $p \in T$, and let us assume that $p \in [xy]$. If $d(p, x) < \varepsilon + m$ or $d(p, y) < \varepsilon + m$, we are done. Otherwise, there is a shortcut vertex x_i such that $d(x_i, p) < \varepsilon$ and a shortcut σ_i, with $x_i \in \sigma_i$ and $L(\sigma_i) \leq m$. Since $[xy]$ is a geodesic, σ_i does not connect two points in $[xy]$ and $d(p, [xz] \cup [yz]) < \varepsilon + m$. Then, by Lemma 3, $\delta(G) \leq \max\{\frac{k}{4}, \varepsilon + m\}$.

Suppose that G is δ-hyperbolic, and consider any cycle that is a geodesic triangle $T = \{x, y, z\}$ with $L(T) \geq 9\delta$. Let $p \in T$, and let us assume, with no loss of generality, that $p \in [xy]$. Since G is δ-hyperbolic, $d(p, [xz] \cup [yz]) \leq \delta$. If $d(p, x), d(p, y) > \delta$, then there is a path γ with $L(\gamma) \leq \delta$ joining p to $[xz] \cup [yz]$. In particular, there is a shortcut $\sigma \subset \gamma$ with $L(\sigma) \leq L(\gamma) \leq \delta$ joining some shortcut vertex $p' \in [xy]$ with $d(p, p') < \delta$ to $[xz] \cup [yz]$. Therefore, if $L([xy]) > 2\delta$, for every point $q \in [xy]$, there is a shortcut vertex $q' \in [xy]$ such that $d_T(q, q') < 2\delta + 1$ associated with a shortcut with length at most δ. Since $L(T) \geq 9\delta$, by triangle inequality, there is at most one side of the triangle with length at most 2δ. Then, for every point p in the triangle, there is a shortcut vertex p' such that $d_T(p, p') < 3\delta + 1$ associated with a shortcut with length at most δ. Thus, it suffices to consider $\varepsilon = 3\delta + 1$, $k = 9\delta$ and $m = \delta$. $\square$

Remark 7. *Notice that in Corollary 3, we obtain that a graph G is quasi-isometric to a tree if and only if all the cycles satisfy a certain property, and Theorem 6 states that the same property, restricted to the cycles that are geodesic triangles, characterizes being hyperbolic.*

The following theorem can be also obtained as a corollary of Theorems 1 and 5. However, the direct proof provides a better bound for the parameter Δ.

Theorem 7. *Given a uniform graph G, if every minimal vertex separator has diameter at most m, then G satisfies (BP) (i.e., G is quasi-isometric to a tree). Moreover, it suffices to take $\Delta = m + 2$.*

Proof. If $m = 0$, it is trivial to see that G is a tree, and it satisfies (BP) with $\Delta = 0$. Assume $m \geq 1$. By Proposition 2, it suffices to check the property for pairs of vertices. Thus, consider any pair of vertices $a, b \in V(G)$, and let c be a midpoint of a geodesic $[ab]$.

If $d(a, b) \leq 2$, then (BP) is trivial with $\Delta' = 1$. Suppose $d(a, b) \geq 3$. Then, there is some vertex v_0 in the interior of $[ab]$ with $d(v_0, c) \leq \frac{1}{2}$. By Lemma 1, since $[ab]$ is a geodesic, there exists a minimal ab-separator S containing v_0. Thus, every ab-path contains a vertex in S, and since $diam(S) \leq m$, every ab-path passes through $N_m(v_0) \subset N_{m+\frac{1}{2}}(c)$. Hence, (BP) is satisfied on the vertices with $\Delta' = m + \frac{1}{2}$, and by Proposition 2, G satisfies (BP) with $\Delta = m + 2$. $\square$

The following example shows that the converse is not true.

Example 2. *Let G be the graph whose vertices are all the pairs (a, b) with either $a \in \mathbb{N}$ and $b = 0$ or $4n + 1 \leq a \leq 4n + 3$ and $1 \leq b \leq n$ for every $n \in \mathbb{N}$, and such that (a, b) is adjacent to (a', b') if and only if either $b = b'$ and $|a' - a| = 1$ or $a = a'$ and $|b' - b| = 1$. See Figure 3.*

Now, notice that $S_n = \{4n + 2, j\}_{0 \leq j \leq n}$ defines a minimal $(4n, 0)(4n + 4, 0)$-separator with diameter n for every $n \in \mathbb{N}$. Therefore, G has minimal ab-separators arbitrarily big. However, to see that G satisfies (BP), consider the map $f : V(G) \to V(G)$ such that $f(i, j) = (4n + 2, j)$ for every $4n + 1 \leq i \leq 4n + 3$ and $1 \leq j \leq n$ and the identity on the rest of the vertices. It is trivial to check that f extends to a $(1, 2)$-quasi-isometry on G where the image is a tree. Therefore, G is quasi-isometric to a tree and satisfies (BP) (and it is ε-densely (k, m)-chordal).

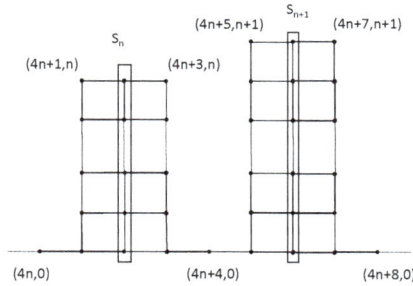

Figure 3. Satisfying the bottleneck property does not imply the existence of minimal vertex separators with uniformly-bounded diameters.

Remark 8. *In the case of uniform graphs, the following theorem can also be obtained as a corollary of Proposition 1 and Theorem 7. Furthermore, it follows from Theorem 3 in [40] and Theorem 5. However, the direct proof provides a better bound for the parameter.*

Theorem 8. *If G is $(k, 1)$-chordal, then G satisfies (BP). Moreover, it suffices to take $\Delta = \frac{k}{4} + \frac{5}{2}$.*

Proof. Consider any pair of vertices a, b, any geodesic $[ab]$ in G and the midpoint c in $[ab]$. If $d(a, b) \leq \frac{k}{2} + 2$, then (BP) is trivially satisfied for $\Delta' = \frac{k}{4} + 1$. Suppose $d(a, b) > \frac{k}{2} + 2$ and that there is an ab-path γ not intersecting $N_{k/4+1}(c)$. Let $a' \in [ac] \subset [ab]$ and $b' \in [cb] \subset [ab]$ such that $d(a', c) = d(c, b') = \frac{k}{4}$. Then, since γ does not intersect $N_{k/4+1}(c)$, there is a cycle C contained in $[ab] \cup \gamma$ such that $[a'b'] \subset C$ and $L(C) \geq k$.

Claim: there is a one-shortcut in C joining a vertex in the interior of $[a'b']$ to a vertex in γ. Since G is $(k, 1)$-chordal, there is a one-shortcut, σ_1, in C. If σ_1 joins a vertex in the interior of $[a'b']$ to a vertex in

γ, we are done. Otherwise, we obtain a new cycle, C_1, such that $[a'b'] \subset C_1$ and, therefore, $L(C_1) \geq k$. Repeating the process, we finally obtain a one-shortcut joining a vertex z_1 in the interior of $[a'b']$ to a vertex z_2 in γ.

Therefore, $d(c, z_2) \leq d(c, z_1) + 1 < \frac{k}{4} + 1$ and $z_2 \in N_{k/4+1}(c)$, leading to a contradiction.

Thus, G satisfies (BP) on the vertices with $\Delta = \frac{k}{4} + 1$, and by Proposition 2, G satisfies (BP) with $\Delta = \frac{k}{4} + \frac{5}{2}$. $\square$

Corollary 4. *If G is $(k, 1)$-chordal, then G is quasi-isometric to a tree.*

Remark 9. *Corollary 4 follows also from Proposition 1 and Corollary 1 in the case of uniform graphs.*

Remark 10. *The converse to Theorem 8 or Corollary 4 is not true. It is immediate to check that the graph from Example 1 is quasi-isometric to a tree through the map sending every cycle C_n to the vertex n.*

Remark 11. *Herein, the gap between being hyperbolic and being quasi-isometric to a tree is shown to depend on which cycles are ε-densely (k, m)-chordal, only geodesic triangles or all of them. Furthermore, we have seen that (BP) characterizes geodesic spaces quasi-isometric to trees. There exist also properties that characterize when a hyperbolic space is quasi-isometric to a tree. Corollary 1.9 in [49] states that two visual hyperbolic geodesic spaces X, Y are quasi-isometric if and only if there is a PQ-symmetric homeomorphism f (where 'PQ' stands for 'power quasi') with respect to any visual metrics between their boundaries (The property of being visual has different names in the literature. For example, it is called "having a pole" in [50,51] or being "almost geodesically complete" in [52].).*

Furthermore, there is a one-to-one correspondence between rooted trees and bounded ultrametric spaces where every tree induces a bounded ultrametric space, and for every bounded ultrametric space X there is a tree whose boundary is X. See [53] or [54].

Thus, a visual hyperbolic space is quasi-isometric to a tree if and only if its boundary is PQ-symmetric homeomorphic to an ultrametric space.

Furthermore, Theorem 1 in [47] states that given a complete geodesic space X with $H_1(X)$ uniformly generated, then X is quasi-isometric to a tree if and only if there is a function $f : X \to \mathbb{R}$ such that f is bornologous and metrically proper on the connected components.

Since any hyperbolic space has uniformly generated H_1, then it follows that for any hyperbolic graph G, G is quasi-isometric to a tree if and only if there is a function $f : G \to \mathbb{R}$ such that f is bornologous and metrically proper on the connected components.

4. Minimal Vertex r-Separators

Definition 11. *Given $r \in \mathbb{N}$, two vertices a and b are r-separated by a subset $S \subset V(G)$ if considering the connected components of $G \setminus S$, G_a and G_b containing a and b respectively, for every pair of vertices $v \in G_a$ and $w \in G_b$, $d(v, w) > r$. If a and b are two vertices r-separated by S, then S is said to be an ab-r-separator.*

Remark 12. *Notice that separated means one-separated.*

Definition 12. *S is a minimal ab-r-separator if no proper subset of S r-separates a and b. Finally, S is a minimal vertex r-separator if it is a minimal r-separator for some pair of vertices.*

Remark 13. *Given any minimal ab-r-separator S, every vertex in S is either adjacent to G_a or G_b. Moreover, if $r \geq 2$, then there are two disjoint subsets S_a and S_b such that $S = S_a \cup S_b$ where the vertices in S_a are adjacent to G_a and the vertices in S_b are adjacent to G_b. Furthermore, for every vertex v in S_a, $d(v, S_b) = r - 1$.*

Lemma 4. *Let G be a uniform graph and $r \geq 2$. Given any geodesic $[ab]$ with $d(a, b) > r$ and two vertices $v_1, v_2 \in [ab]$ distinct from a, b with $d(v_1, v_2) = r - 1$, then there is a minimal ab-r-separator containing $\{v_1, v_2\}$.*

Proof. Suppose $[ab]$ is a geodesic with $d(a,b) > r$. Let us assume that $d(a, v_1) < d(a, v_2)$, and define $\varepsilon_1 = d(a, v_1)$ and $\varepsilon_2 = d(v_2, b)$. Since G is uniform, for every vertex $v \in G$ the set $S(v, \varepsilon)$ is finite for every $\varepsilon \in \mathbb{N}$. Let $S_0 := S(a, \varepsilon_1) \cup S(b, \varepsilon_2)$. It is immediate to check that S_0 is an ab-r-separator and $[ab] \cap S_0 = \{v_1, v_2\}$. Since S_0 is finite, then there is a minimal subset $S \subset S_0$ which is also an ab-r-separator. Finally, since $[ab] \cap S_0 = \{v_1, v_2\}$, $v_i \in S$ for $i = 1, 2$. $\square$

Theorem 9. *Let G be a uniform graph and $r \geq 2$. If every minimal vertex r-separator has diameter at most m with $m \leq r$, then G is $(r + \frac{1}{2})$-densely $(2r + 2, r)$-chordal.*

Proof. Let C be any cycle with $L(C) \geq 2r + 2$, and let x_1 be any vertex in C. Then, consider two vertices a, b in C such that $d_C(a, b) = r + 1$, $d_C(a, x_1) = 1$ and $d_C(x_1, b) = r$. Let γ_1 and γ_2 be the two independent paths joining a and b defined by C, and assume $x_1 \in \gamma_1$. Consider $x_2 \in \gamma_1$ with x_2 between x_1 and b such that $d_C(x_1, x_2) = r - 1$ (and $d_C(x_2, b) = 1$).

If γ_1 is not a geodesic, then there is a shortcut with length at most r and a shortcut vertex v such that $d_C(x_1, v) < r$.

If γ_1 is a geodesic, by Lemma 4, there exists a minimal ab-r-separator S containing x_1, x_2. Then, there exist $y_1, y_2 \in \gamma_2 \cap S$, with y_1 between a and y_2, such that $d_C(y_1, y_2) \geq r - 1$, $d_C(a, y_1) \geq 1$ and $d_C(y_2, b) \geq 1$. Since $diam(S) \leq m$, then $d(x_1, y_2) \leq m$. Since $d_C(x_1, y_2) \geq r + 1 \geq m + 1$, there is a shortcut σ in C joining x_1 and y_2 with $L(\sigma) \leq m \leq r$ and with an associated shortcut vertex v such that $d_C(x_1, v) < m \leq r$.

Thus, for every vertex x_1, there is a shortcut vertex v such that $d_C(x_1, v) < r$, and therefore, shortcut vertices define a $(r + \frac{1}{2})$-dense subset in C. $\square$

Theorem 10. *Let G be a uniform graph and $r \geq 2$. If for every minimal ab-r-separator S either S_a or S_b has diameter at most m, then G is ε-densely $(k, \frac{k}{2})$-chordal with $k = 2m + 2r + 2$ and $\varepsilon = \max\{\frac{m+1}{2} + r, m + \frac{1}{2}\}$.*

Proof. Let C be any cycle with $L(C) \geq 2m + 2r + 2$ and x_1 be any vertex in C. Then, consider two vertices a, b in C such that $d_C(a, b) = m + r + 1$, $d_C(a, x_1) = \frac{m+1}{2}$ and $d_C(x_1, b) = \frac{m+1}{2} + r$ if m is odd, and $d_C(a, x_1) = \frac{m}{2} + 1$ and $d_C(x_1, b) = \frac{m}{2} + r$ if m is even. Let γ_1 and γ_2 be the two independent paths joining a and b defined by C, and assume $x_1 \in \gamma_1$. Consider $x_2 \in \gamma_1$ with x_2 between x_1 and b such that $d_C(x_1, x_2) = r - 1$ (and therefore, $d_C(x_2, b) \geq \frac{m}{2} + 1 > \frac{m}{2}$).

If γ_1 is not a geodesic, then there is a shortcut with length at most $m + r + 1$ and a shortcut vertex v such that $d_C(x_1, v) < \frac{m}{2} + r$.

If γ_1 is a geodesic, consider S the minimal ab-r-separator containing x_1, x_2 built in the proof of Lemma 4, and let us assume, without loss of generality, that S_a has diameter at most m. Then, by construction, there exists $y_1 \in \gamma_2 \cap S$ such that $d_{\gamma_2}(a, y_1) \geq d(a, y_1) = d(a, x_1) \geq \frac{m+1}{2}$. Since $diam(S_a) \leq m$, then $d(x_1, y_1) \leq m$. However, $d_C(x_1, y_1) \geq \min\{m + 1, d_{\gamma_1}(x_1, b) + d(b, y_1)\} = m + 1$, and therefore, there is a shortcut σ in C joining x_1 and y_1 with $L(\sigma) \leq m$. Moreover, there is a shortcut vertex v such that $d_C(x_1, v) < m$.

Thus, for every vertex x_1, there is a shortcut vertex v with $d_C(x_1, v) < \min\{\frac{m}{2} + r, m\}$, and therefore, shortcut vertices define an ε-dense subset in C with $\varepsilon = \max\{\frac{m+1}{2} + r, m + \frac{1}{2}\}$. $\square$

Then, from Theorems 5 and 10, we can obtain immediately the following:

Corollary 5. *Let G be a uniform graph and $r \geq 2$. If for every minimal ab-r-separator S either S_a or S_b has diameter at most m, then G satisfies (BP), i.e., G is quasi-isometric to a tree.*

Furthermore, from Theorems 3, 9 and 10, we obtain:

Corollary 6. *Let G be a uniform graph and $r \geq 2$. If every minimal vertex r-separator has diameter at most m with $m \leq r$, then G is δ-hyperbolic. Moreover, $\delta(G) \leq 2r + \frac{1}{2}$.*

Corollary 7. *Let G be a uniform graph and $r \geq 2$. If for every minimal ab-r-separator S either S_a or S_b has diameter at most m, then G is δ-hyperbolic. Moreover, $\delta(G) \leq \max\{\frac{3m+3}{2} + 2r, 2m + r + \frac{3}{2}\}\}$.*

5. Neighbor Separators

Given a set S in a graph G, let $N_r(S) := \{x \in G \mid d(x, S) \leq r\}$.

Definition 13. *Given two vertices a, b in a graph $G = (V, E)$ and some $r \in \mathbb{N}$, a set $S \subset V$ is an ab-N_r-separator if a and b are in different components of $G \setminus N_r(S)$. S is an ab-neighbor separator if it is an ab-N_r-separator for some r.*

Notice that an ab-separator is just an ab-N_0-separator.

Theorem 11. *G satisfies (BP) if and only if there is a constant $\Delta'' > 0$ such that for every pair of vertices a, b with $d(a, b) \geq 2\Delta'' + 2$ and any geodesic $[ab]$, there exists a vertex $c \in [ab]$ that is an ab-$N_{\Delta''}$-separator.*

Proof. The only if part follows trivially from Proposition 2.

Suppose that for every pair of vertices a, b with $d(a, b) \geq 2\Delta'' + 2$ and any geodesic $[ab]$, there exists a point $c \in [ab]$ that is an ab-$N_{\Delta''}$-separator. Consider any pair of vertices x, y in G, any geodesic $[xy]$ and the midpoint z in $[xy]$.

If $d(x, y) \leq 2\Delta'' + 1$, then (BP) is trivially satisfied on x, y for any $\Delta' \geq \Delta'' + \frac{1}{2}$.

If $d(x, y) \geq 2\Delta'' + 2$, by hypothesis, there is some vertex $z_1 \in [xy]$ such that $N_{\Delta''}(z_1)$ is an xy-$N_{\Delta''}$-separator. If $d(z, z_1) \leq \Delta''$, then it follows that every xy-path intersects $N_{\Delta''}(z_1) \subset N_{2\Delta''}(z)$ and G satisfies (BP) on the vertices for $\Delta' = 2\Delta''$. If $d(z, z_1) > \Delta''$, then we repeat the process with the part of the geodesic, $[xz_1]$ or $[z_1 y]$, containing z. Let us assume, without loss of generality, that $z \in [xz_1]$. Since $d(z, z_1) > \Delta''$ and $d(x, z) > \Delta''$, there is some point $z_2 \in [xz_1]$ that is an xz_1-$N_{\Delta''}$-separator. Since there is a $z_1 y$-path in $G \setminus N_{\Delta''}(z_2)$, z_2 is also an xy-$N_{\Delta''}$-separator. If $d(z, z_2) \leq \Delta''$, we are done. Otherwise, we repeat the process until we obtain some point $z_k \in [xy]$ that is an xy-$N_{\Delta''}$-separator and such that $d(z, z_k) \leq \Delta''$. Therefore, G satisfies (BP) on the vertices for $\Delta' = 2\Delta''$.

Thus, by Proposition 2, G satisfies (BP) with $\Delta = 2\Delta'' + \frac{3}{2}$. □

Corollary 8. *G is quasi-isometric to a tree if and only if there is a constant $\Delta'' > 0$ such that for every pair of vertices a, b with $d(a, b) > \Delta''$ and any geodesic $[ab]$, there exists a vertex $c \in [ab]$ that is an ab-$N_{\Delta''}$-separator.*

Proposition 3. *If G is $(k, 1)$-chordal, then for every pair of vertices a, b, any geodesic $[ab]$ with $d(a, b) \geq \frac{k}{2} + 2$ and every pair of vertices $a', b' \in [ab]$ with $d(\{a', b'\}, \{a, b\}) \geq 2$ and such that $d(a', b') \geq \frac{k}{2} - 2$, $[a'b'] \subset [ab]$ is an ab-N_1-separator. In particular, for every pair of vertices a, b in G with $d(a, b) \geq \frac{k}{2} + 2$, there is a geodesic σ of length $\frac{k}{2} - 2$ or $\frac{k-3}{2}$ such that σ is an ab-N_1-separator.*

Proof. Consider any geodesic $[ab]$ in G with $d(a, b) \geq \frac{k}{2} + 2$ and any pair of vertices $a', b' \in [ab]$ with $d(\{a', b'\}, \{a, b\}) \geq 2$ such that $d(a', b') \geq \frac{k}{2} - 2$. Let a'' be the vertex in $[aa'] \subset [ab]$ adjacent to a' and b'' be the vertex in $[b'b] \subset [ab]$ adjacent to b'. Therefore, $d(a'', b'') \geq \frac{k}{2}$. Suppose that a and b are in the same connected component, A, of $G \setminus N_1([a'b'])$. Clearly, a'' and b'' are adjacent to A. Let γ be a path of minimal length joining a'' and b'' in the subgraph induced by $A \cup \{a'', b''\}$. Therefore, $[a''b''] \cup \gamma$ defines a cycle, C, of length at least k. Since G is $(k, 1)$-chordal, then there is an edge joining two non-adjacent vertices in C. Since $[a''b'']$ is geodesic and γ has minimal length, the edge must join a vertex, $v \in \gamma$ to a vertex in $[a'b']$. Therefore, $v \in N_1([a'b']) \cap A$ leading to a contradiction. □

Definition 14. *A path γ in a graph G is chordal if it has no one-shortcuts in G.*

Proposition 4. *If G is $(k, 1)$-chordal, then for every chordal ab-path σ with $L(\sigma) \geq k$ and every pair of vertices $a', b' \in \sigma$ with $d_\sigma(\{a', b'\}, \{a, b\}) \geq 2$ and such that $d_\sigma(a', b') \geq k - 4$, then the restriction of σ joining a'*

and b', σ', is an ab-N_1-separator. In particular, for every pair of vertices a, b in G joined by a chordal path with length at least k there is a chordal path γ' of length $k - 4$ such that γ' is an ab-N_1-separator.

Proof. Consider any chordal path σ in G with endpoints a, b and $L(\sigma) \geq k$. Consider any pair of vertices $a', b' \in [ab]$ with $d_\sigma(\{a', b'\}, \{a, b\}) \geq 2$ such that $d_\sigma(a', b') \geq k - 4$, and let $\sigma' = [a'b'] \subset [ab]$. Let a'' be the vertex in σ adjacent to a' closer in σ to a and b'' be the vertex in σ adjacent to b' closer in σ to b. Therefore, if σ'' is the restriction of σ joining a'' and b'', then $L(\sigma'') \geq k - 2$. Suppose that a and b are in the same connected component, A, of $G \setminus N_1(\sigma')$. Clearly, a'' and b'' are adjacent to A. Let γ be a path of minimal length joining a'' and b'' in the subgraph induced by $A \cup \{a'', b''\}$. Therefore, $\sigma'' \cup \gamma$ defines a cycle, C, of length at least k. Since G is $(k, 1)$-chordal, then there is an edge joining two non-adjacent vertices in C. Since σ'' is chordal and γ has minimal length, the edge must join a vertex $v \in \gamma$ to a vertex in σ'. Therefore, $v \in N_1(\sigma') \cap A$, leading to a contradiction. $\square$

Proposition 5. *If a graph G satisfies that for some $k, m \in \mathbb{N}$ with $k \geq 4m$, for every geodesic $[ab]$ with $d(a, b) \geq k + 2$ and for every pair of vertices $a', b' \in [ab]$ with $d(\{a', b'\}, \{a, b\}) \geq m + 1$ and such that $d(a', b') \geq k - 2m$, $[a'b'] \subset [ab]$ is an ab-N_m-separator, then G is $(\frac{k}{2} + 2)$-densely $(2k + 4, k + 1)$-chordal.*

Proof. Let C be any cycle with $L(C) \geq 2k + 4$. Let v by any vertex in C and a, b two vertices in C such that $d_C(a, v) = \lfloor \frac{k}{2} \rfloor + 1$ and $d_C(v, b) = \lceil \frac{k}{2} \rceil + 1$, and therefore, $d_C(a, b) = k + 2$. Let γ_1, γ_2 be the two ab-paths defined by the cycle, and let us assume that $v \in \gamma_1$ (and therefore, $L(\gamma_1) \leq L(\gamma_2)$). If there is a shortcut in γ_1, then there is a shortcut in C with length at most $k + 1$ and with a shortcut vertex z such that $d_C(v, z) < \frac{k}{2} + 2$. If there is no shortcut in γ_1, then γ_1 is a geodesic with $d(a, b) = k + 2$. Thus, let $a', b' \in \gamma_1$ with $d(a, a') = m + 1 = d(b', b)$ and $d(a', b') = k - 2m$. Therefore, $[a'b'] \subset \gamma_1$ is an ab-N_m-separator. In particular, there is some vertex w in $\gamma_2 \setminus \{a, b\}$ such that $w \in N_m([a'b'])$, defining a shortcut in C with length at most m and with a shortcut vertex z such that $d_C(v, z) < \frac{k}{2} + 1$. $\square$

Corollary 9. *If a graph G satisfies that for some $k, m \in \mathbb{N}$ with $k \geq 4m$, for every geodesic $[ab]$ with $d(a, b) \geq k + 2$ and for every pair of vertices $a', b' \in [ab]$ with $d(\{a', b'\}, \{a, b\}) \geq m + 1$ and such that $d(a', b') \geq k - 2m$, $[a'b'] \subset [ab]$ is an ab-N_m-separator, then G is quasi-isometric to a tree.*

6. Neighbor Obstructors

Definition 15. *Given two vertices a, b in a graph $G = (V, E)$ and some $r \in \mathbb{N}$, a set $S \subset V$ is ab-N_r-obstructing if for every geodesic γ joining a and b, $\gamma \cap N_r(S) \neq \emptyset$.*

Given any metric space (X, d) and any pair of subsets $A, B \subset X$, let us recall that the Hausdorff metric, d_H, induced by d is:

$$d_H(A_1, A_2) := \max\{\sup_{x \in A_1}\{d(x, A_2)\}, \sup_{y \in A_2}\{d(y, A_1)\}\},$$

or equivalently,

$$d_H(A_1, A_2) := \inf\{\varepsilon > 0 \mid A_1 \subset B(A_2, \varepsilon) \text{ y } A_2 \subset B(A_1, \varepsilon)\}.$$

Definition 16. *In a geodesic metric space (X, d), we say that geodesics are stable if and only if there is a constant $R \geq 0$ such that given two points $x, y \in X$ and any geodesic $[xy]$, then every geodesic σ joining x to y satisfies that $d_H(\sigma, [xy]) \leq R$.*

It is well known that if X is a hyperbolic space, then quasi-geodesics are stable. See, for example, Theorem III.1.7 in [2]. In particular, geodesics are stable in hyperbolic geodesic spaces.

Let $\mathcal{B}$ be the family of cycles that are bigons.

Theorem 12. *Given a graph G, geodesics are stable if and only if there exist constants $\varepsilon > 0$ and $k, m \in \mathbb{N}$ such that G is ε-densely (k, m)-chordal on $\mathcal{B}$.*

Proof. Suppose that G is ε-densely (k, m)-chordal on $\mathcal{B}$. Consider any pair of points x, y and any pair of geodesics, σ_1, σ_2, joining them. Then, for any point $z \in \sigma_1$, either $z \in \sigma_1 \cap \sigma_2$ or there is a cycle $C \subset \sigma_1 \cup \sigma_2$ with $z \in C$. If $L(C) < k$, then $d(z, \sigma_2) < \frac{k}{2}$. If $L(C) \geq k$, then either $d_C(z, \sigma_2) < \varepsilon$ or there is an m-shortcut in C with a shortcut vertex v such that $d_C(z, v) < \varepsilon$, and since σ_1 is geodesic, $d(v, \sigma_2) \leq m$. Thus, if $R = \max\{\frac{k}{2}, \varepsilon + m\}$, $d(z, \sigma_2) < R$ in any case. Hence, $\sigma_1 \subset N_R(\sigma_2)$. The same argument proves that $\sigma_2 \subset N_R(\sigma_1)$, and therefore, $d_H(\sigma_1, \sigma_2) \leq R$.

Suppose that geodesics are stable with constant R. Consider any pair of points x, y with $d(x, y) \geq 2R + 2$ and two xy-geodesics σ_1, σ_2 such that $\sigma_1 \cup \sigma_2$ defines a cycle C. Then, for any point $z \in \sigma_1$ (respectively with σ_2) such that $d_C(z, \sigma_2) > R$ (resp. $d_C(z, \sigma_1) > R$), since $d_H(\sigma_1, \sigma_2) \leq R$, $d(z, \sigma_2) \leq R$ (resp. $d(z, \sigma_1) \leq R$), and there is a strict R-shortcut in C with a shortcut vertex v such that $d_C(v, z) < R$. Thus, shortcut vertices are $(2R + 1)$-dense in C and G is $(2R + 1)$-densely $(4R + 4, R)$-chordal on $\mathcal{B}$. $\square$

Definition 17. *In a graph G, we say that geodesics between vertices are stable if and only if there is a constant $R \geq 0$ such that given two vertices $a, b \in G$ and any geodesic $[ab]$, then every geodesic σ joining a to b satisfies that $d_H(\sigma, [ab]) \leq R$.*

Proposition 6. *Given a graph G, geodesics between vertices are stable if and only if there is some constant $k \in \mathbb{N}$ so that for every pair of vertices a, b with $d(a, b) \geq 2k + 2$, every geodesic $[ab]$ and every vertex $v \in [ab]$ such that $d(v, \{a, b\}) > k$, then v is an ab-N_k-obstructing vertex.*

Proof. Suppose that geodesics between vertices are stable with constant R. Then, given any two vertices $a, b \in G$ with $d(a, b) \geq 2R + 2$ and any geodesic $[ab]$, every geodesic σ joining a to b satisfies that $d_H(\sigma, [ab]) \leq R$. Thus, for every vertex $v \in [ab]$ there is some vertex $w \in \sigma$ such that $d(v, w) \leq R$. Suppose $v \in [ab]$ with $d(v, \{a, b\}) > R$. Hence, v is an ab-N_R-obstructing vertex.

Now, suppose that for every pair of vertices a, b with $d(a, b) \geq 2k + 2$, every geodesic $[ab]$ and every vertex $v \in [ab]$ with $d(v, \{a, b\}) > k$, then v is an ab-N_k-obstructing vertex. Consider any pair of vertices $a, b \in G$ and any pair of ab-geodesics σ_1, σ_2. If $d(a, b) < 2k + 2$, then it is trivial to check that $d_H(\sigma_1, \sigma_2) < k + 1$. Suppose $d(x, y) \geq 2k + 2$. Then, for every vertex $v \in \sigma_1$ such that $d(v, \{a, b\}) > k$, $\sigma_2 \cap N_k(v) \neq \varnothing$. Therefore, it follows immediately that $\sigma_1 \subset N_{k+1/2}(\sigma_2)$. The same argument proves that $\sigma_2 \subset N_{k+1/2}(\sigma_1)$, and therefore, $d_H(\sigma_1, \sigma_2) < k + 1$. $\square$

Let $\mathcal{B}_0$ be the family of cycles that are bigons defined by two geodesics between vertices.

Proposition 7. *If G is $\frac{k}{4}$-densely (k, m)-chordal on $\mathcal{B}_0$, then for every pair of vertices a, b with $d(a, b) \geq \frac{k}{2} + 4$, every geodesic $[ab]$ and every vertex v_0 such that $d(v_0, \{a, b\}) \geq \frac{k}{4} + 1$, v_0 is an ab-N_k-obstructing vertex. In particular, $[ab]$ contains an ab-N_k-obstructing vertex.*

Proof. Consider any pair of vertices a, b with $d(a, b) \geq \frac{k}{2} + 4$, any geodesic $[ab]$ and any vertex $v_0 \in [ab]$ with $d(v_0, \{a, b\}) \geq \frac{k}{4} + 1$. Let a' be the vertex in $[av_0] \subset [ab]$ with $d(a', v_0) = \lceil \frac{k}{4} \rceil$ and b' be the vertex in $[v_0 b] \subset [ab]$ with $d(v_0, b') = \lceil \frac{k}{4} \rceil$. Therefore, $d(a', b') \geq \frac{k}{2}$, $a' \neq a$ and $b' \neq b$.

If there is no geodesic joining a to b disjoint from $N_{k/4}(v_0)$, we are done.

Suppose there is some geodesic γ_0 joining a to b such that $\gamma_0 \cap N_{k/4}(v_0) = \varnothing$. Then, $[ab] \cup \gamma_0$ contains a cycle C (with possibly $C = [ab] \cup \gamma_0$) composed by two geodesics $\gamma_1 = [a''b'']$ with $[a'b'] \subset [a''b''] \subset [ab]$ and $\gamma_2 \subset \gamma_0$ joining also a'' to b''. Clearly, $L(C) \geq k$. Since G is $\frac{k}{4}$-densely (k, m)-chordal on $\mathcal{B}_0$, then there is a strict shortcut σ with $L(\sigma) \leq m$ joining two vertices in C with a shortcut vertex v_1 in $N_{k/4}(v_0)$. Furthermore, since γ_1 and γ_2 are geodesics, then σ joins v_1 to a vertex v_2 in $\gamma_2 \subset \gamma_0$. Therefore, $d(v_2, v_0) \leq m + \frac{k}{4} < k$ (see Remark 2) and $\gamma_0 \cap N_k(v_0) \neq \varnothing$. $\square$

Theorem 13. *Given a graph G, geodesics between vertices are stable if and only if there exist constants $\varepsilon > 0$ and $k, m \in \mathbb{N}$ such that G is ε-densely (k, m)-chordal on $\mathcal{B}_0$.*

Proof. Suppose that G is ε-densely (k, m)-chordal on $\mathcal{B}_0$. By Proposition 7, if $k' = \max\{4\varepsilon, k\}$, then for every pair of vertices a, b with $d(a, b) \geq \frac{k'}{2} + 4$, every geodesic $[ab]$ and every vertex v_0 such that $d(v_0, \{a, b\}) \geq \frac{k'}{4} + 1$, v_0 is an ab-$N_{k'}$-obstructing vertex. Thus, by Proposition 6, geodesics are stable with constant $R = \frac{k'}{4} + 2$.

Let us suppose that geodesics between vertices are stable with constant R. Let a, b be two vertices with $d(a, b) \geq 2R + 2$ and C be a cycle that is a bigon defined by two ab-geodesics, σ_1, σ_2. Therefore, $L(C) \geq 4R + 4$. Consider any vertex $v \in \sigma_1$ (respectively, σ_2) such that $d(v, \{a, b\}) > R$. Then, since geodesics between vertices are stable with parameter R, $v \in N_R(\sigma_2)$ (respectively, σ_1) and there is a strict R-shortcut in C with an associated shortcut vertex w such that $d_C(v, w) < R$, therefore shortcut vertices are $(2R + 1)$-dense in C, and G is $(2R + 1)$-densely $(4R + 4, R)$-chordal on $\mathcal{B}_0$. $\square$

The following example shows that having stable geodesics between vertices does not imply that geodesics are stable.

Example 3. *Consider the family of odd cycles $\{C_{2k+1} : k \in \mathbb{N}\}$, and suppose we fix a vertex v_k in each cycle; we define a connected graph G identifying the family $\{v_k : k \in \mathbb{N}\}$ as a single vertex v. Notice that in G geodesics between vertices are unique. If two vertices belong to the same cycle C_{2k+1}, then the geodesic is contained in the cycle, and it is clearly unique. Otherwise, the geodesic is the union of the two (unique) shortest paths joining the vertices to v. Thus, geodesics between vertices are stable with constant zero.*

Let m_k be the midpoint of an edge in C_{2k+1} such that $d(m_k, v) = k + \frac{1}{2}$. Then, C_{2k+1} is a bigon in G defined by two geodesics, σ_1, σ_2 joining m_k to v and $d_H(\sigma_1, \sigma_2) = \frac{k}{2} + \frac{1}{4}$ with k arbitrarily large.

Remark 14. *Notice that the same property that characterizes being quasi-isometric to a tree (Corollary 3) also characterizes being hyperbolic, when restricted to triangles (Theorem 6), having stable geodesics, when restricted to bigons (Theorem 12), and having stable geodesics between vertices, when restricted to bigons between vertices (Theorem 13).*

Remark 15. *In the context of multi-path routing, (BP) implies that given any nominal path (with minimum cost) joining x and y, then any other path would remain close (at least at some point) to the nominal one. Furthermore, if we consider all paths with minimum cost, the stability of geodesics characterized above implies that every point of any minimal path is close to the nominal one.*

The proof of Proposition 7 can be adapted to prove also the following:

Proposition 8. *If G is $(\frac{k}{4} - m)$-densely (k, m)-chordal on $\mathcal{B}_0$ with $k > 4m$, then for every geodesic $[ab]$ with $d(a, b) \geq \frac{k}{2} + 2$ and every pair of vertices $a', b' \in [ab]$ with $d(\{a', b'\}, \{a, b\}) \geq m + 1$ and such that $d(a', b') \geq \frac{k}{2} - 2m$, $[a'b'] \subset [ab]$ is an ab-N_m-obstructing set. In particular, for every pair of vertices a, b in G with $d(a, b) \geq \frac{k}{2} + 2$, there is a geodesic σ of length $\frac{k}{2} - 2m$ or $\frac{k+1}{2} - 2m$ such that σ is ab-N_m-obstructing.*

Proof. Consider any geodesic $[ab]$ with $d(a, b) \geq \frac{k}{2} + 2$ and any pair of vertices $a', b' \in [ab]$ with $d(\{a', b'\}, \{a, b\}) \geq m + 1$ and $d(a', b') \geq \frac{k}{2} - 2m$. Let a'' be the vertex in $[aa'] \subset [ab]$ with $d(a', a'') = m$ and b'' be the vertex in $[b'b] \subset [ab]$ with $d(b', b'') = m$. Therefore, $d(a'', b'') \geq \frac{k}{2}$.

Suppose that there is some geodesic γ_0 joining a and b such that $\gamma \cap N_m([a'b']) = \emptyset$. Then, $[ab] \cup \gamma_0$ contains a cycle C composed by two geodesics: γ_1 with $[a''b''] \subset \gamma_1 \subset [ab]$ and $\gamma_2 \subset \gamma_0$. Clearly, $L(C) \geq k$. Consider the midpoint c in $[a'b']$. Since G is $(\frac{k}{4} - m)$-densely (k, m)-chordal on $\mathcal{B}_0$, then there is a strict shortcut σ with $L(\sigma) \leq m$ joining two vertices in C with a shortcut vertex v_1 such that $d_C(v_1, c) \leq \frac{k}{4} - m$, and hence, $v_1 \in [a'b']$. Furthermore, since γ_1 and γ_2 are geodesics, then σ joins v_1 to a vertex, v_2, in $\gamma_2 \subset \gamma_0$. Therefore, $d(v_2, [a'b']) \leq m$ and $\gamma_0 \cap N_m([a'b']) \neq \emptyset$, leading to a contradiction. $\square$

Acknowledgments: The author was partially supported by MTM2015-63612P.

Conflicts of Interest: The authors declare no conflict of interest.

References

1. Gromov, M. Hyperbolic groups. In *Essays in Group Theory*; Gersten, S.M., Ed.; Mathematical Science Research Institute Publications; Springer: New York, NY, USA, 1987; Volume 8, pp. 75–263.
2. Bridson, M.; Haefliger, A. *Metric Spaces of Non-Positive Curvature*; Springer: Berlin, Germany, 1999.
3. Burago, D.; Burago, Y.; Ivanov, S. A course in metric geometry. In *Graduate Studies in Mathematics*; AMS: Providence, RI, USA, 2001; Volume 33.
4. Buyalo, S.; Schroeder, V. Elements of Asymptotic Geometry. In *EMS Monographs in Mathematics*; European Mathematical Society: Zürich, Switzerland, 2007.
5. Gyhs, E.; de la Harpe, P. Sur le groupes hyperboliques d'après Mikhael Gromov. In *Progress in Math*; Birkhäuser: Boston, MA, USA, 1990; Volume 83.
6. Väisälä, J. Gromov hyperbolic spaces. *Expos. Math.* **2005**, *23*, 187–231.
7. Bermudo, S.; Rodríguez, J.M.; Rosario, O.; Sigarreta, J.M. Small values of the hyperbolicity constant in graphs. *Discret. Math.* **2016**, *339*, 3073–3084.
8. Bermudo, S.; Rodríguez, J.M.; Sigarreta, J.M. Computing the hyperbolicity constant. *Comput. Math. Appl.* **2011**, *62*, 4592–4595.
9. Carballosa, W.; Pestana, D.; Rodríguez, J.M.; Sigarreta, J.M. Distortion of the hyperbolicity constant of a graph. *Electron. J. Comb.* **2012**, *19*, # P67.
10. Carballosa, W.; Rodríguez, J.M.; Sigarreta, J.M.; Villeta, M. Gromov hyperbolicity of line graphs. *Electron. J. Comb.* **2011**, *18*, # P210.
11. Bermudo, S.; Rodríguez, J.M.; Sigarreta, J.M.; Vilaire, J.-M. Gromov hyperbolic graphs. *Discret. Math.* **2013**, *313*, 1575–1585.
12. Chepoi, V.; Dragan, F.F.; Estellon, B.; Habib, M.; Vaxes, Y. Notes on diameters, centers, and approximating trees of δ-hyperbolic geodesic spaces and graphs. *Electron. Notes Discret. Math.* **2008**, *31*, 231–234.
13. Frigerio, R.; Sisto, A. Characterizing hyperbolic spaces and real trees. *Geom. Dedicata* **2009**, *142*, 139–149.
14. Hästö, P.A. Gromov hyperbolicity of the j_G and $\tilde{j}_G$ metrics. *Proc. Am. Math. Soc.* **2006**, *134*, 1137–1142.
15. Michel, J.; Rodríguez, J.M.; Sigarreta, J.M.; Villeta, M. Hyperbolicity and parameters of graphs. *Ars Comb.* **2011**, *100*, 43–63.
16. Pestana, D.; Rodríguez, J.M.; Sigarreta, J.M.; Villeta, M. Gromov hyperbolic cubic graphs. *Cent. Eur. J. Math.* **2012**, *10*, 1141–1151.
17. Portilla, A.; Rodríguez, J.M.; Sigarreta, J.M.; Vilaire, J.-M. Gromov hyperbolic tessellation graphs. *Util. Math.* **2015**, *97*, 193–212.
18. Portilla, A.; Rodríguez, J.M.; Tourís, E. Gromov hyperbolicity through decomposition of metric spaces II. *J. Geom. Anal.* **2004**, *14*, 123–149.
19. Portilla, A.; Rodríguez, J.M.; Tourís, E. Stability of Gromov hyperbolicity. *J. Adv. Math. Stud.* **2009**, *2*, 77–96.
20. Portilla, A.; Tourís, E. A characterization of Gromov hyperbolicity of surfaces with variable negative curvature. *Publ. Matorsz.* **2009**, *53*, 83–110.
21. Rodríguez, J.M.; Sigarreta, J.M. Bounds on Gromov hyperbolicity constant in graphs. *Proc. Indian Acad. Sci. Math. Sci.* **2012**, *122*, 53–65.
22. Rodríguez, J.M.; Sigarreta, J.M.; Torres-Nuñez, Y. Computing the hyperbolicity constant of a cubic graph. *Int. J. Comput. Math.* **2014**, *91*, 1897–1910.
23. Rodríguez, J.M.; Sigarreta, J.M.; Vilaire, J.-M.; Villeta, M. On the hyperbolicity constant in graphs. *Discret. Math.* **2011**, *311*, 211–219.
24. Sigarreta, J.M. Hyperbolicity in median graphs. *Proc. Indian Acad. Sci. Math. Sci.* **2013**, *123*, 455–467.
25. Tourís, E. Graphs and Gromov hyperbolicity of non-constant negatively curved surfaces. *J. Math. Anal. Appl.* **2011**, *380*, 865–881.
26. Dress, A.; Holland, B.; Huber, K.T.; Koolen, J.H.; Moulton, V.; Weyer-Menkhoff, J. Δ additive and Δ ultra-additive maps, Gromov's trees, and the Farris transform. *Discret. Appl. Math.* **2005**, *146*, 51–73.
27. Dress, A.; Moulton, V.; Terhalle, W. T-theory: An overview. *Eur. J. Comb.* **1996**, *17*, 161–175.

28. Clauset, A.; Moore, C.; Newman, M.E.J. Hierarchical structure and the prediction of missing links in networks. *Nature* **2008**, *453*, 98–101.

29. Krioukov, D.; Papadopoulos, F.; Kitsak, M.; Vahdat, A.; Boguñá, M. Hyperbolic geometry of complex networks. *Phys. Rev. E* **2010**, *82*, 036106, doi:10.1103/PhysRevE.82.036106.

30. Shang, Y. Lack of Gromov-hyperbolicity in small-world networks. *Cent. Eur. J. Math.* **2012**, *10*, 1152–1158.

31. Shang, Y. Non-hyperbolicity of random graphs with given expected degrees. *Stoch. Model.* **2013**, *29*, 451–462.

32. Jonckheere, E.A. Contrôle du traffic sur les réseaux à géométrie hyperbolique—Vers une théorie géométrique de la sécurité l'acheminement de l'information. *J. Eur. Syst. Autom.* **2002**, *8*, 45–60.

33. Jonckheere, E.A.; Lohsoonthorn, P. Geometry of network security. *Proc. Am. Control Conf.* **2004**, *2*, 976–981.

34. Jonckheere, E.A.; Lou, M.; Bonahon, F.; Baryshnikov, Y. Euclidean versus hyperbolic congestion in idealized versus experimental networks. *Int. Math.* **2011**, *7*, 1–27.

35. Sreenivasa Kumar, P.; Veni Madhavan, C.E. Minimal vertex separators of chordal graphs. *Discret. Appl. Math.* **1998**, *89*, 155–168.

36. Blair, J.; Peyton, B. An introduction to chordal graphs and clique trees, Graph Theory and Sparse Matrix Multiplication. In *IMA Volumes in Mathematics and its Applications*; Springer: Berlin, Germany, 1993; Volume 56, pp. 1–29.

37. Brinkmann, G.; Koolen, J.; Moulton, V. On the hyperbolicity of chordal graphs. *Ann. Comb.* **2001**, *5*, 61–69.

38. Wu, Y.; Zhang, C. Hyperbolicity and chordality of a graph. *Electron. J. Comb.* **2011**, *18*, # P43.

39. Bermudo, S.; Carballosa, W.; Rodríguez, J.M.; Sigarreta, J.M. On the hyperbolicity of edge-chordal and path-chordal graphs. *Filomat* **2016**, *30*, 2599–2607.

40. Martínez-Pérez, A. Chordality properties and hyperbolicity on graphs. *Electron. J. Comb.* **2016**, *23*, # P3.51.

41. Dirac, G.A. On rigid circuit graphs. *Abh. Math. Semin. Univ. Hambg.* **1961**, *25*, 71–76.

42. Krithika, R.; Mathew, R.; Narayanaswamy, N.S.; Sadagopan, N. A Dirac-type Characterization of *k*-chordal Graphs. *Discret. Math.* **2013**, *313*, 2865–2867.

43. Anandkumar, A.; Tan, V.; Huang, F.; Willsky, A.S. High-dimensional Gaussian graphical model selection: Walk summability and local separation criterion. *J. Mach. Learn. Res.* **2012**, *13*, 2293–2337.

44. Manning, J.F. Geometry of pseudocharacters. *Geom. Topol.* **2005**, *9*, 1147–1185.

45. Bestvina, M.; Bromberg, K.; Fujiwara, K. Constructing group actions on quasi-trees and applications to mapping class groups. *Publ. Math. l'IHÉS* **2015**, *122*, 1–64.

46. Cashen, C.H. A Geometric Proof of the Structure Theorem for Cyclic Splittings of Free Groups. *Topol. Proc.* **2017**, *50*, 335–349.

47. Martínez-Pérez, A. Real-valued functions and metric spaces quasi-isometric to trees. *Ann. Acad. Sci. Fenn. Math.* **2012**, *37*, 525–538.

48. Rodríguez, J.M.; Tourís, E. Gromov hyperbolicity through decomposition of metric spaces. *Acta Math. Hung.* **2004**, *103*, 53–84.

49. Martínez-Pérez, A. Quasi-isometries between visual hyperbolic spaces. *Manuscr. Math.* **2012**, *137*, 195–213.

50. Cao, J. Cheeger isoperimetric constants of Gromov-hyperbolic spaces with quasi-pole. *Commun. Contemp. Math.* **2000**, *4*, 511–533.

51. Martínez-Pérez, A.; Rodríguez, J.M. Cheeger isoperimetric constant of Gromov hyperbolic manifolds and graphs. *Commun. Contemp. Math.* **2017**, in press, doi:10.1142/S021919971750050X.

52. Bieri, R.; Geoghegan, R. Limit sets for modules over groups on CAT(0) spaces: From the Euclidean to the hyperbolic. *Proc. Lond. Math. Soc.* **2016**, *112*, 1059–1102.

53. Hughes, B. Trees and ultrametric spaces: A categorical equivalence. *Adv. Math.* **2004**, *189*, 148–191.

54. Martínez-Pérez, A.; Morón, M.A. Uniformly continuous maps between ends of ℝ-trees. *Math. Z.* **2009**, *263*, 583–606.

symmetry

MDPI

Article

β-Differential of a Graph

Ludwin A. Basilio [1], Sergio Bermudo [2], Jesús Leaños [1] and José M. Sigarreta [3,*]

[1] Academic Unit of Mathematics, Autonomous University of Zacatecas, Paseo la Bufa, int. Calzada Solidaridad, 98060 Zacatecas, Mexico; ludwin.ali@gmail.com (L.A.B.); jleanos@mate.reduaz.mx (J.L.)
[2] Department of Economics, Quantitative Methods and Economic History, Pablo de Olavide University, Carretera de Utrera Km. 1, 41013 Sevilla, Spain; sbernav@upo.es
[3] Faculty of Mathematics, Autonomous University of Guerrero, Carlos E. Adame 5, Col. La Garita, 39350 Acapulco, Guerrero, Mexico
* Correspondence: josemariasigarretaalmira@yahoo.es; Tel.: +52-744-481-0216

Received: 12 September 2017; Accepted: 26 September 2017; Published: 30 September 2017

Abstract: Let $G = (V, E)$ be a simple graph with vertex set V and edge set E. Let D be a subset of V, and let $B(D)$ be the set of neighbours of D in $V \setminus D$. The *differential* $\partial(D)$ of D is defined as $|B(D)| - |D|$. The maximum value of $\partial(D)$ taken over all subsets $D \subseteq V$ is the *differential* $\partial(G)$ of G. For $\beta \in (-1, \Delta)$, the *β-differential* $\partial_\beta(G)$ of G is the maximum value of $\{|B(D)| - \beta|D| : D \subseteq V\}$. Motivated by an influential maximization problem, in this paper we study the β-differential of G.

Keywords: differential of a graph; domination number

1. Introduction

Social networks, such as Facebook or Twitter, have served as an important medium for communication and information disseminating. As a result of their massive popularity, social networks now have a wide variety of applications in the viral marketing of products and political campaigns. Motivated by its numerous applications, some authors [1–3] have proposed several influential maximization problems, which share a fundamental algorithmic problem for information diffusion in social networks: the problem of determining the best group of nodes to influence the rest. As it was shown in [4], the study of the differential of a graph G, could be motivated from such scenarios. In this work we generalize the notion of differential of a graph and provide new applications. Let us first give some basic notation and then we motivate such a generalization.

Throughout this paper, $G = (V, E)$ is a simple graph of order $n \geq 3$ with vertex set V and edge set E. Let u, v be distinct vertices of V, and let S be a subset of V. We will write $u \sim v$ whenever u and v are adjacent in G. If S is nonempty, then $N_S(v)$ denotes the set of neighbors that v has in S, i.e., $N_S(v) := \{u \in S : u \sim v\}$; the degree of v in S is denoted by $\delta_S(v) := |N_S(v)|$. As usual, $N(v)$ is the set of neighbors that v has in V, i.e., $N(v) := \{u \in V : u \sim v\}$; and $N[v]$ is the closed neighborhood of v, i.e., $N[v] := N(v) \cup \{v\}$. We denote by $\delta(v) := |N(v)|$ the degree of v in G, and by $\delta(G)$ and $\Delta(G)$ the minimum and the maximum degree of G, respectively. The subgraph of G induced by S will be denoted by $G[S]$, and the complement of S in V by $\overline{S}$. Then $N_{\overline{S}}(v)$ is the set of neighbors that v has in $\overline{S} = V \setminus S$. We let $N(S) := \bigcup_{v \in S} N(v)$ and $N[S] := N(S) \cup S$. Finally, we will use $B(S)$ to denote the set of vertices in $\overline{S}$ that have a neighbour in S, and $C(S)$ to denote $\overline{S \cup B(S)}$. Then $\{S, B(S), C(S)\}$ is a partition of V. An *external private neighbor* of $v \in S$ (with respect to S) is a vertex $w \in N(v) \cap \overline{S}$ such that $w \notin N(u)$ for every $u \in S \setminus \{v\}$. The set of all external private neighbors of v is denoted by epn $[v, S]$.

To motivate the notion of β-differential of a graph, assume for a moment that our graph $G = (V, E)$ represents a map of a country, where V is the set of cities of G and E is the set of roads between cities of G. To avoid weights, we could assume that all the cities of G have the same population and have the same importance, and also that all roads have the same length. A supermarket chain wants to build some supermarkets in that country and they are studying which are the best places to do it. For that, they might consider that every supermarket will give service to the own city and the neighboring cities. Moreover, according to some previous studies, the cost of building a new supermarket is $\alpha > 0$ times the benefit that can be obtained by each city in an specific number of years. In consequence, if we consider the unit as the amount of money that we can obtain from a city in that amount of years and we build a supermarket in each vertex of a set $D \subseteq V$, we have that the benefit that we obtain is $|B(D)| + |D| - \alpha|D| = |B(D)| - (\alpha - 1)|D|$, or equivalently, $|B(D)| - \beta|D|$ for $\beta = \alpha - 1$. Such a value is denoted by $\partial_\beta(D)$ and it is called the β-*differential* of D. We are naturally interested in determining the following value:

$$\partial_\beta(G) := \max\{\partial_\beta(D) : D \subseteq V\} = \max\{|B(D)| - \beta|D| : D \subseteq V\}.$$

The number $\partial_\beta(G)$ is the β-*differential* of G. Let Δ be the maximum degree of G. Note that if $v \in V$ has degree Δ, then $\partial_\beta(G) \geq \partial_\beta(\{v\}) = \Delta - \beta$. Thus, if $\beta < \Delta$, there will always be a choice of places giving benefits. On the other hand, if $\beta \geq \Delta$ and $D \subseteq V$, then $\partial_\beta(D) = |B(D)| - \beta|D| \leq \Delta|D| - \beta|D| = (\Delta - \beta)|D| \leq 0$, and hence no set of locations will make benefits. For these reasons, we restrict our study of $\partial_\beta(G)$ to the values of β belonging to $(-1, \Delta)$.

The particular case in which $\beta = 1$ is called the differential of G, and it is usually denoted by $\partial(G)$. The study of $\partial(G)$ together with a variety of other kinds of differentials of a set, started in [5]. In particular, several bounds for $\partial(G)$ were given. The differential of a graph has also been investigated in [4,6–15], and it was proved in [11] that $\partial(G) + \gamma_R(G) = n$, where n is the order of the graph G and $\gamma_R(G)$ is the Roman domination number of G, so every bound for the differential of a graph can be used to get a bound for the Roman domination number. The differential of a set D was also considered in [16], where it was denoted by $\eta(D)$, and the minimum differential of an independent set was considered in [17]. The case of the B-differential of a graph or enclaveless number, defined as $\psi(G) := \max\{|B(D)| : D \subseteq V\}$, was studied in [5,18].

Notice that if G is disconnected, and $G_1, \ldots, G_k$ are its connected components, then $\partial_\beta(G) = \partial_\beta(G_1) + \cdots + \partial_\beta(G_k)$. In view of this, from now on we only consider connected graphs.

Other graph parameters that we will use in this paper are the dominating number and the packing number of G. We recall that a set $S \subseteq V$ is a *dominating set* if every vertex $v \in \overline{S}$ is adjacent to a vertex in S. The *domination number* $\gamma(G)$ is the minimum cardinality among all dominating sets. A *packing* of a graph G is a set of vertices in G that are pairwise at distance more than two. The *packing number* $\rho(G)$ of G is the size of a largest packing in G. An *open packing* of G is a set $S \subseteq V$ such that $N(u) \cap N(v) = \emptyset$ for every two different vertices $u, v \in S$. The *open packing number* $\rho^o(G)$ of G is the size of a largest open packing in G.

2. The Function $f_G(\beta) = \partial_\beta(G)$

Throughout this section Δ denotes the maximum degree of $G = (V, E)$ and $\beta \in (-1, \Delta)$. Clearly, the values of $\partial_\beta(G)$ can be considered as a function $f_G : (-1, \Delta) \to \mathbb{R}$, which is defined as $f_G(\beta) := \partial_\beta(G)$. A subset $D \subseteq V$ satisfying $\partial_\beta(D) = \partial_\beta(G)$ is called a β-∂-*set* or a β-*differential set*. If D has minimum (maximum) cardinality among all β-differential sets, then D is a *minimum* (*maximum*) β-*differential set*. We will write D_β^m (respectively, D_β^M) to indicate that D is a minimum (respectively, maximum) β-differential set. Since the value of $\partial_\beta(G)$ can be achieved by several subsets of V, a natural problem is to determine the properties of such β-differential sets. Our goal in this section is to establish several properties of these sets. In particular, as a consequence of some of them we will show in Theorem 1 that f_G is a continuous function. We have seen before that, if $v \in V$ has maximum degree, then $\partial_\beta(\{v\}) > 0$ for any admissible β. In our next results we continue our study in this direction and we show that the positive value of the β-differential of a subset D of vertices of G will depend on the values of $|D|$ and β.

Proposition 1. *Let $G = (V, E)$ be a graph. If $\beta \in (-1, 1]$ and $k \in \mathbb{N}$ such that $k < \dfrac{n}{\beta+1}$, then there exists a subset $D \subseteq V$ such that $|D| = k$ and $\partial_\beta(D) > 0$.*

Proof 1. If G contains a dominating set $D \subseteq V$ of cardinality k, then $\partial_\beta(D) = |B(D)| - \beta|D| = n - k(\beta + 1) > n - \dfrac{n}{\beta+1}(\beta+1) = 0$, as desired. Now we suppose $\gamma(G) > k$, and consider a maximum matching of G, say $M = \{u_1 v_1, \ldots, u_m v_m\}$. It is known that $m = |M| \geq \gamma(G) > k$. Let $D = \{u_1, \ldots, u_k\}$. If u_m or v_m is adjacent to a vertex in D, then $\partial_\beta(D) = |B(D)| - \beta|D| \geq k + 1 - k = 1 > 0$. Thus we can assume that neither u_m nor v_m is adjacent to any vertex in D. Since G is connected, then at least one of u_m or v_m is adjacent to a vertex of $V \setminus (D \cup \{u_m, v_m\})$. Without loss of generality, we suppose that u_m is adjacent to a vertex $x \in V \setminus (D \cup \{u_m, v_m\})$. If $x \notin \{v_1, \ldots, v_k\}$, then $D' = \{u_1, \ldots, u_{k-1}, u_m\}$ satisfies $\partial_\beta(D') = |B(D')| - \beta|D'| \geq k + 1 - k = 1 > 0$. If $x = v_j$ for some $j \in \{1, \ldots, k\}$, then $D' = \{u_1, \ldots, u_{j-1}, v_j, u_{j+1}, \ldots, u_k\}$ satisfies $\partial_\beta(D') = |B(D')| - \beta|D'| \geq k + 1 - k = 1 > 0$. $\square$

Taking into account that $\frac{n}{\beta+1} \leq \frac{n-\gamma(G)}{\beta}$ when $\beta \in (-1, 1]$, the next proposition shows that the upper bound on the size of D in Proposition 1 cannot be relaxed.

Proposition 2. *Let $G = (V, E)$ be a graph. Every set $D \subseteq V$ such that $|D| \geq \min\left\{\frac{n}{\beta+1}, \frac{n-\gamma(G)}{\beta}\right\}$ satisfies $\partial_\beta(D) \leq 0$.*

Proof 2. Firstly, if $D \subseteq V$ is a set such that $|D| \geq n/(\beta+1)$, then $\beta|D| \geq n - |D| \geq |B(D)|$, consequently, $\partial_\beta(D) \leq 0$. Secondly, if $|D| \geq (n - \gamma(G))/\beta$, then $\beta|D| \geq n - \gamma(G) \geq |B(D)|$, which again implies $\partial_\beta(D) \leq 0$. $\square$

Lemma 1. *If $G = (V, E)$ is a graph and $\beta_1 < \beta_2$, then $\partial_{\beta_1}(G) > \partial_{\beta_2}(G)$.*

Proof 3. If $\beta_1 < \beta_2$, then $|B(D)| - \beta_1|D| > |B(D)| - \beta_2|D|$ for every $D \subseteq V$. Since the number of subsets of V is finite, we conclude that $\partial_{\beta_1}(G) > \partial_{\beta_2}(G)$. $\square$

Proposition 3. *Let $G = (V, E)$ be a graph. If $\beta \notin \mathbb{N}$, then every minimum β-differential set is a maximal β-differential set and every maximum β-differential set is a minimal β-differential set.*

Proof 4. Let D_β^m be a minimum β-differential set and let D be a β-differential set such that $D_\beta^m \subset D$. In such a case, $D \setminus D_\beta^m = \{u_1, \ldots, u_r\} \subseteq B(D_\beta^m) \cup C(D_\beta^m)$ and $\{u_i; z_{i_1}, \ldots, z_{i_{k_i}}\}$ are disjoint stars (not necessarily induced stars) with centers u_i for every $i \in \{1, \ldots, r\}$, such that $z_s \in C(D_\beta^m)$ for every $s \in \{i_j : i \in \{1, \ldots, r\}, j \in \{1, \ldots, k_i\}\}$ and $\sum_{u_i \in B(D_\beta^m)}(-1 + k_i - \beta) + \sum_{u_i \in C(D_\beta^m)}(k_i - \beta) = 0$. Since $\beta \notin \mathbb{N}$, there exists $i \in \{1, \ldots, r\}$ such that $u_i \in B(D_\beta^m)$ and $-1 + k_i - \beta > 0$, or $u_i \in C(D_\beta^m)$ and $k_i - \beta > 0$, thus $\partial_\beta(D_\beta^m \cup \{u_i\}) > \partial_\beta(D_\beta^m)$, a contradiction. Analogously, we can prove that every maximum β-differential set is minimal. $\square$

Now we establish a couple of relationships between dominating sets and β-differential sets of G.

Lemma 2. *Let $G = (V, E)$ be a graph and let A be a dominating set in G. If $D \subseteq V$ with $|D| > |A|$, then $\partial_\beta(D) < \partial_\beta(A)$. In particular,*

$$\partial_\beta(G) = \max\{\partial_\beta(D) : D \subseteq V, |D| \leq \gamma(G)\}.$$

Proof 5. Let $D \subseteq V$ with $|D| > |A|$. Then

$$\partial_\beta(A) = |B(A)| - \beta|A| = |V| - |A| - \beta|A| > |V| - |D| - \beta|D| \geq |B(D)| - \beta|D| = \partial_\beta(D).$$

$\square$

Proposition 4. *Let $G = (V, E)$ be a graph of order n. If $\beta \in [-1, 0]$, then $\partial_\beta(G) = n - (1 + \beta)\gamma(G)$. That is, every β-differential set is a minimum dominating set.*

Proof 6. Let $A, D \subseteq V$ such that A is a dominating set with $\gamma(G) = |A|$ and $\partial_\beta(G) = \partial_\beta(D)$. It is known that $|B(A)| = \max\{|B(S)| : S \subseteq V\}$. If $\beta = 0$ we have that $\partial_0(A) = |B(A)| \geq |B(D)| = \partial_0(D)$, and so $|B(A)| = |B(D)|$. Then $|B(D)| + |A| = n$, or equivalently, $\partial_0(D) = n - \gamma(G)$, as required. Now we suppose that $\beta < 0$. By Lemma 2 we know that $|D| \leq |A|$. If $|D| < |A|$, then

$$\partial_\beta(A) = |B(A)| - \beta|A| \geq |B(D)| - \beta|A| > |B(D)| - \beta|D| = \partial_\beta(D),$$

a contradiction. Finally, since $|D| = |A|$ and $\partial_\beta(D) = |B(D)| - \beta|D| = |B(D)| - \beta|A| \leq |B(A)| - \beta|A| = \partial_\beta(A)$, we have that $|B(D)| = |B(A)|$ and, consequently, that D is a minimum dominating set. $\square$

In view of Proposition 4, unless otherwise stated, from now on we will only consider $\beta > 0$. Note that the trees shown in Figures 1–3, and the 1- and 2-differential sets marked (in black) suggest that if $\beta_1 < \beta_2$ then $|D^M_{\beta_2}| \leq |D^m_{\beta_1}|$. This question will be answered in Lemma 3.

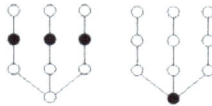

Figure 1. $|D^m_1| = 3$ and $|D^M_2| = 1$.

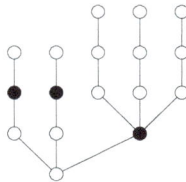

Figure 2. $|D^m_1| = |D^M_2| = 3$.

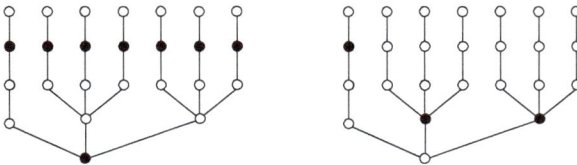

Figure 3. $|D^m_1| = 8$ and $|D^M_2| = 3$.

Lemma 3. *Let $G = (V, E)$ be a graph. If $\beta_1 < \beta_2$, and there is no $D \subseteq V$ such that $\partial_{\beta_1}(G) = \partial_{\beta_1}(D)$ and $\partial_{\beta_2}(G) = \partial_{\beta_2}(D)$, then for every $D_1, D_2 \subseteq V$ such that $\partial_{\beta_1}(G) = \partial_{\beta_1}(D_1)$ and $\partial_{\beta_2}(G) = \partial_{\beta_2}(D_2)$, we have $|D_2| \leq |D_1| - 1$.*

Proof 7. Let D_1 and D_2 be such that $\partial_{\beta_1}(G) = \partial_{\beta_1}(D_1)$ and $\partial_{\beta_2}(G) = \partial_{\beta_2}(D_2)$. By hypothesis $\partial_{\beta_2}(G) \neq \partial_{\beta_2}(D_1)$, so $|B(D_2)| - \beta_2|D_2| > |B(D_1)| - \beta_2|D_1|$. Hence

$$
\begin{aligned}
|B(D_2)| - \beta_1|D_2| &= |B(D_2)| - |B(D_1)| + |B(D_1)| - \beta_1|D_2| + \beta_1|D_1| - \beta_1|D_1| \\
&> \beta_2|D_2| - \beta_2|D_1| + \beta_1|D_1| - \beta_1|D_2| + \partial_{\beta_1}(G) \\
&= (\beta_2 - \beta_1)(|D_2| - |D_1|) + \partial_{\beta_1}(G).
\end{aligned}
$$

Therefore, if $|D_2| - |D_1| \geq 0$, then $|B(D_2)| - \beta_1|D_2| > \partial_{\beta_1}(G)$, a contradiction. $\square$

Lemma 4. *Let $G = (V, E)$ be a graph. If $\beta_1 < \beta_2$, then for every β_1-differential set D_1 and every β_2-differential set D_2 it is satisfied $|D_2| \leq |D_1|$ and $|B(D_2)| \leq |B(D_1)|$.*

Proof 8. By absurdum we suppose that $|D_2| > |D_1|$. Since $\partial_{\beta_2}(D_1) \leq \partial_{\beta_2}(D_2)$, then

$$
|B(D_1)| - \beta_2|D_1| \leq |B(D_2)| - \beta_2|D_2| = |B(D_2)| - \beta_1|D_2| - |D_2|(\beta_2 - \beta_1).
$$

Therefore,

$$
|B(D_1)| - \beta_2|D_1| + |D_2|(\beta_2 - \beta_1) = |B(D_1)| - \beta_1|D_1| + (|D_2| - |D_1|)(\beta_2 - \beta_1) \leq \partial_{\beta_1}(D_2).
$$

Since $(|D_2| - |D_1|)(\beta_2 - \beta_1) > 0$, we have $\partial_{\beta_1}(D_1) = |B(D_1)| - \beta_1|D_1| < \partial_{\beta_1}(D_2)$, a contradiction.

Finally, since $|B(D_2)| - \beta_1|D_2| \leq \partial_{\beta_1}(G) = |B(D_1)| - \beta_1|D_1|$, we have

$$
|B(D_2)| - |B(D_1)| \leq \beta_1(|D_2| - |D_1|) \leq 0.
$$

$\square$

Looking also Figures 1–3, it might be thought that every minimum β-differential set is included in a maximum β-differential set, but this is not true, as we can see in Figure 4, where black vertices sets are the minimum 1-differential set and $\frac{1}{2}$-differential set, respectively, and grey vertices set are the maximum 1-differential set and $\frac{1}{2}$-differential set, respectively.

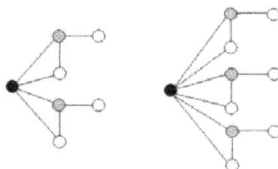

Figure 4. On the left $|D_1^m| = 1$ and $|D_1^M| = 2$, and on the right $|D_{\frac{1}{2}}^m| = 1$ and $|D_{\frac{1}{2}}^M| = 3$.

If β is an irrational number, then $|D_\beta^m| = |D_\beta^M|$, because $|B(D_\beta^M)| - \beta|D_\beta^M| = |B(D_\beta^m)| - \beta|D_\beta^m|$ implies that $\beta(|D_\beta^M| - |D_\beta^m|) = |B(D_\beta^M)| - |B(D_\beta^m)|$.

Proposition 5. *Let $G = (V, E)$ be a graph. If $\beta_1 < \beta_2$ and there exists $D \subseteq V$ such that $\partial_{\beta_1}(G) = \partial_{\beta_1}(D)$ and $\partial_{\beta_2}(G) = \partial_{\beta_2}(D)$, then for every $\beta \in (\beta_1, \beta_2)$ it holds $\partial_\beta(G) = \partial_\beta(D)$ and $|D| = |D_{\beta_2}^M| = |D_{\beta_1}^m|$.*

Proof 9. Let $\beta \in (\beta_1, \beta_2)$ and let $D' \subseteq V$ be a β-differential set, by Lemma 4 we have

$$
|D| \leq |D'| \leq |D| \quad \text{and} \quad |B(D)| \leq |B(D')| \leq |B(D)|,
$$

thus, $\partial_\beta(D) = |B(D)| - \beta|D| = |B(D')| - \beta|D'| = \partial_\beta(G)$. Finally, since D is a β_2-differential set, by Lemma 4, we have $|D| \le |D^m_{\beta_1}|$. Using now that D is also a β_1-differential set, we have $|D| \ge |D^m_{\beta_1}|$. The equality $|D| = |D^M_{\beta_2}|$ can be obtained analogously. $\square$

Theorem 1. *Let $G = (V, E)$ be a graph, then the function $f_G(\beta) = \partial_\beta(G)$ is continuous for every $\beta \in (-1, \Delta)$.*

Proof 10. It follows from Lemma 3 and Proposition 5 that the graphic representation of the function $f_G(\beta)$ is formed by pieces of straight lines with negative slope. That is, there exists a partition $0 < \beta_1 < \beta_2 < \cdots < \beta_{r-1} < \beta_r = \Delta$ of $[0, \Delta]$ such that

$$f_G(\beta) = \begin{cases} n - (1+\beta)\gamma(G) & -1 < \beta \le 0 \\ a_1 - b_1\beta & 0 < \beta \le \beta_1 \\ a_2 - b_2\beta & \beta_1 < \beta \le \beta_2 \\ \vdots & \vdots \\ a_r - b_r\beta & \beta_{r-1} < \beta \le \beta_r \end{cases}$$

where $a_i, b_i \in \mathbb{N}$. Moreover, $b_i \le b_{i-1} - 1$ and $r \le \gamma(G)$. Observe that $f_G(\beta)$ is a continuous function because, if $a_i - b_i\beta_i > a_{i+1} - b_{i+1}\beta_i$, then there exist $\delta > 0$ and $\beta' \in (\beta_i, \beta_i + \delta)$ such that $a_i - b_i\beta' > a_{i+1} - b_{i+1}\beta'$, so $f_G(\beta')$, since it is a maximum, should be equal to $a_i - b_i\beta'$, a contradiction. $\square$

For instance, in the graphs shown in Figure 5 we have $f(\beta) = 4 - 3\beta$ if $\beta \in \left(0, \frac{1}{2}\right]$ and $f(\beta) = 3 - \beta$ if $\beta \in \left(\frac{1}{2}, 3\right]$.

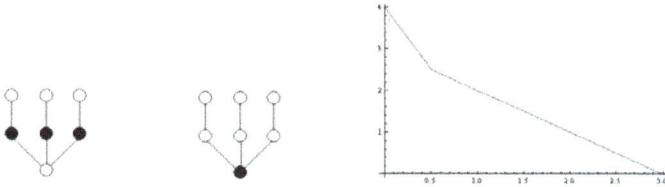

Figure 5. β-differential set when $\beta \in \left(0, \frac{1}{2}\right]$ (on the left) and β-differential set when $\beta \in \left(\frac{1}{2}, 3\right]$ (on the right).

Notice that from the point of view explained in the introduction, if the cost of building a supermarket is $\alpha = \frac{7}{5}$ (that is $\beta = \frac{2}{5}$), it is more profitable to build three supermarket giving service to all the towns. However, if the cost of building a supermarkets is $\alpha = \frac{8}{5}$ (that is $\beta = \frac{3}{5}$), it is more profitable to build only one supermarket, leaving without service to three towns.

Note that there exist another generalizations in graphs using continuous parameters, for instance, α-domination number in [19], where the resulting function is not continuous.

It might be also thought that the intervals where the function is a straight line are big, but there are graphs where these intervals are really small. For instance, if we consider the graph G_r with $3r + 1$ vertices shown in Figure 6, the β-differential set is an unitary set containing the black vertex when $\beta > \frac{1}{r-1}$, and the set containing the grey vertices when $\beta \le \frac{1}{r-1}$.

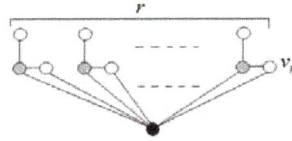

Figure 6. An example of a graph G_r such that $\partial_\beta(G_r) = 1 + 2r - \beta r$ if $\beta \leq \frac{1}{r-1}$, and $\partial_\beta(G_r) = 2r - \beta$ if $\beta > \frac{1}{r-1}$.

As $v_r \in B(D_\beta)$ for every β-differential set D_β in G_r, we can consider a graph G whose vertices are $V(G) = \bigcup_{i=0}^{j} V(G_{r+i})$ and edges $E = \bigcup_{i=0}^{j} E(G_{r+i}) \cup \{v_s v_{s+1} : s \in \{r, \ldots, r+j-1\}\}$. In such a case, the partition of the interval $(0, \Delta)$ for the definition of the piecewise function $f_G(\beta)$ is $0 < \frac{1}{r+j-1} < \frac{1}{r+j-2} < \cdots < \frac{1}{r-1} < \Delta$.

3. Bounds on the β-Differential of a Graph

As we have mentioned in the introduction, $\partial_\beta(G)$ will be the maximum benefit we could obtain if the cost of placing the considered service is $\alpha = \beta + 1$, so it will be interesting to get lower and upper bounds for this benefit.

Proposition 6. *Let $G = (V, E)$ be a graph with order n and maximum degree Δ. Then $\Delta - \beta \leq \partial_\beta(G) \leq n - (1 + \beta)$.*

Proof 11. Let $v \in V$ such that $\delta(v) = \Delta$. Then $\partial(\{v\}) = \Delta - \beta \leq \partial_\beta(G)$. Now, for any β-differential set D we have that

$$\partial_\beta(G) = |B(D)| - \beta|D| \leq n - 1 - \beta|D| \leq n - 1 - \beta.$$

$\square$

Proposition 7. *Let $G = (V, E)$ be a graph with order n and maximum degree Δ. The following properties hold.*

(a) $\partial_\beta(G) = n - (1 + \beta)$ *if and only if $\Delta = n - 1$.*
(b) $\partial_\beta(G) = n - (2 + \beta)$ *if and only if $\Delta = n - 2$.*
(c) *If $\beta > 1$, then $\partial_\beta(G) = n - (3 + \beta)$ if and only if $\Delta = n - 3$.*

Proof 12.   *(a)* If $\Delta = n - 1$, by Proposition 6 we have $n - 1 - \beta \leq \partial_\beta(G) \leq n - 1 - \beta$, then $\partial_\beta(G) = n - 1 - \beta$. If $\partial_\beta(G) = n - 1 - \beta$ and D is a β-differential set, then we have $n - 1 - \beta = |B(D)| - \beta|D| \leq n - 1 - \beta|D|$. Therefore, $|D| \leq 1$, that is, $|D| = 1$ and $|B(D)| = n - 1$, which means that $\Delta = n - 1$.

(b) If $\Delta = n - 2$, by Proposition 6 and (a) we have $n - 2 - \beta \leq \partial_\beta(G) < n - 1 - \beta$. If D is a β-differential set such that $|D| \geq 2$, then $n - 2 - \beta \leq |B(D)| - \beta|D| \leq |B(D)| - 2\beta$, consequently, $n - 2 - \beta \leq |B(D)|$. Since $\beta > 0$, we have $n - 1 \leq |B(D)|$, which is a contradiction. If D is a β-differential set such that $|D| = 1$, by $n - 2 - \beta \leq |B(D)| - \beta|D|$ and (a) we obtain $|B(D)| = n - 2$, so $\partial_\beta(G) = n - (2 + \beta)$. Now, if $\partial_\beta(G) = n - 2 - \beta$, there exists a β-differential set D such that $n - 2 - \beta = |B(D)| - \beta|D| \leq |B(D)| - \beta$, therefore $|B(D)| \geq n - 2$. By (a) we know that $|B(D)| \neq n - 1$, then $|B(D)| = n - 2$ and, using again that $|B(D)| - \beta|D| = n - 2 - \beta$, we conclude that $|D| = 1$, which means that $\Delta = n - 2$.

(c) If $\partial_\beta(G) = n - (3 + \beta)$, there exists a β-differential set D such that $n - (3 + \beta) = |B(D)| - \beta|D| \leq |B(D)| - \beta$, then $|B(D)| \geq n - 3$. By (a) we know that $|B(D)| \neq n - 1$. If $|B(D)| = n - 2$, then we have $|D| \leq 2$ and $|D| = 1 + \frac{1}{\beta}$, which is a contradiction with the fact that $\beta > 1$. Therefore, $|B(D)| = n - 3$ and, consequently, $|D| = 1$, which means that $\Delta = n - 3$. Finally, if $\Delta = n - 3$ and D is a β-differential set, by Proposition 6 we have $n - 3 - \beta \leq |B(D)| - \beta|D| \leq |B(D)| - \beta$,

then $|B(D)| = n - 2$ or $|B(D)| = n - 3$. If $|B(D)| = n - 2$, since $\Delta = n - 3$, we have $|D| = 2$ and $-1 - \beta \leq -\beta|D| = -2\beta$, a contradiction. If $|B(D)| = n - 3$, since $n - 3 - \beta \leq |B(D)| - \beta|D| = n - 3 - \beta|D|$, we have $|D| = 1$ and $\partial_\beta(G) = n - 3 - \beta$.

$\square$

Let us note that, if we consider the path P_5 with five vertices and $\beta < 1$, then we have $\Delta = n - 3$ but $\partial_\beta(P_5) = n - 2 - 2\beta \neq n - (3 + \beta)$.

To characterize the graphs G such that $\partial_\beta(G) = \Delta - \beta$ is much more difficult. A characterization of these graphs when $\beta = 1$ was given in [5], but only for trees. Next we will give some properties that the graphs verifying that equality must satisfy.

Proposition 8. *Let $G = (V, E)$ be a graph with maximum degree Δ. If $\partial_\beta(G) = \Delta - \beta$ and $v \in V$ is such that $\delta(v) = \Delta$, then:*

(a) $\Delta(G[V \setminus N[v]]) \leq \beta$.
(b) $\delta_{\overline{N[v]}}(u) \leq \beta + 1$ *for every $u \in N(v)$.*
(c) $|N_{\overline{N[v]}}(A)| + |N_{N(v) \setminus A}(A)| \leq \Delta - 1 + \beta(|A| - 1)$ *for every $A \subseteq N(v)$.*

Proof 13. We suppose that $\partial_\beta(G) = \Delta - \beta$ and we take any vertex $v \in V$ such that $\delta(v) = \Delta$. If $\Delta(G[V \setminus N[v]]) > \beta$, there exist $\{u, u_1, \ldots, u_j\} \subseteq V \setminus N[v]$ such that $u \sim u_i$ for every $i \in \{1, \ldots, j\}$ and $j > \beta$. In such a case, $\partial_\beta(\{v, u\}) = \Delta + j - 2\beta > \Delta - \beta$, a contradiction. If there exists $u \in N(v)$ such that $\delta_{\overline{N[v]}}(u) > \beta + 1$, then $\partial_\beta(\{v, u\}) > \Delta - 1 + \beta + 1 - 2\beta = \Delta - \beta$, a contradiction. If there exists $A \subseteq N(v)$ such that $|N_{\overline{N[v]}}(A)| + |N_{N(v) \setminus A}(A)| > \Delta - 1 + \beta(|A| - 1)$, then $\partial_\beta(A) = |N_{\overline{N[v]}}(A)| + |N_{N(v) \setminus A}(A)| + 1 - \beta|A| > \Delta - \beta$, a contradiction. $\square$

Note that conditions (a)–(c) in the above Proposition are not enough to guarantee that $\partial_\beta(G) = \Delta - \beta$. The graph G shows in Figure 7 satisfies these three conditions but $\partial_2(G) = 4 > \Delta - 2$.

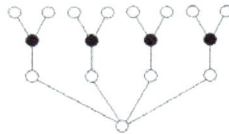

Figure 7. An example with $\beta = 2$ and $\partial_2(G) = 4$.

Proposition 9. *Let $G = (V, E)$ be a graph with order n and maximum degree Δ. If $\beta \in (0, 1)$ and $\partial_\beta(G) = \Delta - \beta$, then $n \leq 2\Delta + 1$. Moreover, if $\beta < \frac{1}{\Delta - 1}$, then $n \leq 2\Delta$.*

Proof 14. The first statement is directly obtained by Proposition 8. Assume that $\beta < \frac{1}{\Delta - 1}$ and $n = 2\Delta + 1$. If $\delta(v) = \Delta$, then $\partial_\beta(N(v)) = \Delta + 1 - \beta\Delta > \Delta - \beta$, a contradiction. $\square$

Another lower and upper bound is shown in the following lemma.

Lemma 5. *Let $G = (V, E)$ be a graph with order n. Then*

$$n - (1 + \beta)\gamma(G) \leq \partial_\beta(G) \leq n - \gamma(G) - \beta.$$

Proof 15. For any set of vertices D it is known that $|B(D)| \leq n - \gamma(G)$. Therefore, for any β-differential set D we have

$$\partial_\beta(G) = |B(D)| - \beta|D| \leq n - \gamma(G) - \beta|D| \leq n - \gamma(G) - \beta.$$

Finally, if D is a minimum dominating set, then $n - (1 + \beta)\gamma(G) = \partial_\beta(D) \leq \partial_\beta(G)$. $\square$

Next, we will characterize all trees attaining the upper bound given in this lemma. For that, we will need the following result. We recall that a *wounded spider* is a graph that results by subdividing at most $m - 1$ edges of the complete bipartite graph $K_{1,m}$.

Lemma 6 ([20]). *If $G = (V, E)$ is a tree, then $\gamma(G) = n - \Delta$ if only if G is a wounded spider.*

Theorem 2. *If G is a tree of order n, then $\partial_\beta(G) = n - \gamma(G) - \beta$ if only if G is a wounded spider.*

Proof 16. Assume that $\partial_\beta(G) = n - \gamma(G) - \beta$, and let D be a β-differential set of G. Since $|B(D)| - \beta|D| = n - \gamma(G) - \beta$ and we know that $|B(D)| \leq n - \gamma(G)$, we deduce that $D = \{v\}$ for some $v \in V$, and $\delta(v) = n - \gamma(G)$. Therefore, $\delta(v) = \Delta$ and, by Lemma 6 we have that G is a wounded spider. If G is a wounded spider, again by Lemma 6 we have that $\Delta = n - \gamma(G)$, so $\partial_\beta(G) \geq \Delta - \beta \geq n - \gamma(G) - \beta$. Finally, using Lemma 5 we conclude that $\partial_\beta(G) = n - \gamma(G) - \beta$. $\square$

Proposition 10. *If $G = (V, E)$ is a graph with minimum degree δ. Then,*

(a) *if $\beta \in (0, \delta - 1)$, then $\partial_\beta(G) \geq \rho^o(G)(\delta - \beta - 1)$,*
(b) *if $\beta \in (0, \delta)$, then $\partial_\beta(G) \geq \rho(G)(\delta - \beta)$.*

Proof 17. (a) Let S be a maximum open packing in G. If $u \in S$ then $\delta_{\overline{[S]}}(u) \geq \delta - 1$, and so $\partial_\beta(S) \geq |S|(\delta - 1) - \beta|S| = \rho^o(G)(\delta - \beta - 1)$. The proof of (b) is analogous. $\square$

Proposition 11. *Let $G = (V, E)$ be a graph with maximum degree Δ. If $\beta \in [1, \Delta)$, then $\partial_{\beta-1}(G) - \gamma(G) \leq \partial_\beta(G) \leq \partial_{\beta-1}(G) - 1$.*

Proof 18. On one hand

$$
\begin{aligned}
\partial_{\beta-1}(G) &= \max\{|B(D)| - \beta|D| + |D| : D \subseteq V\} \\
&\geq \max\{|B(D)| - \beta|D| + 1 : D \subseteq V\} \\
&= \max\{|B(D)| - \beta|D| : D \subseteq V\} + 1 = \partial_\beta(G) + 1.
\end{aligned}
$$

On the other hand, if D is a $(\beta - 1)$-differential set, then $|D| \leq \gamma(G)$ and

$$
\partial_{\beta-1}(G) = |B(D)| - \beta|D| + |D| \leq |B(D)| - \beta|D| + \gamma(G) \leq \partial_\beta(G) + \gamma(G).
$$

$\square$

Proposition 12. *Let K_n, P_n and C_n be the complete, path and cycle graph of order n and let $S_{n,m}$ and $K_{n,m}$ be the double star and the bipartite complete graph of orders $n + m + 2$ and $n + m$ respectively. Then*

$$\partial_\beta(K_n) = \partial_\beta(W_n) = n - 1 - \beta.$$
$$\partial_\beta(P_n) = \partial_\beta(C_n) = \begin{cases} \lfloor \frac{n}{3} \rfloor (2 - \beta) + 1 - \beta & \text{if } \beta \in (0,1) \text{ and } n \equiv 2 \text{ (mod 3)} \\ \lfloor \frac{n}{3} \rfloor (2 - \beta) & \text{otherwise.} \end{cases}$$

If $m \geq n$

$$\partial_\beta(K_{n,m}) = \begin{cases} m + n - 2(1 + \beta) & \text{if } 0 < \beta < n - 2 \\ m - \beta & \text{if } \beta \geq n - 2. \end{cases}$$
$$\partial_\beta(S_{n,m}) = \begin{cases} m + n - 2\beta & \text{if } 0 < \beta < n - 1 \\ m + 1 - \beta & \text{if } \beta \geq n - 1. \end{cases}$$

Proof 19. $\partial_\beta(K_n) = \partial_\beta(W_n) = n - 1 - \beta$ follows immediately from Proposition 6. Let $V(P_n) = V(C_n) = \{u_1, \ldots, u_n\}$ with $n = 3k$ or $n = 3k + 1$. Let $D = \{u_2, u_5, \ldots, u_{3\lfloor \frac{n}{3} \rfloor - 1}\}$ then

$\partial_\beta(D) = \lfloor \frac{n}{3} \rfloor (2 - \beta)$. Since any other set has β-differential less than or equal to $\partial_\beta(D)$, then $\partial_\beta(P_n) = \partial_\beta(C_n) = \lfloor \frac{n}{3} \rfloor (2 - \beta)$. Similarly, we can check the other cases. $\square$

Lemma 7. *Let* $G = (V, E)$ *be a graph. If* D *is a minimum (respectively, maximum) β-differential set of* G, *then* $|B(D)| \geq (\lfloor \beta \rfloor + 1) |D|$ *(respectively,* $|B(D)| \geq \lceil \beta \rceil |D|$*).*

Proof 20. If D is a minimum β-differential set, then for every $v \in D$, the number k of vertices in $B(D)$ which are adjacent to v but not to any $w \in D \setminus \{v\}$, that means that they are private neighbors of v with respect to D, must satisfy $k > \beta$, and so $k \geq \lfloor \beta \rfloor + 1$. If we consider the same situation when D is a maximum β-differential set, it must be satisfied $k \geq \beta$, that is, $k \geq \lceil \beta \rceil$. $\square$

Observe that $\lfloor \beta \rfloor + 1 = \lceil \beta \rceil$ when $\beta \notin \mathbb{N}$.

Proposition 13. *Let* $G = (V, E)$ *be a graph. If* D *is a β-differential set of* G, *then* $(\lceil \beta \rceil - \beta) |D| \leq \partial_\beta(G)$. *Moreover, if* D *is a minimum β-differential set, then* $(\lfloor \beta \rfloor - \beta + 1) |D| \leq \partial_\beta(G)$.

Proof 21. It is enough to prove the first statement for a maximum β-differential set. By Lemma 7 we have $\lceil \beta \rceil |D| \leq |B(D)|$, so $|D| (\lceil \beta \rceil - \beta) \leq \partial_\beta(G)$. If D is a minimum β-differential set, Again by Lemma 7 we have $(\lfloor \beta \rfloor + 1) |D| \leq |B(D)|$, so $(\lfloor \beta \rfloor - \beta + 1) |D| \leq \partial_\beta(G)$. $\square$

Theorem 3. *Let* $G = (V, E)$ *be a graph with maximum degree* Δ.

(i) *If* $\beta \in (0, 1]$, *then* $\frac{(\Delta - \beta)\partial(G)}{\Delta - 1} \leq \partial_\beta(G)$.
(ii) *If* $\beta \in (1, \Delta)$, *then* $\frac{(\lfloor \beta \rfloor - \beta + 1)\partial(G)}{\lfloor \beta \rfloor} \leq \partial_\beta(G)$.

Proof 22. (i) Let D be a 1-differential set of G. Since $\partial(G) \leq (\Delta - 1)|D|$, we have

$$
\begin{aligned}
\partial_\beta(G) &\geq |B(D)| - \beta |D| = |B(D)| - |D| + (1 - \beta)|D| = \partial(G) + (1 - \beta)|D| \\
&\geq \partial(G) + \frac{(1 - \beta)\partial(G)}{\Delta - 1} = \frac{(\Delta - \beta)\partial(G)}{\Delta - 1}.
\end{aligned}
$$

(ii) Let D be a 1-differential set of G. Since $1 - \beta < 0$, by Proposition 13 we have

$$
\begin{aligned}
\partial_\beta(G) &\geq |B(D)| - \beta |D| = |B(D)| - |D| + (1 - \beta)|D| = \partial(G) + (1 - \beta)|D| \\
&\geq \partial(G) + \frac{(1 - \beta)\partial_\beta(G)}{(\lfloor \beta \rfloor - \beta + 1)},
\end{aligned}
$$

then $\left(1 - \frac{(1 - \beta)}{(\lfloor \beta \rfloor - \beta + 1)}\right) \partial_\beta(G) \geq \partial(G)$ or, equivalently $\partial_\beta(G) \geq \frac{(\lfloor \beta \rfloor - \beta + 1)\partial(G)}{\lfloor \beta \rfloor}$. $\square$

Theorem 4. *Let* $G = (V, E)$ *be a graph of order n and minimum degree δ, and let* $\beta \in (0, \delta)$. *Then,*

$$
\partial_\beta(G) \geq \left(\frac{\lfloor \beta \rfloor - \beta + 1}{\beta \lfloor \beta \rfloor + 2 \lfloor \beta \rfloor + 2} \right) n.
$$

Proof 23. Let D be a minimum β-differential set. Since $\beta < \delta$ we have that every vertex in $C(D)$ has at least one neighbor in $B(D)$, that is, $B(D)$ is a dominating set. On one hand, since $\partial_\beta(G) \geq$

$|B(B(D))| - \beta|B(D)| = |D| + |C(D)| - \beta|B(D)|$, we have $|C(D)| \le \partial_\beta(G) + \beta|B(D)| - |D| = (1 + \beta)\partial_\beta(G) + (\beta^2 - 1)|D|$. Now, using that, by Proposition 13, $(\lfloor \beta \rfloor - \beta + 1)|D| \le \partial_\beta(G)$, we have

$$
\begin{aligned}
n &= |D| + |B(D)| + |C(D)| \le (\beta + 1)\partial_\beta(G) + |B(D)| + \beta^2|D| \\
&= (\beta + 2)\partial_\beta(G) + (\beta^2 + \beta)|D| \le (\beta + 2)\partial_\beta(G) + \left(\frac{\beta^2 + \beta}{\lfloor \beta \rfloor - \beta + 1} \right)\partial_\beta(G) \\
&= \left(\frac{(\beta + 2)(\lfloor \beta \rfloor - \beta + 1) + \beta^2 + \beta}{\lfloor \beta \rfloor - \beta + 1} \right)\partial_\beta(G) = \left(\frac{\beta\lfloor \beta \rfloor + 2\lfloor \beta \rfloor + 2}{\lfloor \beta \rfloor - \beta + 1} \right)\partial_\beta(G).
\end{aligned}
$$

$\square$

Note that (i) in Theorem 3 is attained for any graph with order n and maximum degree $\Delta = n - 1$. On the other hand, (ii) is attained in any double star, like the one shown in Figure 8, when $\beta \in \mathbb{N}$ and $r = s = 1 + \beta$.

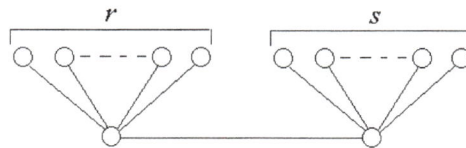

Figure 8. This graph show that the bound (ii) in Theorem 3 is attained when $\beta \in \mathbb{N}$ and $r = s = 1 + \beta$.

On the other hand, notice that, if $\beta \ge \delta$ it is not possible to give a bound similar to the one given in Theorem 4. For instance, it fails for the graph shown in Figure 9 with $\delta = 2, \beta = 3$ and $k \ge 29$, where C_k represents a cycle of k vertices, we have $n = 6 + k$ and $\partial_3(G) = 2$.

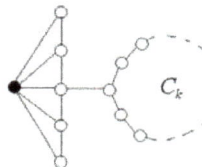

Figure 9. This graph show that Theorem 4 can fail when $\beta > \delta$.

Theorem 5. *Let $G = (V, E)$ be a graph of order n, size m and maximum degree Δ. Then*

$$
\partial_\beta(G) \ge \frac{(2m - n\lfloor \beta \rfloor)(\lfloor \beta \rfloor - \beta + 1)}{\Delta(\lfloor \beta \rfloor + 2) + 1}.
$$

Proof 24. We note that if D is a minimum β-differential set of G, then the following properties hold:

(1) $|D| \le \dfrac{\partial_\beta(G)}{\lfloor \beta \rfloor - \beta + 1}$.

(2) If $v \in B(D)$, then $\delta_{C(D)}(v) \le \lfloor \beta \rfloor + 1$.

(3) If $v \in C(D)$, then $\delta_{C(D)}(v) \le \lfloor \beta \rfloor$.

Let r be the number of edges from $B(D)$ to $C(D)$. Then from (3) and (2) we have

$$
\left(\sum_{u \in C(D)} \delta(u) \right) - |C(D)|\lfloor \beta \rfloor \le r \le |B(D)|(\lfloor \beta \rfloor + 1).
$$

Therefore,

$$
\begin{aligned}
2m \;&\leq\; |D|\Delta + |B(D)|\Delta + \sum_{u \in C(D)} \delta(u) \leq |D|\Delta + |B(D)|\Delta + |B(D)|(\lfloor \beta \rfloor + 1) + |C(D)|\lfloor \beta \rfloor \\
&=\; |D|\Delta + |B(D)|(\Delta + 1) + (n - |D|)\lfloor \beta \rfloor \\
&=\; |D|(\Delta - \lfloor \beta \rfloor) + \partial_\beta(G)(\Delta + 1) + \beta |D|(\Delta + 1) + n\lfloor \beta \rfloor \\
&=\; (\Delta - \lfloor \beta \rfloor + \beta(\Delta + 1))|D| + \partial_\beta(G)(\Delta + 1) + n\lfloor \beta \rfloor \\
&\leq\; \left(\frac{\Delta - \lfloor \beta \rfloor + \beta(\Delta + 1)}{\lfloor \beta \rfloor - \beta + 1} \right) \partial_\beta(G) + \partial_\beta(G)(\Delta + 1) + n\lfloor \beta \rfloor \\
&=\; \left(\frac{\Delta - \lfloor \beta \rfloor + \beta(\Delta + 1) + (\Delta + 1)(\lfloor \beta \rfloor - \beta + 1)}{\lfloor \beta \rfloor - \beta + 1} \right) \partial_\beta(G) + n\lfloor \beta \rfloor \\
&=\; \left(\frac{\Delta(\lfloor \beta \rfloor + 2) + 1}{\lfloor \beta \rfloor - \beta + 1} \right) \partial_\beta(G) + n\lfloor \beta \rfloor.
\end{aligned}
$$

In consequence, $\partial_\beta(G) \geq \frac{(2m - n\lfloor \beta \rfloor)(\lfloor \beta \rfloor - \beta + 1)}{\Delta(\lfloor \beta \rfloor + 2) + 1}$. $\quad\square$

We present now a technical lemma which will be used in the proof of Theorem 6.

Lemma 8. *If $\beta \in (0, \delta)$, then*

$$
\left(\frac{\lceil \beta \rceil (1 + \lfloor \beta \rfloor)}{\delta - \lfloor \beta \rfloor} + 2 + \lfloor \beta \rfloor \right) \left(\frac{\Delta - \beta}{1 + \lfloor \beta \rfloor - \beta} + 1 \right) > 2 + \Delta(2 + \lfloor \beta \rfloor).
$$

Proof 25. We write $\beta = k + \frac{\alpha}{10}$, where $k \in \mathbb{N}$ and $\alpha \in [0, 10)$, so the inequality is

$$
\left(\frac{(k + 1)^2}{\delta - k} + 2 + k \right) \left(\frac{\Delta - k - \frac{\alpha}{10}}{1 - \frac{\alpha}{10}} + 1 \right) > 2 + \Delta(2 + k)
$$

or, equivalently,

$$
\left(\frac{(k + 1)^2 + (2 + k)(\delta - k)}{\delta - k} \right) \left(\frac{10\Delta - 10k - 2\alpha + 10}{10 - \alpha} \right) > 2 + \Delta(2 + k).
$$

Since $h_1(\delta) := \frac{(k+1)^2 + (2+k)(\delta - k)}{\delta - k}$ is decreasing in δ and $h_2(\alpha) = \frac{10\Delta - 10k - 2\alpha + 10}{10 - \alpha}$ is increasing in α, we have

$$
\begin{aligned}
&\left(\frac{(k + 1)^2 + (2 + k)(\delta - k)}{\delta - k} \right) \left(\frac{10\Delta - 10k - 2\alpha + 10}{10 - \alpha} \right) \\
&\qquad \geq \left(\frac{(k + 1)^2 + (2 + k)(\Delta - k)}{\Delta - k} \right) \left(\frac{10\Delta - 10k + 10}{10} \right) \\
&\qquad = \left(\frac{(k + 1)^2 + (2 + k)(\Delta - k)}{\Delta - k} \right) (\Delta - k + 1) \\
&\qquad = (k + 1)^2 + (2 + k)(\Delta - k) + \frac{(k + 1)^2}{\Delta - k} + (2 + k) \\
&\qquad = k(k + 2) + 1 + (2 + k)\Delta - (2 + k)k + \frac{(k + 1)^2}{\Delta - k} + (2 + k) \\
&\qquad = 1 + \Delta(2 + k) + \frac{(k + 1)^2}{\Delta - k} + (2 + k) > 2 + \Delta(2 + k).
\end{aligned}
$$

$\square$

Theorem 6. *Let* $G = (V, E)$ *be a graph of order* n, *minimum degree* δ *and maximum degree* Δ. *Let* $\beta < \delta$ *and* $h(k) := \left(\frac{\lceil \beta \rceil (1 + \lfloor \beta \rfloor)}{\delta - \lfloor \beta \rfloor} + 2 + \lfloor \beta \rfloor \right) \left(\frac{\Delta - \beta}{1 + \lfloor \beta \rfloor - \beta} + k + 1 \right)$, *where* $k \in \mathbb{N}$. *If* $n \geq h(k)$, *then*

$$\partial_\beta(G) \geq \Delta - \beta + (k+1)(\lfloor \beta \rfloor - \beta + 1).$$

Proof 26. We proceed by induction on k. For $k = 0$ we suppose that $n \geq h(0)$ and take $v \in V$ such that $\delta(v) = \Delta$. If there exists $u \in B(\{v\})$ with $\delta_{C(\{v\})}(u) \geq \lfloor \beta \rfloor + 2$, then for $D = \{v, u\}$ we obtain $\partial_\beta(D) \geq \Delta - 1 + \lfloor \beta \rfloor + 2 - 2\beta = \Delta - \beta + (\lfloor \beta \rfloor - \beta + 1)$. Therefore, we can assume $\delta_{C(\{v\})}(u) \leq \lfloor \beta \rfloor + 1$ for every $u \in B(\{v\})$. Note that if there exists a $w \in C(\{v\})$ such that $N(w) \cap B(\{v\}) = \varnothing$, then

$$\partial(\{v, w\}) \geq \Delta + \delta - 2\beta = \Delta - \beta + \delta - \beta \geq \Delta - \beta + (\lfloor \beta \rfloor - \beta + 1),$$

because $\delta \geq \lfloor \beta \rfloor + 1$. If we assume that $N(w) \cap B(\{v\}) \neq \varnothing$ for every $w \in C(\{v\})$, and hence $|C(\{v\})| \leq \Delta(1 + \lfloor \beta \rfloor)$. From Lemma 8 it follows

$$n = 1 + \Delta + |C(\{v\})| \leq 1 + \Delta(2 + \lfloor \beta \rfloor) < h(0),$$

contradicting the hypothesis. Now, we suppose that the theorem is true for k and $n \geq h(k+1)$. Let $\mathcal{M}$ be the collection of all β-differential sets of G such that every $D \in \mathcal{M}$ satisfies that every vertex $v \in D$ has at least $\lfloor \beta \rfloor + 1$ external private neighbors with respect to D. That is, $|\text{epn}[v, D]| \geq \lfloor \beta \rfloor + 1$. Let $D' \in \mathcal{M}$ with maximum cardinality. Since $n \geq h(k+1) \geq k$, by induction hypothesis we know that $\partial_\beta(D') \geq \Delta - \beta + (k+1)(\lfloor \beta \rfloor - \beta + 1)$. Moreover, as $|B(D')| \geq (\lfloor \beta \rfloor + 1)|D'|$, we also have $\partial_\beta(G) \geq (\lfloor \beta \rfloor - \beta + 1)|D'|$.

If there exists $w \in C(D')$ such that $\delta_{C(D')}(w) > \lfloor \beta \rfloor$, then we have

$$\partial(D' \cup \{w\}) \geq \Delta - \beta + (k+1)(\lfloor \beta \rfloor - \beta + 1) + \lfloor \beta \rfloor - \beta + 1 = \Delta - \beta + (k+2)(\lfloor \beta \rfloor - \beta + 1)$$

and we are done. Therefore, we suppose that for every $w \in C(D')$ it is satisfied $\delta_{C(D')}(w) \leq \lfloor \beta \rfloor$. If m' is the number of edges in $G[C(D')]$, then

$$m' \leq \frac{(n - |D'| - |B(D')|)\lfloor \beta \rfloor}{2}.$$

We suppose that there exists $v \in D'$ and $u \in B(\{v\})$ such that $\delta_{C(D')}(u) \geq 1 + \beta$. If $|\text{epn}[v, D']| = \lfloor \beta \rfloor + 1$, then $D'' = (D' \setminus \{v\}) \cup \{u\}$ gives a β-differential bigger than $\partial_\beta(D')$, which is impossible. If $|\text{epn}[v, D']| > \lfloor \beta \rfloor + 1$, then $D'' = D' \cup \{u\} \in \mathcal{M}$ contradicting the choice of D'. Thus, we can assume that $\delta_{C(D')}(u) < 1 + \beta$ for any $u \in B(\{v\})$ and $v \in D'$, that is, $\delta_{C(D')}(u) \leq \lceil \beta \rceil$ for any $u \in B(\{v\})$ and $v \in D'$.

Let r be the number of edges between $B(D')$ and $C(D')$. Then $r \leq \lceil \beta \rceil |B(D')| = \lceil \beta \rceil (\partial_\beta(G) + \beta|D'|)$. Hence,

$$m' \geq \frac{(n - |D'| - |B(D')|)\delta - r}{2} \geq \frac{(n - |D'| - |B(D')|)\delta - \lceil \beta \rceil (\partial_\beta(G) + \beta|D'|)}{2},$$

consequently,

$$\frac{(n - |D'| - |B(D')|)\delta - \lceil \beta \rceil (\partial_\beta(G) + \beta|D'|)}{2} \leq \frac{(n - |D'| - |B(D')|)\lfloor \beta \rfloor}{2}$$

or, equivalently,

$$n \leq \frac{\lceil \beta \rceil}{\delta - \lfloor \beta \rfloor}(\partial_\beta(G) + \beta|D'|) + \partial_\beta(G) + (\beta + 1)|D'|.$$

Finally, using that $|D'| \leq \frac{\partial_\beta(G)}{\lfloor \beta \rfloor - \beta + 1}$, we obtain

$$
\begin{aligned}
n &\leq \frac{\lceil \beta \rceil}{\delta - \lfloor \beta \rfloor}(\partial_\beta(G) + \beta|D'|) + \partial_\beta(G) + (\beta+1)|D'| \\
&\leq \frac{\lceil \beta \rceil}{\delta - \lfloor \beta \rfloor}\left(\partial_\beta(G) + \frac{\beta \partial_\beta(G)}{\lfloor \beta \rfloor - \beta + 1}\right) + \partial_\beta(G) + \frac{(\beta+1)\partial_\beta(G)}{\lfloor \beta \rfloor - \beta + 1} \\
&= \frac{\lceil \beta \rceil}{\delta - \lfloor \beta \rfloor}\left(\frac{(1 + \lfloor \beta \rfloor)\partial_\beta(G)}{\lfloor \beta \rfloor - \beta + 1}\right) + \frac{(2 + \lfloor \beta \rfloor)\partial_\beta(G)}{\lfloor \beta \rfloor - \beta + 1} \\
&= \left(\frac{\lceil \beta \rceil(1 + \lfloor \beta \rfloor)}{\delta - \lfloor \beta \rfloor} + 2 + \lfloor \beta \rfloor\right)\frac{\partial_\beta(G)}{\lfloor \beta \rfloor - \beta + 1}
\end{aligned}
$$

and, as

$$
\left(\frac{\lceil \beta \rceil(1 + \lfloor \beta \rfloor)}{\delta - \lfloor \beta \rfloor} + 2 + \lfloor \beta \rfloor\right)\left(\frac{\Delta - \beta + (k+2)(\lfloor \beta \rfloor - \beta + 1)}{\lfloor \beta \rfloor - \beta + 1}\right) = h(k+1) \leq n,
$$

we conclude that $\partial_\beta(G) \geq \Delta - \beta + (k+2)(\lfloor \beta \rfloor - \beta + 1)$. $\square$

Acknowledgments: We thank both referees for their valuable and constructive comments. This paper was supported in part by two grants from AEI and FEDER (MTM2016-78227-C2-1-P and MTM2015-69323-REDT), and a grant from CONACYT (FOMIX-CONACyT-UAGro 249818), Mexico. First author's work was also partially supported by Plan Nacional I+D+I Grant MTM2015-70531 and Junta de Andalucía FQM-260.

Author Contributions: These authors contributed to equally to this work.

Conflicts of Interest: The authors declare no conflict of interest.

References

1. Kempe, D.; Kleinberg, J.; Tardos, E. Maximizing the spread of influence through a social network. In Proceedings of the Ninth ACM SIGKDD International Conference on Knowledge Discovery and Data Mining, New York, NY, USA, 24–27 August 2003; pp. 137–146, doi:10.1145/956750.956769.

2. Kempe, D.; Kleinberg, J.; Tardos, E. Influential nodes in a diffusion model for social networks. In Proceedings of the 32nd international conference on Automata, Languages and Programming, Lisbon, Portugal, 11–15 July 2005; pp. 1127–1138.

3. Haynes, T.W.; Hedetniemi, S.; Slater, P.J. *Domination in Graphs: Advanced Topics*; Taylor and Francis: Didcot, UK, 1998.

4. Bermudo, S.; Fernau, H. Lower bound on the differential of a graph. *Discret. Math.* **2012**, *312*, 3236–3250, doi:10.1016/j.disc.2012.07.021.

5. Mashburn, J.L.; Haynes, T.W.; Hedetniemi, S.M.; Hedetniemi, S.T.; Slater, P.J. Differentials in graphs. *Util. Math.* **2006**, *69*, 43–54.

6. Basilio, L.A.; Bermudo, S.; Sigarreta, J.M. Bounds on the differential of a graph. *Util. Math.* **2015**, in press.

7. Bermudo, S. On the Differential and Roman domination number of a graph with minimum degree two. *Discret. Appl. Math.* doi:10.1016/j.dam.2017.08.005.

8. Bermudo, S.; De la Torre, L.; Martín-Caraballo, A.M.; Sigarreta, J.M. The differential of the strong product graphs. *Int. J. Comput. Math.* **2015**, *92*, 1124–1134, doi:10.1080/00207160.2014.941359.

9. Bermudo, S.; Fernau, H. Computing the differential of a graph: Hardness, approximability and exact algorithms. *Discret. Appl. Math.* **2014**, *165*, 69–82, doi:10.1016/j.dam.2012.11.013.

10. Bermudo, S.; Fernau, H. Combinatorics for smaller kernels: The differential of a graph. *Theor. Comput. Sci.* **2015**, *562*, 330–345, doi:10.1016/j.tcs.2014.10.007.

11. Bermudo, S.; Fernau, H.; Sigarreta, J.M. The differential and the Roman domination number of a graph. *Appl. Anal. Discret. Math.* **2014**, *8*, 155–171, doi:10.2298/AADM140210003B.

12. Bermudo, S.; Rodríguez, J.M.; Sigarreta, J.M. On the differential in graphs. *Util. Math.* **2015**, *97*, 257–270.

13. Pushpam, P.R.L.; Yokesh, D. Differential in certain classes of graphs. *Tamkang J. Math.* **2010**, *41*, 129–138, doi:10.5556/j.tkjm.41.2010.664.

14. Sigarreta, J.M. Differential in cartesian product graphs. *Ars Comb.* **2016**, *126*, 259–267.

15. Hernández-Gómez, J.C. Differential and operations on graphs. *Int. J. Math. Anal.* **2015**, *9*, 341–349, doi:10.12988/ijma.2015.411344.

16. Goddard, W.; Henning, M.A. Generalised domination and independence in graphs. *Congr. Numer.* **1997**, *123*, 161–171.

17. Zhang, C.Q. Finding critical independent sets and critical vertex subsets are polynomial problems. *SIAM J. Discret. Math.* **1990**, *3*, 431–438, doi:10.1137/0403037.

18. Slater, P.J. Enclaveless sets and MK-systems. *J. Res. Nat. Bur. Stand.* **1977**, *82*, 197–202.

19. Dahme, F.; Rautenbach, D.; Volkmann, L. Some remarks on α-domination. *Discuss. Math. Gr. Theory* **2004**, *24*, 423–430, doi:10.7151/dmgt.1241.

20. Domke, G.S.; Dumbar, J.E.; Markus, L.R. Gallai-type theorems and domination parameters. *Discret. Math.* **1997**, *167–168*, 237–248, doi:10.1016/S0012-365X(97)00231-8.

symmetry

Article

Mathematical Properties on the Hyperbolicity of Interval Graphs

MDPI

Juan Carlos Hernández-Gómez [1,*] , Rosalío Reyes [2], José Manuel Rodríguez [2]
and José María Sigarreta [1]

1 Faculty of Mathematics, Autonomous University of Guerrero, Carlos E. Adame 54, La Garita,
   Acapulco 39650, Mexico; josemariasigarretaalmira@hotmail.com
2 Department of Mathematics, Carlos III University of Madrid, Avenida de la Universidad 30, 28911 Leganés,
   Spain; khanclawn@hotmail.com (R.R.); jomaro@math.uc3m.es (J.M.R.)
* Correspondence: jcarloshg@gmail.com

Received: 3 October 2017; Accepted: 27 October 2017; Published: 1 November 2017

Abstract: Gromov hyperbolicity is an interesting geometric property, and so it is natural to study it in the context of geometric graphs. In particular, we are interested in interval and indifference graphs, which are important classes of intersection and Euclidean graphs, respectively. Interval graphs (with a very weak hypothesis) and indifference graphs are hyperbolic. In this paper, we give a sharp bound for their hyperbolicity constants. The main result in this paper is the study of the hyperbolicity constant of every interval graph with edges of length 1. Moreover, we obtain sharp estimates for the hyperbolicity constant of the complement of any interval graph with edges of length 1.

Keywords: interval graphs; indifference graphs; euclidean graphs; geometric graphs; Gromov hyperbolicity; geodesics

MSC: Primary 05C75; Secondary 05C12

1. Introduction

The focus of the first works on Gromov hyperbolic spaces were finitely generated groups [1]. Initially, the main application of hyperbolic spaces were the automatic groups (see, e.g., [2]). This concept appears also in some algorithmic problems (see [3] and the references therein). Besides, they are useful in the study of secure transmission of information on the internet [4].

In [5], the equivalence of the hyperbolicity of graphs and negatively curved surfaces was proved. The study of hyperbolic graphs is a topic of increasing interest (see, e.g., [4–28] and the references therein).

If $\gamma : [a, b] \to X$ is a continuous curve in the metric space (X, d), γ is a *geodesic* if $L_X(\gamma|_{[t,s]}) = d(\gamma(t), \gamma(s)) = |t - s|$ for every $t, s \in [a, b]$. We say that X is a *geodesic metric space* if, for every $x, y \in X$, there exists a geodesic in X joining them. Let us denote by $[xy]$ any geodesic joining x and y (this notation is very convenient although it is ambiguous, recall that we do not assume uniqueness of geodesics). Consequently, any geodesic metric space is connected.

$G = (V(G), E(G))$ will denote a non-trivial $(V(E) \neq \varnothing)$ simple graph such that we have defined a length function, denoted by L_G or L, on the edges $L_G : E(G) \to \mathbb{R}_+$; the length of a path $\eta = \{e_1, e_2, \ldots, e_k\}$ is defined as $L_G(\eta) = \sum_{i=1}^{k} L_G(e_i)$. We assume that $\ell(G) := \sup\{L_G(e) \mid e \in E(G)\} < \infty$. In order to consider a graph G as a geodesic metric space, identify (by an isometry $\mathcal{I}$) any edge $uv \in E(G)$ with the interval $[0, L_G(uv)]$ in the real line; then, the real interval $[0, L_G(uv)]$ is isometric to the edge uv (considered as a graph with a single edge). If $x, y \in uv$ and η_{xy} denotes the segment contained in uv joining x and y, we define the length of η_{xy} as $L_G(\eta_{xy}) = |\mathcal{I}(x) - \mathcal{I}(y)|$. Thus, the points in G are the vertices $u \in V(G)$ and, in addition, the points in the interior of any edge $uv \in E(G)$.

We denote by d_G or d the natural distance of the graph G. If x, y belong to different connected components of G, then let us define $d_G(x, y) = \infty$. In Section 3, we just consider graphs with every edge of length 1. Otherwise, if a graph G has edges with different lengths, then we also assume that it is locally finite. These properties guarantee that any connected component of G is a geodesic metric space.

If X is a geodesic metric space and $x_1, x_2, x_3 \in X$, the union of three geodesics $[x_1 x_2]$, $[x_2 x_3]$ and $[x_3 x_1]$ is a *geodesic triangle* that will be denoted by $T = \{x_1, x_2, x_3\}$ and we will say that x_1, x_2 and x_3 are the vertices of T; we can also write $T = \{[x_1 x_2], [x_2 x_3], [x_3 x_1]\}$. The triangle T is δ-*thin* if any side of T is contained in the δ-neighborhood of the union of the two other sides. Let us denote by $\delta(T)$ the sharp thin constant of the geodesic triangle T, i.e., $\delta(T) := \inf\{\delta \geq 0 \mid T \text{ is } \delta\text{-thin}\}$. We say that the space X is δ-*hyperbolic* if every geodesic triangle in X is δ-thin. Let us define:

$$\delta(X) := \sup\{\delta(T) \mid T \text{ is a geodesic triangle in } X\}.$$

The geodesic metric space X is *hyperbolic* if it is δ-hyperbolic for some $\delta \geq 0$; then, X is hyperbolic if and only if $\delta(X) < \infty$. If Y is the union of geodesic metric spaces $\{Y_i\}_{i \in I}$, we define its hyperbolicity constant by $\delta(Y) := \sup_{i \in I} \delta(Y_i)$, and we say that Y is hyperbolic if $\delta(Y) < \infty$.

To relate hyperbolicity with other properties of graphs is an interesting problem. The papers [6,9,28] prove, respectively, that chordal, k-chordal and edge-chordal graphs are hyperbolic; these results are improved in [23]. In addition, several authors have proved results on hyperbolicity for some particular classes of graphs (see, e.g., [21,29–31]).

A geometric graph is a graph in which the vertices or edges are associated with geometric objects. Two of the main classes of geometric graphs are Euclidean graphs and intersection graphs. A graph is *Euclidean* if the vertices are points in $\mathbb{R}^n$ and the length of each edge connecting two vertices is the Euclidean distance between them (this makes a lot of sense with the cities and roads analogy commonly used to describe graphs). An *intersection graph* is a graph in which each vertex corresponds with a set, and two vertices are connected by an edge if and only if their corresponding sets have non-empty intersection. In this paper, we work with interval graphs (a class of intersection graphs) and indifference graphs (a class of Euclidean graphs).

We say that G is an *interval graph* if it is the intersection graph of a family of intervals in $\mathbb{R}$: there is a vertex for each interval in the family, and an edge joins two vertices if and only if the their corresponding intervals intersect. Usually, we consider that every edge of an interval graph has length 1, but we also consider interval graphs whose edges have different lengths. It is well-known that interval graphs are always chordal graphs [32,33]. The complements of interval graphs also have interesting properties: they are comparability graphs [34], and the comparability relations are the interval orders [32]. The theory of interval graphs was developed focused on its applications by researchers at the RAND Corporation's mathematics department (pp. ix–10, [35]).

An *indifference graph* is an interval graph whose vertices correspond to a set of intervals with length 1, and the length of the corresponding edge to two unit intervals that intersect is the distance between their midpoints. In addition, we can see an indifference graph as an Euclidean graph in $\mathbb{R}$ constructed by taking the vertex set as a subset of $\mathbb{R}$ and two vertices are connected by an edge if and only if they are within one unit from each other. Since it is a Euclidean graph, the length of each edge connecting two vertices is the Euclidean distance between them. Indifference graphs possess several interesting properties: connected indifference graphs have Hamiltonian paths [36]; an indifference graph has a Hamiltonian cycle if and only if it is biconnected [37]. In the same direction, we consider indifference graphs since for these graphs we can remove one of the hypothesis of a main theorem on interval graphs (compare Theorem 8 and Corollary 6).

We would like to mention that Ref. [38] collects very rich results, especially those concerning path properties, about interval graphs and unit interval graphs. It is well-known that interval graphs (with a very weak hypothesis) and indifference graphs are hyperbolic. One of the main results in this paper is Theorem 8, which provides a sharp upper bound of the hyperbolicity constant of interval graphs verifying a very weak hypothesis. This result allows for obtaining bounds for the hyperbolicity constant of every indifference graph (Corollary 6) and the hyperbolicity constant of every interval graph with edges of length 1 (Corollary 7). Moreover, Theorem 10 provides sharp bounds for the hyperbolicity constant of the complement of any interval graph with edges of length 1. Note that it is not usual to obtain such precise bounds for large classes of graphs. The main result in this paper is Theorem 9, which allows for computing the hyperbolicity constant of every interval graph with edges of length 1, by using geometric criteria.

2. Previous Results

We collect some previous results that will be useful along the paper.

A *cycle* is a path with different vertices, unless the last vertex, which is equal to the first one.

Lemma 1. *([39] Lemma 2.1) Let us consider a geodesic metric space X. If every geodesic triangle in X that is a cycle is δ-thin, then X is δ-hyperbolic.*

Corollary 1. *In any geodesic metric space X,*

$$\delta(X) = \sup \left\{ \delta(T) \mid T \text{ is a geodesic triangle that is a cycle} \right\}.$$

Recall that a *chordal graph* is one in which all cycles of four or more vertices have a chord, which is an edge that is not part of the cycle but connects two vertices of the cycle.

If C is a cycle in G and $v \in V(G)$, we denote by $\deg_C(v)$ the degree of the vertex v in the subgraph Γ induced by $V(C)$ (note that Γ could contain edges that are not contained in C, and thus it is possible to have $\deg_C(v) > 2$).

Lemma 2. *([9] Lemma 2.2) Consider a chordal graph G and a cycle C in G with $a, v, b \in C \cap V(G)$ and $av, vb \in E(G)$. If $ab \notin E(G)$, then $\deg_C(v) \geq 3$.*

Corollary 2. *Consider a cycle C in a chordal graph G and v_1, v_2, v_3 consecutive vertices in C. If $\deg_C(v_2) = 2$, then $v_1 v_3 \in E(G)$. Consequently, if C has at least four vertices, then $\deg_C(v_1) \geq 3$ and $\deg_C(v_3) \geq 3$.*

Let $J(G)$ be the set of vertices and midpoints of edges in G. Consider the set $\mathbb{T}_1$ of geodesic triangles T in G that are cycles and such that the three vertices of the triangle T belong to $J(G)$, and denote by $\delta_1(G)$ the infimum of the constants λ such that every triangle in $\mathbb{T}_1$ is λ-thin.

Theorem 1. *([40] Theorem 2.5) For every graph G with edges of length 1, we have $\delta_1(G) = \delta(G)$.*

The next result will narrow the possible values for the hyperbolicity constant.

Theorem 2. *([40] Theorem 2.6) If G is a hyperbolic graph with edges of length 1, then $\delta(G)$ is a multiple of $1/4$.*

Theorem 3. *([40] Theorem 2.7) If G is a hyperbolic graph with edges of length 1, then there exists a geodesic triangle $T \in \mathbb{T}_1$ such that $\delta(T) = \delta(G)$.*

In the following theorems, we study the graphs G with $\delta(G) < 1$.

Theorem 4. *([41] Theorem 11) If G is a graph with edges of length 1 with $\delta(G) < 1$, then we have either $\delta(G) = 0$ or $\delta(G) = 3/4$. Furthermore,*

- $\delta(G) = 0$ *if and only if G is a tree.*
- $\delta(G) = 3/4$ *if and only if G is not a tree and every cycle in G has length 3.*

Corollary 3. *A graph G with edges of length 1 satisfies $\delta(G) \geq 1$ if and only if there exists a cycle in G with length at least 4.*

In order to characterize from a geometric viewpoint the interval graphs with hyperbolicity constant 1, we need the following result, which is a direct consequence of Theorems 2 and 4, and ([7] Theorem 4.14).

Theorem 5. *Let G be any graph with edges of length 1. We have $\delta(G) = 1$ if and only if $\delta(G) \notin \{0, 3/4\}$ and, for every cycle C in G and every $x, y \in C \cap J(G)$, we have $d(x, y) \leq 2$.*

Theorems 4 and 5 have the following consequences.

Corollary 4. *Let G be any graph with edges of length 1. We have $\delta(G) \leq 1$ if and only if, for every cycle C in G and every $x, y \in C \cap J(G)$, we have $d(x, y) \leq 2$.*

By Theorems 2 and 4, and ([7] Theorems 4.14 and 4.21), we have the following result.

Theorem 6. *Let G be any graph with edges of length 1. If there exists a cycle in G with $p, q \in V(G)$ and $d(p, q) \geq 3$, then $\delta(G) \geq 3/2$.*

We will also need this last result.

Theorem 7. *([41] Theorem 30) If G is any graph with edges of length 1 and n vertices, then $\delta(G) \leq n/4$.*

3. Interval Graphs and Hyperbolicity

Given a cycle C in an interval graph G, let $\{v_1, \ldots, v_k\}$ be the vertices in G with

$$C = v_1 v_2 \cup \cdots \cup v_{k-1} v_k \cup v_k v_1.$$

Denote by $\{I_1, \ldots, I_k\}$ the corresponding intervals to $\{v_1, \ldots, v_k\}$. If $I_j = [a_j, b_j]$, then let us define the *minimal interval of C* as the interval $I_{j_1} = [a_{j_1}, b_{j_1}]$ with $a_{j_1} \leq a_j$ for every $1 \leq j \leq k$ and $b_{j_1} > b_j$ if $a_j = a_{j_1}$ with $1 \leq j \leq k$ and $j \neq j_1$, and the *maximal interval of C* as the interval $I_{j_2} = [a_{j_2}, b_{j_2}]$ with $b_{j_2} \geq b_j$ for every $1 \leq j \leq k$ and $a_{j_2} < a_j$ if $b_j = b_{j_2}$ with $1 \leq j \leq k$ and $j \neq j_2$. If $i \in \mathbb{Z} \setminus \{1, 2, \ldots, k\}$, $1 \leq j \leq k$ and $i \equiv j \pmod{k}$, then we define $v_i := v_j$ and $I_i := I_j$.

If H is a subgraph of G and $w \in V(H)$, we denote by $\deg_H(w)$ the degree of the vertex w in the subgraph induced by $V(H)$.

For any graph G,

$$\operatorname{diam} V(G) := \sup \{d_G(v, w) \mid v, w \in V(G)\},$$
$$\operatorname{diam} G := \sup \{d_G(x, y) \mid x, y \in G\},$$

i.e., $\operatorname{diam} V(G)$ is the diameter of the set of vertices of G, and $\operatorname{diam} G$ is the diameter of the whole graph G (recall that in order to have a geodesic metric space, G must contain both the vertices and the points in the interior of any edge of G).

The following result is well-known.

Lemma 3. *For any geodesic triangle T in a graph G, we have $\delta(T) \leq (\operatorname{diam} T)/2 \leq L(T)/4$.*

Corollary 5. *The inequalities*

$$\delta(G) \leq \frac{1}{2} \operatorname{diam} G \leq \frac{1}{2} \left(\operatorname{diam} V(G) + \ell(G) \right)$$

hold for every graph G.

A graph G is *length-proper* if every edge is a geodesic. A large class of length-proper graphs are the graphs with edges of length 1. Another important class of length-proper graphs are the following geometric graphs: consider a discrete set V in an Euclidean space (or in a metric space) where we consider two points connected by an edge if some criterium is satisfied. If we define the length of an edge as the distance between its vertices, then we obtain a length-proper graph.

It is well-known that every interval graph is chordal. Hence, every length-proper interval graph is hyperbolic. The following result is one of the main theorems in this paper, since it provides a sharp inequality for the hyperbolicity constant of any length-proper interval graph. Recall that $\ell(G) := \sup \{L_G(e) \mid e \in E(G)\}$.

Theorem 8. *Every length-proper interval graph G satisfies the sharp inequality*

$$\delta(G) \leq \frac{3}{2} \ell(G).$$

Proof. Consider a geodesic triangle $T = \{x, y, z\}$ that is a cycle in G and $p \in [xy]$. Assume first that T satisfies the following property:

$$\text{if } a, b \in V(G) \cap [xy] \text{ and } ab \in E(G), \text{ then } ab \subseteq [xy]. \tag{1}$$

Consider the consecutive vertices $\{v_1, \ldots, v_k\}$ in the cycle T, and their corresponding intervals $\{I_1, \ldots, I_k\}$. As before, we denote by I_{j_1} and I_{j_2} the minimal and maximal intervals, respectively.

If $k < 4$, then $L(T) \leq 3\ell(G)$ and Lemma 3 gives:

$$d(p, [xz] \cup [zy]) \leq \frac{1}{4} L(T) \leq \frac{3}{4} \ell(G). \tag{2}$$

Assume now that $k \geq 4$.

Case (A). Assume that $p \in V(G)$. Let $a, b \in V(G)$ with $ap, bp \in E(G)$ and $ap \cup bp \subset T$.

Case $(A.1)$. If $ab \notin E(G)$, then Lemma 2 gives $\deg_T(p) \geq 3$, and there exists $q \in V(G) \cap T$ with $pq \in E(G)$ such that pq is not contained in T. By (1), $q \in [xz] \cup [zy]$ and so:

$$d(p, [xz] \cup [zy]) \leq d(p, q) = L(pq) \leq \ell(G). \tag{3}$$

Case $(A.2)$. If $ab \in E(G)$, then ab is not contained in T, since T is a cycle and $k \geq 4$. By (1), $\{a, b\}$ is not contained in $[xy]$, and:

$$d(p, [xz] \cup [zy]) \leq \max \{d(p, a), d(p, b)\} = \max \{L(pa), L(pb)\} \leq \ell(G). \tag{4}$$

Case (B). Assume that $p \notin V(G)$. Let $a, b \in V(G)$ with $p \in ab \subset T$ and $d(p, a) \leq L(ab)/2 \leq \ell(G)/2$. Corollary 2 gives that we have $\deg_T(a) \geq 3$ or $\deg_T(b) \geq 3$.

Case $(B.1)$. Assume that $\deg_T(a) \geq 3$.

Case $(B.1.1)$. If $a \notin [xy]$, then:

$$d(p, [xz] \cup [zy]) \leq d(p, a) \leq \frac{1}{2} \ell(G). \tag{5}$$

Case $(B.1.2)$. Assume that $a \in [xy]$. Since $\deg_T(a) \geq 3$, there exists $q \in V(G) \cap T$ with $aq \in E(G)$ such that aq is not contained in T. By (1), $q \in [xz] \cup [zy]$ and so:

$$d(p, [xz] \cup [zy]) \leq d(p, a) + d(a, [xz] \cup [zy]) \leq d(p, a) + d(a, q)$$
$$= d(p, a) + L(aq) \leq \frac{1}{2} \ell(G) + \ell(G) = \frac{3}{2} \ell(G). \tag{6}$$

Case (B.2). Assume that $\deg_T(a) = 2$ and $\deg_T(b) \geq 3$. Let $\alpha \neq b$ with $\alpha \in V(G)$, $\alpha a \in E(G)$ and $\alpha a \subset T$. Corollary 2 gives that we have $\alpha b \in E(G)$. By (1), we have that $\{\alpha, b\}$ is not contained in $[xy]$, and:

$$d(p, [xz] \cup [zy]) \leq \max \{d(p, \alpha), d(p, b)\} \leq \max \{d(p, a) + d(a, \alpha), d(p, b)\}$$

$$\leq \max \left\{ \frac{1}{2}\ell(G) + \ell(G), \ell(G) \right\} = \frac{3}{2}\ell(G). \tag{7}$$

Inequalities (2)–(7) give in every case $d(p, [xz] \cup [zy]) \leq 3\ell(G)/2$.

Consider now a geodesic triangle $T = \{x, y, z\} = \{[xy], [xz], [yz]\}$ that does not satisfy property (1). We are going to obtain a new geodesic γ joining x and y such that the geodesic triangle $T' = \{\gamma, [xz], [yz]\}$ satisfies (1).

Let us define inductively a finite sequence of geodesics $\{g_0, g_1, g_2, \ldots, g_r\}$ joining x and y in the following way:

If $j = 0$, then $g_0 := [xy]$.

Assume that $j \geq 1$. If the geodesic triangle $\{g_{j-1}, [xz], [yz]\}$ satisfies (1), then $r = j - 1$ and the sequence stops. If $\{g_{j-1}, [xz], [yz]\}$ does not satisfy (1), then there exists $a, b \in V(G) \cap [xy]$ such that $ab \in E(G)$ and ab is not contained in $[xy]$. Denote by $[ab]$ the geodesic joining a and b contained in g_{j-1}. Let us define $g_j := (g_{j-1} \setminus [ab]) \cup ab$. Note that $g_j \cap V(G) \subset g_{j-1} \cap V(G)$ and $|g_j \cap V(G)| < |g_{j-1} \cap V(G)|$.

Since $|g_j \cap V(G)| < |g_{j-1} \cap V(G)|$ for any $j \geq 1$, this sequence must finish with some geodesic g_r such that the geodesic triangle $T' := \{g_r, [xz], [yz]\}$ satisfies (1). Thus, define $\gamma := g_r$. Hence,

$$g_r \cap V(G) \subset g_{r-1} \cap V(G) \subset \cdots \subset g_1 \cap V(G) \subset g_0 \cap V(G),$$

and so $\gamma \cap V(G) \subset [xy] \cap V(G)$.

Let us consider $p \in [xy] \subset T$.

If $p \in \gamma \subset T'$, then, by applying the previous argument to the geodesic triangle T', we obtain $d(p, [xz] \cup [zy]) \leq 3\ell(G)/2$. Assume that $p \notin \gamma$.

Since $\gamma \cap V(G) \subset [xy] \cap V(G)$, there exist $v, w \in \gamma \cap V(G)$ with $vw \in E(G)$ such that, if $[vw]$ denotes the geodesic joining v and w contained in $[xy]$, then:

$$p \in [vw], \qquad [vw] \cap vw = \{v, w\}.$$

Since vw and $[vw]$ are geodesics, we have $L(vw) = L([vw])$. Thus, we can define $p' \in \gamma$ as the point in vw with $d(p', v) = d(p, v)$ and $d(p', w) = d(p, w)$. By applying the previous argument to p' and T', we obtain $d(p', [xz] \cup [zy]) \leq 3\ell(G)/2$. Since p' belongs to the edge vw, we have $d(p', [xz] \cup [zy]) = d(p', v) + d(v, [xz] \cup [zy])$ or $d(p', [xz] \cup [zy]) = d(p', w) + d(w, [xz] \cup [zy])$. By symmetry, we can assume that $d(p', [xz] \cup [zy]) = d(p', v) + d(v, [xz] \cup [zy])$. Since $d(p', v) = d(p, v)$, we have:

$$d(p, [xz] \cup [zy]) \leq d(p, v) + d(v, [xz] \cup [zy]) = d(p', v) + d(v, [xz] \cup [zy]) = d(p', [xz] \cup [zy]) \leq \frac{3}{2}\ell(G).$$

Finally, Corollary 1 gives $\delta(G) \leq 3\ell(G)/2$.

Proposition 1 below shows that the inequality is sharp. $\square$

Note that, if we remove the hypothesis $\ell(G) < \infty$, then there are non-hyperbolic length-proper interval graphs: if Γ is any graph such that every cycle in Γ has exactly three vertices and $\sup\{L(C) \mid C$ is a cycle in $\Gamma\} = \infty$, then Γ is a non-hyperbolic chordal graph. Some of these graphs Γ are length-proper interval graphs.

Recall that every indifference graph is an Euclidean graph. Hence, every indifference graph G is a length-proper graph and $\ell(G) \leq 1$.

Theorem 8 has the following direct consequence.

Corollary 6. *Every indifference graph G satisfies the inequality:*

$$\delta(G) \leq \frac{3}{2}\ell(G) \leq \frac{3}{2}.$$

4. Interval Graphs with Edges of Length 1

Along this section, we just consider graphs with edges of length 1. This is a very usual class of graphs. Note that every graph G with edges of length 1 is a length-proper graph with $\ell(G) = 1$.

The goal of this section is to compute the precise value of the hyperbolicity constant of every interval graph with edges of length 1 (see Theorem 9). We wish to emphasize that it is unusual to be able to compute the hyperbolicity constant of every graph in a large class of graphs. Let us start with a direct consequence of Theorem 8.

Corollary 7. *Every interval graph G with edges of length 1 satisfies the inequality:*

$$\delta(G) \leq \frac{3}{2}.$$

First of all, we characterize the interval graphs with edges of length 1 and $\delta(G) = 3/2$ in Proposition 1 below. Furthermore, Proposition 1 shows that the inequality in Theorem 8 is sharp.

Let G be an interval graph. We say that G has the $(3/2)$-*intersection property* if there exists two disjoint intervals I' and I'' corresponding to vertices in a cycle C in G such that there is no corresponding interval I to a vertex in G with $I \cap I' \neq \emptyset$ and $I \cap I'' \neq \emptyset$.

Proposition 1. *An interval graph G with edges of length 1 satisfies $\delta(G) = 3/2$ if and only if G has the $(3/2)$-intersection property.*

Proof. Assume that G has the $(3/2)$-intersection property. Thus, there exist two disjoint corresponding intervals I' and I'' to vertices in a cycle C in G such that there is no corresponding interval I to a vertex in G with $I \cap I' \neq \emptyset$ and $I \cap I'' \neq \emptyset$. If v' and v'' are the corresponding vertices to I' and I'', respectively, then $v', v'' \in C$ and $d(v', v'') \geq 3$. Thus, Theorem 6 gives $\delta(G) \geq 3/2$ and, since $\delta(G) \leq 3/2$ by Corollary 7, we conclude $\delta(G) = 3/2$.

Assume now that G does not have the $(3/2)$-intersection property. Seeking for a contradiction, assume that $\delta(G) = 3/2$. By Theorem 3, there exist a geodesic triangle $T = \{x, y, z\}$ that is a cycle in G and $p \in [xy]$ such that $d(p, [xz] \cup [zy]) = \delta(T) = \delta(G) = 3/2$ and $x, y, z \in J(G)$. Since $d(p, \{x, y\}) \geq d(p, [xz] \cup [zy]) = 3/2$, we have $d(x, y) \geq 3$. Since G does not have the $(3/2)$-intersection property, for each two disjoint corresponding intervals I' and I'' to vertices in the cycle T, there exists a corresponding interval I to a vertex in G with $I \cap I' \neq \emptyset$ and $I \cap I'' \neq \emptyset$. If v' and v'' are the corresponding vertices to I' and I'', respectively, then $v', v'' \in T$ and $d(v', v'') = 2$. We conclude that $\operatorname{diam}(T \cap V(G)) \leq 2$ and $\operatorname{diam} T \leq 3$. Since $d(x, y) \geq 3$ with $x, y \in J(G)$, we have $\operatorname{diam}(T \cap V(G)) = 2$, $\operatorname{diam} T = 3$, $d(x, y) = 3$, $L([xy])/2 = d(p, x) = d(p, y) = d(p, [xz] \cup [zy]) = \delta(T) = \delta(G) = 3/2$ and p is the midpoint of $[xy]$. Thus $x, y \in J(G) \setminus V(G)$ and $p \in V(G)$. If $x \in x_1 x_2 \in E(G)$ and $y \in y_1 y_2 \in E(G)$, then $d(\{x_1, x_2\}, \{y_1, y_2\}) = 2$. Let $I_{x_1}, I_{x_2}, I_{y_1}, I_{y_2}, I_p$ be the corresponding intervals to the vertices x_1, x_2, y_1, y_2, p, respectively. We can assume that $x_1, y_1 \in [xy]$ and thus $I_{x_1} \cap I_p \neq \emptyset$ and $I_{y_1} \cap I_p \neq \emptyset$ since $d(x_1, y_1) = 2$, $I_{x_1} \cap I_{y_1} = \emptyset$. Thus, there exists $\zeta \in I_p \setminus (I_{x_1} \cup I_{y_1})$. Since $[xy] \cap V(G) = \{x_1, p, y_1\}$ and T is a cycle containing x_1, p, y_1, by continuity, there exists a corresponding interval J to a vertex $v \in ([xz] \cup [zy]) \cap V(G)$ with $\zeta \in J$. Thus, $pv \in E(G)$ and $3/2 = d(p, [xz] \cup [zy]) \leq d(p, v) = 1$, which is a contradiction. Hence, $\delta(G) \neq 3/2$. $\square$

Corollary 7 and Theorems 2 and 4 give that $\delta(G) \in \{0, 3/4, 1, 5/4, 3/2\}$ for every interval graph G with edges of length 1. Proposition 1 characterizes the interval graphs with edges of length 1 and

$\delta(G) = 3/2$. In order to characterize the interval graphs with the other values of the hyperbolicity constant, we need some definitions.

Let G be an interval graph.

We say that G has the *0-intersection property* if, for every three corresponding intervals I', I'' and I''' to vertices in G, we have $I' \cap I'' \cap I''' = \emptyset$.

G has the *(3/4)-intersection property* if it does not have the 0-intersection property and for every four corresponding intervals I', I'', I''' and I'''' to vertices in G we have $I' \cap I'' \cap I''' = \emptyset$ or $I' \cap I'' \cap I'''' = \emptyset$.

By a *couple* of intervals in a cycle C of G, we mean the union of two non-disjoint intervals whose corresponding vertices belong to C. We say that G has the *1-intersection property* if it does not have the 0 and (3/4)-intersection properties and, for every cycle C in G, each interval and a couple of corresponding intervals to vertices in C are not disjoint.

One can check that every chordal graph that has a cycle with length of at least four has a cycle with length four and, since this cycle has a chord, it also has a cycle with length three.

Next, we provide a characterization of the interval graphs with hyperbolicity constant 0. It is well-known that these are the caterpillar trees, see [42], but we prefer to characterize them by the 0-intersection property in Proposition 2 below, since it looks similar to the other intersection properties.

Proposition 2. *An interval graph G with edges of length 1 satisfies $\delta(G) = 0$ if and only if G has the 0-intersection property.*

Proof. By Theorem 4, $\delta(G) = 0$ if and only if G is a tree. Since every interval graph is chordal, G is not a tree if and only if it contains a cycle with length 3, and this last condition holds if and only if there exist three corresponding intervals I', I'' and I''' to vertices in G with $I' \cap I'' \cap I''' \neq \emptyset$. Hence, G has a cycle if and only if it does not have the 0-intersection property. □

Proposition 3. *An interval graph G with edges of length 1 satisfies $\delta(G) = 3/4$ if and only if G has the (3/4)-intersection property.*

Proof. By Theorem 4, $\delta(G) = 3/4$ if and only if G is not a tree and every cycle in G has length 3. Proposition 2 gives that G is not a tree if and only if G does not have the 0-intersection property. Therefore, it suffices to show that every cycle in G has length 3, if and only if for every four corresponding intervals I', I'', I''' and I'''' to vertices in G, we have $I' \cap I'' \cap I''' = \emptyset$ or $I' \cap I'' \cap I'''' = \emptyset$.

Since every interval graph is chordal, G has a cycle with length at least 4 if and only if it has a cycle C with length 4 and this cycle has at least a chord.

Assume first that there exists such a cycle C. If I', I'', I''', I'''' are the corresponding intervals to the vertices in C and I', I'' corresponds to vertices with a chord, and then $I' \cap I'' \cap I''' \neq \emptyset$ and $I' \cap I'' \cap I'''' \neq \emptyset$.

Assume now that there are corresponding intervals I', I'', I''', I'''' to the vertices v', v'', v''', v'''' in G with $I' \cap I'' \cap I''' \neq \emptyset$ and $I' \cap I'' \cap I'''' \neq \emptyset$. Thus, $v'v''', v''v'' \in E(G)$ and $v'v''', v''v'''' \in E(G)$, and so $v'v''' \cup v'''v'' \cup v''v'''' \cup v''''v'$ is a cycle in G with length 4. □

Proposition 4. *An interval graph G with edges of length 1 satisfies $\delta(G) = 1$ if and only if G has the 1-intersection property.*

Proof. By Theorem 5, $\delta(G) = 1$ if and only if $\delta(G) \notin \{0, 3/4\}$, and, for every cycle C in G and every $x, y \in C \cap J(G)$, we have $d(x, y) \leq 2$. Propositions 2 and 3 give that $\delta(G) \notin \{0, 3/4\}$ if and only if G does not have the 0 and (3/4)-intersection properties. Therefore, it suffices to show that for every cycle C in G, we have $d(x, y) \leq 2$ for every $x, y \in C \cap J(G)$ if and only if each interval and couple of corresponding intervals to vertices in C are not disjoint.

Fix a cycle C in G. Each interval and couple of corresponding intervals to vertices in C are not disjoint if and only if $d(x, y) \leq 3/2$ for every $x \in C \cap V(G)$ and $y \in C \cap (J(G) \setminus V(G))$. Since every point in $C \cap (J(G) \setminus V(G))$ has a point in $C \cap V(G)$ at distance $1/2$, this last condition is equivalent to $d(x, y) \leq 2$ for every $x, y \in C \cap J(G)$. $\quad \square$

Finally, we collect the previous geometric characterizations in the following theorem. Note that the characterization of $\delta(G) = 5/4$ in Theorem 9 is much simpler than the one in [7]. Recall that to characterize the graphs with hyperbolicity $3/2$ is a very difficult task, as it was shown in ([7] Remark 4.19).

Theorem 9. *Every interval graph G with edges of length 1 is hyperbolic and $\delta(G) \in \{0, 3/4, 1, 5/4, 3/2\}$. Furthermore,*

- $\delta(G) = 0$ *if and only if G has the 0-intersection property.*
- $\delta(G) = 3/4$ *if and only if G has the $(3/4)$-intersection property.*
- $\delta(G) = 1$ *if and only if G has the 1-intersection property.*
- $\delta(G) = 5/4$ *if and only if G does not have the $0, 3/4, 1$ and $(3/2)$-intersection properties.*
- $\delta(G) = 3/2$ *if and only if G has the $(3/2)$-intersection property.*

Complement of Interval Graphs

The *complement* $\overline{G}$ of the graph G is defined as the graph with $V(\overline{G}) = V(G)$ and such that $e \in E(\overline{G})$ if and only if $e \notin E(G)$. Recall that, for every disconnected graph G, we define $\delta(G)$ as the supremum of $\delta(G_i)$, where G_i varies in the set of connected components of G.

We consider that the length of the edges of every complement graph is 1.

If Γ is a subgraph of G, we consider in Γ the *inner metric* obtained by the restriction of the metric in G, that is:

$$d_\Gamma(v, w) := \inf \{ L(\gamma) \mid \gamma \subset \Gamma \text{ is a continuous curve joining } v \text{ and } w \} \geq d_G(v, w).$$

Note that the inner metric d_Γ is the usual metric if we consider the subgraph Γ as a graph.

Since the complements of interval graphs belong to the class of comparability graphs [34], it is natural to also study the hyperbolicity constant of complements of interval graphs. In order to do it, we need some preliminary results and the following technical lemma.

Lemma 4. *Let G be an interval graph with edges of length 1, $V(G) = \{v_1, \ldots, v_r\}$ and corresponding intervals $\{I_1, \ldots, I_r\}$. We have $\operatorname{diam} V(G) = 2$ if and only if there exists an interval I_i with $I_j \cap I_i \neq \varnothing$ for every $1 \leq j \leq r$ and $\operatorname{diam} V(G') \geq 2$, where G' is the corresponding interval graph to $\{I_1, \ldots, I_r\} \setminus I_i$. Furthermore, if this is the case, then $\delta(\overline{G}) = \delta(\overline{G'})$.*

Proof. Assume that $\operatorname{diam} V(G) = 2$. Let $[a_j, b_j] = I_j$ for $1 \leq j \leq r$. Consider integers $1 \leq i_1, i_2 \leq r$ satisfying:

$$b_{i_1} \leq b_j, \qquad a_j \leq a_{i_2}, \qquad \text{for every } 1 \leq j \leq r. \tag{8}$$

Since $\operatorname{diam} V(G) = 2$, we have $b_{i_1} < a_{i_2}$. Thus, $d_G(v_{i_1}, v_{i_2}) = 2$ and there exists i with $v_i v_{i_1}, v_i v_{i_2} \in E(G)$. Hence, $I_{i_1} \cap I_i \neq \varnothing$ and $I_{i_2} \cap I_i \neq \varnothing$. Thus, (8) gives $I_j \cap I_i \neq \varnothing$ for every $1 \leq j \leq r$, and we deduce $d_G(v_j, v_i) \leq 1$ for every $1 \leq j \leq r$.

Seeking for a contradiction assume that $\operatorname{diam} V(G') \leq 1$. Thus, $d_G(v_j, v_{j'}) \leq d_{G'}(v_j, v_{j'}) \leq 1$ for every $1 \leq j, j' \leq r$ with $j, j' \neq i$. Furthermore, we have proved $d_G(v_j, v_i) \leq 1$ for every $1 \leq j \leq r$. Therefore, $d_G(v_j, v_{j'}) \leq 1$ for every $1 \leq j, j' \leq r$ and we conclude $\operatorname{diam} V(G) \leq 1$, which is a contradiction. Hence, $\operatorname{diam} V(G') \geq 2$.

The converse implication is well-known.

Finally, since $v_j v_i \in E(G)$ for every $1 \leq j \leq r$ with $j \neq i$, we have $\overline{G} = \{v_i\} \cup \overline{G'}$ and:

$$\delta(\overline{G}) = \max \{ \delta(\{v_i\}), \delta(\overline{G'}) \} = \max \{ 0, \delta(\overline{G'}) \} = \delta(\overline{G'}).$$

$\square$

Note that it is not usual to obtain such close lower and upper bounds for a large class of graphs. Some inequalities are not difficult to prove; the most difficult cases are the upper bound when diam $V(G) = 2$ (recall that this is the more difficult case in the study of the complement of a graph), and the lower bound when diam $V(G) \geq 4$.

Theorem 10. *Let G be any interval graph.*

- *If* diam $V(G) = 1$, *then* $\delta(\overline{G}) = 0$.
- *If* $2 \leq$ diam $V(G) \leq 3$, *then* $0 \leq \delta(\overline{G}) \leq 2$.
- *If* diam $V(G) \geq 4$, *then* $5/4 \leq \delta(\overline{G}) \leq 3/2$.

Furthermore, the lower bounds on $\delta(\overline{G})$ *are sharp.*

Proof. If diam $V(G) = 1$, then G is a complete graph. Thus, $\overline{G}$ is a union of isolated vertices and $\delta(\overline{G}) = 0$.

Let us prove now the upper bounds.

It is well-known that if diam $V(G) \geq 3$, then $\overline{G}$ is connected and diam $V(\overline{G}) \leq 3$. Therefore, Corollary 5 gives $\delta(\overline{G}) \leq 2$.

If diam $V(G) \geq 4$, then ([43] Theorem 2.14) gives $\delta(\overline{G}) \leq 3/2$.

Assume now that diam $V(G) = 2$. By Lemma 4, there exists an interval graph G' with $|V(G')| = |V(G)| - 1$, diam $V(G') \geq 2$ and $\delta(\overline{G}) = \delta(\overline{G'})$. Let us define inductively a finite sequence of interval graphs $\{G^{(0)}, G^{(1)}, G^{(2)}, \ldots, G^{(k)}\}$ with:

$$\delta(\overline{G^{(0)}}) = \delta(\overline{G^{(1)}}) = \delta(\overline{G^{(2)}}) = \cdots = \delta(\overline{G^{(k)}}),$$

$$|V(G^{(j)})| = |V(G^{(j-1)})| - 1, \qquad \text{for } 0 < j \leq k,$$

$$\text{diam } V(G^{(j)}) \geq 2, \qquad \text{for } 0 \leq j \leq k,$$

in the following way:

If $j = 0$, then $G^{(0)} := G$.

If $j = 1$, then $G^{(1)} := G'$.

Assume that $j > 1$. If diam $V(G^{(j-1)}) \geq 3$, then $k = j - 1$ and the sequence stops. If diam $V(G^{(j-1)}) = 2$, then Lemma 4 provides an interval graph $(G^{(j-1)})'$ with:

$$|V((G^{(j-1)})')| = |V(G^{(j-1)})| - 1, \quad \text{diam } V((G^{(j-1)})') \geq 2, \quad \delta(\overline{G^{(j-1)}}) = \delta(\overline{(G^{(j-1)})'}),$$

and we define $G^{(j)} := (G^{(j-1)})'$.

Since $|V(G^{(j)})| = |V(G^{(j-1)})| - 1$ for $0 < j \leq k$ and the diameter of a graph with just a vertex is 0, this sequence must finish with some graph $G^{(k)}$ satisfying diam $V(G^{(k)}) \geq 3$. Thus,

$$\delta(\overline{G}) = \delta(\overline{G^{(0)}}) = \delta(\overline{G^{(1)}}) = \cdots = \delta(\overline{G^{(k)}}) \leq 2.$$

We prove now that $\delta(\overline{G}) \geq 5/4$ if diam $V(G) \geq 4$. Let us fix any graph G with diam $V(G) \geq 4$. Thus, there exists a geodesic $[v_0 v_4] = v_0 v_1 \cup v_1 v_2 \cup v_2 v_3 \cup v_3 v_4$ in G. If Γ is the subgraph of $\overline{G}$ induced by $\{v_0, v_1, v_2, v_3, v_3, v_4\}$, then $E(\Gamma) = \{v_0 v_2, v_0 v_3, v_0 v_4, v_1 v_3, v_1 v_4, v_2 v_4\}$. Consider the cycle $C := v_0 v_2 \cup v_2 v_4 \cup v_4 v_1 \cup v_1 v_3 \cup v_3 v_0$ in Γ. If p is the midpoint of $v_0 v_2$, then $d_\Gamma(v_1, p) = 5/2$ and so Corollary 4 gives $\delta(\Gamma) > 1$. Therefore, Theorem 2 gives $\delta(\Gamma) \geq 5/4$. Since Γ is an induced subgraph of $\overline{G}$, if g is a path in $\overline{G}$ joining v_i and v_j ($0 \leq i, j \leq 4$) and g is not contained in Γ, then $L_{\overline{G}}(g) \geq 2$. Since diam$_{\overline{G}} V(\Gamma) = 2$, we have $d_\Gamma(v_j, v_j) = d_{\overline{G}}(v_j, v_j)$ for every $0 \leq i, j \leq 4$; consequently, $d_\Gamma(x, y) = d_{\overline{G}}(x, y)$ for every $x, y \in \Gamma$, i.e., Γ is an isometric subgraph of $\overline{G}$. Hence, the geodesic triangles in Γ are also geodesic triangles in $\overline{G}$, and we have $\delta(\overline{G}) \geq \delta(\Gamma) \geq 5/4$.

Let us show now that the lower bounds on $\delta(\overline{G})$ are sharp. Recall that the path graph with n vertices P_n is a graph with $V(P_n) = \{v_1, v_2, \ldots, v_n\}$ and $E(P_n) = \{v_1 v_2, v_2 v_3, \ldots, v_{n-1} v_n\}$.

Consider the path graph with four vertices $G = P_4$. Since $\overline{G} = P_4$, we have diam $V(G) = 3$ and $\delta(\overline{G}) = 0$.

Consider the path graph with five vertices $G = P_5$. Since diam $V(G) = 4$, we have $\delta(\overline{G}) \geq 5/4$. Note that $\overline{G}$ has five vertices and thus Theorem 7 gives $\delta(\overline{G}) \leq 5/4$. Hence, we conclude $\delta(\overline{G}) = 5/4$. □

Corollary 8. *If G is any interval graph with edges of length 1, then*

$$\delta(G)\,\delta(\overline{G}) \leq \begin{cases} 0, & \text{if } \operatorname{diam} V(G) = 1, \\ 3, & \text{if } 2 \leq \operatorname{diam} V(G) \leq 3, \\ 9/4, & \text{if } \operatorname{diam} V(G) \geq 4. \end{cases}$$

Note that we can not improve the trivial lower bound $\delta(G)\delta(\overline{G}) \geq 0$, since it is attained if G is any tree.

Corollary 9. *If G is any interval graph with edges of length 1, then*

$$\delta(G) + \delta(\overline{G}) \leq \begin{cases} 3/2, & \text{if } \operatorname{diam} V(G) = 1, \\ 7/2, & \text{if } 2 \leq \operatorname{diam} V(G) \leq 3, \\ 3, & \text{if } \operatorname{diam} V(G) \geq 4. \end{cases}$$

In addition, $\delta(G) + \delta(\overline{G}) \geq 5/4$ for every graph G with diam $V(G) \geq 4$.

5. Conclusions

Gromov hyperbolicity is an interesting geometric property, and so it is natural to study it in the context of geometric graphs. In this work we deal with interval and indifference graphs, which are important classes of intersection and Euclidean graphs, respectively. It is well-known that interval graphs (with a very weak hypothesis) and indifference graphs are hyperbolic. One of our main results is Theorem 8, which provides a sharp upper bound of the hyperbolicity constant of interval graphs verifying a very weak hypothesis. This result allows for obtaining bounds for the hyperbolicity constant of every indifference graph (Corollary 6) and the hyperbolicity constant of every interval graph with edges of length 1 (Corollary 7). Moreover, Theorem 10 provides sharp bounds for the hyperbolicity constant of the complement of any interval graph with edges of length 1. Note that it is not usual to obtain such precise bounds for large classes of graphs. Our main result is Theorem 9, which provides the hyperbolicity constant of every interval graph with edges of length 1, by using geometric criteria.

Acknowledgments: This paper was supported in part by a grant from CONACYT (FOMIX-CONACyT-UAGro 249818), México and by two grants from the Ministerio de Economía y Competitividad, Agencia Estatal de Investigación (AEI) and Fondo Europeo de Desarrollo Regional (FEDER) (MTM2016-78227-C2-1-P and MTM2015-69323-REDT), Spain. We would like to thank the referees for their careful reading of the manuscript and several useful comments that have helped us to improve the presentation of the paper.

Author Contributions: Juan Carlos Hernández-Gómez, Rosalío Reyes, José Manuel Rodríguez and José María Sigarreta contributed equally with the ideas and writing of this paper.

Conflicts of Interest: The authors declare no conflict of interest.

References

1. Gromov, M. Hyperbolic groups. In *Essays in Group Theory*; Gersten, S.M., Ed.; Springer: New York, NY, USA, 1987; Volume 8, pp. 75–263.
2. Oshika, K. *Discrete Groups*; AMS Bookstore: Providence, RI, USA, 2002.
3. Krauthgamer, R.; Lee, J.R. Algorithms on Negatively Curved Spaces. In Proceedings of the Foundations of Computer Science (FOCS'06), Berkeley, CA, USA, 21–24 October 2006.
4. Jonckheere, E.A. Contrôle du traffic sur les réseaux à géométrie hyperbolique–Vers une théorie géométrique de la sécurité l'acheminement de l'information. *J. Eur. Syst. Autom.* **2002**, *8*, 45–60. (In French)
5. Tourís, E. Graphs and Gromov hyperbolicity of non-constant negatively curved surfaces. *J. Math. Anal. Appl.* **2011**, *380*, 865–881.
6. Bermudo, S.; Carballosa, W.; Rodríguez, J.M.; Sigarreta, J.M. On the Hyperbolicity of Edge-Chordal and Path-Chordal Graphs. *Filomat* **2016**, *30*, 2599–2607.
7. Bermudo, S.; Rodríguez, J.M.; Rosario, O.; Sigarreta, J.M. Small values of the hyperbolicity constant in graphs. *Discret. Math.* **2016**, *339*, 3073–3084.
8. Boguñá, M.; Papadopoulos, F.; Krioukov, D. Sustaining the Internet with Hyperbolic Mapping. *Nature Commun.* **2010**, *1*, 1–19.
9. Brinkmann, G.; Koolen, J.; Moulton, V. On the hyperbolicity of chordal graphs. *Ann. Comb.* **2001**, *5*, 61–69.
10. Carballosa, W. Gromov hyperbolicity and convex tessellation graph. *Acta Math. Hungarica* **2017**, *151*, 24–34.
11. Chalopin, J.; Chepoi, V.; Papasoglu, P.; Pecatte, T. Cop and Robber Game and Hyperbolicity. *SIAM J. Discret. Math.* **2014**, *28*, 1987–2007.
12. Chepoi, V.; Dragan, F.F.; Estellon, B.; Habib, M.; Vaxès, Y. Notes on diameters, centers, and approximating trees of δ-hyperbolic geodesic spaces and graphs. *Electron. Notes Discret. Math.* **2008**, *31*, 231–234.
13. Chepoi, V.; Dragan, F.F.; Estellon, B.; Habib, M.; Vaxès, Y.; Xiang, Y. Additive Spanners and Distance and Routing Labeling Schemes for Hyperbolic Graphs. *Algorithmica* **2012**, *62*, 713–732.
14. Cohen, N.; Coudert, D.; Lancin, A. On computing the Gromov hyperbolicity. *ACM J. Exp. Algortm.* **2015**, *20*, 18.
15. Fournier, H.; Ismail, A.; Vigneron, A. Computing the Gromov hyperbolicity of a discrete metric space. *J. Inf. Process. Lett.* **2015**, *115*, 576–579.
16. Frigerio, R.; Sisto, A. Characterizing hyperbolic spaces and real trees. *Geom. Dedic.* **2009**, *142*, 139–149.
17. Jonckheere, E.A.; Lohsoonthorn, P. Geometry of network security. *Am. Control Conf.* **2004**, *6*, 111–151.
18. Jonckheere, E.A.; Lohsoonthorn, P.; Ariaesi, F. Upper bound on scaled Gromov-hyperbolic delta. *Appl. Math. Comput.* **2007**, *192*, 191–204.
19. Jonckheere, E.A.; Lohsoonthorn, P.; Bonahon, F. Scaled Gromov hyperbolic graphs. *J. Graph Theory* **2007**, *2*, 157–180.
20. Kiwi, M.; Mitsche, D. A Bound for the Diameter of Random Hyperbolic Graphs. In *ANALCO*; Society for Industrial and Applied Mathematics: Philadelphia, PA, USA, 2015; pp. 26–39, doi:10.1137/1.9781611973761.3.
21. Koolen, J.H.; Moulton, V. Hyperbolic Bridged Graphs. *Eur. J. Comb.* **2002**, *23*, 683–699.
22. Krioukov, D.; Papadopoulos, F.; Kitsak, M.; Vahdat, A.; Boguñá, M. Hyperbolic geometry of complex networks. *Phys. Rev. E* **2010**, *82*, 036106.
23. Martínez-Pérez, A. Chordality properties and hyperbolicity on graphs. *Electron. J. Comb.* **2016**, *23*, P3.51.
24. Martínez-Pérez, A. Generalized Chordality, Vertex Separators and Hyperbolicity on Graphs. *Symmetry* **2017**, *9*, 199, doi:10.3390/sym9100199.
25. Mitsche, D.; Pralat, P. On the Hyperbolicity of Random Graphs. *Electron. J. Comb.* **2014**, *21*, P2.39.
26. Shang, Y. Lack of Gromov-hyperbolicity in small-world networks. *Cent. Eur. J. Math.* **2012**, *10*, 1152–1158.
27. Shang, Y. Non-hyperbolicity of random graphs with given expected degrees. *Stoch. Models* **2013**, *29*, 451–462.
28. Wu, Y.; Zhang, C. Chordality and hyperbolicity of a graph. *Electron. J. Comb.* **2011**, *18*, P43.
29. Calegari, D.; Fujiwara, K. Counting subgraphs in hyperbolic graphs with symmetry. *J. Math. Soc. Jpn.* **2015**, *67*, 1213–1226.
30. Eppstein, D. Squarepants in a tree: sum of subtree clustering and hyperbolic pants decomposition. In *Proceedings of the Eighteenth Annual ACM-SIAM Symposium on Discrete Algorithms*; Society for Industrial and Applied Mathematics: Philadelphia, PA, USA, 2007; pp. 29–38.

31. Li, S.; Tucci, G.H. Traffic Congestion in Expanders, (p, δ)-Hyperbolic Spaces and Product of Trees. *arXiv* **2013**, arXiv:1303.2952.

32. Fishburn, P.C. Interval orders and interval graphs: A study of partially ordered sets. In *Wiley-Interscience Series in Discrete Mathematics*; John Wiley and Sons: New York, NY, USA, 1985.

33. Golumbic, M.C. *Algorithmic Graph Theory and Perfect Graphs*; Academic Press: New York, NY, USA, 1980.

34. Gilmore, P.C.; Hoffman, A.J. A characterization of comparability graphs and of interval graphs. *Can. J. Math.* **1964**, *16*, 539–548.

35. Cohen, J.E. Food webs and niche space. In *Monographs in Population Biology 11*; Princeton University Press: Princeton, NJ, USA, 1978.

36. Bertossi, A.A. Finding Hamiltonian circuits in proper interval graphs. *Inf. Process. Lett.* **1983**, *17*, 97–101.

37. Panda, B.S.; Das, S.K. A linear time recognition algorithm for proper interval graphs. *Inf. Process. Lett.* **2003**, *87*, 153–161.

38. Li, P.; Wu, Y. Spanning connectedness and Hamiltonian thickness of graphs and interval graphs. *Discret. Math. Theor. Comput. Sci.* **2015**, *16*, 125–210.

39. Rodríguez, J.M.; Tourís, E. Gromov hyperbolicity through decomposition of metric spaces. *Acta Math. Hung.* **2004**, *103*, 107–138.

40. Bermudo, S.; Rodríguez, J.M.; Sigarreta, J.M. Computing the hyperbolicity constant. *Comput. Math. Appl.* **2011**, *62*, 4592–4595, doi:10.1016/j.camwa.2011.10.041.

41. Michel, J.; Rodríguez, J.M.; Sigarreta, J.M.; Villeta, M. Hyperbolicity and parameters of graphs. *Ars Comb.* **2011**, *100*, 43–63.

42. Jürgen, E. Extremal interval graphs. *J. Graph Theory* **1993**, *17*, 117–127.

43. Hernández, J.C.; Rodríguez, J.M.; Sigarreta, J.M. On the hyperbolicity constant of circulant networks. *Adv. Math. Phys.* **2015**, *2015*, 1–11.

![symmetry logo] *symmetry*

MDPI

Article

Analytical Treatment of Higher-Order Graphs: A Path Ordinal Method for Solving Graphs

Hala Kamal [1,2,*], Alicia Larena [3] and Eusebio Bernabeu [1]

[1] Department of Optics, Facultad de Ciencias Físicas, Universidad Complutense de Madrid, Ciudad Universitaria, E-28040 Madrid, Spain; ebernabeu@fis.ucm.es
[2] Department of Physics, Faculty of Science, Ain Shams University, 1156 Cairo, Egypt
[3] Department of Chemical Enginering, E.T.S. de Ingenieros Industriales, Universidad Politécnica de Madrid, C/José Gutiérrez Abascal, 2, E-28006 Madrid, Spain; alicia.larena@upm.es
* Correspondence: hkamal@ucm.es

Received: 13 October 2017; Accepted: 17 November 2017; Published: 22 November 2017

Abstract: Analytical treatment of the composition of higher-order graphs representing linear relations between variables is developed. A path formalism to deal with problems in graph theory is introduced. It is shown how paths in the composed graph representing individual contributions to variables relation can be enumerated and represented by ordinals. The method allows for one to extract partial information and gives an alternative to classical graph approach.

Keywords: flow graph; matrix algebra; linear equations; cascade graph; graph order; path ordinal; path set; path set diagram

1. Introduction

A flow graph is a graphical representation of a system of linear equations. It is introduced by Euler [1], this notion is especially useful in simplifying the treatment of certain linear problems arising e.g., in optical systems [2], classical and quantum field theory [3,4], and network theory [5], just to mention some relevant examples [6]. While this approach is worthy for 2×2 systems, for higher-order arrangements it becomes cumbersome. In consequence, to introduce an alternative treatment to solve these higher-order composition graph problems seems to be a relevant task. In this way, some significant contributions were presented earlier [7,8].

Flow graphs are applicable to several fields, such as System of Systems (SoS) implementations. Moreover, higher order graph reduction method could be used as a tool in optimizing the design of SoS, such as, obtaining self-managed smart grids, creating communication networks between all of the possible nodes of systems, setting up a secure transport and auxiliary routes of transportation in real time, managing the energy distribution around systems, permitting flexible and optimized manufacturing, or in financial and business flux analysis.

Flow graph algebra represents a set of linear equations in terms of a complex graph. Through the basic rules, this graph can be reduced to a simpler equivalent form called the "residual graph". For higher order graphs, there are several paths connecting the input nodes with the output ones, where it results to be difficult to follow a particular trajectory. For this reason, except for in the simplest cases, it is more practical to use numerical methods. Nevertheless, other features of flow graphs are still useful.

Here, we propose a new didactic and intuitive tool to solve graphs in any dimension without reducing them by the conventional rules. The new approach is called Path Ordinal Method (POM). The result is equivalent to a matrix product or graph reduction. However, the utility of the presented method arises in the simplicity of predicting such product graphically by means of a simple calculus table, as well as finding the impact of a certain parameter upon others without solving the entire graph.

The plan of this work is as follows. Section 2.1 is devoted to give a brief overview of the flow graph algebra and the basic reduction rules. In Section 2.2 higher-order graph composition is treated. Each possible path in the graph is defined by an ordinal and its trajectory is characterized by a "path-set". A new method to solve graphs of any order is introduced showing the way to extract partial information from the composition. Finally, in Section 2.3 we present an application to 3×3 matrix composition to demonstrate the validity of the developed method.

2. Materials and Methods

2.1. Flow Graphs

Graphs are geometrical structures that can represent linear equations. They relate magnitudes (variables) by graphic interconnections, following a few rules. A variable is represented by a small circle, called a "node". White and black colors are used to indicate the orientation of the nodes, which is analogous to the sides of the equation, in standard algebra: black nodes are "sources", that is, the input variables one has to handle to obtain the output variables called "sinks", which are indicated by white nodes. The line connecting two nodes is called "branch" and the corresponding label is termed "transmittance", which indicates that the relation between the interconnected variables. Furthermore, if this transmittance is not specified for a branch, it will be understood that it has the value 1. Branches with transmittance zero are not drawn. Figure 1 shows some flow graph representation of linear equations.

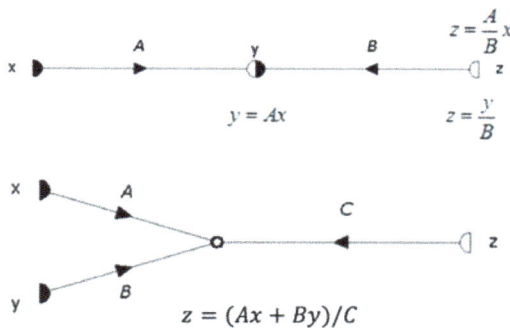

Figure 1. Elementary graph-algebra operations.

The order of a graph is the smallest number of sources or sinks in the graph. Besides, a cascade graph is the results of the composition of several graphs of the same order. Moreover, there is a mutual relation between matrix representation and flow graphs. For example, the algebraic relation between the following two-dimensional vectors,

$$\begin{pmatrix} x_2 \\ y_2 \end{pmatrix} = \begin{pmatrix} A & B \\ C & D \end{pmatrix} \begin{pmatrix} x_1 \\ y_1 \end{pmatrix}, \tag{1}$$

can be represented by the second-order graph as in Figure 2.

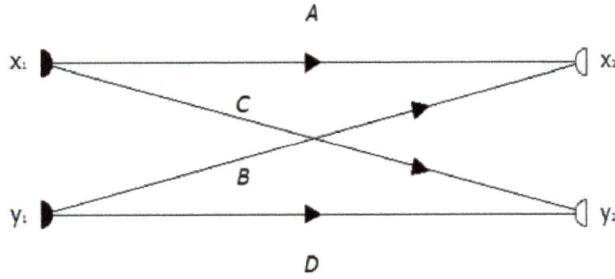

Figure 2. A second order graph representing Equation (1). The homologous of the vector parameters are the nodes while the homologous of the matrix elements are the branches transmittance.

Matrix multiplication can be solved through its alternative graphical representation. Figure 3 shows a composition of n graphs of order two: the result is a $2 \times n$ cascade graph.

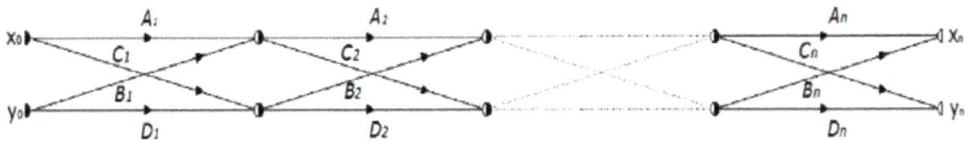

Figure 3. A cascade graph composed of n graphs of order two, attached side by side. Each individual graph represents a 2×2 matrix.

Which is equivalent to the following matrix expression of Equation (2),

$$\begin{pmatrix} x_2 \\ y_2 \end{pmatrix} = \begin{pmatrix} A_n & B_n \\ C_n & D_n \end{pmatrix} \cdots \begin{pmatrix} A_2 & B_2 \\ C_2 & D_2 \end{pmatrix} \begin{pmatrix} A_1 & B_1 \\ C_1 & D_1 \end{pmatrix} \begin{pmatrix} x_0 \\ y_0 \end{pmatrix}. \tag{2}$$

For a cascade flow graph there are two ways to proceed. First, by using the five basic algebraic rules namely: addition, product, transmission, suck up node, and self-loop elimination, where an equivalent simpler graph is obtained. The second one is the Mason's rule, recommended when we are only interested in one of the output variables as a function of one of the input variables.

2.2. Graph Composition and Path Characterization

The analysis of bulky systems made of several elements implies the composition of higher-order graphs, which turns to be complicated. In this section, we propose a general method to obtain the equivalent matrix, as well as the residual graph directly from the individual elements. Also, the influence of a certain input parameters upon an output one could be obtained without solving the whole graph.

We define two graphs: the cascade graph representing the whole system and the individual graph corresponding to any arbitrary element of the system. Consider, for example, the cascade graph of Figure 4.

The input variables are the vector:

$$\overrightarrow{x}_0 = (x_{01}, x_{02}, x_{03}, \ldots, x_{0k}, \ldots, x_{0m}),$$

and the output variables are represented by the vector:

$$\overrightarrow{x}_n = (x_{n1}, x_{n2}, x_{n3}, \ldots, x_{nk}, \ldots, x_{nm}),$$

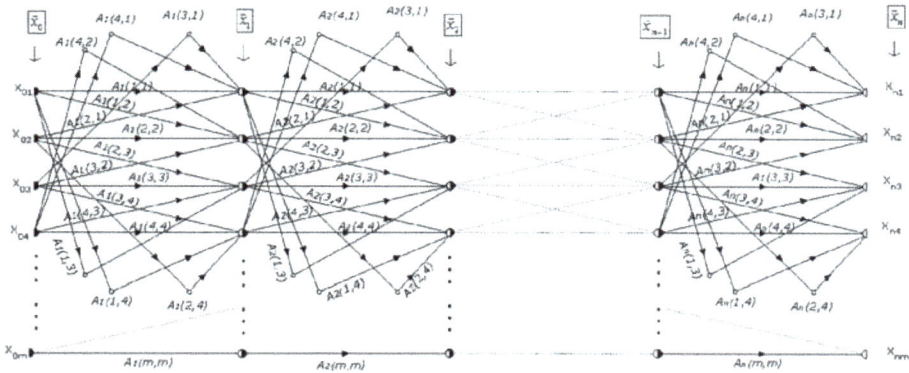

Figure 4. A $m \times n$ cascade graph.

This cascade graph is composed of n graphs, attached side by side, each one of order m. In consequence, the total number of possible paths connecting the input nodes to the output ones is:

$$N_{m,n} = m^{n+1}.$$

Now, let us consider the jth constituent of the cascade graph as sketched in Figure 5.

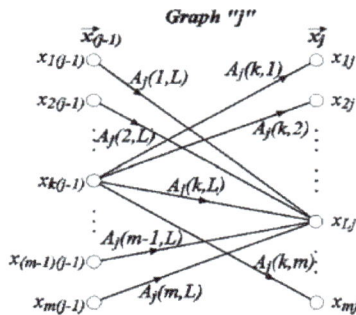

Figure 5. The jth graph of order m inside the cascade graph of Figure 4. $A_j(k,L)$ represents the transmittance of the branch connecting the nodes x_k and x_L.

The total number of possible paths is m^2. This individual graph is defined by the incoming nodes vector $\vec{x}_{j-1}$ and the outgoing nodes vector $\vec{x}_j$, where j takes the values $j = 1, 2, 3, \ldots, n$.

2.2.1. Ordinal of a Path and Path Value

An arbitrary path with an ordinal i ($1 \leq i \leq N_{m,n}$), connecting any node in the input vector with another node in the output vector, is characterized by a set of numbers $\{\theta_{ij}\}$ that we will call "path-set", which defines the trajectory of the path.

$$\{\theta_{ij}\} = \{\theta_{i0}, \theta_{i1}, \theta_{i2}, \theta_{i3}, \ldots \ldots \ldots, \theta_{in}\}$$

These θ_{ij} can take any of the values $1 \leq \theta_{ij} \leq m$. If $\theta_{ij} = k$ this means that the ith path passes through the kth node of the jth vector, x_{jk}. An example of a path-set is illustrated in Figure 6.

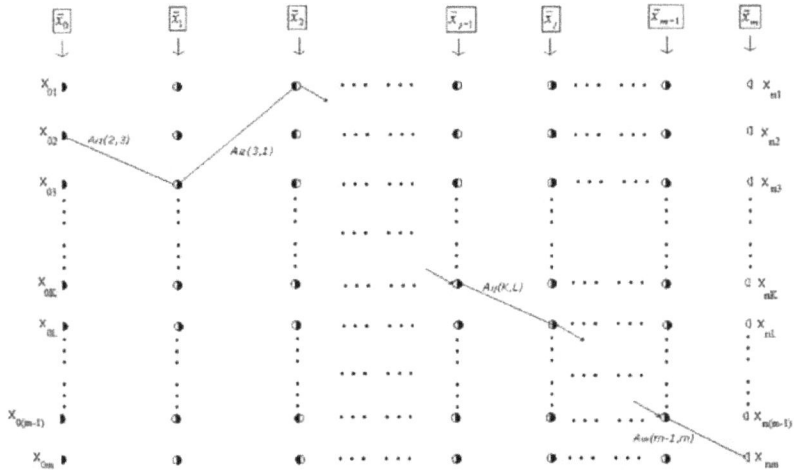

Figure 6. An example illustrating a path i. The trajectory of the path corresponds to the path-set $\{\theta_{ij}\} = \{2,3,1,\ldots,k,L,\ldots,m-1,m\}$ where $x_{mn} = P_i x_{20}$ and $P_i = (A_1(2,3)\cdot A_1(3,1)\cdot \ldots A_j(k,L)\cdot \ldots A_n(m-1,m))$.

The value of any possible path P_i can be seen as the product of the transmittances corresponding to each branch along the path

$$P_i = \prod_{j=1}^{n} A_{ij}(\theta_{i(j-1)}, \theta_{ij}), \tag{3}$$

where A_{ij} is the transmittance of the branch in the jth graph within the path i (see Figure 5).

Separating the contribution of the first and the last graph we get:

$$P_i = A_{i1}(\theta_{i0}, \theta_{i1}) \left[\prod_{j=2}^{n} A_{ij}(\theta_{i(j-1)}, \theta_{ij}) \right] A_{in}(\theta_{i(n-1)}, \theta_{in}), \tag{4}$$

The path value P_i that starts in an arbitrary node x_{0k} in the input vector $\vec{x}_0$ and reaches the output vector $\vec{x}_n$ in any arbitrary node x_{nL}, is given by

$$P_i^{kL} = A_{i1}(k, \theta_{i1}) \left[\prod_{j=2}^{n} A_{ij}(\theta_{i(j-1)}, \theta_{ij}) \right] A_{in}(\theta_{i(n-1)}, L), \tag{5}$$

So, each path is defined by two items, the path ordinal i and the path value P_i^{kL}, where both are associated to a "path-set".

Defining a path sequence, the first path ($i = 1$) will start from the node x_{01} and will end at the node x_{n1}, the second starts from the node x_{02} and ends at the node x_{n1}, the kth will start from the node x_{0k} till the path m is reached, which starts from the node x_{0m} and ends at the node x_{n1}.

When considering the output vector $\vec{x}_n$ as it is composed of m outgoing nodes. The total number of paths $N_{m,n}$ is divided into m groups each has m^n paths. The first group ends at the node x_{1n} where ($1 \leq i \leq m^n$) and the second group ends at the node x_{n2} where ($1 + m^n \leq i \leq 2m^n$). As a consequence, all of the paths that end at the node x_{nL} have the path ordinals within the limit ($1 + (L-1)m^n \leq i \leq Lm^n$). On the other hand, all of the paths that start from the node x_{0k}, according to the path sequence, have the ordinals $i = k, k+m, k+2m, \ldots$.

Thus, for a $m \times n$ cascade graph, there are m^{n-1} paths connecting an output node with an input one. These paths that start from an input node x_{0k} and ends at an output node x_{nL}, have the path ordinals $i = k + (L-1)m^n, k + m + (L-1)m^n, \dots, k - m + Lm^n$.

The contribution of the source x_{0k} to the sink x_{nL}, can be expressed as the summation of all the paths that start from x_{0k} and end at x_{nL} as follows,

$$x_{nL}^{(x_{0k})} = x_{0k} \sum_{r=1}^{m^{n-1}} P_{k-m+(L-1)m^n+rm}^{kL}. \tag{6}$$

Similarly, the total contribution of the input vector nodes $\vec{x}_0$ to the sink x_{nL} is given by,

$$x_{nL} = \sum_{k=1}^{m} x_{0k} \sum_{r=1}^{m^{n-1}} P_{k-m+(L-1)m^n+rm}^{kL}. \tag{7}$$

Calling T^{kL} to the summation of all the path values that start from x_{0k} and end at x_{nL}

$$T^{kL} = \sum_{r=1}^{m^{n-1}} P_{k-m+(L-1)m^n+rm}^{kL}, \tag{8}$$

So, the sink x_{nL} can be expressd as,

$$x_{nL} = \sum_{k=1}^{m} T^{kL} x_{0k}. \tag{9}$$

Hence, for a $m \times n$ graph, the contribution of all the sources to all of the sinks, can be represented by the equation:

$$
\begin{pmatrix}
x_{n1} \\
x_{n2} \\
\vdots \\
x_{nL} \\
\vdots \\
x_{n(m-1)} \\
x_{nm}
\end{pmatrix}
=
\begin{pmatrix}
T^{11} & T^{21} & \cdots & T^{k1} & \cdots & T^{m1} \\
T^{12} & T^{22} & \cdots & T^{k2} & \cdots & T^{m2} \\
\vdots & \vdots & \ddots & \vdots & \vdots \\
T^{1L} & T^{2L} & \cdots & T^{kL} & \cdots & T^{mL} \\
\vdots & \vdots & & \vdots & \ddots & \vdots \\
T^{2(m-1)} & T^{2(m-1)} & \cdots & T^{k(m-1)} & \cdots & T^{m(m-1)} \\
T^{1m} & T^{2m} & \cdots & T^{km} & \cdots & T^{mm}
\end{pmatrix}
\begin{pmatrix}
x_{01} \\
x_{02} \\
\vdots \\
x_{0L} \\
\vdots \\
x_{0(m-1)} \\
x_{0m}
\end{pmatrix}. \tag{10}
$$

According to Equation (3) for all of the path-sets representing such trajectories, only θ_{i0} and θ_{in} are defined, with the values k and L, respectively. Now, the goal is to define the trajectory of an arbitrary path, i.e., to evaluate the set of numbers $\{\theta_{ij}\}$.

2.2.2. Determination of the Characteristic Path Set

As it is mentioned before, for a $m \times n$ cascade graph, there are m groups of paths, of which, each is composed m^n paths. Each group reaches an output node. Accordingly, the group of paths that reaches an arbitrary node x_{nL} in the output vector $\vec{x}_n$ has the path ordinals within the range $(1 + (L-1)m^n \leq i \leq Lm^n)$.

Proceeding to calculate the path set. For any path of ordinal i, if the path ordinal i is subtracted by one and then divided by the number of paths that reach an output node (m^n), we get:

$$i - 1 = C_n m^n + R_n,$$

where m^n is de divisor, C_n is the quotient ($0 \leq C_n < m$), and R_n is the reminder ($0 \leq R_n < m^n$). So, the last element of the path-set θ_{in} can be expressed as:

$$\theta_{in} = C_n + 1.$$

Similarly the penultimate element of the path set is calculated by considering a cascade graph of $n - 1$ graphs each of order m. The total number of paths corresponding to such graph is m^{n-1} paths. Now, the path ordinal becomes:

$$i = R_n + 1.$$

To determine the node $x_{(n-1)L}$ that the path ends at, following the previous procedure, the path ordinal is subtracted by one and then divided by m^{n-1}, so we have:

$$R_n = C_{n-1}m^{n-1} + R_{n-1} \Rightarrow \theta_{i(n-1)} = C_{n-1} + 1,$$

where $0 \leq C_{n-1} < m$ and $0 \leq R_{n-1} < m_{n-1}$. Iterating the same procedure, we finally get:

$$i - 1 = C_n m^n + C_{n-1}m^{n-1} + \cdots + C_1 m + R_1. \tag{11}$$

On account of this:

$$\{\theta_{ij}\} = \{R_1 + 1, C_1 + 1, C_2 + 1, \ldots\ldots, C_n + 1\}.$$

Thus, we conclude that, for a given path-ordinal i the corresponding path-set $\{\theta_{ij}\}$ can be determined as follows:

I   The path ordinal is subtracted by one.
II   Then, it is divided by m for n-times.
III   Finally, one is added to the remainders of the division, $R_1, C_1, C_2, \ldots, C_n$.

A scheme illustrating the calculation of the path-set is shown in the next diagram Figure 7.

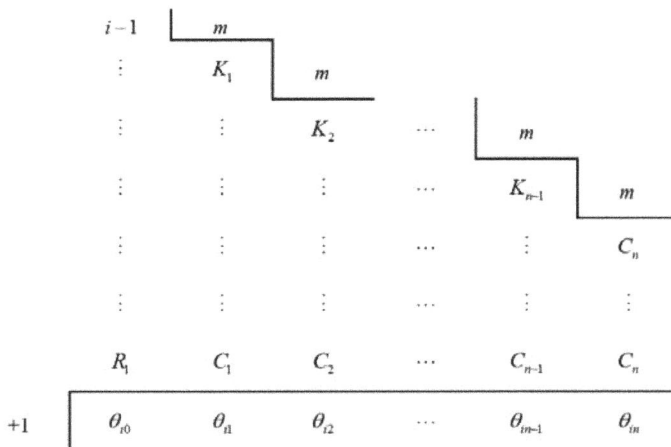

Figure 7. The Path Set Diagram (PSD). Starting from a path ordinal i, its corresponding characteristic path set $\{\theta_{ij}\} = \{\theta_{i0}, \theta_{i1}, \theta_{i2}, \theta_{i3}, \ldots\ldots, \theta_{in}\}$ is calculated as follows: firstly, the path ordinal is subtracted by one, then it is divided by m n-times, finally one is added to the remainders of the division.

The utility of the Path Set Diagram (PSD) is crucial in the application of Equations (7)–(9). In the next section, we discuss a simple and explicit example to illustrate how the Path Ordinal Method (POM) works.

2.3. Examples and Concluding Remarks

Consider an arbitrary system that is composed of two elements, each one is represented by a 3×3 matrix. Starting from the physical scheme of the system, the flow graph is formed by attaching side by side the graph corresponding to each element. The result is a 3×2 cascade graph. The cascade graph of the problem is shown the Figure 8.

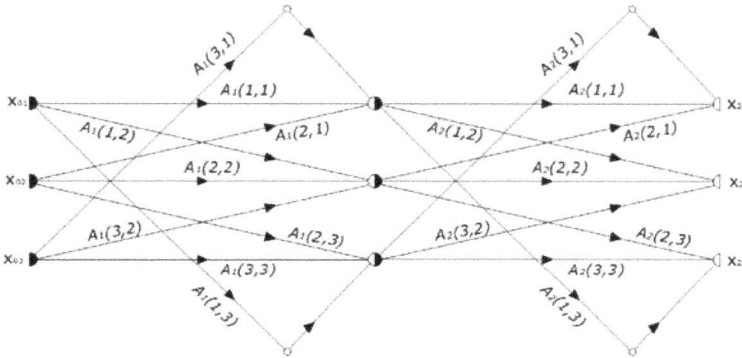

Figure 8. A cascade graph resulting from the composition of two graphs of order 3.

The problem can be solved either by matrix multiplication as in Equation (12) or through conventional graph reduction rules.

$$
\begin{pmatrix} x_{21} \\ x_{22} \\ x_{23} \end{pmatrix} = \begin{pmatrix} A_2(1,1) & A_2(1,1) & A_2(1,1) \\ A_2(1,1) & A_2(1,1) & A_2(1,1) \\ A_2(1,1) & A_2(1,1) & A_2(1,1) \end{pmatrix} \begin{pmatrix} A_2(1,1) & A_2(1,1) & A_2(1,1) \\ A_2(1,1) & A_2(1,1) & A_2(1,1) \\ A_2(1,1) & A_2(1,1) & A_2(1,1) \end{pmatrix} \begin{pmatrix} x_{01} \\ x_{02} \\ x_{03} \end{pmatrix} \tag{12}
$$

$$
\begin{pmatrix} x_{21} \\ x_{22} \\ x_{23} \end{pmatrix} = \begin{pmatrix} T^{11} & T^{21} & T^{31} \\ T^{12} & T^{22} & T^{32} \\ T^{13} & T^{23} & T^{33} \end{pmatrix} \begin{pmatrix} x_{01} \\ x_{02} \\ x_{03} \end{pmatrix}
$$

However, we will proceed to get the residual graph as well as the equivalent matrix by applying the POM. We denote the equivalent matrix as:

$$
\begin{pmatrix} T^{11} & T^{21} & T^{31} \\ T^{12} & T^{22} & T^{32} \\ T^{13} & T^{23} & T^{33} \end{pmatrix}.
$$

Using the path ordinal formalism we will find the partial contribution of an input parameter, as well as the total solution.

2.3.1. The Contribution of and Input Parameter to an Output Parameter

If one is interested to know the effect of an input parameter on an output one, it is not necessary to build the whole matrix or to reduce the whole graph. When considering the path sequence, the contribution of e.g., the input parameter x_{03} to the output parameter x_{22} is given by the matrix element T^{32}. When applying Equation (8), we get:

$$T^{32} = \sum_{r=1}^{3} P^{32}_{3-3+(3-1)3^2+3r} = \sum_{r=1}^{3} P^{32}_{9+3r} = P^{32}_{12} + P^{32}_{15} + P^{32}_{18}.$$

Accordingly, there are three paths that connect both nodes. These paths have the ordinals 12, 15, and 18. The path-values of the above equation are calculated by specifying, firstly, the path-set corresponding to each trajectory.

For the paths of ordinals 12, 15, and 18, the corresponding path sets are obtained by means of the PSD as follows:

$$
\begin{array}{lll}
12-1 = \begin{array}{c|cc} 11 & 3 & \\ : & 3 & 3 \\ : & & 1 \\ : & : & : \\ 2 & 0 & 1 \\ \hline 3 & 1 & 2 \end{array}
&
15-1 = \begin{array}{c|cc} 14 & 3 & \\ : & 4 & 3 \\ : & & 1 \\ : & : & : \\ 2 & 1 & 1 \\ \hline 3 & 2 & 2 \end{array}
&
18-1 = \begin{array}{c|cc} 17 & 3 & \\ : & 5 & 3 \\ : & & 1 \\ : & : & : \\ 2 & 2 & 1 \\ \hline 3 & 3 & 2 \end{array}
\\
+1 & +1 & +1
\end{array}
$$

$$\{\theta_{12,j}\} = \{3,1,2\} \quad \{\theta_{15,j}\} = \{3,1,2\} \quad \{\theta_{12,j}\} = \{3,1,2\}$$

For completeness, the corresponding graph-trajectories according to the above calculations appear in Figure 9.

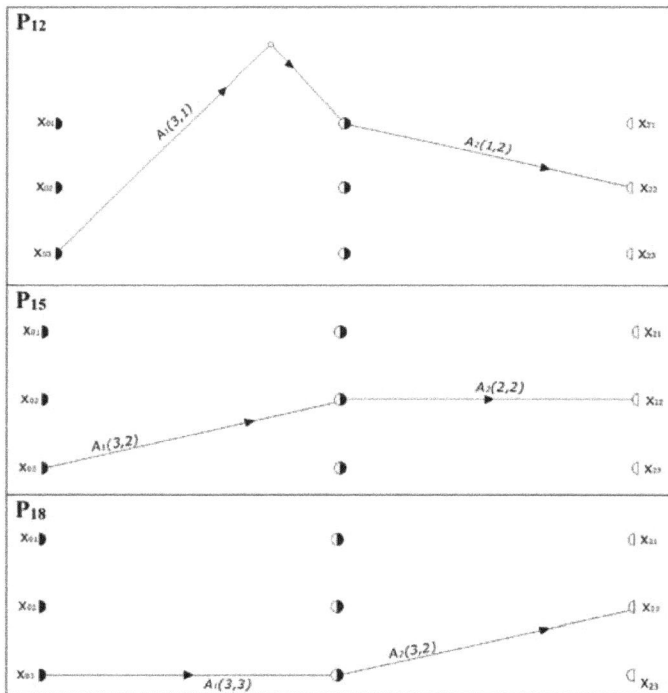

Figure 9. The three possible graph paths P_{12}, P_{15}, and P_{18}, connecting the input node x_{03} to the output node x_{22} according to the path sets calculated by the Path Set Diagram (PSD).

Hence, we get,

$$T^{32} = P_{12} + P_{15} + P_{18} = A_1(3,1)A_2(1,2) + A_1(3,2)A_2(2,2) + A_1(3,3)A_2(3,2). \tag{13}$$

We emphasize that, by means of the POM and as an alternative to the classical approach, we have simply obtained a matrix element without solving the whole matrix product or reducing the whole graph. This is precisely one of the applications of the formalism.

2.3.2. The Total Solution

The cascade graph is solved by specifying the equivalent matrix of the system. In consequence, the residual graph can be drawn easily. According to Equation (10) and recalling Equation (12), the algebraic expression representing the system is

$$\begin{pmatrix} x_{21} \\ x_{22} \\ x_{23} \end{pmatrix} = \begin{pmatrix} T^{11} & T^{21} & T^{31} \\ T^{12} & T^{22} & T^{32} \\ T^{13} & T^{23} & T^{33} \end{pmatrix} \begin{pmatrix} x_{01} \\ x_{02} \\ x_{03} \end{pmatrix}. \tag{14}$$

The equivalent matrix elements are calculated through three steps, each is represented within a table. The path-sets corresponding to each possible path is illustrated in Table 1, while Table 2 gives the path values corresponding to the path sets of Table 1.

Table 1. The path-sets corresponding to the 3 × 2 cascade graph. The table is divided vertically into three parts, each part represents the paths that reach the output nodes x_{21}, x_{22} and x_{23}, respectively.

i	θ_{i0}	θ_{i1}	θ_{i2}	i	θ_{i0}	θ_{i1}	θ_{i2}	i	θ_{i0}	θ_{i1}	θ_{i2}
1	1	1	1	10	1	1	2	19	1	1	3
2	2	1	1	11	2	1	2	20	2	1	3
3	3	1	1	12	3	1	2	21	3	1	3
4	1	2	1	13	1	2	2	22	1	2	3
5	2	2	1	14	2	2	2	23	2	2	3
6	3	2	1	15	3	2	2	24	3	2	3
7	1	3	1	16	1	3	2	25	1	3	3
8	2	3	1	17	2	3	2	26	2	3	3
9	3	3	1	18	3	3	2	27	3	3	3

Table 2. The path values corresponding to the path-sets of Table 1.

P_i	Path Value	P_i	Path Value	P_i	Path Value
P_1	$A_1(1,1)^*A_2(1,1)$	P_{10}	$A_1(1,1)^*A_2(1,2)$	P_{19}	$A_1(1,1)^*A_2(1,3)$
P_2	$A_1(2,1)^*A_2(1,1)$	P_{11}	$A_1(2,1)^*A_2(1,2)$	P_{20}	$A_1(2,1)^*A_2(1,3)$
P_3	$A_1(3,1)^*A_2(1,1)$	P_{12}	$A_1(3,1)^*A_2(1,2)$	P_{21}	$A_1(3,1)^*A_2(1,3)$
P_4	$A_1(1,2)^*A_2(2,1)$	P_{13}	$A_1(1,2)^*A_2(2,2)$	P_{22}	$A_1(1,2)^*A_2(2,3)$
P_5	$A_1(2,2)^*A_2(2,1)$	P_{14}	$A_1(2,2)^*A_2(2,2)$	P_{23}	$A_1(2,2)^*A_2(2,3)$
P_6	$A_1(3,2)^*A_2(2,1)$	P_{15}	$A_1(3,2)^*A_2(2,2)$	P_{24}	$A_1(3,2)^*A_2(2,3)$
P_7	$A_1(1,3)^*A_2(3,1)$	P_{16}	$A_1(1,3)^*A_2(3,2)$	P_{25}	$A_1(1,3)^*A_2(3,3)$
P_8	$A_1(2,3)^*A_2(3,1)$	P_{17}	$A_1(2,3)^*A_2(3,2)$	P_{26}	$A_1(2,3)^*A_2(3,3)$
P_9	$A_1(3,3)^*A_2(3,1)$	P_{18}	$A_1(3,3)^*A_2(3,2)$	P_{27}	$A_1(3,3)^*A_2(3,3)$

The matrix elements are calculated according to Equation (8) and represented in Table 3.

Table 3. The matrix elements corresponding to the example, according to 3 × 2 graph analysis.

$T^{11} = P_1 + P_4 + P_7$	$T^{21} = P_2 + P_5 + P_8$	$T^{31} = P_3 + P_6 + P_9$
$T^{12} = P_{10} + P_{13} + P_{16}$	$T^{22} = P_{11} + P_{14} + P_{17}$	$T^{32} = P_{12} + P_{15} + P_{18}$
$T^{13} = P_{19} + P_{22} + P_{25}$	$T^{23} = P_{20} + P_{23} + P_{26}$	$T^{33} = P_{21} + P_{24} + P_{27}$

Obtaining the matrix elements, which are homologues to the graph transmittances, the residual graph could be drawn easily.

In summary, what we expect to have accomplished is to work out a new didactic and useful tool to solve graphs of different dimensions as an alternative to reducing them by the conventional rules.

3. Results and Discussion

A new didactic, simple, and intuitive tool is developed. The POM is applicable to any type of problems that could be raised with the usual matrix algebra or graphs of any order, contributing an alternative and powerful treatment that allows for treating multitude of problems in physics that nowadays are approached by means of standard matrix treatment or flow graph algebra.

The aptitude of the method to treat as an independent form, each of the contributions of the different components of the input and output vectors, is especially useful in problems of Physics in which one is interested in knowing the impact of certain input parameter of the problem on others. Also, the utility of the method could be observed in problems with higher order matrix compositions or higher order graphs.

The POM states that; for any arbitrary $m \times n$ cascade graph represented by the input variables vector

$$\vec{x}_0 = (x_{01}, x_{02}, x_{03}, \ldots, x_{0k}, \ldots, x_{0m}), \tag{15}$$

and the output variables vector

$$\vec{x}_n = (x_{n1}, x_{n2}, x_{n3}, \ldots, x_{nk}, \ldots, x_{nm}), \tag{16}$$

there exist $N_{m,n} = m^{n+1}$ possible paths connecting the input nodes with the output ones (Figure 4). These paths are defined by an ordinal ($1 \leq i \leq N_{m,n}$), which, as a consequence, is attached to a characteristic Path-Set that determines the path along the graph and a Path-Value that is considered as the product of the transmittances corresponding to each branch along the path. Once the path values are calculated, the transmittances of the branches of the residual graph are calculated through Equation (8), which are homologues to the matrix elements representing the system. For better organization, simplicity, and in order to avoid calculation mistakes, we suggest that all of the calculations to be put in tables. Tables 4 and 5, and Figure 10 summarizes the process.

Table 4. A general form of a table used to calculate all the possible paths of a $m \times n$ cascade graph. $Aj(k,L)$ represents the transmittance of the branch connecting the node $X(j-1)k$ and Xjk in the jth graph.

Path Ordinal P_i	Path Set $\{\theta_{ij}\} = \{\theta_{i0}, \theta_{i1}, \theta_{i2}, \ldots\ldots\ldots, \theta_{in}\}$	Path Value $P_i = \prod_{j=1}^{n} A_{ij}(\theta_{i(j-1)}, \theta_{ij})$
1	$\{1, 1 \ldots \ldots \ldots \ldots \ldots \ldots, 1\}$	$P_1 = A_1(1,1)^* \ldots \ldots \ldots \ldots \ldots \ldots A_n(1,1)$
$\vdots$	$\vdots$	$\vdots$
m^n	$\{m, m \ldots \ldots \ldots \ldots \ldots, 1\}$	$P_{m^n} = A_1(m,m)^* \ldots \ldots \ldots \ldots A_n(m,1)$
$\vdots$	$\vdots$	$\vdots$
m^{n+1}	$\{m, m \ldots \ldots \ldots \ldots \ldots, m\}$	$P_{m^{n+1}} = A_1(m,m)^* \ldots \ldots \ldots \ldots A_n(m,m)$

Table 5. A table illustrating the value of each transmittance in the residual graph as a sum of its corresponding path-values.

$T^{11} = \sum_{r=1}^{m^{n-1}} P_{1-m+r.m}$	$\cdots$	$T^{k1} = \sum_{r=1}^{m^{n-1}} P_{k-m+r.m}$	$\cdots$	$T^{m1} = \sum_{r=1}^{m^{n-1}} P_{r.m}$	
$\vdots$	$\cdots$	$\vdots$	$\cdots$	$\vdots$	
$T^{1L} = \sum_{r=1}^{m^{n-1}} P_{1-m+(L-1).m^n+r.m}$	$\cdots$	$T^{kL} = \sum_{r=1}^{m^{n-1}} P_{k-m+(L-1).m^n+r.m}$	$\cdots$	$T^{mL} = \sum_{r=1}^{m^{n-1}} P_{(L-1).m^n+r.m}$	
$\vdots$	$\cdots$	$\vdots$	$\cdots$	$\vdots$	
$T^{1m} = \sum_{r=1}^{m^{n-1}} P_{1-m+(m-1).m^n+r.m}$	$\cdots$	$T^{km} = \sum_{r=1}^{m^{n-1}} P_{k-m+(m-1).m^n+r.m}$	$\cdots$	$T^{mm} = \sum_{r=1}^{m^{n-1}} P_{(m-1).m^n+r.m}$	

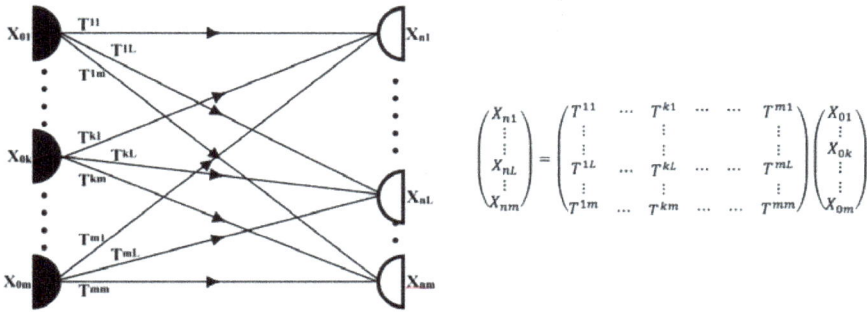

Figure 10. The residual graph and its equivalent matrix form.

In addition, a useful feature of the POM is its practicability in special problems when the impact of a certain parameter upon others is of our interest. Taking into account the path order, the contribution of a source x_{0k} to a sink x_{nL} is given by the transmittance T^{kL} of the residual graph. Which is the summation of its corresponding path values in agreement with Equation (8). These path values can be calculated directly by specifying their corresponding path sets by means of the PSD illustrated in Figure 7, i.e., by applying the POM the contribution of a source to a sink can be calculated easily without solving the entire problem.

The fields of application of this tool spread to all of those linear problems that are treated in physics by means of matrix algebra or flow graphs, being especially effective in the simplification of some specific calculations possessing composition of several high order matrices or bulky graphs.

Clear examples of applications could be fields as: matrix optics in asymmetric systems, matrix treatments in quantum mechanics and quantum theory of fields, treatments of dielectric multilayers, analysis of tensor mechanical properties of materials, optical networks, classic mechanics formulations, polarization and depolarization problems in optics, fluids dynamics, acoustic, and in general, any problem that holds linear relations and sets the stage for a matrix treatment. The POM was applied to 3-layers dielectric system [9].

Finally, other industrial applications could be the simulation in designing System of Systems focused on communication networks, multimodal traffic control, energy distribution systems, multi-site industrial manufacture or emergency management, among others.

Acknowledgments: This work has been funded by the European Commission within the 7th Framework Program and the project "ReBorn–Innovative re-use of modular equipment based on integrated factory design", Grant Agreement No. 609223. Also, below the support of SEGVAUTO-TRIES-CM Program P2013-MIT-2713. The authors would like to express their gratitude to Eduardo Martin-Martinez for his preliminary help at the beginning of this research.

Author Contributions: Hala Kamal analyzed the problem, done all the mathematical calculations, derived the achieved method and wrote the paper. Eusebio Bernabeu directed the work pointing out the tools to be used and revised the paper. Alicia Larena oriented the selection of some input/output parameters in relation to possible realistic applications in future works. She has done a critical analysis of the manuscript.

Conflicts of Interest: The authors declare no conflict of interest.

References

1. Euler, L. The Seven Bridges of Konigsberg. *Comment. Acad. Sci. Imp. Petropolitanae* **1741**, *8*, 128–140.
2. Wang, S.; Zhao, D. *Matrix Optics*; Zhejiang University Press: Hangzhou, China, 2000.
3. Feynman, R.P. *Quantum Electrodynamics*; Perseus: New York, NY, USA, 1998.
4. Peskin, M.E.; Schroeder, D.V. *Quantum Field Theory*; Perseus: New York, NY, USA, 1995.
5. Weissman, Y. *Optical Network Theory*; Artech: New York, NY, USA, 1992.
6. Bondy, J.A.; Murty, U.S.R. *Graph Theory with Applications*; North-Holland: Oxford, UK, 1976.

7. Coates, C.L. Flow—Graph Solutions of Linear Algebric Equations. *IRE Trans. Circuit Theory* **1959**, *6*, 170–187. [CrossRef]

8. Chen, W.-K. On Flow graph Solutions of Linear Algebric Equations. *SIAM J. Appl. Math.* **1967**, *15*, 136–142. [CrossRef]

9. Kamal, H.; Bernabeu, E. High order graph formalism for multilayer structures. *Opt. Int. J. Light Electron Opt.* **2016**, *127*, 1384–1390. [CrossRef]

symmetry

MDPI

Article

Graphical Classification in Multi-Centrality-Index Diagrams for Complex Chemical Networks

Yasutaka Mizui, Tetsuya Kojima, Shigeyuki Miyagi and Osamu Sakai *

Electronic Systems Engineering, The University of Shiga Prefecture, 2500 Hassakacho, Hikone, Shiga 522-8533, Japan; oh23ymizui@ec.usp.ac.jp (Y.M.); oh23tkojima@ec.usp.ac.jp (T.K.); miyagi.s@e.usp.ac.jp (S.M.)
* Correspondence: sakai.o@e.usp.ac.jp; Tel.: +81-749-28-8382

Received: 22 October 2017; Accepted: 5 December 2017; Published: 9 December 2017

Abstract: Various sizes of chemical reaction network exist, from small graphs of linear networks with several inorganic species to huge complex networks composed of protein reactions or metabolic systems. Huge complex networks of organic substrates have been well studied using statistical properties such as degree distributions. However, when the size is relatively small, statistical data suffers from significant errors coming from irregular effects by species, and a macroscopic analysis is frequently unsuccessful. In this study, we demonstrate a graphical classification method for chemical networks that contain tens of species. Betweenness and closeness centrality indices of a graph can create a two-dimensional diagram with information of node distribution for a complex chemical network. This diagram successfully reveals systematic sharing of roles among species as a semi-statistical property in chemical reactions, and distinguishes it from the ones in random networks, which has no functional node distributions. This analytical approach is applicable for rapid and approximate understanding of complex chemical network systems such as plasma-enhanced reactions as well as visualization and classification of other graphs.

Keywords: chemical reaction network; centrality index; statistical analysis; random graph

1. Introduction

Graph theory provides for us a graphical approach to a system containing various elements with connections between them [1,2]. Chemical reaction networks are one such system, and small and simple reaction systems are visualized in linear or small graphs with sufficient understanding of reaction procedures [3–5]. On the other hand, the networks of protein and metabolic systems in a biological cell are quite complicated due to their numbers of nodes (at least, more than 1000), so that not only visualization as a graph but also statistical properties such as degree distributions are representative for characterizing their complexity [6,7].

Recently, we performed graph visualization for plasma-enhanced chemical reactions [8,9]. In low-temperature reactive plasma, high-energy electrons trigger a number of simultaneous dissociations of mother molecules, and its chemistry is more complex than other chemical systems in artificial environments for chemical plants [10–12]. After definitions of a node (for one species) and an edge (for each reaction) for a display in a graph, centrality indices of nodes derived from the graph work as representatives of chemical roles in the system, such as agents, intermediates, and products. However, except for such microscopic points of view, approaches have not been accomplished for describing macroscopic properties of graphs for chemical reaction networks that contain several tens of species.

Such relatively small-sized chemical network systems create rich outputs despite limited numbers of nodes and edges in a graph. For instance, information processing in biochemical reactions revealed collective behaviors that can be interpreted using interactions among analogical spins, leading to similar features to electronic information processing or mechanical systems [13,14]. Another example of graphical network approaches for medium-sized chemical complexity was on numerical calculations of rate equations in plasma-enhanced chemical reactions, and the calculated results were visualized in reaction pathways in a graph to summarize complicated time evolutions of densities of species [15]. In comparison with the previous achievements based on numerical finite-difference methods [10–12,15], direct visualization of reactions based on graph theory is applied here, and we focus on graphical classification of nodes or species on statistical aspects that are missing in our previous studies [8,9].

In this study, for such systems as medium-sized chemical reaction networks, which are neither so small as several numbers of reactions nor too large with more than 1000 species, we demonstrate suitable macroscopic measures by graphical diagrams based on multi-centrality indices. Using such diagrams, semi-statistical properties with confirmation of systematic and global structures in the corresponding system can be deduced, even if the total number of nodes is limited to less than 100, as well as identifications of roles in each species, such as agents, intermediates and products. This method of classification of species in macroscopic points of view provides us with insight and understandings about predictions of global properties of chemical reactions for approximate designs of upcoming chemical reactors and rapid selection of mother chemicals for products when accurate computer aided designs are not available. In Section 2, using reactions in silane and methane plasmas, we show the diagrams with axes of betweenness and closeness centrality indices. Using such a diagram, we can understand both macroscopic and microscopic properties of graphs with tens of nodes. In Section 3, we compare these two examples with random graphs, and discuss the validities of this graphical characterization.

2. Analysis of Reaction Networks in Plasma Chemistry

As examples, we use two plasma-enhanced chemical networks reported in [10,11]. We performed some analyses of methane plasma [8] and silane plasma [9] using some centrality indices, but they did not include macroscopic measures of complex chemical networks. Here, we proceed to study them using the same reaction systems to obtain their macroscopic and microscopic properties simultaneously. Table A1 in Appendix A shows active species in silane plasma with temporary indicator numbers, where we use reaction sets in [11]. There are a wide variety of species that originate initially from two species: SiH_4 and Ar. The number of species or nodes in the corresponding graph is 58, and that of reactions or edges is 222. The reason why we can observe such rich diversity is based on high-energy electrons whose energy spreads up to 20 eV with electron temperature of 1–5 eV; kinetic energy with 1–20 eV induces most of reactions of decomposition, dissociation and ionization [16]. Almost all reactions are bi-molecule, and we treat all reactions as irreversible ones.

To convert a chemical reaction into a graph, when we handle the following reaction,

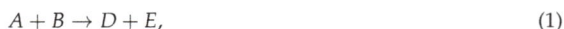

$$A + B \rightarrow D + E, \tag{1}$$

and use the names of species in reactions as node indicators, we set directed edges from node A to node D, from node A to node E, from node B to node D and from node B to node E, as shown in Figure 1a. When one performs graph representation for an underlying system of interest, elements of the system becomes nodes and interactions between elements are displayed as edges. For chemical reactions, as interpreted in [3], species becomes nodes, and not a simple co-existence but an agent-product relation is suitable for a directed edge, leading to a form of representation in Figure 1a. In general chemical reactions for inorganic molecules, since roles of agents, intermediates and products are fairly clear with energy consumption as driving forces, regular graphs or graphs with frequent cycles are less important, and monotone graphs or trees with several nodes are sufficient in many cases [3–5]. However, in reactions enhanced by energetic electrons in plasma, cycles appear with their various sizes with increasing values of clustering coefficients [8], which makes a role of each species more complicated.

After completion of an edge list for a graph, we calculate several centrality indices. In this study, we use betweenness centrality index C_b and closeness centrality index C_c. C_b is defined as [2]:

$$C_b(i) = \sum_{\substack{j,k \\ j \neq k \neq i}} \frac{g_{jk}(i)}{g_{jk}}, \tag{2}$$

where i, j and k are species/node indicator numbers and g_{jk} is number of the shortest paths between nodes j and k. $g_{jk}(i)$ indicates number of g_{jk} passing through node i. C_c is defined as [2]:

$$C_c(i) = \frac{1}{\sum_j d_{ij}}, \tag{3}$$

where d_{ij} is distance or edge numbers from node i to j along the shortest path. Since we consider directed graphs, no path may exist from node i to j; in such a case, d_{ij} is defined as number of all nodes. This fact indicates that, although C_c represents one of the topological aspects, it includes a measure of agents in chemical reactions. In our previous studies [8,9], we calculated simplified PageRank values [17], which represent information on roles of species, i.e., in a microscopic point of view. In this study, we put more emphasis on analysis of topological and statistical properties of graphs, and C_b and C_c are suitable for this purpose.

Figure 1b shows the graph of reactions with species in silane plasma. Since the number of nodes is not large, we can identify an individual species. On the other hand, the number of edges is pretty large, and we cannot trace all of them. For complex networks with huge size, one can recognize neither of them, and they are mainly analyzed using statistical properties (i.e., Power-law tails in degree distributions [18]) from a macroscopic point of view, neglecting identification of microscopic roles of each species. In our case, however, the number of nodes is insufficient to obtain smooth statistical trends. Figure 2 shows the in-degree and out-degree distribution. Nodes scatter broadly in both sides of the dashed line that shows equal numbers of the in-degrees and the out-degrees, which indicates both agents, with larger values of the out-degree, and products are present in balance. When we carry out searches for global information in Figure 2, although it indicates that this degree distribution is not the Poisson one, we cannot clarify its statistical characteristic at this moment.

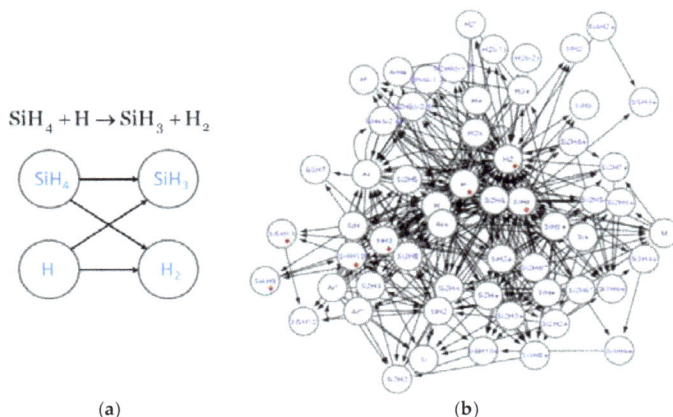

(a)   (b)

Figure 1. Graph for chemical reaction network in silane plasma. (**a**) Example of formation of nodes and edges for one typical reaction; and (**b**) whole structure of graph. Nodes represent species in chemical reactions, and edges start from agents and ends at products of each reaction. Chemical reactions are listed in Ref. [11]. Closed red diamonds indicate species displayed in Figures 2 and 3.

Symmetry **2017**, *9*, 309

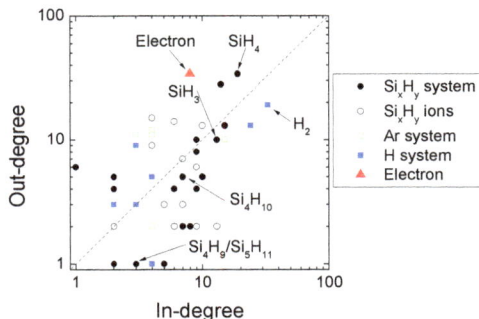

Figure 2. Distributions of in-degree and out-degree of species (or nodes) in silane plasma displayed in Figure 1. Inset dashed line indicates equal cases of in-degrees and out-degrees.

Here, we propose a diagram with C_b and C_c to analyze a macroscopic property of a graph and to obtain microscopic roles of species simultaneously, as shown in Figure 3 for the graph in Figure 1. In general, both C_b and C_c show centrality indices, and a node closer to the center has a higher value, and positive correlation is roughly expected between them. However, this diagram includes various types of information, as described below. For instance, nodes Si_4H_9 and Si_5H_{11} are located in the periphery region of Figure 1. Both of them are in the low-C_b area in Figure 3, but their positions in the diagram are different. Si_4H_9 has three directed paths from other species and another to Si_4H_{10}. The degree, the total number of edges, is limited to 4 as shown in Figure 2, and we can recognize the edges and the nodes around Si_4H_9 in Figure 1. Consequently, Si_4H_9 is not a significant species that affects many reactions, and it is located in the low-C_b area in Figure 3, while C_c of Si_4H_9 is in the middle range. Si_5H_{11} also has three directed paths from other species and one directed path to another species, Si_5H_{12}. It is in the low-C_b area with very low C_c. The difference of locations between Si_4H_9 and Si_5H_{11} is destination of out-degree edges; Si_4H_{10} is more active in reactions than Si_5H_{12} in our model, and we can clarify such a point using this diagram. Note that both of them have the same statistical values of degrees, and they are in the same position in the degree distribution that is a conventional classification in theory of complex network [2].

From a macroscopic point of view, this diagram displays semi-statistical properties of the graph as a visual classification tool. The overall distribution of data points displayed here is fairly uniform along the $\log(C_b)$ axis, and it covers the range in 4 orders of magnitude of C_b. The Si_xH_y system, which includes main species coming from the mother gas, SiH_4, is distributed throughout the entire range. The H system, which also originates from SiH_4, scatters in all ranges. The Ar system is in a less important area as C_b values. In the area with high-value C_b, stable species such as SiH_4 and H_2 exist, and electrons are also one of the highest-value species both in C_b and C_c. SiH_3, which is the most important precursor for Si thin film deposition using silane plasma, is also in this high-value region. In the case of the linear C_b axis shown in the inset of Figure 3, data points are concentrated around 0, and a few of them scatter in the area of high C_b, which is not a sufficient visualization of the graph. A logarithmic plot for C_b is a key manner for graphs for reaction paths and visual classifications of chemical reactions, in particular, in plasma-enhanced chemistry.

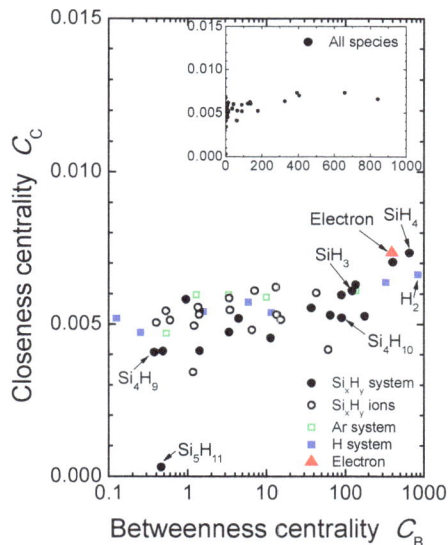

Figure 3. Diagram of betweenness-closeness centrality indices of reaction network in silane plasma displayed in Figure 1. Inset is diagram in linear scales.

Another fact that we can deduce from this diagram is rough estimation on differences coming from roles of species. Values of C_c, which are shown along the vertical axis, become high when the out-degree of a given species is larger than its in-degree (see Figure 2 to confirm these correspondences). For instance, C_c of SiH$_4$ is larger than that of H$_2$: the ratio, in-degree/out-degree of SiH$_4$, is 19/34, while that of H$_2$ is 33/19. Electrons have also unbalanced values: the ratio is 8/34. These facts indicate that species mainly working as agents (on the left-hand side of chemical reactions, such as A and B in Reaction (1), having larger out-degrees) are in the upper area of C_c. Products in chemical reactions tend to be in the lower area.

Figure 4 shows another example of chemical reactions, CH$_4$ system in methane plasma. In [8], we treat species except ions, but here we include all species listed in Ref. [10], shown in Table A2 in Appendix A. The $\log(C_b) - C_c$ diagram for the graph in Figure 4 is shown in Figure 5, which is quite similar to that in Figure 3; the area of $\log(C_b)$ is quite wide in comparison with the width of the C_c range. Also, most of the species that have been so far pointed out on their importance of roles are located in the higher-C_b range. In a microscopic point of view, we can also find similar points to the case of silane plasma. For instance, CH$_4$ and CH$_3$ are in the similar locations to SiH$_4$ and SiH$_3$ in Figure 3, respectively. Our previous report [8] in which the centrality indices similar to PageRank [17] indicate importance of CH$_3$, and the result here is consistent with the one in Ref. [8]. There are a few points that are different from the case of silane plasma, such as the location of electrons between Figures 3 and 5, although such a feature may be a factor coming from each specific system.

Two results shown here indicate that complex reactions in plasma chemistry can be visualized in a $\log(C_b) - C_c$ diagram, and both macroscopic and microscopic properties are derived from it. In particular, since plasma chemistry includes various levels of roles as well as wide range of contribution frequencies to reactions, the range of node distributions along the $\log(C_b)$ axis is quite wide. C_b, given by Equation (2), includes information as reaction connection between species, while C_c indicates rather simple information about a location in a network.

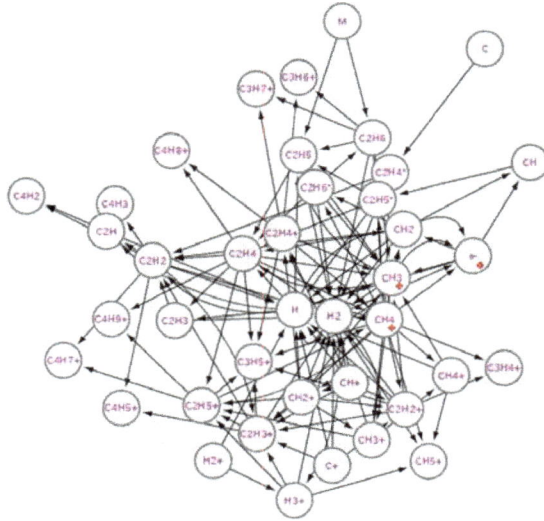

Figure 4. Graph for chemical reaction network in methane plasma. Nodes represent species in chemical reactions, and edges start from agents and ends at products of each reaction. Chemical reactions are listed in Ref. [10]. Closed red diamonds indicate species displayed in Figure 4.

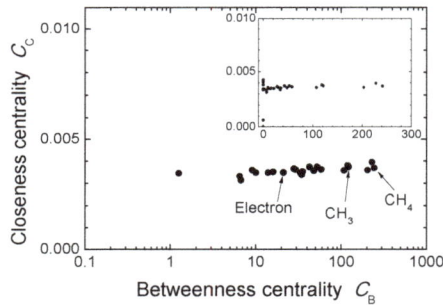

Figure 5. Diagram of betweenness-closeness centrality indices of reaction network in methane plasma displayed in Figure 3. Inset is diagram in linear scales.

3. Discussion

Figures 3 and 5 show wide-range distributions of nodes on the $\log(C_b)$ axis, but it might arise from simple randomness that also exists in random graphs. Here we compare such tendencies to those in arbitrary random graphs.

We fixed the numbers of nodes and edges to the ones in Figure 1, and created random graphs with directed edges in computation. Figure 6 shows three examples, and we cannot see any common points in allocation of roles for specific species. In all $\log(C_b) - C_c$ diagrams shown in Figure 7, the values C_b and C_c for the nodes are around a certain area. In particular, the values of $\log(C_b)$ is in a range with approximately one order of magnitude except for a few nodes. This is attributed to the fact that degrees of random graphs are in the Poisson distributions [18].

To compare spectra in the $\log(C_b)$ scale, the data in Figures 3, 5 and 7 are summarized as histograms of cumulative probabilities or relative densities P in Figure 8. Values of the nodes in random graphs are localized around 10^2, and almost no changes among these three plots. This is attributed to the fact that degrees of random graphs are in the Poisson distribution [18], and these graphs have certain averaged profiles of parameters with some deviations. On the other hand, the spectrum of silane plasma chemistry with the same numbers of the nodes and the edges is quite broad. The case of methane plasma chemistry shows similar tendencies. This comparison implies that networks of complex plasma chemistry include self-arranged systematic roles in the constituent species. The roles of agents and products are partially distinguished from scattering along the C_c axis in the diagram. A wide range of $\log(C_b)$ indicates a number of levels of intermediate roles from one species to another along successive sequential reactions, where the levels are from inevitable functions in reactions to less-frequent contributions for sub-products creations, etc.

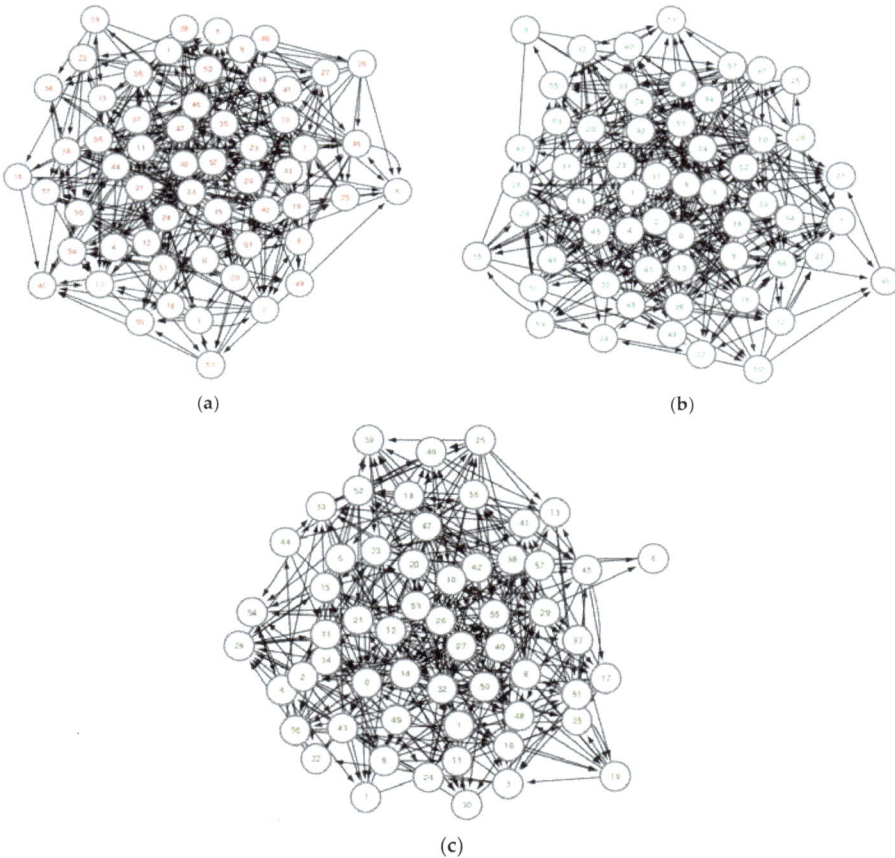

(a)

(b)

(c)

Figure 6. Random graphs created randomly in computation (a–c).

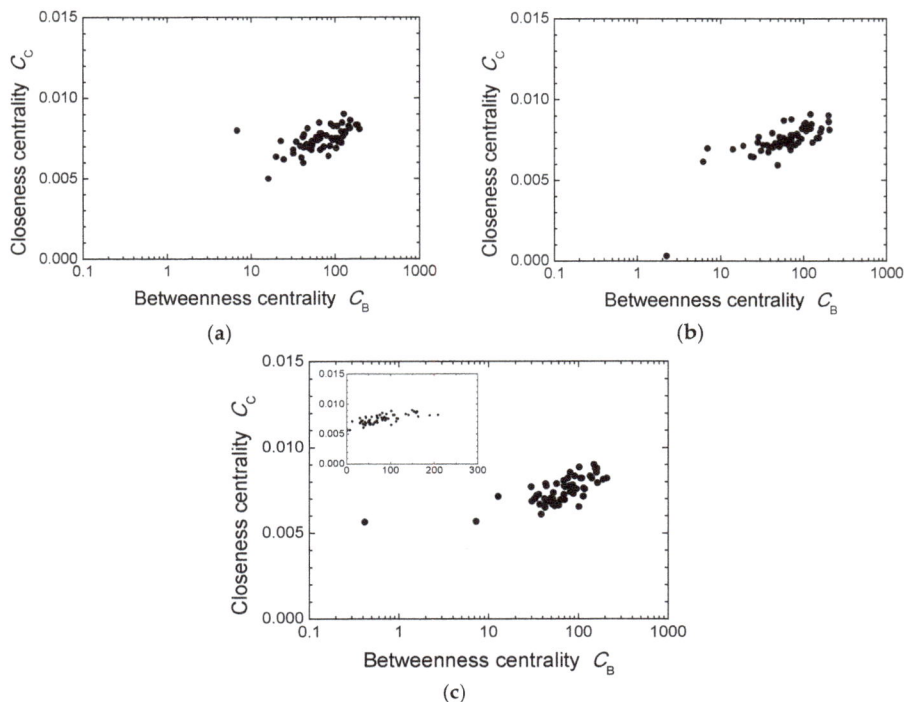

Figure 7. Diagrams of betweenness-closeness centrality indices. (**a**–**c**) corresponds to (**a**–**c**) in Figure 6, respectively. Inset is diagram in linear scales.

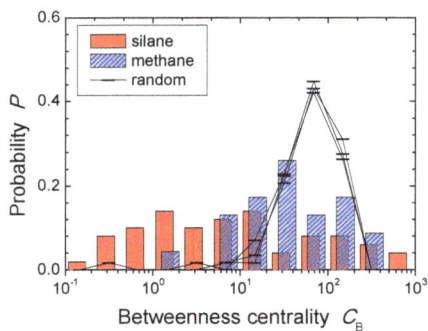

Figure 8. Histograms of cumulative probabilities or relative densities as function of betweenness centrality index. Data points are from Figures 3, 5 and 7.

The above descriptions are based on the data set on silane and methane plasma chemistry. If another sufficient set of reactions is available, one can perform similar data analysis on complex chemical networks. As shown here, approximate estimations of roles on species and global balance in the reaction system are beneficial for various purposes such as rapid analysis of robustness of a reaction system against impurity mixing. Another feasible and interesting contribution is selection of species for monitoring an industrial chemical reactor; behaviors of the limited number of the species detectable on the edge of reaction space can predict the ongoing entire system working on by comparing values of indices with the detected signals.

This kind of graph analysis using diagrams composed of multi-centrality indices is applicable for analyses of other networks. In particular, networks that include less than 100 nodes are suitable for this method to investigate both macroscopic topologies and microscopic identifications of nodes. This is also applicable for larger networks to clarify roles of nodes by specifying their positions in a rough distribution of node data in a diagram. For instance, other physical and chemical processes can be analyzed by this approach when they are described in rate equations, since chemical reactions are given in one type of rate equation and successfully interpreted in this study.

4. Conclusions

A diagram with axes of multi-centrality indices works to clarify both macroscopic structures of the graph and microscopic features such as roles of a species. Plasma-enhanced chemical reactions, in which tens of chemical reactions take place in parallel, are analyzed using a diagram based on C_b and C_c indices. The derived diagram indicates, from its macroscopic distribution of node values, that nodes have a wide variety of roles that are quite different from random classification. It also shows some roles of specific nodes by their locations, such as roles of products, agents, and intermediates. For plasma chemistry, this study provides us a measure for statistical properties for tens or hundreds of species in reaction systems, and for graph analysis, this analysis will become a practical and applicable method for understanding such medium-sized graphs.

Acknowledgments: One of the authors (Osamu Sakai) thanks Professor Tomoyuki Murakami at Seikei University, Professor Mark J. Kushner at University of Michigan, and Professor Tsuyoshi Akiyama at The University of Shiga Prefecture for their useful comments in this study. This study was partly supported by Grant-in-Aid for Scientific Research from the Japanese Ministry of Education, Culture, Sports, Science and Technology, Japan (JSPS KAKENHI Grant Number JP16K13711).

Author Contributions: Yasutaka Mizui performed main analysis shown here and wrote the paper by drawing figures. Tetsuya Kojima performed numerical calculation for random graphs. Shigeyuki Miyagi prepared for calculation environments and gave advices on analysis procedures including error corrections. Osamu Sakai conceived and designed the research activities, and wrote the paper.

Conflicts of Interest: The authors declare no conflict of interest.

Appendix A

The chemical reactions we analyze in this study are in two sets: silane (SiH_4) and methane (CH_4) plasma chemical reactions. All species treated in Figures 1–3 are shown in Table A1, and those in Figures 4 and 5 are in Table A2; they are listed in [10,11], respectively.

Table A1. List of species in silane plasma chemical reactions [11]. Except charge indicators "+" and "−", other symbols like "*" and "v" mean various electronic and mechanical excitation levels of species. M indicates arbitrary species except ions.

Number	Species	Number	Species
0	Ar	31	Si_2H_6 **
1	Ar *	32	$Si_2H_6^+$
2	Ar **	33	$Si_2H_7^+$
3	Ar^+	34	$Si_3H_4^+$
4	ArH^+	35	$Si_3H_6^+$
5	e^-	36	Si_3H_7
6	H	37	Si_3H_8
7	H *	38	Si_4H_{10}
8	H^+	39	$Si_4H_2^+$
9	H_2	40	$Si_4H_6^+$
10	$H_2(v(1))$	41	$Si_4H_8^+$
11	$H_2(v(2))$	42	Si_4H_9

Table A1. *Cont.*

Number	Species	Number	Species
12	H_2 *	43	$Si_5H_{10}^+$
13	H_2^+	44	Si_5H_{11}
14	H_3^+	45	Si_5H_{12}
15	M	46	$Si_5H_4^+$
16	Si	47	SiH
17	Si^+	48	SiH^+
18	Si_2H^+	49	SiH_2
19	Si_2H_2	50	SiH_2^-
20	$Si_2H_2^+$	51	SiH_2^+
21	Si_2H_3	52	SiH_3
22	$Si_2H_3^+$	53	SiH_3^-
23	Si_2H_4	54	SiH_3^+
24	$Si_2H_4^+$	55	SiH_4
25	Si_2H_5	56	$SiH_4(v(1,3))$
26	$Si_2H_5^+$	57	$SiH_4(v(2,4))$
27	Si_2H_6		
28	$Si_2H_6(v(1,3))$		
29	$Si_2H_6(v(2,4))$		
30	Si_2H_6 *		

Table A2. List of species in methane plasma chemical reactions [10]. Except charge indicator "+", other symbols like "*" mean various electronic and mechanical excitation levels of species. M indicates arbitrary species except ions.

Number	Species	Number	Species
1	CH_4	21	CH_3^+
2	H	22	CH_4^+
3	CH_3	23	$C_2H_2^+$
4	CH_2	24	$C_2H_3^+$
5	C_2H_6 *	25	$C_2H_4^+$
6	M	26	$C_2H_5^+$
7	CH	27	H_2^+
8	C_2H_5 *	28	H_3^+
9	C	29	e^-
10	C_2H_4 *	30	C_4H_3
11	C_2H_6	31	C_4H_2
12	C_2H_5	32	CH_5^+
13	C_2H_4	33	$C_3H_4^+$
14	C_2H_3	34	$C_3H_5^+$
15	C_2H_2	35	$C_4H_5^+$
16	C_2H	36	$C_4H_8^+$
17	C^+	37	$C_3H_6^+$
18	CH^+	38	$C_3H_7^+$
19	H_2	39	$C_4H_7^+$
20	CH_2^+	40	$C_4H_9^+$

References

1. Fox, J. *Proceedings of the Symposium on Generalized Networks*; Polytechnic Press: New York, NY, USA, 1966.
2. Kolaczyk, E.D. *Statistical Analysis of Network Data: Methods and Models*; Springer: Berlin, Germany, 2009.
3. Temkin, O.N.; Zeigarnik, A.V.; Bonchev, D. *Chemical Reaction Networks*; CRC Press: Boca Raton, FL, USA, 1996.
4. Leenheer, P.D.; Angeli, D.; Sontag, E.D. Monotone chemical reaction networks. *J. Math. Chem.* **2007**, *41*, 295–314. [CrossRef]
5. Gorban, A.N.; Yablonsky, G.S. Extended detailed balance for systems with irreversible reactions. *Chem. Eng. Sci.* **2011**, *66*, 5388–5399. [CrossRef]

6. Jeong, H.; Tombor, B.; Albert, R.; Oltvai, Z.N.; Barabasi, A.-L. The large-scale organization of metabolic networks. *Nature* **2000**, *407*, 651–654. [PubMed]
7. Albert, R.; Jeong, H.; Barabasi, A.-L. Error and attack tolerance of complex networks. *Nature* **2000**, *406*, 378–382. [CrossRef] [PubMed]
8. Sakai, O.; Nobuto, K.; Miyagi, S.; Tachibana, K. Analysis of weblike network structures of directed graphs for chemical reactions in methane plasmas. *AIP Adv.* **2015**, *5*, 107140. [CrossRef]
9. Mizui, Y.; Nobuto, K.; Miyagi, S.; Sakai, O. Complex reaction network in Silane Plasma chemistry. In *Complex Networks VIII*; Springer International Publishing: Cham, Switzerland, 2017; pp. 135–140.
10. Tachibana, K.; Nishida, M.; Harima, H.; Urano, Y. Diagnostics and modelling of a methane plasma used in the chemical vapour deposition of amorphous carbon films. *J. Phys. D* **1984**, *17*, 1727–1742. [CrossRef]
11. Kushner, M.J. A model for the discharge kinetics and plasma chemistry during plasma enhanced chemical vapor deposition of amorphous silicon. *J. Appl. Phys.* **1988**, *63*, 2532–2551. [CrossRef]
12. Murakami, T.; Niemi, K.; Gans, T.; O'Connell, D.; Graham, W.G. Chemical kinetics and reactive species in atmospheric pressure helium-oxygen plasmas with humid-air impurities. *Plasma Sources Sci. Technol.* **2013**, *22*, 015003. [CrossRef]
13. Agliari, E.; Barra, A.; Bartolucci, S.; Galluzzi, A.; Guerra, F.; Moauro, F. Parallel processing in immune networks. *Phys. Rev. E* **2013**, *87*, 42701. [CrossRef] [PubMed]
14. Agliari, E.; Barra, A.; Schiavo, L.D.; Moro, A. Complete integrability of information processing by biochemical reactions. *Sci. Rep.* **2016**, *6*, 36314. [CrossRef] [PubMed]
15. Bie, C.D.; Dijk, J.; Bogaerts, A. The dominant pathways for the conversion of methane into oxygenates and syngas in an atmospheric pressure dielectric barrier discharge. *J. Phys. Chem. C* **2015**, *119*, 22331–22350. [CrossRef]
16. Lieberman, M.A.; Lichtenberg, A.J. *Principles of Plasma Discharges and Material Processing*; John Wiley and Sons: New York, NY, USA, 1994.
17. Brin, S.; Page, L. The anatomy of a large-scale hypertextual web search engine. *Comput. Netw. ISDN Syst.* **1998**, *30*, 107–117. [CrossRef]
18. Albert, R.; Barabasi, A.-L. Statistical mechanics of complex networks. *Rev. Mod. Phys.* **2002**, *74*, 47–97. [CrossRef]

symmetry

MDPI

Article

Reconstructing Damaged Complex Networks Based on Neural Networks

Ye Hoon Lee [1] and Insoo Sohn [2],*

[1]   Department of Electronic and IT Media Engineering, Seoul National University of Science and Technology, Seoul 01811, Korea; y.lee@snut.ac.kr
[2]   Division of Electronics and Electrical Engineering, Dongguk University, Seoul 04620, Korea
*   Correspondence: isohn@dongguk.edu; Tel.: +82-2-2260-8604

Received: 2 October 2017; Accepted: 9 December 2017; Published: 9 December 2017

Abstract: Despite recent progress in the study of complex systems, reconstruction of damaged networks due to random and targeted attack has not been addressed before. In this paper, we formulate the network reconstruction problem as an identification of network structure based on much reduced link information. Furthermore, a novel method based on multilayer perceptron neural network is proposed as a solution to the problem of network reconstruction. Based on simulation results, it was demonstrated that the proposed scheme achieves very high reconstruction accuracy in small-world network model and a robust performance in scale-free network model.

Keywords: network reconstruction; neural networks; small world networks; scale free networks

1. Introduction

Complex networks have received growing interest from various disciplines to model and study the network topology and interaction between nodes within a modeled network [1–3]. One of the important problems that are actively studied in the area of network science is the robustness of a network under random failure of nodes and intentional attack on the network. For example, it has been found that scale-free networks are more robust compared to random networks against random removal of nodes, but are more sensitive to targeted attacks [4–6]. Furthermore, many approaches have been proposed to optimize conventional networks against random failure and intentional attack compared to the conventional complex networks [7–10]. However, these approaches have been mainly concentrated on designing network topology based on various optimization techniques to minimize the damage to the network. So far, no work has been reported on techniques to repair and recover the network topology after the network has been damaged. Numerous solutions have been proposed on reconstruction of spreading networks based on spreading data [11–13], but the problem on the reconstruction of damaged networks due to random and targeted attacks has not been addressed before.

Artificial neural networks (NNs) have been applied to solve various problems in complex systems due to powerful generalization abilities of NNs [14]. Some of the applications where NNs have been successfully used are radar waveform recognition [15], image recognition [16,17], indoor localization [18,19], and peak-to-average power reduction [20,21]. In this paper, we propose a novel network reconstruction method based on the NN technique. To the best of our knowledge, this work is the first attempt to recover the network topology after the network has been damaged and also the first attempt to apply NN technique for complex network optimization. We formulate the network reconstruction problem as an identification of a network structure based on a much reduced amount of link information contained in the adjacency matrix of the damaged network. The problem is especially challenging due to (1) very large number of possible network configurations on the order of 2^{N^2}, where N is the number of nodes and (2) very small number of node interaction information due to node removals. We simplify the problem by the following assumptions (1) average number of real connections of a network is much smaller than all the possible link configurations and (2) link information of M undamaged networks are available. Based on these assumptions, we chose the multiple-layer perceptron neural network (MLPNN), which is one of the frequently used NN techniques, as the basis of our method. We evaluate the performance of the proposed method based on simulations in two classical complex networks (1) small-world network and (2) scale-free networks, generated by Watts and Strogatz model and Barabási and Albert model, respectively.

The rest of the paper is organized as follows. Section 2 describes the small-world network model and scale-free network, followed by the network damage model. In Section 3, we propose the reconstruction method based on NN technique. In Section 4, we present the numerical results of the proposed method, and conclusions are given Section 5.

2. Model

2.1. Small-World Network

Small-world network is an important network model with low average path length and high clustering coefficient [22–24]. Small-world networks have homogeneous network topology with a degree distribution approximated by the Poisson distribution. A small-world network is created based on a regular lattice such a ring of N nodes, where each node is connected to J nearest nodes. Next, the links are randomly rewired to one of N nodes in the network with probability p. The rewiring process is repeated for all the nodes $n = 1 \ldots N$. By controlling the rewiring probability p, the network will interpolate between a regular lattice ($p = 0$) to a random network ($p = 1$). Figure 1 shows the detailed algorithm for small-world network construction in pseudocode format.

```
1: Initialization:
2:      Set the total number of nodes N
3:      Set the rewiring probability p
4:      Set the node degree K
5: end initialization
6:
7: for each node n ∈ (1, N)
8:      Connect to nearest K nodes.
9: end for
10:
11: for each node n ∈ (1, N)
12:     for each link k ∈ (1, K)
13:         Generate a random number r ∈ (0, 1)
14:         if r < p
15:             Select a node randomly m ∈ (1, N)
16:             Rewire to node m
17:         end if
18:     end for
19: end for
```

Figure 1. Small-world network algorithm.

2.2. Scale-Free Network

Scale-free networks, such as Barabási and Albert (BA) network, are evolving networks that have degree distribution following power-law model [4–6]. Scale-free networks consist of a small number of nodes with very high degree of connections and rest of the nodes with low degree connections. A scale-free network starts the evolution process with a small number of m_0 nodes. Next, a new node is introduced to the network and attaches to m existing nodes with high degree k. The process consisting of new node introduction and preferential attachment is repeated until a network with $N = t + m_0$ nodes has been constructed. Figure 2 shows the detailed algorithm for scale-free network construction in pseudocode format.

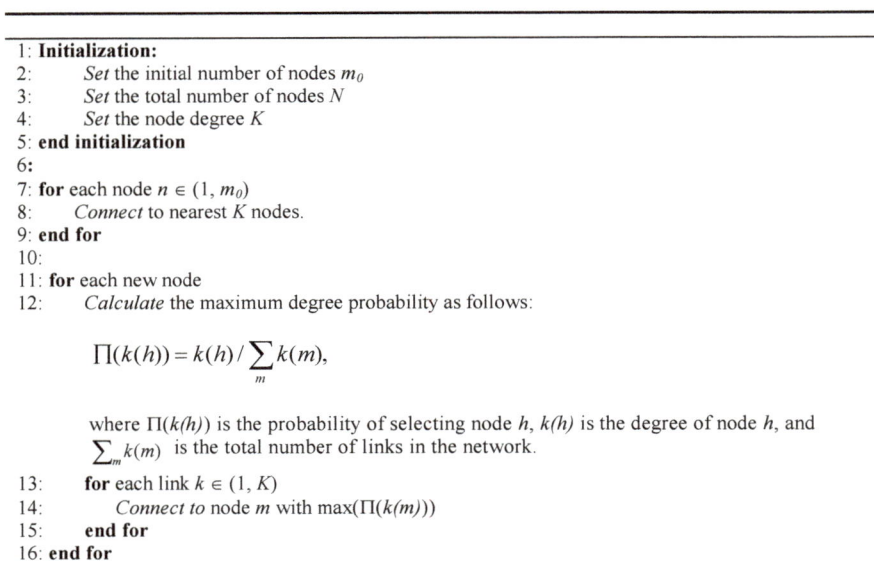

1: **Initialization:**
2:      *Set* the initial number of nodes m_0
3:      *Set* the total number of nodes N
4:      *Set* the node degree K
5: **end initialization**
6:
7: **for** each node $n \in (1, m_0)$
8:      *Connect* to nearest K nodes.
9: **end for**
10:
11: **for** each new node
12:      *Calculate* the maximum degree probability as follows:

$$\prod(k(h)) = k(h) / \sum_m k(m),$$

      where $\prod(k(h))$ is the probability of selecting node h, $k(h)$ is the degree of node h, and $\sum_m k(m)$ is the total number of links in the network.
13:      **for** each link $k \in (1, K)$
14:          *Connect to* node m with max($\prod(k(m))$)
15:      **end for**
16: **end for**

Figure 2. Scale-free network algorithm.

2.3. Network Damage Model

We simulate the damage process on complex networks by considering the random attack model. In the random attack model, nodes are randomly selected and removed. Note that when a node is removed, all the links connected to that node are also removed [25]. To evaluate the performance of the proposed reconstruction method, the difference in the number of links between the original network and the reconstructed network is used and represented as probability of reconstruction error, P_{RE}, that is defined as follows:

$$P_{RE} = \frac{N_{L,diff}}{N_L}, \tag{1}$$

where N_L is the total number of existing links in the complex network that were damaged due to random attack and $N_{L,diff}$ is the total number of links in the reconstructed network that are different from the links in the original network before any node removal. For example, let us assume that the number of nodes $N = 4$ and the node pair set in the original network is equal to $E_o = \{(1, 2), (1, 4), (2, 3), (3, 4)\}$. If the reconstructed network has node pair set as $E_r = \{(1, 2), (1, 3), (1, 4), (2, 3)\}$, then $N_{L,diff} = 2$ and $N_L = 4$, giving us $P_{RE} = 0.5$. Note that the estimated links in the reconstructed network that were not in the original network, in additions to links that were not reproduced, are all counted as errors.

3. Reconstruction Method

3.1. Neural Network Model

Neural networks are important tools that are used for system modeling with good generalization properties. We propose to use MLPNN employing backpropagation based supervised learning in this work. A MLPNN has three types of layers: An input layer, an output layer, and multiple hidden layers in between the input and output layer as shown in Figure 3. The input layer receives input data and is passed to the neurons or units in the hidden layer. The hidden layer units are nonlinear activation function of the weighted sum of inputs from the previous layer. The output of the *j*th unit in the hidden layer *O*(*j*) can be represented as [26]

$$A(j) = \sum_{i=1}^{m} w(j,i)\,O(i) - U(j),$$ (2)

$$O(j) = \Im(A(j)) = \frac{1}{1+e^{-A(j)}},$$ (3)

where $A(j)$ is the activation input to *j*th hidden layer unit, $w(i,j)$ is the weights from unit i to j, $O(j)$ is the input to unit j, $U(j)$ is the threshold of unit j, and $\Im(\bullet)$ is a nonlinear activation function such as sigmoid function, as shown in Equation (3), hardlimit function, radial basis function, and triangular function. The number of units in the output layer is equal to the dimension of the desired output data format. The weights on the network connections are adjusted using training input data and desired output data until the mean square error (MSE) between them are minimized. To implement the MPLNN, the *feedforwardnet* function provided by the MATLAB Neural Network Toolbox was utilized. Additionally, the MPLNN weights were trained using the Levenberg-Marguardt algorithm [27].

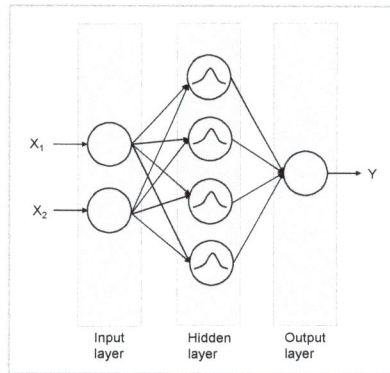

Figure 3. Multiple-layer perceptron neural network (MLPNN) model.

3.2. Neural Network Based Method

To reconstruct the network topology of a damaged complex network due to random attack, we apply MLPNN as a solution to solve the complex network reconstruction problem. One of the key design issues in MLPNN is the training process of weights on the network connections such that the MLPNN is successfully configured to reconstruct damaged networks. Usually, a complex network topology is represented by an adjacency matrix describing interactions of all the nodes in the network. However, an adjacency matrix is inappropriate as training input data for MLPNN training process due to its complexity. Thus, we define a link list (LL) that contains binary elements representing existence of node pairs among all possible combination of node pairs $\begin{pmatrix} N \\ 2 \end{pmatrix}$, where N is the number of nodes in the network. For example, for a network with $N = 4$, possible node pair set is equal to, $E = \{(1, 2), (1, 3), (1, 4), (2, 3), (2, 4), (3, 4)\}$. If a network to be reconstructed has four nodes and four links given by $E = \{(1, 2), (1, 4), (2, 3), (3, 4)\}$, then $LL = [1\ 0\ 1\ 1\ 0\ 1]$, where 1 s represents the existence of the four specific links among six possible ones. To obtain the training input data, M networks are damaged by randomly removing f percent of N nodes in a network. As shown in Figure 4, the proposed method consists of the following modules: Adjacency matrix of damaged network input module, adjacency matrix to link list transformation module, MLPNN module, and network index to adjacency matrix transformation module. In the second module, the adjacency matrices of the damaged networks are pre-processed into LLs that can be entered into the MLPNN. Note that the input dimension of the MLPNN is equal to the dimension of the training input data format. Thus, input dimension of the MLPNN is equal to $\binom{N}{2}$. As for the desired output data, which is the output of the MLPNN module, binary sequence numbers are used to represent the indices of the original complex networks that have been damaged. The number of MLPNN output will depend on the number of training networks used to train the MLPNN. For example, eight binary outputs will be sufficient to represent 256 complex networks. Based on the training input data and desired output data, representing network topology of different complex networks, the goal of the MLPNN is to be able to identify, reconstruct, and produce node pair information of the original network, among numerous networks used to train the neural network. The detailed training algorithm of MLPNN is described in Figure 5.

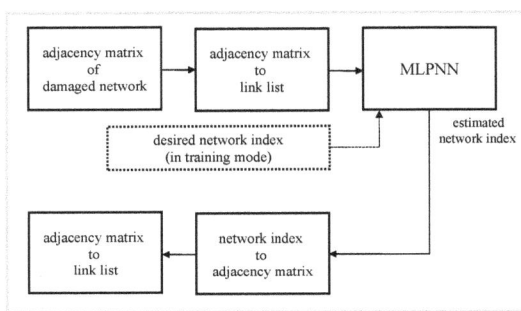

Figure 4. MLPNN based reconstruction method.

```
 1: Initialization:
 2:      Generate a set of M complex networks.
 3:      Set number of hidden layers.
 4:      Set number of neurons in each hidden layers.
 5:      Define activation function in hidden layers.
 6:      Set number of training iteration Iteration Num
 7: end Initialization
 8:
 9:    for i = 1 : Iteration Num
10:         Randomly select a complex networks m ∈ (1, M).
11:         Apply random attack procedure to the complex network m.
12:         Obtain adjacency matrix for the damaged complex networks m.
13:         Transform adjacency matrix into LL.
14:         InputTrainSeq = LL(m).
15:         DesOutputTrainSeq = dec2bin(m).
16: end for
17:
18: while estimation error > threshold do
19:      for i = 1 : Iteration Num
20:           MLPNN (InputTrainSeq) = MLPNN_OUT.
21:           Mean Squrare Error (MLPNN_OUT, DesOutputTrainSeq).
22:           Update weights w in each layer using LMS algorithm.
23:      end for
24: end while
```

Figure 5. MLPNN training algorithm.

4. Performance Evaluations

4.1. Simulation Environment

We study and evaluate the proposed reconstruction method based on the probability of reconstruction error P_{RE} described in Section 2. For the network damage model, we assume random attack process, where nodes are randomly removed with attached links. The MLPNN used in our method has two hidden layers with 64 neurons in the first layer and four neurons in the second layer. The nonlinear activation function in the hidden layer is chosen to be triangular activation function. The number of inputs to the MLPNN depends on the number of nodes in the network. To train and test the MLPNN, using complex networks with $N = 10$, $N = 30$, and $N = 50$, the number of inputs are set equal to the possible number of node pair combinations, which are 45, 435, and 1225, respectively. As for the number of outputs, eight are chosen to represent maximum number of 256 complex networks. The training input and output data patterns are randomly chosen from LL of M damaged complex networks with different percentage f of failed nodes out of total N nodes and corresponding indices of the complex networks.

4.2. Small-World Network

To evaluate the performance of the proposed method in small-world network model, the network is implemented based on the algorithm described in Figure 1. Figure 6 and Table 1 shows the reconstruction error probability as a function of percentage of random node failure f. Furthermore, we study the influence of the number of node on the network reconstruction performance with $N = 10$, $N = 30$, and $N = 50$. The initial degree K of the network is set to two and the links are randomly rewired with probability $p = 0.15$. One can see that with the increase in the number of node failures, the reconstruction performance deteriorates for all different N, but for $f = 0.1$, P_{RE} is less than 0.35 and for $f = 0.5$, P_{RE} is less than 0.5. In another words, the proposed method can reconstruct almost close to 70% of the network topology for 10% node failures and more than 60% of the network topology for 50% node failures. Note that lower reconstruction error probability is observed for larger number of nodes. The reason for this results is due to the higher dimension of input data to the MLPNN, e.g., 1225 for $N = 50$. Furthermore, from the figure, we observe that P_{RE} is less than what one might expect for the case where most of the nodes are destroyed, e.g., $f = 0.7$. This phenomenon is due to the large number of overlap in the node connections in LL among the M damaged networks due to small rewiring probability p. To study how the rewiring probability affects the reconstruction accuracy, simulations are performed with $p = 0.3$, $p = 0.5$, and $p = 0.7$, as shown in Figure 7 and Table 2. From the figure, we can see that there is a significant deterioration in performance in reconstruction accuracy with increase in rewiring probability p. This is because the small-world network topology becomes increasingly disordered with increase in rewiring probability and results in decrease in ability of the proposed method to reproduce the original network topology.

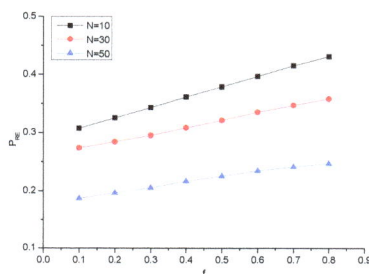

Figure 6. Probability of reconstruction error for small-world networks with $N = 10$, $N = 30$, $N = 50$, $M = 10$, and $p = 0.15$.

Table 1. Probability of reconstruction error for small-world networks with $N = 10$, $N = 30$, $N = 50$, $M = 10$, and $p = 0.15$.

N/f	0.1	0.2	0.3	0.4	0.5	0.6	0.7	0.8
10	0.307	0.325	0.343	0.361	0.379	0.397	0.415	0.431
30	0.273	0.284	0.295	0.308	0.321	0.335	0.347	0.358
50	0.186	0.196	0.205	0.216	0.225	0.234	0.241	0.246

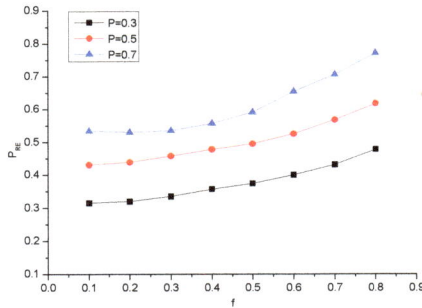

Figure 7. Probability of reconstruction error for small-world networks with $p = 0.3$, $p = 0.5$, $p = 0.7$, $M = 10$, and $N = 50$.

Table 2. Probability of reconstruction error for small-world networks with $p = 0.3$, $p = 0.5$, $p = 0.7$, $M = 10$, and $N = 50$.

P/f	0.1	0.2	0.3	0.4	0.5	0.6	0.7	0.8
0.3	0.315	0.319	0.334	0.356	0.373	0.399	0.431	0.477
0.5	0.430	0.438	0.457	0.477	0.494	0.524	0.566	0.616
0.7	0.534	0.529	0.534	0.556	0.590	0.653	0.705	0.770

However, even in the case of high rewiring probability $p = 0.5$, 50% of links can be successfully estimated. In Figure 8 and Table 3, we study the influence of the number of networks M that were used to train and test the MLPNN on the reconstruction performance. The number of nodes N is assumed to be 50 and the rewiring probability p is set to 0.5. It can be observed from the figure that there is a small degradation in performance with increase in M, but, P_{RE} remains less than 0.3.

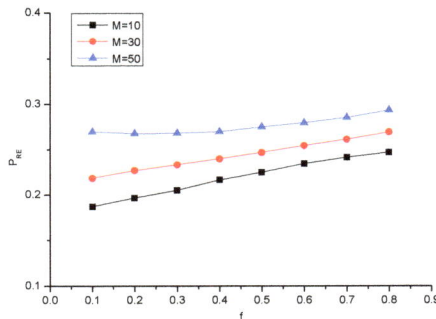

Figure 8. Probability of reconstruction error for small-world networks with $M = 10$, $M = 30$, $M = 50$, $N = 50$, and $p = 0.15$.

Table 3. Probability of reconstruction error for small-world networks with $M = 10$, $M = 30$, $M = 50$, $N = 50$, and $p = 0.15$.

M/f	0.1	0.2	0.3	0.4	0.5	0.6	0.7	0.8
10	0.186	0.196	0.205	0.216	0.225	0.234	0.241	0.246
30	0.218	0.226	0.233	0.239	0.246	0.253	0.260	0.268
50	0.26	0.267	0.267	0.269	0.274	0.279	0.285	0.293

4.3. Scale-Free Network

The proposed method is also evaluated in scale-free network model that is generated using the algorithm described in Figure 2. Figure 9 and Table 4 compares the reconstruction error probability for different number of nodes $N = 10$, $N = 30$, and $N = 50$. The initial number of nodes m_0 was set to two and the node degree $K = 2$ for the preferential attachment process. Figure 9 shows that the reconstruction accuracy in scale-free network model is significantly lower compared to the small-world network. The reason for the poor performance is that the network topologies of M scale-free networks are more complex compared to the small-world network models. Furthermore, the links in LL between the M damaged networks do not overlap as much as in the small-world network models. In Figure 10 and Table 5, the reconstruction error probability performance with $N = 30$ and $m_0 = 2$, for different number of networks $M = 10$, $M = 30$, and $M = 50$, is shown. Compared to the small-world network environment, the reconstruction error probability values are quite high even in low percentage of node failures for $M = 30$ and $M = 50$. Due to the complex topology of the scale-free network model, increase in M affects the link estimation ability of the MLPNN. Finally, Figure 11 and Table 6 shows the reconstruction accuracy performance with different initial number of nodes in constructing scale-free network model. One can observe that there is a small difference in reconstruction performance for high percentage of node failures, regardless of the initial number of nodes. This is because the link estimation difficulty is almost equal to the MLPNN, even if they have different degree distributions, when the number of hubs remains the same in scale-free networks.

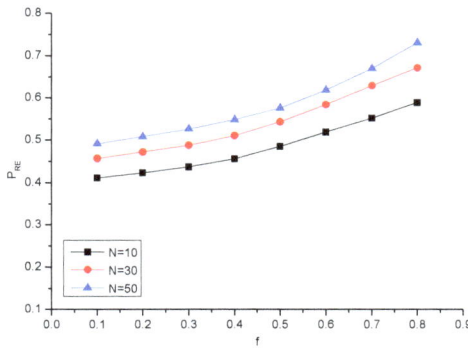

Figure 9. Probability of reconstruction error for scale-free networks with $N = 10$, $N = 30$, $N = 50$, $M = 10$, and $m_0 = 2$.

Table 4. Probability of reconstruction error for scale-free networks with $N = 10$, $N = 30$, $N = 50$, $M = 10$, and $m_0 = 2$.

N/f	0.1	0.2	0.3	0.4	0.5	0.6	0.7	0.8
10	0.411	0.423	0.437	0.455	0.484	0.518	0.551	0.588
30	0.455	0.471	0.487	0.510	0.542	0.582	0.628	0.671
50	0.490	0.507	0.525	0.547	0.575	0.618	0.669	0.730

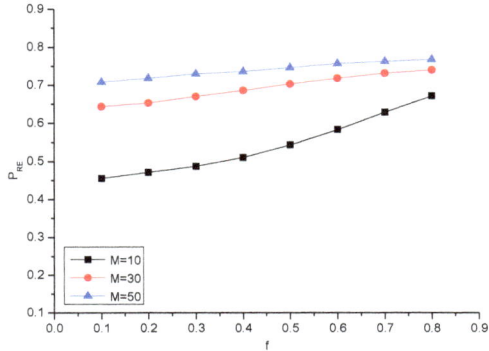

Figure 10. Probability of reconstruction error for scale-free networks with $M = 10$, $M = 30$, $M = 50$, $N = 30$, and $m_0 = 2$.

Table 5. Probability of reconstruction error for scale-free networks with $M = 10$, $M = 30$, $M = 50$, $N = 30$, and $m_0 = 2$.

M/f	0.1	0.2	0.3	0.4	0.5	0.6	0.7	0.8
10	0.455	0.471	0.487	0.510	0.542	0.582	0.628	0.671
30	0.643	0.652	0.669	0.685	0.702	0.717	0.730	0.738
50	0.708	0.718	0.729	0.735	0.745	0.755	0.761	0.766

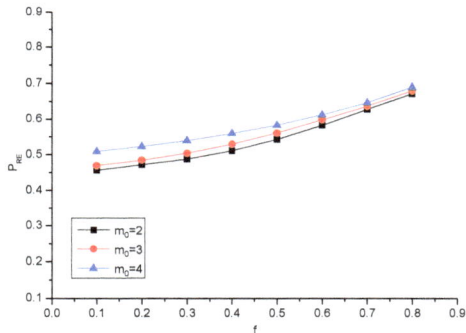

Figure 11. Probability of reconstruction error for scale-free networks with $m_0 = 2$, $m_0 = 3$, $m_0 = 4$, $N = 30$, and $M = 10$.

Table 6. Probability of reconstruction error for scale-free networks with $m_0 = 2$, $m_0 = 3$, $m_0 = 4$, $N = 30$, and $M = 10$.

m_0/f	0.1	0.2	0.3	0.4	0.5	0.6	0.7	0.8
2	0.455	0.471	0.487	0.510	0.542	0.582	0.628	0.671
3	0.468	0.483	0.503	0.528	0.559	0.597	0.636	0.679
4	0.508	0.522	0.530	0.558	0.582	0.612	0.646	0.689

5. Conclusions

In this paper, we proposed a new method that efficiently reconstructs the topology of the damaged complex networks based on NN technique. To the best of our knowledge, our proposed method is the first known attempt in the literature to recover the network topology after the network has been damaged and also the first known application of the NN technique for complex network optimization. The main purpose of our work was to design a NN solution based on known damaged network topology for accurate reconstruction. The proposed reconstruction method was evaluated based on the probability of reconstruction error in small-world network and scale-free network models. From simulation results, the proposed method was able to reconstruct around 70% of the network topology for 10% node failures for small-world networks and around 50% of the network topology for 10% node failures for scale-free networks. Important topics that need to be considered in the future work is to develop a new link list that can represent both unidirectional and bidirectional link information and various performance metric needs to be developed that can be used to provide deeper understanding on the network reconstruction performance.

Acknowledgments: This research was supported by Basic Science Research Program through the National Research Foundation of Korea (NRF) funded by the Ministry of Education (2017R1D1A1B03035522).

Author Contributions: Insoo Sohn conceived, designed the experiments, and wrote the paper. Ye Hoon Lee analyzed data and designed the experiments.

Conflicts of Interest: The authors declare no conflict of interest.

References

1. Wang, X.F.; Chen, G. Complex networks: Small-world, scale-free and beyond. *IEEE Circuits Syst. Mag.* **2003**, *3*, 6–20. [CrossRef]
2. Newman, M. *Networks: An Introduction*; Oxford University Press: Oxford, UK, 2010.
3. Sohn, I. Small-world and scale-free network models for IoT systems. *Mob. Inf. Syst.* **2017**, *2017*. [CrossRef]
4. Barabási, A.; Albert, R. Emergence of scaling in random networks. *Science* **1999**, *286*, 509–512. [PubMed]
5. Albert, R.; Barabási, A. Statistical mechanics of complex networks. *Rev. Mod. Phys.* **2002**, *74*, 47–97. [CrossRef]
6. Barabási, A.L. Scale-free networks: A decade and beyond. *Science* **2009**, *325*, 412–413. [CrossRef] [PubMed]
7. Crucitti, P.; Latora, V.; Marchiori, M.; Rapisarda, A. Error and attack tolerance of complex networks. *Phys. A Stat. Mech. Appl.* **2004**, *340*, 388–394. [CrossRef]
8. Tanizawa, T.; Paul, G.; Cohen, R.; Havlin, S.; Stanley, H.E. Optimization of network robustness to waves of targeted and random attacks. *Phys. Rev. E* **2005**, *71*, 1–4. [CrossRef] [PubMed]
9. Schneider, C.M.; Moreira, A.A.; Andrade, J.S.; Havlin, S.; Herrmann, H.J. Mitigation of malicious attacks on networks. *Proc. Natl. Acad. Sci. USA* **2011**, *108*, 3838–3841. [CrossRef] [PubMed]
10. Ash, J.; Newth, D. Optimizing complex networks for resilience against cascading failure. *Phys. A Stat. Mech. Appl.* **2007**, *380*, 673–683. [CrossRef]
11. Clauset, A.; Moore, C.; Newman, M.E.J. Hierarchical structure and the prediction of missing links in networks. *Nature* **2008**, *453*, 98–101. [CrossRef] [PubMed]
12. Guimerà, R.; Sales-Pardo, M. Missing and spurious interactions and the reconstruction of complex networks. *Proc. Natl. Acad. Sci. USA* **2010**, *106*, 22073–22078. [CrossRef] [PubMed]
13. Zeng, A. Inferring network topology via the propagation process. *J. Stat. Mech. Theory Exp.* **2013**. [CrossRef]
14. Haykin, S. *Neural Networks: A Comprehensive Foundation*; Macmillan: Basingstoke, UK, 1994.
15. Zhang, M.; Diao, M.; Gao, L.; Liu, L. Neural networks for radar waveform recognition. *Symmetry* **2017**, *9*. [CrossRef]
16. Lawrence, S.; Giles, C.L.; Tsoi, A.C.; Back, A.D. Face recognition: A convolutional neural-network approach. *IEEE Trans. Neural Netw.* **1997**, *8*, 98–113. [CrossRef] [PubMed]
17. Lin, S.H.; Kung, S.Y.; Lin, L.J. Face recognition/detection by probabilistic decision-based neural network. *IEEE Trans. Neural Netw.* **1997**, *8*, 114–132. [PubMed]
18. Fang, S.H.; Lin, T.N. Indoor location system based on discriminant-adaptive neural network in IEEE 802.11 environments. *IEEE Trans. Neural Netw.* **2008**, *19*, 1973–1978. [CrossRef] [PubMed]
19. Sohn, I. Indoor localization based on multiple neural networks. *J. Inst. Control Robot. Syst.* **2015**, *21*, 378–384. [CrossRef]
20. Sohn, I. A low complexity PAPR reduction scheme for OFDM systems via neural networks. *IEEE Commun. Lett.* **2014**, *18*, 225–228. [CrossRef]

21. Sohn, I.; Kim, S.C. Neural network based simplified clipping and filtering technique for PAPR reduction of OFDM signals. *IEEE Commun. Lett.* **2015**, *19*, 1438–1441. [CrossRef]

22. Watts, D.; Strogatz, S. Collective dynamics of 'small-world' network. *Nature* **1998**, *393*, 440–442. [CrossRef] [PubMed]

23. Kleinberg, J.M. Navigation in a small world. *Nature* **2000**, *406*, 845. [CrossRef] [PubMed]

24. Newman, M.E.J. Models of the small world. *J. Stat. Phys.* **2000**, *101*, 819–841. [CrossRef]

25. Holme, P.; Kim, B.J.; Yoon, C.N.; Han, S.K. Attack vulnerability of complex networks. *Phys. Rev. E* **2002**, *65*, 1–14. [CrossRef] [PubMed]

26. Johnson, J.; Picton, P. How to train a neural network: An introduction to the new computational paradigm. *Complexity* **1996**, *1*, 13–28. [CrossRef]

27. Marquardt, D.W. An algorithm for least-squares estimation of nonlinear parameters. *J. Soc. Ind. Appl. Math.* **1963**, *11*, 431–441. [CrossRef]

symmetry

MDPI

Article

Analyzing Spatial Behavior of Backcountry Skiers in Mountain Protected Areas Combining GPS Tracking and Graph Theory

Karolina Taczanowska [1,*], Mikołaj Bielański [2], Luis-Millán González [3], Xavier Garcia-Massó [4] and José L. Toca-Herrera [5]

[1] Institute of Landscape Development, Recreation and Conservation Planning, Department of Landscape, Spatial and Infrastructure Sciences, University of Natural Resources and Life Sciences (BOKU), 1190 Vienna, Austria

[2] Department of Tourism and Recreation, University School of Physical Education, 31-571 Cracow, Poland; mikolaj.bielanski@awf.krakow.pl

[3] Department of Physical Education and Sport, University of Valencia, 46010 Valencia, Spain; luis.m.gonzalez@uv.es

[4] Departament de Didàctica de l'Expressió Musical, Plàstica i Corporal, University of Valencia, 46010 Valencia, Spain; xavier.garcia@uv.es

[5] Institute for Biophysics, Department of Nanobiotecnology, University of Natural Resources and Life Sciences (BOKU), 1190 Vienna, Austria; jose.toca-herrera@boku.ac.at

* Correspondence: karolina.taczanowska@boku.ac.at; Tel.: +43-1-47654-85324

Received: 15 November 2017; Accepted: 11 December 2017; Published: 14 December 2017

Abstract: Mountain protected areas (PAs) aim to preserve vulnerable environments and at the same time encourage numerous outdoor leisure activities. Understanding the way people use natural environments is crucial to balance the needs of visitors and site capacities. This study aims to develop an approach to evaluate the structure and use of designated skiing zones in PAs combining Global Positioning System (GPS) tracking and analytical methods based on graph theory. The study is based on empirical data (n = 609 GPS tracks of backcountry skiers) collected in Tatra National Park (TNP), Poland. The physical structure of the entire skiing zones system has been simplified into a graph structure (structural network; undirected graph). In a second step, the actual use of the area by skiers (functional network; directed graph) was analyzed using a graph-theoretic approach. Network coherence (connectivity indices: β, γ, α), movement directions at path segments, and relative importance of network nodes (node centrality measures: degree, betweenness, closeness, and proximity prestige) were calculated. The system of designated backcountry skiing zones was not evenly used by the visitors. Therefore, the calculated parameters differ significantly between the structural and the functional network. In particular, measures related to the actually used trails are of high importance from the management point of view. Information about the most important node locations can be used for planning sign-posts, on-site maps, interpretative boards, or other tourist infrastructure.

Keywords: protected areas; tourism; tourist mobility; backcountry skiing; outdoor recreation; GPS tracking; graph theory; graph connectivity; centrality measures; network analysis; network

1. Introduction

Protected areas (PAs) play a crucial role in the conservation of vulnerable mountain ecosystems [1]. Depending on the nature conservation regime, they may also have social functions, such as provisioning space for recreation, research, and educational purposes [2]. Mountain PAs frequently serve as attractive tourist destinations, where multifunctional use requires effective management strategies [3]. Therefore, understanding the way people use natural environments is crucial to balance the needs of visitors and site capacities [4,5].

Network science methods, largely based on graph theory, are gaining in importance in social, physical, medical, and many other disciplines [6,7]. In recent years an increased interest in network analytics can also be observed within tourism domain [8,9]. This analytical approach contributes to a better understanding of the structure and the behavior of the whole system, where various types of relationships (social, economic, operational, informational, spatial, etc.) may be investigated [9]. Mobility of tourists is one example of application context, where empirical travel data fits well into the graph theoretic analytical framework [9,10].

Spatial behavior of visitors in outdoor leisure areas can be documented in many different ways [11]. Data on tourists' mobility can be obtained using traditional survey methods [12,13], trip diaries, or map sketches [14], and since early 2000 also via automatic tracking devices [15,16]. Within the last two decades, Global Positioning System (GPS) tracking has become a well-established method to collect the exact movement trajectories of visitors [15,17–20]. The raw data comprises series of recorded trackpoints (pairs of geographic positions and associated time stamps), termed GPS tracks. GPS tracking data is typically used to analyze route parameters of individual visitors or a collective distribution of visitors in the study area [21–23]. Studies on outdoor recreation in protected areas often use the Geographic Information Systems (GIS) framework for conducting spatial analysis [22]. However, dedicated network analytic approaches investigating the structural and functional properties of recreational/tourism systems are underrepresented in comparison to other analysis methods [9].

TNP is known for its outstanding alpine landscape and is characterized by high biodiversity [24]. Recent rapid development of a new winter recreational activity—backcountry skiing—fosters new management challenges, related to an effective visitor guidance within the protected area [25]. Therefore, this study aims to develop a methodology to evaluate the structure and use of newly designated skiing zones in TNP combining GPS tracking of visitors and analytical methods based on graph theory.

2. Materials and Methods

2.1. Study Area

The study is based on empirical data (n = 609 GPS tracks of backcountry skiers) collected in the western part of Tatra National Park (TNP), Poland. The Tatra Mountains are situated in Central Eastern Europe and are the highest mountain range of the Carpathians. The national border between Poland and Slovakia runs along the main mountain ridge, dividing the protected area into two neighboring national parks. The Polish part of TNP covers an area of 21,197 ha [24], where recreational activities, due to nature conservation objectives, are restricted to designated zones [26]. Due to the rapid development of backcountry skiing in the Tatras within the last decade, TNP management has recently introduced a system of designated skiing zones (Figure 1), covering 13% of the TNP area [25]. Backcountry skiers can move up or down over snow-covered terrain without the necessity of using ski lifts nor prepared ski pistes. Ascents are possible with special skins fixed on visitors' skis. The backcountry skiing traffic within TNP is estimated at 10,000 visits per winter season [27]. Most of the visitor entries (approximately 70%) concentrate at Kuźnice trailhead, whereas other entry points, such as Chochołowska, Kościeliska, and Białka Valleys, are less frequently used [27].

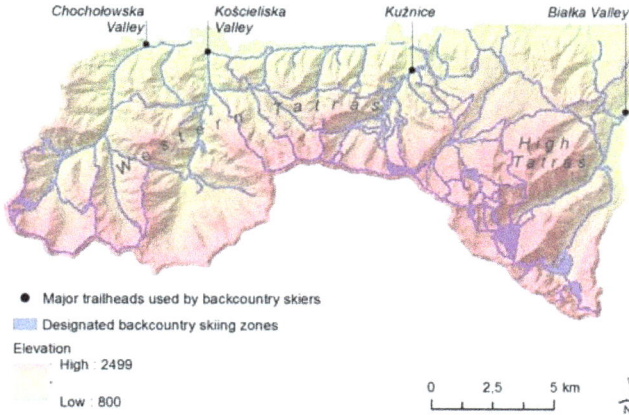

Figure 1. Characteristics of the study area: designated backcountry skiing zones within the border of Tatra National Park (TNP), Poland.

2.2. Data

2.2.1. Inventory of Skiing Zones System

The exact delineation of skiing zones system was obtained from the TNP management and stored within the Geographic Information Systems (GIS) as a vector dataset (Supplementary Materials Data S2. Ski touring area maps).

2.2.2. Mobility Data of Backcountry Skiers

In order to collect trip itineraries of backcountry skiers in TNP, the on-site GPS tracking method was applied. One hundred and three GPS loggers (Hollux M-1000C) were distributed by trained staff at major TNP entry points during sampling days throughout the winter seasons of 2012 and 2013. The GPS loggers were distributed simultaneously during the sampling days at three chosen entrances (Kuznice, Koscieliska, Chocholowska). The choice of these points was based on the pilot study [25]. The chosen entrances provide access for the ski tourers to central and western parts of the park. Each ski tourer entering TNP (at the chosen entry points) was approached and asked to carry a GPS device for the duration of his or her trip. The response rate was 87% and the total sample size reached 609 backcountry skiers participating in the study. Thirty-six GPS tracks were partially incomplete due to low battery level and/or device failure, and were therefore excluded from further analysis. Therefore, 573 complete tracks were used as a basis for final calculations (Supplementary Materials Data S1. Tracks).

2.3. Data Analysis

The general procedure of data analysis consisted of the following steps:

- GPS data pre-processing
- Creation of the structural network (undirected graph)
- Creation of the functional network (directed graph)
- Quantification of network connectivity indices
- Calculation of centrality measures of network nodes.

The calculations were made in MATLAB (version R2012b) [28] and Pajek software (version 5.01) [29].

2.3.1. Pre-Processing of GPS Data

GPS data was stored in GPX data format and pre-processed. It was necessary to clear some spatial artefacts caused by GPS signal reflection observed while starting up a device or location in unsuitable terrain conditions (e.g., deep mountain valleys, indoor usage of GPS logger in mountain huts or cable car station). GPS loggers were configured to record trackpoints every 50 meters (or every 120 seconds in the absence of movement). Taking into account these conditions, all anomalous data (e.g., displacements > 200 m) were identified and quantified. GPS tracks having signal errors above 10% were not considered for further analysis. However, tracks presenting smaller levels of error (<10%) were corrected using 1-D median filtering (medfilt1 Matlab function). In addition, the first five records from each GPS track were deleted due to the instability of the GPS device immediately after activation.

2.3.2. Creating the Structural Network

The physical structure of the entire skiing zones system was simplified into a graph structure (structural network; undirected graph). In order to build this structure, we used the geometric planes of the ski area in ESRI Shapefile (Supplementary Materials Date S2. Ski touring area maps). Trailheads as well as crossing points of skiing areas were used as nodes of the constructed network (all of them were labeled as a function of their west-east location, from 1 to 93). The node number 1 corresponded to the location situated most west, while number 93 referred to the node located most east. The ski paths between nodes were converted into network links (edges of the graph). In this way, the structure of the graph consisted of 93 vertices and 133 edges. The graph structure was saved as plaintext in Pajek file format for further analysis. Figures 2 and 3 present the resulting graph, based on the network of designated skiing zones in TNP. The height of the nodes (Figure 3) was calculated using SRTM Elevation Data [30].

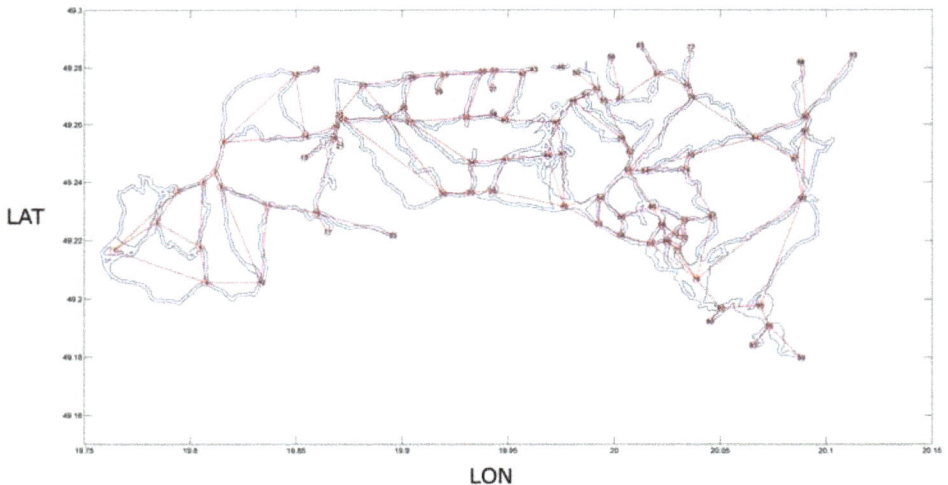

Figure 2. Translation of the designated backcountry skiing areas system into a structural network (undirected graph). The geographic coordinates (longitude and latitude) were defined in WGS84 spatial reference system and are expressed in decimal degrees.

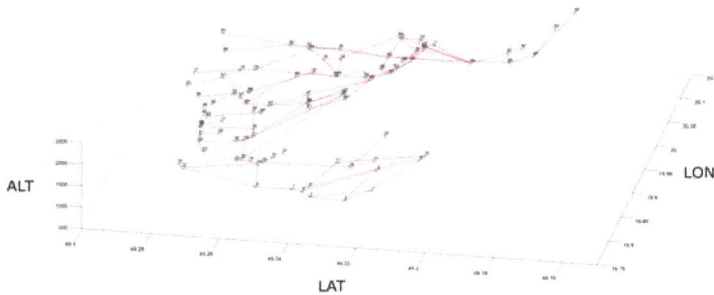

Figure 3. Three-dimensional visualization of the structural skiing network (undirected graph). The geographic coordinates (longitude and latitude) were defined in WGS84 spatial reference system and are expressed in decimal degrees. Altitude is expressed in meters above sea level (m.a.s.l.). The elevation data was used only for visualization purposes.

2.3.3. Creating the Functional Network

The actual use of the area by skiers (GPS records of visitors' trip itineraries) was assigned to the network (functional network; directed graph). In order to allocate visits to specific node locations, a buffer of 200 m for every node was defined. However, after discussion with the experts of the national park, the buffers of 21 nodes were additionally adapted to specific local conditions (the buffer radius was reduced or extended). Intersections of GPS trackpoints with node buffer zones were used to establish a relation between visitor routes and the network. Figure 4 shows an example of a recorded skiing route (a) and its trip assignment to the network (b). Assignment of all GPS tracks resulted in 2420 arcs, linking network nodes. This data was stored as plaintext and analyzed with Pajek software (version 5.01) [31]. The programming code utilized to build the network can be found in the Supplementary Materials Code S3.

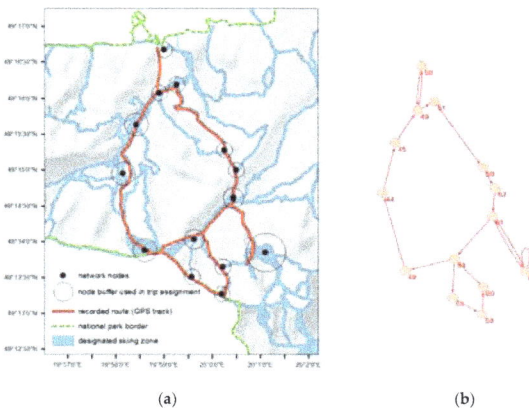

(a)   (b)

Figure 4. Example of a skiing route. (**a**) Recorded GPS track of a TNP visitor; (**b**) representation of the corresponding directed graph G = (V,E) where the vertices are V = {44, . . . ,66} and the edges are E = {(50, 49), (49, 51), (51, 59), (59,62), (62, 61), (61, 66), (66, 61), (61,54), (54, 53), (53, 58), (58, 60), (60, 54), (54, 48), (48, 44), (44, 45), (45, 49), (49, 50)}. In the final graph the loop, (66, 66) was deleted, since a property of directed graphs is vi ≠ vj.

2.3.4. Calculating Network Connectivity Indices

Network connectivity indices quantify the coherence of a network. In order to compare the coherence of the entire structural network and the actually used one (functional network), the following measures were calculated:

- Kansky index $\beta = \frac{e}{v}$ where e is the number of edges, v is the number of vertices (the higher the value of β, the more coherent the network);
- Kansky index $\gamma = e/(3*(v-2))$, defining the ratio of the existing number of edges (e) to the maximum possible number of edges resulting from the number of vertices (v). The γ value ranges from 0 to 1, where the value of 1 indicates a completely connected network.
- Kansky index $\alpha = \frac{\mu}{2v-5}$, where μ is a cyclomatic measure calculated as $\mu = e - v + p$, where p is the number of isolated subgraphs. An α value of 1 indicates a completely meshed network, and 0 indicates a very simple network.

A more detailed description of the indices can be found in Rodrigue et al. [10].

2.3.5. Calculating Node Centrality Measures

In order to investigate the relative importance of nodes within the system of designated skiing zones, the following centrality measures were computed: degree, input degree, output degree, degree all, weighted input degree, weighted output degree, weighted degree all, closeness, betweenness, and proximity prestige. The calculations were made for both the structural and the functional network, whereas in functional network the nodes without activity were not taken into account. Table 1 gives an overview of the calculated centrality measures.

Table 1. Overview of node centrality measures used in the study.

Centrality Measure	Description	Mathematical Equation
Input degree	Number of edges entering to vertex i.	$d_i^+ = \sum_{j\in G} x_{ij}$ x_{ij} signal the position between node i and node j.
Output degree	Number of edges leaving vertex i.	$d_i^- = \sum_{j\in G} x_{ij}$
Degree (all)	Total number of edges connected to the vertex i.	$D_i = d^+ + d^-$
Closeness	Inverse sum of distances from a given vertex to all other vertices in the graph.	$C_i = \sum_{j\ \neq i\in G} (d(i,j))^{-1}$ where $d(i,j)$ is a topological distance between vertices i and j.
Betweenness	A number of times a vertex is crossed by shortest paths in the graph.	$B(i) = \sum_{j<k} g_{jk}(i)/g_{jk}$ where g_{jk} is the number of geodesics connecting jk, and $g_{jk}(i)$ is the number that geodesics i is on.
Proximity prestige	Expresses the influence domain of a vertex by the average distance from all vertices in the influence domain. Pp value of 0 indicates that node i is unreachable; whereas Pp = 1 if all nodes are directly connected to node i.	$Pp_i = \frac{Ii}{g-1} \sum_{j\ \neq i\in G} (d(i,j)/Ii)^{-1}$

Equations based on References [10,32].

3. Results

3.1. General Characteristics of the Structural and Functional Networks

Figure 5 depicts the overall structure of the skiing network. The constructed structural network consisted of 93 nodes connected by 133 links. Additionally, one loop (node 46), which was not included in the link quantification, can be also observed. The average degree was 2.86. The longest network distance was established between nodes 1 and 83 (with a total of 20 steps between both), while the number of unreachable pairs was 184.

To build the functional network, all previously defined 93 nodes were considered. However, the collected GPS data indicated that 30 nodes were not visited. The functional network had 175 links connecting 63 nodes. The average degree of the connected network was 3.76, and the longest distance occurred between the nodes 1 and 17 (with 17 steps between them). Figure 6 illustrates the functional skiing network. Both network datasets were stored in Pajek software (Supplementary Materials File S4).

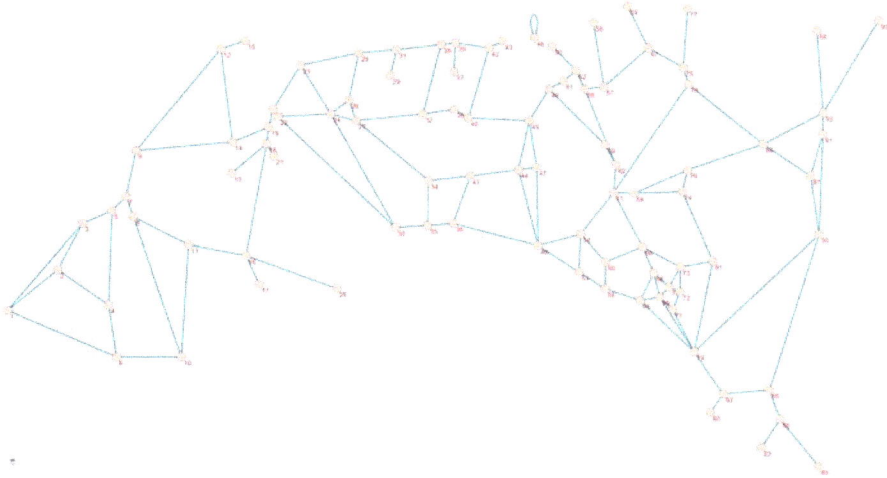

Figure 5. Structural network (undirected graph) depicting the designated backcountry skiing system in TNP. The network is composed of 93 nodes and 133 links.

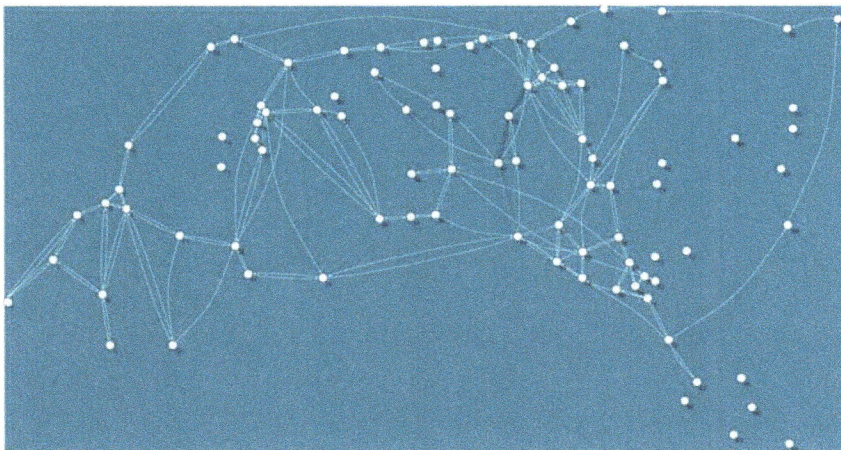

Figure 6. Functional network (directed graph) of the designated backcountry skiing system; graph based on the recorded visitors' trip itineraries. Arrows indicate movement direction; color scale (grey scale) illustrates the intensity of skiing traffic. A darker color means higher use intensity.

3.2. Network Coherence (Connectivity Indices)

The existing network of designated skiing zones had moderate connectivity ($\beta = 1.43$; $\gamma = 0.49$; $\alpha = 0.23$). Using the network systems typology introduced by Taaffe and Gauthier [33], the investigated network can be classified as a lattice network type. Lattice networks are characterized by a moderate density of edges relative to the number of nodes [33].

The actual recreational use concentrated only on a part of the designated skiing network. Taking into account all 93 nodes of the structural network and the documented visitors' movement between nodes, we observed that the connectivity indices of the functional network were slightly lower ($\beta = 1.25$; $\gamma = 0.42$; $\alpha = 0.13$). Although 30 out of the total 93 nodes (22%) were not visited at all, several additional connections between node locations were identified outside of the officially designated skiing network. This can be confirmed by the additional calculation of connectivity indices for the connected graph only (63 interconnected nodes used by visitors). These results show higher connectivity in comparison to the structural network ($\beta = 1.84$; $\gamma = 0.63$; $\alpha = 0.54$).

3.3. Relative Importance of Network Nodes (Centrality Measures)

The results show a large heterogeneity of node centrality measures within the investigated network. Each of the applied parameters generated slightly different outputs, indicating the relative importance of nodes within the skiing network of TNP.

All nodes' centrality values, calculated separately for the structural and the functional networks, can be found in Supplementary Materials Table S1. Figure 7 shows a graphic representation of calculated centrality measures in both networks. Results of the 10 most important nodes, considering "all degree" values, for both networks (structural and functional) are listed in Tables 2 and 3.

The degree values were obtained for a structural network ranging from 1 to 6, and the average was 2.86. Nodes located near the northern border of the study area served as entrance locations and usually had very low degree values. Low degree values were also a typical feature of distant mountain peaks. The most interconnected nodes were located in the central part of the study area; the node with the highest degree was located in the Valley of Five Lakes. Weighted degree measures, calculated for the functional network, reflected the actual intensity of use at particular node locations. It can be seen that the majority of skiing traffic concentrates in the Kuznice-Kasprowy Wierch area. The highest weighted degree values were observed at nodes 45 and 49 (located lower in the valley, close to the Kuznice entrance point). Other nodes were less frequently used (low weighted degree values). The directions of use were balanced at most node locations; the disproportion of in and out degree values was observed at some trailheads, such as node 49 (Kuznice) and the neighboring node 45.

Closeness centrality is based on the network distance between nodes. In the structural network, nodes located in the central part of the study area had higher closeness measures, whereas nodes placed in the most western part of the park had the lowest closeness measures. Closeness calculations made for the functional network exposed the importance of the Kuznice-Kasprowy Wierch area. Nodes located in the western and the eastern parts of the study area were considered less important. Proximity prestige measures calculated for the directed network (functional network) showed equal the closeness values.

Table 2. Structural network node centrality measures of the 10 nodes with the highest "all degree" values.

Node Number	Degree	Closeness	Betweenness
79	6	0.14	0.14
24	5	0.15	0.08
48	5	0.18	0.43
61	5	0.17	0.30
69	5	0.14	0.02
16	4	0.11	0.12
18	4	0.12	0.17
27	4	0.15	0.09
44	4	0.17	0.09
45	4	0.15	0.14

Table 3. Functional network node centrality and prestige measures of the 10 nodes with the highest "weighted all degree" values.

Node Number	Weighted Input Degree	Weighted Output Degree	Weighted All Degree	Closeness	Betweenness	Proximity Prestige
45	350	532	882	0.19	0.08	0.19
49	351	149	500	0.20	0.13	0.20
48	197	194	391	0.19	0.06	0.19
44	152	149	301	0.18	0.01	0.17
61	109	90	199	0.17	0.10	0.17
62	93	100	193	0.15	0.05	0.15
41	86	84	170	0.17	0.04	0.17
54	82	88	170	0.17	0.06	0.17
7	79	78	157	0.12	0.07	0.12
51	56	83	139	0.17	0.02	0.17

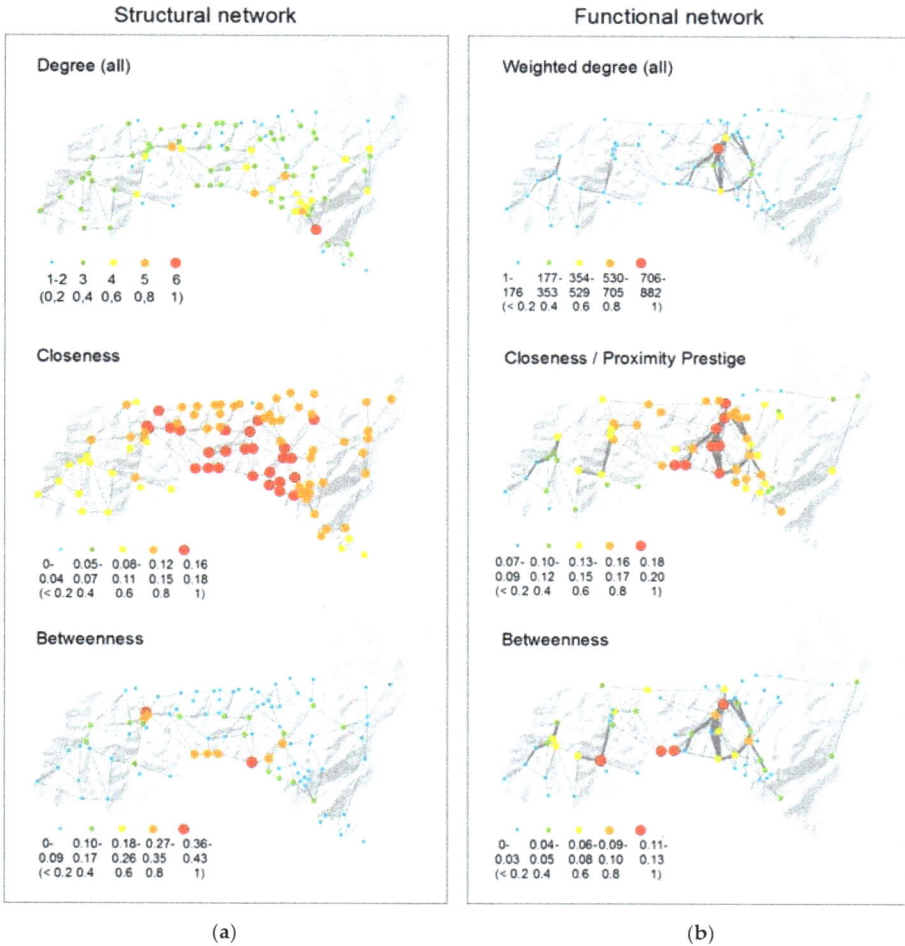

Figure 7. Node centrality measures in the structural (**a**) and functional networks (**b**) of designated backcountry skiing zones in TNP.

Betweenness centrality provides information about the number of geodesics in the network crossing particular nodes. In the structural network, the highest betweenness measures were observed at node 48 (the summit of Kasprowy Wierch) and at the beginning of the Koscieliska Valley, as well as at several nodes located along the main ridge of the Tatra Mountains. The functional network exposed the importance of the Kuznice entrance (node 49), Ornak mountain hut (node 16), and Czerwone Wierchy peaks (nodes 30 and 33).

4. Discussion and Conclusions

4.1. General Meaning of the Findings

The main contribution of the study is the application of graph theory to analyze a recreational system considering not only the physical structure of the environment, but also its actual use (visitor mobility data obtained via GPS tracking).

Typically, GPS tracks are analyzed within Geographic Information Systems (GIS) and focus on calculations of (track)point densities illustrating the intensity of recreational use [17,20,22,23], intersections with other thematic layers such as nature protection zones and wildlife habitats [20], or movement parameters of individual visitors [21,23,34]. Also, recent approaches of visual analytics dedicated to movement data, such as V-analytics [35,36], have gained importance and are frequently used in the field of spatial planning and transportation [35,37]. Typically, GPS tracks analyzed within a GIS framework are not constrained to network structures [35], although network analysis is an important approach used in geoinformatics [38].

Network analytic tools, although highly recognized in other disciplines such as sociology, economics, transportation, computer science, and medicine [6], are still underrepresented in the domain of recreation research [39]. Most of the studies related to outdoor recreation using graph theoretic analysis methods focus on the physical structures of recreation systems [40–42]. Even in the tourism sector (an important branch of the world economy), scientists underline the necessity for tourism scholars to refine and improve their analysis methods and tools, which also refers to network analytics [9].

Our study builds upon our previous work [39], applying graph theoretical approach to analyze the structure and function of a recreational system, using as an example a network of designated backcountry skiing zones.

4.2. Comparison of the Structural and Functional Networks in the Study Area

Our results show that the existing network of designated skiing zones is not evenly used by backcountry skiers. The main objective of the structure created by the park management is to minimize visitors' impact on nature [26]. However, skiers' major motivation is to reach an attractive peak and descent [43]. Although general connectivity indices of the structural and functional networks remained similar, a closer look at the spatial distribution of the use intensity (number of visits per network link or weighted degree measures at node locations) revealed large heterogeneity in network use. The region of Kuznice-Kasprowy Wierch turned out to be the most heavily used area, whereas other locations were less frequently visited.

When we analyzed the centrality measures of network nodes in depth, some remarkable differences were observed. The largest differences refer to degree and weighted degree measures as they directly reflect the intensity of tourist traffic. A high degree centrality value in the structural network (the existence of many skiing routes around a specified node) does not always imply higher use intensity at the specified location. For instance, a well-connected node (79) is located in the distant Valley of Five Lakes at a higher elevation. Therefore, although it has a high degree value, it is not that frequently used. In contrary, nodes 49 and 45, located close to the most popular trailhead, Kuznice, have medium degree values in the structural network, but attract most of the skiing visits in the park. For both networks only one node (48—Kasprowy Wierch peak) was similarly important, which is the

Symmetry **2017**, *9*, 317

central point with the top cable car station. It can be reached by ski-tourers from many directions, even in bad weather conditions with a high level of avalanche risk. Moreover, this peak (Kasprowy Wierch) is also surrounded by the groomed pistes for downhill skiing. These factors explain the high popularity of this node location.

The closeness parameter is generally similar in the structural and functional networks. However, there are some nodes that differ in use. Nodes 36 and 53 have a high closeness value in the structural network. These two points are also important in the functional network, as they are usually ski tour destinations and, after reaching them, the skiers turn back and descend. It is also remarkable that nodes with low closeness values were generally less frequently used, or even not visited at all.

The betweenness values showed us that in the structural network the nodes 48, 20, and 19 have very high values, while in the functional network the maximum intermediation values are in nodes 33, 49, 30, and 16. The areas close to the peaks are those used by skiers as transit areas (e.g., nodes: 30, 33), while the areas close to the forest or to the entrances (e.g., nodes: 19, 20) are the most important points according to the structure.

The comparison of the structural and functional properties of the skiing network showed that recreational use only partly depends on the network structure (e.g., distribution of use in the Kuznice-Kasprowy Wierch region). The overall popularity of specific network locations is possibly caused by additional external factors, such as accessibility from tourist resorts or the difficulty of ski tours related to local terrain and snow conditions [25]. A further reason for strong differences between the two mentioned networks is the origin of the structural network, which was primarily designed for summer hiking [26]. The winter network was created only recently [25], and according to the legal regulations it was obligatory to base it on the summer hiking trial network [44].

4.3. Limitations of the Proposed Methodology

The main strength of this study is a parallel analysis of the structural and functional properties of the system of designated skiing zones in TNP. Collected mobility data (GPS tracks) were assigned to the existing network structure and several measures characterizing the overall network as well as the relative importance of node locations were investigated in order to better understand how the park space is being used by backcountry skiers.

Nevertheless, there is still room for methodological improvements. One of the limitations of this study is related to the use of GPS loggers to collect the mobility data of backcountry skiers. It is possible that skiers may have changed their behaviors by feeling observed. This problem is a subject of discussion in recreation monitoring studies using GPS tracking to document visitor behavior in protected areas [19,23,39]. However, so far there is no empirical confirmation of this thesis. Moreover, the accuracy of GPS records can be influenced by mountain terrain conditions (e.g., lower accuracy in narrow valleys, forested areas) [45], which is another limitation of this data collection technique.

Another discussion point refers to the sampling strategy used during the distribution of GPS loggers. GPS devices were distributed simultaneously at major trailheads of the national park by trained staff, which required a demanding logistic strategy and was challenging in winter conditions. Therefore, the study concentrated on the most important entrance locations and some minor trailheads were disregarded. We analyzed only the three main entry points, excluding the Bialka Valley due to its lower importance from the management perspective and high cost of data collection campaign (far away from Zakopane). In the future this entrance might be a case of its own for researching backcountry skiing traffic accessing the highest peaks of the Tatra Mountains.

As a consequence, their surroundings were underrepresented in the overall spatial distribution of skiing traffic, which possibly could affect the calculations of the functional network measures. In order to overcome these limitations, in the future, technological advances should enable anonymous tracking of visitors' electronic devices (such as mobile phones) at desirable resolutions [15,46,47].

Nevertheless, so far systematic tracking via distributed GPS loggers is one of the most reliable methods used for visitor monitoring in outdoor leisure settings. It is also worth mentioning that the total number of backcountry skiers entering TNP at the investigated trailheads [27] corresponds with the proportions of the collected GPS tracks.

Backcountry skiing is an example of a semi-open recreational system [22], where people can move along designated trails, but they also can leave a marked track and move freely over snow-covered terrain. Assigning recorded visitors' trip itineraries to a network structure is very successful in cases when visitors follow designated trails. It is also possible when visitors are crossing network nodes and are moving partly off-trail (the presence at particular node locations enables establishing a network link). Yet, the most challenging situation from the network analytics perspective occurs when a visitor completely leaves the system of designated skiing zones and moves freely out of the skiing network. Such type of spatial behavior cannot be documented with the methodology proposed in this paper. Extending the number of network nodes, including all potential off-trail destinations (such as unconnected mountain passes or peaks) in the network could possibly overcome this problem and will be considered in future studies. Visitors who do not cross the specified node locations cause "untightens" of the investigated network system. This problem is particularly visible by comparing the directed degree measures of network nodes (in-degree and out-degree). In transit locations, those values should be practically the same, but in the presented results in- and out-degree values sometimes differed from each other. Large differences in directed degree measures were observed in nodes 49 and 45. This fact is caused mainly by another reason: the data pre-processing algorithm deleting the first five records of each track, which were mainly located around the entrance node 49 (Kuznice). In this way, the network link "49, 45" was not considered in several skiing itineraries.

From the network analytics point of view, we only calculated well-established parameters such as network connectivity (β, γ, α indices) and basic node centrality measures (degree, closeness, betweenness, and proximity prestige). Future work could use other calculations [6,8,10,29,32,48] that could extend our analysis. Next to the topological properties of the network, geographic dimension could be additionally considered by using spatial graphs [49]. In this way metric distances between nodes and terrain steepness could be included in distance calculations and accessibility measures.

4.4. Meaning of the Findings for Visitor Management in Protected Areas

The presented results deliver comprehensive information about the structural properties of the designated skiing network in TNP and its actual use over the winter season. Outcomes of the study can be practically used for evaluating the newly introduced regulations concerning backcountry skiing in this protected area and for supporting management decisions related to the strategic allocation of infrastructure and provisioning purposive tourist information.

Although designated skiing zones cover 13% of the total park area, most of the skiing traffic concentrated in one particular region: Kuznice-Kasprowy Wierch. Basically, the management strategy of TNP requires some improvements in the guidance of winter visitors. The first important aspect is the lack of signage dedicated to backcountry skiers in many strategic locations, such as the Kuznice trailhead and its surroundings, as well as the Kasprowy Wierch peak. The presented results indicated node locations with the highest centrality measures (especially those calculated for the functional network) which could be used as a background for further decisions concerning the locations of information boards and signposts in TNP. Node locations with high betweenness values should be especially carefully investigated, as their strategic position enables directing visitor flow to other regions of the national park. Experiences from other protected areas underline the importance of signposting for effective visitor guidance in protected areas [50,51]. It would also be desirable to promote certain backcountry skiing destinations through written information (e.g., maps, informative leaflets), participative events, and discussions in order to avoid skiing traffic in the most vulnerable nature conservation zones [50].

Symmetry **2017**, *9*, 317

We conclude that that the newly designated system of backcountry skiing zones, largely based on the summer hiking trail network and the existing groomed pistes of downhill skiing, needs detailed functional analysis and partial adaptation in order to conform with winter recreation requirements in addition to nature protection objectives. The presented study based on graph theory is a first step to evaluate the whole winter recreation system, considering not only its structural properties but also its function.

Supplementary Materials: The following will be available online at www.mdpi.com/2073-8994/9/12/317/s1, Data: S1. Maps: Tracks; S2. Ski touring area maps; Code: S3. Code MATLAB; Files: S4. Pajek; Tables S1. Centrality parameters.

Acknowledgments: This study was supported by a research grant of the University of Valencia, Spain (Ayudas para estancias de investigacion: UV-INV EPDI15-276152) and the bilateral Scientific and Technical Cooperation Programme between Austria and Poland (S&T Cooperation/WTZ grant number PL01/2016). The authors wish to thank the Tatra National Park management, especially the director—Szymon Ziobrowski—for support, organizational help, provisioning of GPS loggers, and agreement to provide open access to the collected data. We also would like to thank TNP volunteers and the students from the Department of Tourism and Recreation, University School of Physical Education in Cracow, Poland involved in data collection.

Author Contributions: K.T., L.-M.G. and J.L.T.-H. came across the idea of linking GPS tracking data and graph theoretic analytical approaches in the field of outdoor recreation research. K.T. and L.-M.G. designed the general methodology and decided on network analytic methods being used. M.B. conceived, designed, and coordinated the field experiments (GPS tracking of backcountry skiers) in Tatra National Park and provided his expertise and critical comments at all stages of study design, data management, data processing, analysis, and interpretation of the results; L.-M.G. and X.G.-M. analyzed the data in MATLAB and Pajek; K.T. did calculations and results visualizations in GIS. K.T. wrote the paper, L.-M.G. and X.G.-M. contributed to the methodological part of the manuscript and the results description, L.-M.G. and M.B. contributed additionally to the discussion and conclusions section; J.L.T.-H. provided substantial advice at the conceptual and analytical stage of the study along with constructive comments on data presentation, discussion of the results, as well as the overall preparation of the manuscript.

Conflicts of Interest: The authors declare no conflict of interest. The founding sponsors had no role in the design of the study; in the collection, analyses, or interpretation of data; in the writing of the manuscript, and in the decision to publish the results.

References

1. IUCN. Red List of Ecosystems. Available online: https://www.iucn.org/about/union/commissions/cem/cem_work/tg_red_list/ (accessed on 28 November 2017).
2. Dudley, N. *Guidelines for Applying Protected Area Management Categories*; International Union for Conservation of Nature (IUCN): Gland, Switzerland, 2013.
3. Newsome, D.; Moore, S.A.; Dowling, R.K. *Natural Area Tourism: Ecology, Impacts and Management*, 2nd ed.; Newsome, D., Moore, S.A., Dowling, R.K., Eds.; Channel View Publications: Bristol, UK; Tonawnda, NY, USA; North York, ON, Canada, 2012; ISBN 978-1-84541-381-1.
4. Bell, S. *Design for Outdoor Recreation*; Taylor & Francis: Abingdon, UK, 2008; ISBN 978-1-134-10804-6.
5. Cole, D.N.; Daniel, T.C. The science of visitor management in parks and protected areas: From verbal reports to simulation models. *J. Nat. Conserv.* **2003**, *11*, 269–277. [CrossRef]
6. Barabási, A.-L.; Pósfai, M. *Network Science*; Cambridge University Press: Cambridge, UK, 2016; ISBN 978-1-107-07626-6.
7. Newman, M. *Networks: An Introduction*; Oxford University Press: Oxford, UK, 2010; ISBN 978-0-19-920665-0.
8. Scott, N.; Baggio, R.; Cooper, C. *Network Analysis and Tourism: From Theory to Practice*; Channel View Publications: Clevedon, UK; Buffalo, NY, USA, 2008; ISBN 978-1-84541-087-2.
9. Baggio, R. Network science and tourism—The state of the art. *Tour. Rev.* **2017**, *72*, 120–131. [CrossRef]
10. Rodrigue, J.-P.; Comtois, C.; Slack, B. *The Geography of Transport Systems*, 3rd ed.; Routledge: New York, NY, USA, 2013; ISBN 978-0-415-82254-1.
11. Cessford, G.; Muhar, A. Monitoring options for visitor numbers in national parks and natural areas. *J. Nat. Conserv.* **2003**, *11*, 240–250. [CrossRef]
12. D'Agata, R.; Gozzo, S.; Tomaselli, V. Network analysis approach to map tourism mobility. *Qual. Quant.* **2013**, *47*, 3167–3184. [CrossRef]

13. Stimson, R. (Ed.) *Handbook of Research Methods and Applications in Spatially Integrated Social Science*; Edward Elgar Publishing: Cheltenham, UK, 2014; ISBN 978-0-85793-297-6.

14. Arnberger, A.; Hinterberger, B. Visitor monitoring methods for managing public use pressures in the Danube Floodplains National Park, Austria. *J. Nat. Conserv.* **2003**, *11*, 260–267. [CrossRef]

15. Shoval, N.; Isaacson, M. Tracking tourists in the digital age. *Ann. Tour. Res.* **2007**, *34*, 141–159. [CrossRef]

16. Thimm, T.; Seepold, R. Past, present and future of tourist tracking. *J. Tour. Future* **2016**, *2*, 43–55. [CrossRef]

17. Beeco, J.A.; Hallo, J.C.; Brownlee, M.T.J. GPS Visitor Tracking and Recreation Suitability Mapping: Tools for understanding and managing visitor use. *Landsc. Urban Plan.* **2014**, *127*, 136–145. [CrossRef]

18. Meijles, E.W.; de Bakker, M.; Groote, P.D.; Barske, R. Analysing hiker movement patterns using GPS data: Implications for park management. *Comput. Environ. Urban Syst.* **2014**, *47*, 44–57. [CrossRef]

19. D'Antonio, A.; Monz, C. The influence of visitor use levels on visitor spatial behavior in off-trail areas of dispersed recreation use. *J. Environ. Manag.* **2016**, *170*, 79–87. [CrossRef] [PubMed]

20. Rupf, R.; Wyttenbach, M.; Köchli, D.; Hediger, M.; Lauber, S.; Ochsner, P.; Graf, R. Assessing the spatio-temporal pattern of winter sports activities to minimize disturbance in capercaillie habitats. *J. Prot. Mt. Areas Res.* **2011**, *3*, 23–32. [CrossRef]

21. Orellana, D.; Wachowicz, M. Exploring patterns of movement suspension in pedestrian mobility. *Geogr. Anal.* **2011**, *43*, 241–260. [CrossRef] [PubMed]

22. Beeco, J.A.; Brown, G. Integrating space, spatial tools, and spatial analysis into the human dimensions of parks and outdoor recreation. *Appl. Geogr.* **2013**, *38*, 76–85. [CrossRef]

23. Taczanowska, K.; Muhar, A.; Brandenburg, C. Potential and limitations of GPS tracking for monitoring spatial and temporal aspects of visitor behaviour in recreational areas. In Proceedings of the Fourth International Conference on Monitoring and Management of Visitor Flows in Recreational and Protected Areas, Montecatini Terme, Italy, 14–19 October 2008; Raschi, A., Trampetti, S., Eds.; CNR-Ibimet: Firenze, Italy, 2008; pp. 451–456.

24. TNP—Tatra National Park Tatra National Park. Available online: http://tpn.pl/poznaj (accessed on 1 November 2017).

25. Bielański, M. Skitouring in Tatra National Park and Its Environmental Impacts. Ph.D. Thesis, University School of Physical Education, Cracow, Poland, 2013.

26. TNP—Tatra National Park Decree Nr 3/2017 of the Tatra National Park Director from 23.02.2017 on Hiking, Bicycling and Skiing in the Area of the Tatra National Park. (Zarzadzenie nr 3/2017 Dyrektora Tatrzanskiego Parku Narodowego z dnia 23 lutego 2017 roku w sprawie ruchu pieszego, rowerowego oraz uprawiania narciarstwa na terenie Tatrzańskiego Parku Narodowego) 2017.

27. Adamski, P.; Bielański, M. *Analiza Danych Dotyczących Ruchu Wejściowego Narciarzy Skiturowych w Tatrzańskim Parku Narodowym, w Sezonach Zimowych 2013 Oraz 2013/2014 (Report)*; Tatra National Park: Zakopane, Poland, 2014; p. 23.

28. *MATLAB*; Mathworks Inc.: Natick, MA, USA, 2012.

29. Mrvar, A.; Batagelj, V. Analysis and visualization of large networks with program package Pajek. *Complex Adapt. Syst. Model.* **2016**, *4*, 6. [CrossRef]

30. United States Geological Survey. *United States Geological Survey SRTM Elevation Data 2017*; United States Geological Survey: Reston, VA, USA.

31. Batagelj, V.; Mrvar, A. Pajek—Analysis and Visualization of Large Networks. In *Graph Drawing Software*; Jünger, M., Ed.; Springer: Berlin, Germany, 2003; pp. 77–103.

32. Wasserman, S.; Faust, K. *Social Network Analysis: Methods and Applications*; Cambridge University Press: Cambridge, UK, 1994; ISBN 978-1-139-78861-8.

33. Taaffe, E.J.; Gauthier, H.L. *Geography of Transportation*; Prentice-Hall foundations of economic geography series; Prentice-Hall: Englewood Cliffs, NJ, USA, 1973; ISBN 0133513955.

34. Olliff, T.; Legg, K.; Kaeding, B. *Effects of Winter Recreation on Wildlife of the Greater Yellowstone Area: A Literature Review and Assessment*; Report to the Greater Yellowstone Coordinating Committee; National Park Service: Yellowstone National Park, WY, USA, 1999; p. 315.

35. Andrienko, N.; Andrienko, G. Visual analytics of movement: An overview of methods, tools and procedures. *Inf. Vis.* **2013**, *12*, 3–24. [CrossRef]

36. Andrienko, N.; Andrienko, G. V-Analytics. Available online: http://geoanalytics.net/V-Analytics/ (accessed on 9 December 2017).

37. Burian, J.; Zajickova, L.; Popelka, S.; Rypka, M. Spatial Aspects of Movement of Olomuc and Ostrava Citizens. In *SGEM2016 Conference Proceedings*; 2016; Volume III, pp. 439–446.

38. Fischer, M. GIS and Network Analysis. In *Spatial Analysis and GeoComputation*; Springer: Berlin/Heidelberg, Germany, 2006; pp. 43–60. ISBN 978-3-540-35729-2.

39. Taczanowska, K.; González, L.-M.; Garcia-Massó, X.; Muhar, A.; Brandenburg, C.; Toca-Herrera, J.-L. Evaluating the structure and use of hiking trails in recreational areas using a mixed GPS tracking and graph theory approach. *Appl. Geogr.* **2014**, *55*, 184–192. [CrossRef]

40. Kolodziejczyk, K. Hiking Trails for Tourists in the "Chelmy" Landscape Park—Assessment of Their Route and Infrastructure Development. *Pol. J. Sport Tour.* **2011**, *18*, 324–329. [CrossRef]

41. Styperek, J. Tourist hiking trails in Polish National Parks. *Turyzm* **2001**, *11*, 25–37. (In Polish)

42. Li, W.; Ge, X.; Liu, C. Hiking trails and tourism impact assessment in protected area: Jiuzhaigou Biosphere Reserve, China. *Environ. Monit. Assess.* **2005**, *108*, 279–293. [CrossRef] [PubMed]

43. Derezińska, M. Pochwala skialpinizmu, czyli o filozofii zacisnietych zebow. In *Tatry*; Tatra National Park: Zakopane, Poland, 2007; p. 30. (In Polish)

44. Sejm of the Republic of Poland. *Nature Conservation Act (Ustawa z dnia 16 Kwietnia 2004 r. o Ochronie Przyrody)*; Sejm of the Republic of Poland: Warsaw, Poland, 2004.

45. Hofmann-Wellenhof, B.; Lichtenegger, H.; Collins, J. *Global Positioning System—Theory and Practice*; Springer: New York, NY, USA, 2001; ISBN 978-3-7091-6199-9.

46. Shoval, N. Tracking technologies and urban analysis. *Cities* **2008**, *25*, 21–28. [CrossRef]

47. Korpilo, S.; Virtanen, T.; Lehvävirta, S. Smartphone GPS tracking—Inexpensive and efficient data collection on recreational movement. *Landsc. Urban Plan.* **2017**, *157*, 608–617. [CrossRef]

48. Luke, D.A. *A User's Guide to Network Analysis in R*, 1st ed.; Springer: New York, NY, USA, 2016; ISBN 978-3-319-23882-1.

49. Fall, A.; Fortin, M.-J.; Manseau, M.; O'Brien, D. Spatial graphs: Principles and applications for habitat connectivity. *Ecosystems* **2007**, *10*, 448–461. [CrossRef]

50. Sterl, P.; Eder, R.; Arnberger, A. Exploring factors in influencing the attitude of on-site ski mountaineers towards the ski touring management measures of the Gesäuse National Park. *J. Prot. Mt. Areas Res.* **2006**, *2/1*, 31–38. [CrossRef]

51. Zeidenitz, C.; Mosier, H.J.; Hunziker, M. Outdoor recreation: From analysing motivations to furthering ecologically responsible behaviour. *For. Snow Landsc. Res.* **2007**, *81*, 175–190.

symmetry

MDPI

Article

Efficient Location of Resources in Cylindrical Networks

José Juan Carreño [1], José Antonio Martínez [2] and María Luz Puertas [3],*

[1] Department of Applied Mathematics for Information and Communication Technologies, Universidad Politécnica de Madrid, Calle Alan Turing s\n, 28031 Madrid, Spain; jjcc@etsisi.upm.es
[2] Department of Computer Science, Universidad de Almería, Carretera Sacramento s\n, 04120 Almería, Spain; jmartine@ual.es
[3] Department of Mathematics, Universidad de Almería, Carretera Sacramento s\n, 04120 Almería, Spain
* Correspondence: mpuertas@ual.es; Tel.: +34-950-015-463

Received: 4 December 2017; Accepted: 8 January 2018; Published: 10 January 2018

Abstract: The location of resources in a network satisfying some optimization property is a classical combinatorial problem that can be modeled and solved by using graphs. Key tools in this problem are the domination-type properties, which have been defined and widely studied in different types of graph models, such as undirected and directed graphs, finite and infinite graphs, simple graphs and hypergraphs. When the required optimization property is that every node of the network must have access to exactly one node with the desired resource, the appropriate models are the efficient dominating sets. However, the existence of these vertex sets is not guaranteed in every graph, so relaxing some conditions is necessary to ensure the existence of some kind of dominating sets, as efficient as possible, in a larger number of graphs. In this paper, we study independent $[1,2]$-sets, a generalization of efficient dominating sets defined by Chellali et al., in the case of cylindrical networks. It is known that efficient dominating sets exist in very special cases of cylinders, but the particular symmetry of these graphs will allow us to provide regular patterns that guarantee the existence of independent $[1,2]$-sets in every cylinder, except in one single case, and to compute exact values of the optimal parameter, the independent $[1,2]$-number, in cylinders of selected sizes.

Keywords: cartesian product of graphs; efficient domination; tropical matrix algebra

MSC: 05C69; 05C85

1. Introduction

Graphs have been used, since their informal beginnings, as a model to represent complex networks, to describe different properties within them and to find solutions to diverse problems, such as optimal routes or the best location of resources. Formally, a *finite graph* is a pair $G = (V, E)$, where V is a non-empty finite set, whose elements are called *vertices*, and $E \subseteq V \times V$ is a set of unordered pairs of vertices called *edges*. If uv is an edge, we say that u and v are *neighbors*. The *distance* between two vertices u and v in a graph is the number of edges in a shortest path connecting them and it is denoted by $d(u,v)$. The *open neighborhood* of a vertex u in a graph G is $N(u) = \{v \in V(G): uv \in E(G)\}$ and the *closed neighborhood* is $N[u] = N(u) \cup \{u\}$. The *degree* of a vertex is the cardinality of its open neighborhood.

Domination-type properties in graphs, formally defined in the late 1950s [1] and early 1960s [2], are a good example of how these models can help to represent real world problems and to provide optimal solutions to them. A *dominating set* in a graph G is a vertex set S such that every vertex not in S has at least one neighbor in S. This classical definition provides a model of resource location in a network, in such a way that every node of the network has access to such a resource.

Once this model is formulated, it is an immediate question to find the best possible distribution, which results in the definition of the *domination number* $\gamma(G)$ of a graph G, which is the minimum cardinality of a dominating set of G. An extensive compilation of results on this subject can be found in [3], as well as some of its applications such as the planning of school bus routes, the design of computer communication networks, the location of broadcasting stations, social networks modeling and land surveying.

Following this classic pattern, a large number of variants have been defined, which pay attention to different aspects. For instance, k-domination requests at least k neighbors in S for the vertices not in the k-dominating set S and locating-domination asks for $N(u) \cap S \neq N(v) \cap S$ for every pair of vertices $u, v \in V(G) \setminus S$. Another interesting variation consists of considering dominating sets with additional properties, such as connectedness or independence. A vertex set is *connected* if every two vertices in it can be joined by a sequence of edges consisting of vertices in the set. This model is useful to distribute resources in a network that need to be connected to each other. The opposite point of view is independence. A vertex set is *independent* if no pair of vertices in it are neighbors. Independent dominating sets have been widely studied, as they provide a model of resource location in cases where such resources should be placed far away from each other.

The most precise way to dominate a graph is the so-called efficient domination. A vertex set S is an *efficient dominating set* [4], or *perfect code* [5,6], if S is independent and every vertex not in S has a unique neighbor in S. The idea behind this definition is keeping the domination of each vertex to the minimum, so that vertices in S are dominated just by themselves, and vertices not in S are dominated just once by vertices in S. Clearly, this definition provides a desirable form of domination, which becomes even more interesting with the repeatedly rediscovered result that ensures every efficient dominating set is a minimum dominating set [3]. However, the main problem with efficient dominating sets is that their existence is not guaranteed in every graph. There are well-known graph families that have no efficient dominating sets and, in these cases, a number of relaxations of conditions are possible in order to obtain a dominating set as efficient as possible. In this paper, we focus on one of these relaxed forms, called *independent* $[1,2]$-*sets* and defined in [7]. The lower level of requirement of these sets consists of allowing at most two neighbors in the set, for vertices not in it, while keeping independence. Although existence is not guaranteed either in this case, the lesser requirement leads one to think that the family of graphs that possess such sets is larger than in the case of efficient domination. The *independent* $[1,2]$-*number* of a graph G was also defined in [7] as the minimum cardinality of an independent $[1,2]$-set, if such sets exist in G, and it is denoted by $i_{[1,2]}(G)$. The following general relationship among the three mentioned domination parameters can easily be deduced from definitions. Every graph G satisfies the first inequality, while the last one is true for graphs admitting an independent $[1,2]$-set.

$$\gamma(G) \leq i(G) \leq i_{[1,2]}(G) \tag{1}$$

Given two graphs, G and H, the *Cartesian product* of them is the graph $G \square H$ with vertex set $V(G \square H) = V(G) \times V(H)$ and edge set defined as follows: $(u_1, v_1)(u_2, v_2) \in E(G \square H)$ if and only if $u_1 u_2 \in E(G), v_1 = v_2$ or $u_1 = u_2, v_1 v_2 \in E(H)$. Cartesian product graphs play an interesting role in the domination-type properties, due in part to the well-known Vizing's Conjecture [8], which states $\gamma(G \square H) \geq \gamma(G)\gamma(H)$ for every two graphs G and H. It was formulated in 1968 and is still open.

We study the particular case of *cylindrical networks* $C_m \square P_n$, which are the Cartesian product of a cycle C_m and a path P_n, and our interest comes from the known fact that they have no efficient dominating set, except in one particular case [9]. On the other hand, studying the existence of independent $[1,2]$-sets in cylinders was proposed as an open problem in [7]. Unlike in the case of *grids* $P_m \square P_n$, the Cartesian product of two paths, where the domination number is completely computed [10], the domination number of the cylinder is unknown in the general case, while formulas for $\gamma(C_m \square P_n)$ have recently been obtained for $m \leq 11$ [11] and $m \leq 30$ [12]. This makes cylinders a graph family of interest for domination-type properties, in which there is still much to study.

The rest of the paper is organized as follows. In Section 2, we prove that all cylinders have an independent $[1,2]$-set, except the single case of the Cartesian product of the cycle with 5 vertices and the path with 2 vertices, which has no such set. We use the symmetry of the cylindrical graphs to provide regular models of independent $[1,2]$-sets that, in addition, will give an upper bound for the independent $[1,2]$-number. We will also prove that this upper bound is indeed the exact value, in some small cases. In Section 3, we present a modification of a dynamic programming algorithm, originally developed for computing the domination number of grids, to provide information about the independent $[1,2]$-number in cylinders of selected sizes. Depending on the size of the cylinder, this modified algorithm computes the exact value of the parameter or just an upper bound for it. In the latter case we combine these results with an appropriate lower bound, in order to obtain the desired exact value. Finally, in Sections 4 and 5 we present and discuss the experimental results obtained with the above mentioned algorithm. All graphs that appear in this paper are finite, simple and undirected. For undefined general concepts of graph theory, we refer to [13].

2. Dominating a Cylinder as Efficiently as Possible

We devote this section to studying the existence of independent $[1,2]$-sets in cylinders of any size. As we mentioned before, a cylinder $C_m \square P_n$ is the Cartesian product of a cycle C_m with $m \geq 3$ vertices and a path P_n with $n \geq 2$ vertices. We will not consider the smallest case $C_m \square P_1 = C_m$ because all vertices in a cycle have degree two, so that every independent dominating set is trivially an independent $[1,2]$-set.

We present some regular patterns that provide independent $[1,2]$-sets in every cylinder, except in the case $C_5 \square P_2$, where no such sets exist. The key point of this graph family is the particular symmetry of cylinders, which allows one to replicate small pieces in order to cover the whole graph.

The interest in obtaining independent $[1,2]$-sets in cylinders lies in the known fact that just a particular case of them have an efficient dominating set, as we recall in the following proposition.

Proposition 1. *[9] The cylinder $C_m \square P_n$ has an efficient dominating set if and only if $m \equiv 0 \pmod 4$ and $n = 2$.*

We begin our study about the behaviour of independent $[1,2]$-sets in cylinders proving that $C_5 \square P_2$ has no such sets. This will eventually be the unique case of a cylinder failing this property.

Proposition 2. *The cylinder $C_5 \square P_2$ has no independent $[1,2]$-set.*

Proof. Denote $V(C_5 \square P_2)$ by $\{(u_i, k): 1 \leq i \leq 5, 1 \leq k \leq 2\}$ and suppose on the contrary that $S \subseteq V(C_5 \square P_2)$ is an independent $[1,2]$-set. Suppose that there exist at least two vertices in S sharing the second coordinate. By the symmetry of the graph, we may assume that $(u_i, 1), (u_j, 1) \in S$.

Note that $d((u_i, 1), (u_j, 1)) \leq 2$, so it must be $d((u_i, 1), (u_j, 1)) = 2$, because S is independent. By symmetry, we may assume that $u_i = u_1$ and $u_j = u_3$. Then $(u_1, 2), (u_2, 1), (u_3, 2) \notin S$, by independence, so $(u_2, 2) \in S$ in order to be dominated. However, this means that $(u_2, 1)$ has three neighbors in S, which is not possible (see Figure 1). Therefore, if $(u_i, 1) \in S$ then $(u_{i'}, 1) \notin S$ for every $i' \neq i$. In the same way, if $(u_j, 2) \in S$ then $(u_{j'}, 2) \notin S$ for every $j' \neq j$. Finally, $|S| \leq 2$, but $3 = \gamma(C_5 \square P_2) \leq i_{[1,2]}(C_5 \square P_2)$, a contradiction. $\square$

$$(u_1, 1) \quad (u_2, 1) \quad (u_3, 1)$$

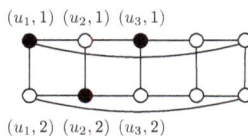

$$(u_1, 2) \quad (u_2, 2) \quad (u_3, 2)$$

Figure 1. $C_5 \square P_2$ has no independent $[1,2]$-set.

We now focus on the rest of the cylinders with $n = 2$, where we provide an example of an independent $[1,2]$-set in each size and we also obtain the exact value of the independent $[1,2]$-number.

Proposition 3. *Let $m \geq 3$ be an integer. Then, the cylinder $C_m \square P_2$ has an independent $[1,2]$-set if and only if $m \neq 5$. Moreover,*

$$i_{[1,2]}(C_m \square P_2)) = i(C_m \square P_2) = \begin{cases} \lceil \frac{m}{2} \rceil & \text{if } m \equiv 0,3 \ (\text{mod } 4), \\ \lceil \frac{m}{2} \rceil + 1 & \text{if } m \equiv 1,2 \ (\text{mod } 4), \ m \neq 5. \end{cases}$$

Proof. By Proposition 2, we know that $C_5 \square P_2$ has no independent $[1,2]$-set. It is shown in [14] that $i(C_m \square P_2) = \lceil \frac{m}{2} \rceil$, if $m \equiv 0,3 \ (\text{mod } 4)$, and $i(C_m \square P_2) = \lceil \frac{m}{2} \rceil + 1$, if $m \equiv 1,2 \ (\text{mod } 4)$. For every integer $m \neq 5$ we will construct an independent $[1,2]$-set, with $i(C_m \square P_2)$ vertices, so we will obtain that $i_{[1,2]}(C_m \square P_2) \leq i(C_m \square P_2)$.

If $m \equiv 0 \ (\text{mod } 4)$, then $m = 4r \geq 4$ and consider the set of black vertices in Figure 2a. The basic block can be repeated r times to obtain an independent $[1,2]$-set S of $C_{4r} \square P_2$. Each basic block contains 2 vertices of S, so $|S| = 2r = \frac{m}{2} = \lceil \frac{m}{2} \rceil = i(C_m \square P_2)$. Note that S is also an efficient dominating set.

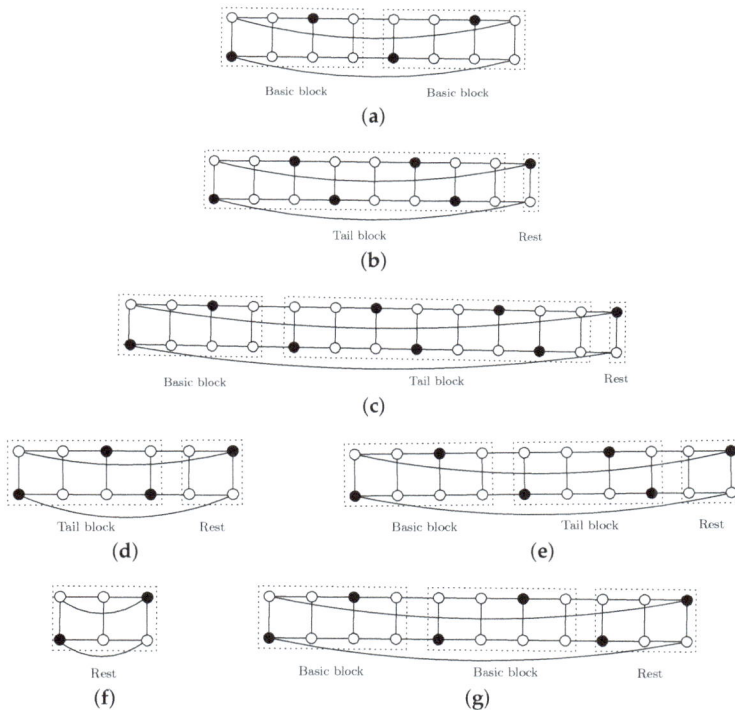

Figure 2. Regular patterns for an independent $[1,2]$-set (black vertices) in $C_m \square P_2$, $m \neq 5$: (**a**) $m \equiv 0$ (mod 4), $m \geq 4$; (**b**) $m = 9$; (**c**) $m \equiv 1$ (mod 4), $m \geq 13$; (**d**) $m = 6$; (**e**) $m \equiv 2$ (mod 4), $m \geq 10$; (**f**) $m = 3$; (**g**) $m \equiv 3$ (mod 4), $m \geq 7$.

If $5 < m \equiv 1 \ (\text{mod } 4)$, then $m = 4r + 1 \geq 9$ and $r \geq 2$. We repeat the basic block $r - 2$ times, add the tail block and the rest (see Figure 2b for the case $r = 2, m = 9$ and Figure 2c for the general case

$r \geq 3$). We obtain an independent $[1,2]$-set S (black vertices) with $2(r-2)+6 = 2r+2 = \frac{m-1}{2}+2 = \frac{m+1}{2}+1 = \lceil \frac{m}{2} \rceil + 1 = i(C_m \square P_2)$ vertices.

If $m \equiv 2 \pmod 4$, then $m = 4r+2 \geq 6$ and $r \geq 1$. We add $r-1$ copies of the basic block, the tail block for this case and the appropriate rest (see Figure 2d for the case $r = 1, m = 6$ and Figure 2e for the general case $r \geq 2$) and we obtain an independent $[1,2]$-set with $2(r-1)+4 = 2r+2 = \frac{m}{2}+1 = \lceil \frac{m}{2} \rceil + 1 = i(C_m \square P_2)$ vertices.

If $m \equiv 3 \pmod 4$, then $m = 4r+3 \geq 3$ and $r \geq 0$. We construct the independent $[1,2]$-set by adding r copies of the basic block to the rest for this size, as is shown in Figure 2f if $r = 0, m = 3$ and in Figure 2g if $r \geq 1$. We obtain an independent $[1,2]$-set with $2r+2 = \frac{m+1}{2} = \lceil \frac{m}{2} \rceil = i(C_m \square P_2)$ vertices.

Once we have proven the existence of an independent $[1,2]$-set in $C_m \square P_2$ ($m \neq 5$), with size $i(C_m \square P_2)$, the desired equality now comes from Equation (1). $\square$

Remark 1. *It is known that* $\gamma(C_m \square P_2) = \lceil \frac{m}{2} \rceil + 1$ *if* $m \equiv 2 \pmod 4$ *and* $\gamma(C_m \square P_2) = \lceil \frac{m}{2} \rceil$ *if* $m \equiv 3 \pmod 4$ *[11]. In these cases, no efficient dominating set exists and Proposition 3 shows that minimum independent $[1,2]$-sets play a similar role to such sets, in the sense that they provide the most efficient way of dominating these cylinders and they are at the same time minimum dominating sets.*

In the same way, we obtained the value of $i_{[1,2]}$ in the case n is as small as possible, we now study the case with the smallest cycle, that is $C_3 \square P_n$. Here, we also supply an example of an independent $[1,2]$-set that proves to be minimum. Henceforth, we will say that $C_m \square P_n$ has m rows and n columns. Each row is a path with n vertices and each column is a cycle with m vertices. We numerate rows from top to bottom and we numerate columns from left to right.

Proposition 4. *Let* $n \geq 2$ *be an integer. Then the cylinder* $C_3 \square P_n$ *has an independent* $[1,2]$-*set. Moreover,*

$$i_{[1,2]}(C_3 \square P_n) = i(C_3 \square P_n) = n.$$

Proof. We just need to prove that $C_3 \square P_n$ has an independent $[1,2]$-set with $i(C_3 \square P_n)$ elements. In Figure 3, we show a regular pattern for such a set, for any value of n. Clearly, this set has n elements, one in each column, so $i_{[1,2]}(C_3 \square P_n)) \leq n = i(C_3 \square P_n)$ (for the last equality see [14]). $\square$

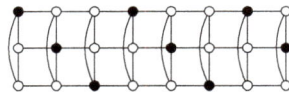

Figure 3. Regular pattern for an independent $[1,2]$-set in $C_3 \square P_n$.

Remark 2. *We would like to point out that the value of the domination number of cylinder* $C_3 \square P_n$ *that appears in [12],* $\gamma(C_3 \square P_n) = \lceil \frac{3n+1}{3} \rceil$ *is not correct. The correct one is* $\gamma(C_3 \square P_n) = \lceil \frac{3n}{4} \rceil + 1$ *if* $n \equiv 0 \pmod 4$ *and* $\gamma(C_3 \square P_n) = \lceil \frac{3n}{4} \rceil$ *otherwise, and it can be found in [11]. In Figure 4a, we show a minimum dominating set of* $C_3 \square P_8$ *with* $\lceil \frac{24}{4} \rceil + 1 = 7$ *vertices and, in Figure 4b, a dominating set of* $C_3 \square P_9$ *with* $\lceil \frac{27}{4} \rceil = 7$ *vertices. None of them are independent sets; indeed,* $\gamma(C_3 \square P_n) = i(C_3 \square P_n)$ *if and only if* $1 \leq n \leq 4$.

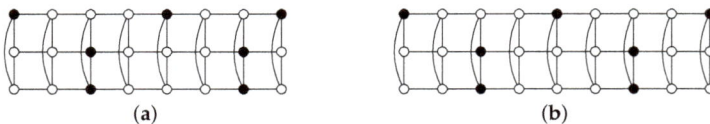

(a)

(b)

Figure 4. $\gamma(C_3 \square P_n) < i(C_3 \square P_n)$, for every $n \geq 5$: (a) $\gamma(C_3 \square P_8) = 7$; (b) $\gamma(C_3 \square P_9) = 7$.

Once we have studied the cases with the smallest values of m and n, we now focus on the general case. Our target is to prove that every cylinder $C_m \square P_n$, with $m \geq 4$ and $n \geq 3$ has an independent $[1,2]$-set. To this end we construct regular patterns that can be replicated in order to cover all the cases. These patterns will also provide an upper bound of $i_{[1,2]}$. We think that it could be possible to separately study some other small cases, for instance $m = 4$ or $m = 5$, to obtain the exact values of the independent $[1,2]$-number. However, we now prefer to provide general constructions, even if they are not minimum ones, that can be used in cylinders of any size, in order to prove the general existence of independent $[1,2]$-sets. On the other hand, we will compute the exact value of $i_{[1,2]}$ for a number of small cases in the following sections. We divide our study into two results, one for odd paths and another for even paths. We begin with the odd case.

Theorem 1. *Let $m \geq 4$ be an integer and let $n \geq 3$ be an odd integer. Then the cylinder $C_m \square P_n$ has an independent $[1,2]$-set. Moreover in this case*

$$i_{[1,2]}(C_m \square P_n) \leq \left\lfloor \frac{m(n+1)}{4} \right\rfloor .$$

Proof.

Case 1: $m = 2r \geq 4$.

Using $n = 2k + 1 \geq 3$, consider the pattern described in Figure 5, where set S consists of black vertices. There are no vertices of S in the even columns. Regarding odd columns, we begin with vertices in odd positions in the first one and then we alternate with vertices in even positions. Clearly S is an independent $[1,2]$-set in $C_{2r} \square P_{2k+1}$. Moreover, in each odd column there are r vertices of S, so

$$|S| = r(k+1) = \frac{m}{2} \cdot \frac{n+1}{2} = \frac{m(n+1)}{4} = \left\lfloor \frac{m(n+1)}{4} \right\rfloor .$$

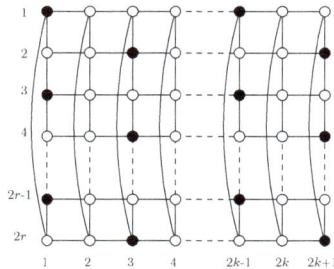

Figure 5. Regular pattern (black vertices) for an independent $[1,2]$-set in $C_{2r} \square P_{2k+1}$.

Case 2: $m = 2r + 1 \geq 5$ and $n = 4k + 1 \geq 5$.

Consider the pattern described in Figure 6. Here, we use a basic block with four columns, which we repeat k times, and a rest with one column.

In the basic block, vertices in S in the first column are in even positions, in the second column the unique vertex in S is the last one. In the third column, vertices in S are the ones in the odd positions, except the first and the last ones. Finally, in column number four, we pick just the first vertex. Therefore, there are $r + 1 + (r - 1) + 1 = 2r + 1$ vertices of S in each basic block (see Figure 6).

The last column contains r vertices in S which are in even positions (see Figure 6). Clearly, S is an independent $[1,2]$-set of $C_{2r+1} \square P_{4k+1}$ and moreover,

$$|S| = (2r + 1)k + r = m \cdot \frac{n-1}{4} + \frac{m-1}{2} = \frac{m(n+1) - 2}{4} = \left\lfloor \frac{m(n+1)}{4} \right\rfloor .$$

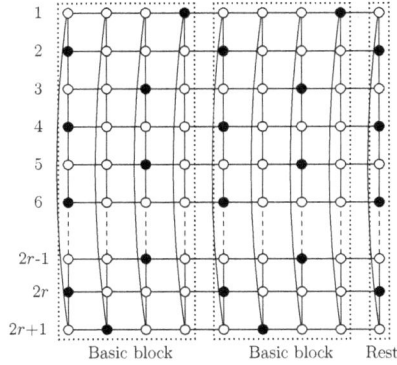

Figure 6. Regular pattern (black vertices) for an independent $[1,2]$-set in $C_{2r+1}\square P_{4k+1}$.

Case 3: $m = 2r + 1 \geq 5$ and $n = 4k + 3 \geq 3$.

We now take the pattern in Figure 7. Again, we repeat the same basic block as before k times, and we add a rest with three columns. If $k = 0$ and $n = 3$, we just consider the rest with three columns. In any case, the resulting set S is an independent $[1,2]$-set of $C_{2r+1}\square P_{4k+3}$. We know that a basic block contains $2r + 1$ vertices of S and note that the rest contains $r + 1 + r$ vertices of S, therefore

$$|S| = (2r+1)k + (2r+1) = (2r+1)(k+1) = m \cdot \frac{n+1}{4} = \left\lfloor \frac{m(n+1)}{4} \right\rfloor. \quad \square$$

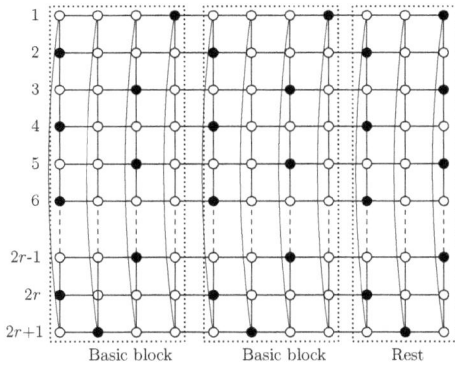

Figure 7. Regular pattern (black vertices) for an independent $[1,2]$-set in $C_{2r+1}\square P_{4k+3}$.

We now complete the study of the existence of independent $[1,2]$-sets in cylinders with the following theorem, covering the remaining case, that is, when n is even.

Theorem 2. *Let $m \geq 4$ be an integer and let $n \geq 4$ be an even integer. Then, the cylinder $C_m\square P_n$ has an independent $[1,2]$-set. Moreover, in this case,*

$$i_{[1,2]}(C_m\square P_n) \leq \begin{cases} \left\lfloor \dfrac{m}{3} \right\rfloor n & \text{if } m \equiv 0,1 \pmod 3, \\[12pt] \left\lfloor \dfrac{m}{3} \right\rfloor n + \dfrac{n}{2} & \text{if } m \equiv 2 \pmod 3. \end{cases}$$

Proof.

Case 1: $m = 3r$.

We consider here the pattern described in Figure 8 that provides an independent $[1, 2]$-set S, in $C_{3r} \square P_n$. Note that each column contains r vertices of S, so

$$|S| = rn = \frac{m}{3}n = \left\lfloor \frac{m}{3} \right\rfloor n.$$

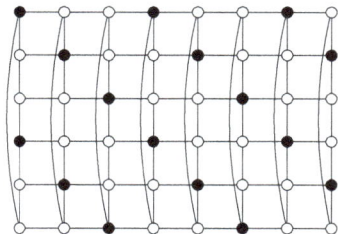

Figure 8. Regular pattern (black vertices) for an independent $[1, 2]$-set in $C_{3r} \square P_n$.

Case 2: $m = 3r + 1 \geq 4$.

To describe the pattern we use here, we begin with a single column. Vertices in S in such a column are those in positions a multiple of three plus one, except the vertex in position $m = 3r + 1$, so the column contains r vertices in S. We call the vertex in position one in the column the mark vertex, and, by construction, the mark vertex belongs to S (see Figure 9a, mark vertex with a square). We repeat this distribution of vertices in S in the second column, but rotating the cycle in such a way that the mark vertex is in the position two units smaller (modulo m) (see Figure 9b, the arrow shows rotation). The set S is an independent $[1, 2]$-set in this pair of columns.

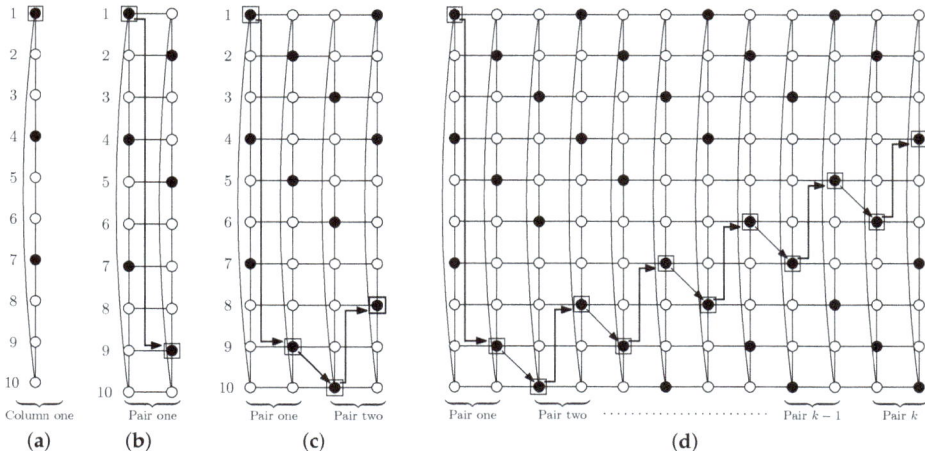

Figure 9. Construction of a regular pattern for an independent $[1, 2]$-set in $C_{3r+1} \square P_{2k}$: (**a**) first column; (**b**) a pair of columns; (**c**) two pairs of columns; (**d**) k pairs of columns.

Consider now a new pair of columns with the distribution of vertices in S described above. We join it to the preceding pair, and, in this case, we rotate the cycles so that the mark vertex is in position one unit larger (modulo m) (see Figure 9c, the arrows show rotations). Again, we obtain an independent $[1,2]$-set in the resulting cylinder. Repeating this operation as many times as necessary, we obtain an independent $[1,2]$-set S in $C_{3r+1}\square P_{2k}$ (see Figure 9d, the arrows show rotations).

We show in Figure 9 an example with ten rows, but the construction also works for smaller cases (four and seven rows), and for bigger cases. Regarding the cardinality of S, note that there are r vertices of S in each column, so

$$|S| = rn = \frac{m-1}{3}n = \left\lfloor \frac{m}{3} \right\rfloor n.$$

Case 3: $m = 3r + 2 \geq 5$.

We begin here with two types of columns. In Type A columns, vertices in S are in positions multiple of three plus one, except in position $3m+1$; in addition we include vertex in position $3m$ (see Figure 10a). We call the vertex in position one in the column the Type A mark vertex, and, by construction, this vertex belongs to S. Note that there are $r+1$ vertices of S in type A columns.

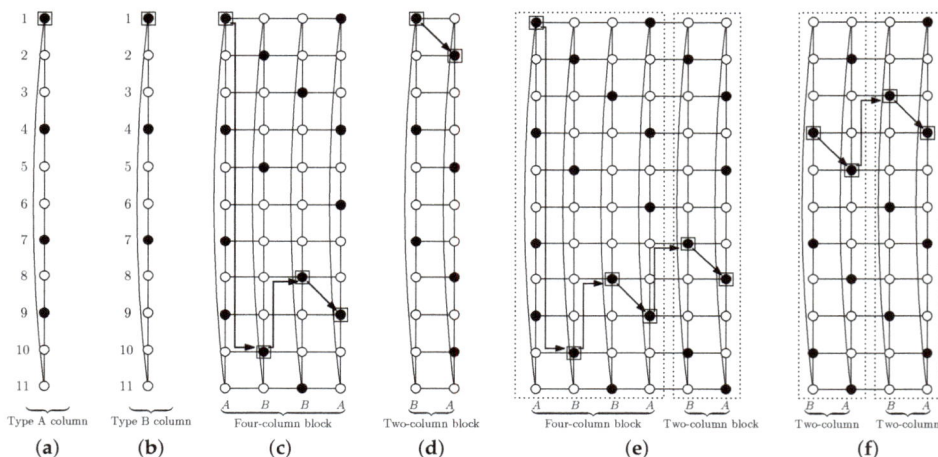

Figure 10. Construction of a regular pattern for an independent $[1,2]$-set in $C_{3r+2}\square P_{2k}$: (**a**) Type A column; (**b**) Type B column; (**c**) four-column block; (**d**) two-column block; (**e**), (**f**) joining blocks.

In Type B columns, vertices in S are in positions multiple of three plus one, except in position $3m+1$. We do not add any other vertex here (see Figure 10b). We call the vertex in position one in the column the Type B mark vertex, which belongs to S. In a Type B column there are r vertices in S.

Rules to join both types of columns, in order to obtain the desired independent $[1,2]$-set, are the following. We first need a four-column block, of types A, B, B and A (in this order), where each column is rotated, in reference to the previous one, until placing mark vertices as shown in Figure 10c. Note that vertices in S in this block are an independent $[1,2]$-set in the block.

We now make a two-column block by placing a Type B column and a Type A column (in this order), in such a way that mark vertices are in positions shown in Figure 10d.

We place a two-column block after another block (with four or two columns), rotating the second block in such a way that its mark vertex is in positions shown in Figure 10e,f.

The desired independent $[1,2]$-set S is obtained with one initial four-column block and attaching to it as many two-column blocks as necessary, to obtain the cylinder $C_m\square P_{2k}$ (note that $2k \geq 4$) (see Figure 11). Note that, although vertices in S in a two-column block are not a dominating set by themselves, as can be seen in Figure 10d, when we attach such a block to one four-column block following the rotation rules as described above, we obtain an independent $[1,2]$-set S (see Figure 10e). After this first operation, adding a new two-column block keeps independent $[1,2]$-domination (see Figure 11). We show in this figure the case $m = 11$, but the same pattern works with smaller cases $m = 5$ and $m = 8$, and also with larger ones.

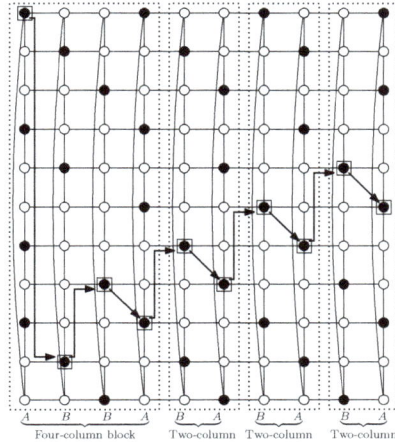

Figure 11. Regular pattern (black vertices) for an independent $[1,2]$-set in $C_{3r+2} \square P_{2k}$.

The four-column block contains two type A and two type B columns, so there are $2(r+1) + 2r = 4r + 2$ vertices of S in the block. On the other hand, two-column blocks consist of one type A and one type B column, so they have $(r+1) + r = 2r + 1$ vertices in S. Finally, S contains one four-column block and $k - 2$ two-column blocks, so:

$$|S| = (4r+2) + (2r+1)(k-2) = (2r+1)k = (2\frac{m-2}{3} + 1)\frac{n}{2} = \frac{m-2}{3}n + \frac{n}{2} = \left\lfloor \frac{m}{3} \right\rfloor n + \frac{n}{2}. \quad \square$$

3. Computing the Independent [1, 2]-Number in Cylinders

Having proven that every cylinder, except $C_5 \square P_2$, has an independent $[1,2]$-set, in this section, we present an algorithm to compute the independent $[1,2]$-number in cylinders of small sizes. This is an adaptation of the algorithm presented in [15], to compute the domination number of grids. A modified version of this algorithm also appears in [16,17]. Although the steps of the algorithm are quite similar to the original one, we prefer to completely described it, in order to fix notation and to point out the differences. We focus on cylinders $C_m \square P_n$ with $m \geq 4$, because the case $m = 3$ was solved in Proposition 4.

The main tool of the algorithm is the $(min, +)$ matrix multiplication. This is the standard matrix multiplication for the semi-ring of tropical numbers in the min convention (see [18]), that is, the usual multiplication is replaced by sum, whereas the usual sum is replaced by minimization. Therefore, the $(min, +)$ product of an $m \times n$ matrix A and an $n \times p$ matrix B is computed by the formula $(A \boxtimes B)_{i,j} = \min\limits_{1 \leq k \leq n} (a_{i,k} + b_{k,j})$. Moreover, the $(min, +)$ product of a matrix A and a scalar c is computed by $(c \boxtimes A)_{i,j} = c + a_{i,j}$.

The other key point of the algorithm is the identification of columns of the cylinder with words in the alphabet $\{0, 1, 2, 3\}$, in such a way that the number in each vertex describes its behaviour regarding independent $[1,2]$-domination. We next describe how we will do this. Let $C_m \square P_n$ be a cylinder and let $S \subseteq V(C_m \square P_n)$ be an independent $[1,2]$-set. We identify each vertex u with an element of the set $\{0, 1, 2, 3\}$, following these rules:

$u = 0$ if $u \in S$;
$u = 1$ if u has exactly one neighbor in S in its column or in the previous one;
$u = 2$ if u has exactly two neighbors in S in its column or in the previous one;
$u = 3$ if u has no neighbors in S in its column or in the previous one.

By definition, each vertex equal to 2 has at least one neighbor in S in its column and each vertex equal to 3 has a unique neighbor in S, which is in the following column. Every vertex in the cylinder is in exactly one of the preceding situations, so all of them can be identified with a unique number. Therefore each column can be seen as a word of length m in the alphabet $\{0, 1, 2, 3\}$, where the first and last letters are consecutive, because columns in the cylinder $C_m \square P_n$ are cycles with m vertices. The property of set S being an independent $[1, 2]$-set provides a number of restrictions in words that can appear, which we list below. In all the restrictions, we consider that the first and last letters of the word $P (\equiv$ column$)$ are consecutive, because each column is a cycle. If P is a word associated to an independent $[1, 2]$-set S, then P does not contain any of the following sequences:

(a) $00, 22, 33, 112, 211, 212, 213, 321, 1111, 1113, 3111, 3113$
(b) $03, 30, 010, 121, 123, 321, 323$
(c) $12021, 12023, 32021, 32023$

Note that any sequence of List (a) in a column implies that in the column, or in the previous one, or in the following one, there are two consecutive vertices of S, so it is not possible. Sequences in list (b) are not allowed because of the rules for associating each vertex with its number. Finally, sequences of list (c) do not appear in any column because each vertex has at most two neighbors in S and in all cases, such a sequence implies that in the previous column there is a vertex (next to the zero) with three neighbors in S. A word of length m in the alphabet $\{0, 1, 2, 3\}$ not containing any of the sequences in Lists (a), (b) and (c) is called a *correct word*.

We also need to bear in mind that the first and last columns of the cylinder play a different role than interior ones, because they are placed next to a unique column, not between two columns. A word in the first column satisfies that every vertex equal to 1 has a neighbor in the column (just one) equal to 0, and every vertex equal to 2 has both neighbors in the column equal to 0. A correct word satisfying these conditions is called an *initial word*. On the other hand, the word in the last column does not contain any vertex equal to 3 and we call correct words satisfying this property *final words*.

Some restrictions regarding rows also arise. They can be described with rules that describe where a column P can be placed in the following position of column Q. In order to collect these restrictions, we denote letters in words P and Q as $P = p_1 p_2 \ldots p_m$ and $Q = q_1 q_2 \ldots q_m$ and in the following cases, indices are taken modulo m,

(i)   if $q_i = 0$, then either $\left\{ p_i = 2 \text{ and } \{p_{i-1} \neq 0 \text{ or } p_{i+1} \neq 0, \text{ but not both are non-zero}\} \right\}$ or $\left\{ p_i = 1 \text{ and } p_{i-1} \neq 0 \text{ and } p_{i+1} \neq 0 \right\}$;

(ii) if $q_i = 1$, then either $p_i = 0$ or $\left\{ p_i = 1 \text{ and } \{p_{i-1} \neq 0 \text{ or } p_{i+1} \neq 0, \text{ but not both are non-zero}\} \right\}$ or $\left\{ p_i = 2 \text{ and } p_{i-1} = 0 \text{ and } p_{i+1} = 0 \right\}$ or $p_i = 3$;

(iii) if $q_i = 2$, then either $\left\{ p_i = 1 \text{ and } \{p_{i-1} \neq 0 \text{ or } p_{i+1} \neq 0, \text{ but not both are non-zero}\} \right\}$ or $p_i = 3$;

(iv) if $q_i = 3$, then $p_i = 0$.

If words P and Q satisfy Conditions (i) to (iv), we say that P *can follow* Q. Clearly, each independent $[1, 2]$-set in the cylinder $C_m \square P_n$ can be identified with a unique ordered list $P_1, P_2, \ldots, P_n$, of n correct words of length m in the alphabet $\{0, 1, 2, 3\}$, such that

1. P_1 is initial;
2. P_{i+1} can follow P_i, for $i \in \{1, 2, \ldots n - 1\}$;
3. P_n is final.

We denote such ordered lists as (m, n)-lists. An ordered list satisfying properties 1 and 2 is called an (m, n)-quasilist.

To compute the independent $[1, 2]$-number of $C_m \Box P_n$, that is, the minimum number of zeros among all (m, n)-lists, we define the following vector. Denote by c_m the cardinality of the set $\{P_1, P_2, \ldots, P_{c_m}\}$ of correct words of length m. The vector $X^1 = (\alpha_1, \alpha_2, \ldots, \alpha_{c_m})$, of length c_m, is defined as follows:

$$\alpha_i = \begin{cases} \text{number of zeros of } P_i & \text{if } P_i \text{ is initial,} \\ +\infty & \text{otherwise.} \end{cases}$$

We also define the matrix $A_{c_m \times c_m} = (a_{i,j})$, a square matrix of size c_m such that

$$a_{i,j} = \begin{cases} \text{number of zeros of } P_i & \text{if } P_i \text{ can follow } P_j, \\ +\infty & \text{otherwise.} \end{cases}$$

By multiplying vector X^1 and matrix A, with the $(min, +)$ matrix multiplication, we obtain a new vector $X^2 = A \boxtimes X^1 = (\beta_1, \beta_2, \ldots, \beta_{c_m})$, and, clearly, every finite entry

$$\beta_i = \min_{1 \leq j \leq c_m} (\alpha_j + a_{i,j}) < +\infty$$

represents the minimum number of zeros among all $(m, 2)$-quasilists having P_i as its second word. We can repeat this process as many times as we need, so if $X^n = A \boxtimes X^{(n-1)} = (\delta_1, \delta_2, \ldots, \delta_{c_m})$, then every finite entry $\delta_i < +\infty$, represents the minimum number of zeros among all (m, n)-quasilists having P_i as its last word. Having in mind that an (m, n)-quasilist whose last word is final is an (m, n)-list, we obtain

$$i_{[1,2]}(C_m \Box P_n) = \min\{\delta_i : P_i \text{ is a final word}\}. \tag{2}$$

Moreover, algebraic properties of $(min, +)$ matrix multiplication give the following result, which is similar to Theorem 2.2 of [16].

Proposition 5. *Let $m \geq 3$ be an integer and suppose that there exist $n_0, c, d > 0$ such that $X^{n_0 + d} = c \boxtimes X^{n_0}$. Then $X^{n+d} = c \boxtimes X^n$ for every $n \geq n_0$, and moreover*

$$i_{[1,2]}(C_m \Box P_{n+d}) = i_{[1,2]}(C_m \Box P_n) + c.$$

Proof. We prove by induction that $X^{n+d} = c \boxtimes X^n$ for every $n \geq n_0$. On the one hand, $X^{n_0+d} = c \boxtimes X^{n_0}$ by hypothesis. Assume now that $X^{(n-1)+d} = c \boxtimes X^{(n-1)}$ for $n - 1 \geq n_0$. Then, properties of the $(min, +)$ matrix algebra and the inductive hypothesis give

$$X^{n+d} = A \boxtimes X^{(n-1)+d} = A \boxtimes (c \boxtimes X^{(n-1)}) = c \boxtimes (A \boxtimes X^{(n-1)}) = c \boxtimes X^n.$$

Therefore, if $X^{n+d} = (\epsilon_1, \epsilon_2, \ldots, \epsilon_{c_m})$ and $X^n = (\delta_1, \delta_2, \ldots, \delta_{c_m})$, then $\epsilon_i = \delta_i + c$ for every index i. In particular, ϵ_i is finite if and only if δ_i is finite. Finally this gives

$$\begin{aligned} i_{[1,2]}(C_m \Box P_{n+d}) &= \min\{\epsilon_i : P_i \text{ is a final word}\} \\ &= \min\{\delta_i + c : P_i \text{ is a final word}\} \\ &= \min\{\delta_i : P_i \text{ is a final word}\} + c \\ &= i_{[1,2]}(C_m \Box P_n) + c. \quad \Box \end{aligned}$$

Results presented in this section provide the following algorithm to compute the independent $[1, 2]$-number of $C_m \Box P_n$. It is an adaptation of the algorithm presented in [15], to compute the domination number of grids. A modified version of this algorithm also appears in [16,17]. On the one hand, the algorithm computes the value of $i_{[1,2]}(C_m \Box P_n)$ for fixed and small enough (m, n). On the other hand, given a fixed m, it computes a list of vectors $X^1, X^2, \ldots, X^r$ and it tries to find the recurrence relationship on the hypothesis of Proposition 5.

If a recurrence is found by Algorithm 1, Proposition 5 provides a finite difference equation

$$i_{[1,2]}(C_m \Box P_{n+d}) - i_{[1,2]}(C_m \Box P_n) = c, \text{ for } n \geq n_0.$$

with d boundary values $i_{[1,2]}(C_m \Box P_s)$, for $n_0 \leq s \leq n_0 + d - 1$. The solution of this equation is

$$i_{[1,2]}(C_m \Box P_n) = \left\lceil \frac{cn + \beta}{d} \right\rceil, \text{ for } n \geq n_0,$$

where the value of β depends on the boundary values. Remaining values of $i_{[1,2]}(C_m \Box P_n)$, that is for $2 \leq n \leq n_0 - 1$, have also been computed by the algorithm, so finding the recurrence means that the independent $[1, 2]$-number of $C_m \Box P_n$ is completely computed.

Algorithm 1: Recurrence for the Independent $[1, 2]$-number in Cylinders

---

   Input: integers $m \geq 4$ and $r \geq 2$
   Output: $n_0, c, d > 0$ such that $X^{n_0+d} = c \boxtimes X^{n_0}$ and $i_{[1,2]}(C_m \Box P_n)$ for $2 \leq n \leq n_0 + d - 1$, or NO RECURRENCE FOUND
1 compute the ordered set of all correct words on length m;
2 compute vector X^1;
3 compute matrix A;
4 use $(min, +)$ matrix multiplication to obtain $X^2, X^3, \ldots, X^r$;
5 **if** $(X^{n_0+d} = c \boxtimes X^{n_0}$ *for some* n_0, d, c==**True**) **then**
6   use Equation (2) to compute $i_{[1,2]}(C_m \Box P_s)$ for $2 \leq s \leq n_0 + d - 1$;
7   return n_0, d, c and $i_{[1,2]}(C_m \Box P_s)$ for $2 \leq s \leq n_0 + d - 1$;
8 **else**
9   return NO RECURRENCE FOUND;
10 **end**

However, the existence of such a recurrence relationship is not guaranteed, as was mentioned in Section 6 of [15]. Some sufficient conditions on an arbitrary matrix A, in the $(min, +)$ matrix algebra, are known that ensure such a recurrence exists. We recall the following definition from [15]. A matrix A is *irreducible* if there exists an integer K such that for every $k \geq K$, matrix A^k has no infinite entries, and this is analogous to the definition of irreducible matrix in regular matrix algebra. Theorem 6.3 of [15] states that, if A is an irreducible matrix, then the recurrence needed to apply Proposition 5 occurs.

The following characterization is well known; see, for instance, [19]. Considering matrix A as the adjacency matrix of a directed graph (there is an arc from j to i if and only if $a_{i,j} < +\infty$), A is irreducible if and only if such a directed graph is strongly connected. In particular, a matrix such that every entry in the ith row is equal to $+\infty$ is not irreducible, because no arc arrives to i. When this situation happens, the recurrence needed in Proposition 5 could not occur.

This is the main difference between the computation of the independent $[1, 2]$-number in cylinders and the previous cases where this algorithm has been used (see [15–17]). In our case, matrices defined using the properties of independent $[1, 2]$-sets have an increasing number of rows with no finite entry, when m gets larger. In fact, we just found the desired recurrence in cases $m = 4$ and $m = 5$. If $m \geq 6$ our strategy consists of modifying the first step of the algorithm. We consider a reduced collection $\mathcal{C}$ of correct words instead of having all of them, so we obtain an auxiliary parameter $f_m(n)$ that represents the minimum number of zeros among all (m, n)-lists with words in $\mathcal{C}$, if there exists at least one of such (m, n)-list. Then, we use the algorithm to look for the recurrence relationship, but using just correct words in subset $\mathcal{C}$ and, in the case it is found, the auxiliary function obtained satisfies

$i_{[1,2]}(C_m \square P_n) \leq f_m(n)$. The final step consists of combining this upper bound with an appropriate lower bound, in order to localize the independent $[1,2]$-number in an interval as small as possible.

4. Experimental Results

In this section, we present the results obtained with Algorithm 1, for values of m between 4 and 15. The link to the source code, in programming language C, to perform all the operations described in the algorithm can be found in the Supplementary Materials. Notice that vector X^1 is computed using its definition, but vectors $X^2, X^3, \ldots, X^r$ are obtained by successively multiplying matrix A and vector X^1, with the $(min, +)$ matrix multiplication. We do this operation by means of an adaptation of the CSPARSE library [20]. This library provides a fast method for multiplying sparse matrices with the usual product, and we have adapted it for the $(min, +)$ matrix product. Therefore, we have obtained a library that efficiently multiplies sparse matrices, also with this product, as can be seen in Table 1, where we show that execution times for computing the first 100 vectors are shorter than times for the computation of the matrix. We have also included computation times for the set of correct words, but not for vector X^1, because it is less that one second in all cases.

Table 1. Execution times with only one core, Intel(R) Core(TM) i7-3632QM at 2.20 GHz processor.

m	4^m = Words of Length m in Alphabet $\{0,1,2,3\}$	c_m = Number of Correct Words	Execution Times		
			Correct Words	A	$X^r, 2 \leq r \leq 100$
4	256	20	< 1 s.	< 1 s.	< 1 s.
5	1024	35	< 1 s.	< 1 s.	< 1 s.
6	4096	79	< 1 s.	< 1 s.	< 1 s.
7	16384	154	< 1 s.	< 1 s.	< 1 s.
8	65536	332	< 1 s.	< 1 s.	< 1 s.
9	262144	666	< 1 s.	< 1 s.	< 1 s.
10	1048576	1389	< 1 s.	< 1 s.	< 1 s.
11	4194304	2849	< 1 s.	2.6 s.	< 1 s.
12	16777216	5891	2.2 s.	11.8 s.	1.9 s.
13	67108864	12116	8.1 s.	53.2 s.	3.7 s.
14	268435456	25008	31.1 s.	4 m. 3 s.	6.9 s.
15	1073741824	51509	2 m. 5 s.	18 m. 25 s.	13.3 s.

4.1. Cases $m = 4$ and $m = 5$

In both cases, we have found the recurrence needed to apply Proposition 5, and we have completely computed the independent $[1,2]$-number of cylinders $C_m \square P_n$, for $n \geq 2$. Data obtained with Algorithm 1 and finite difference equations appear in Table 2.

Table 2. Recurrence values for reduced sets of correct words.

m	n_0	d	c	Finite Difference Equation	Boundary Values	Rest of Values
4	3	2	2	$i_{[1,2]}(C_4 \square P_{n+2}) - i_{[1,2]}(C_4 \square P_n) = 2, n \geq 3$	$i_{[1,2]}(C_4 \square P_3) = 3$ $i_{[1,2]}(C_4 \square P_4) = 4$	$i_{[1,2]}(C_4 \square P_2) = 2$
5	4	1	1	$i_{[1,2]}(C_5 \square P_{n+1}) - i_{[1,2]}(C_5 \square P_n) = 1, n \geq 4$	$i_{[1,2]}(C_5 \square P_4) = 6$	$i_{[1,2]}(C_5 \square P_3) = 5$

The solution of the finite difference equation gives the formula of the independent $[1,2]$-number. In both cases, the independent $[1,2]$-number agrees with the domination number (see [12]); therefore, independent $[1,2]$-sets provide the most efficient way to dominate these cylinders.

$$i_{[1,2]}(C_4 \square P_n) = n, \text{ for } n \geq 2$$

$$i_{[1,2]}(C_5 \Box P_n) = n + 2, \text{ for } n \geq 3.$$

4.2. Cases $6 \leq m \leq 10$

In these cases, matrices obtained with the algorithm have a large number of rows with all entries equal to $+\infty$, so they are not irreducible. This means the recurrence is not guaranteed and, in fact, we have not found it in the first 100 vectors. Clearly, it could happen that the recurrence relationship exists for some $n_0 \geq 100$, but instead of keeping on looking for a recurrence in larger values, we prefer a different approach.

We remove a group of correct words and we use Algorithm 1 again, but considering the remaining word subset instead of the complete list of correct words computed in step 1. The criterion for removing words takes into account formulas for the independent domination number $i(C_m \Box P_n) = \lceil \frac{cn+\beta}{d} \rceil$ obtained in [14], and we look for similar formulas for the independent $[1,2]$-number. We fixed vector X^{20} because it is small enough to quickly use the algorithm and it gives positive results in cases we are considering. For each m, we take the integer d in formula $i(C_m \Box P_n) = \lceil \frac{cn+\beta}{d} \rceil$ and we study the pair of vectors X^{20}, X^{20+d}. Their entries in the ith position, $X^{20}(i), X^{20+d}(i)$, are in one of the following situations:

1. both are infinite;
2. both are finite, then we compute the difference $X^{20+d}(i) - X^{20}(i)$;
3. one of them is finite and the other one is infinite, then we say that they are non-comparable.

If a recurrence is not found, then either there are non-comparable pairs or differences of finite entries are not equal, or both. In Table 3, we show values of differences found and if there are non-comparable words. Recall that, for each m, the set of correct words $\{P_1, P_2, \ldots, P_{c_m}\}$ is ordered and it has c_m elements, and the size of vector X^n is c_m. We now take the integer c in $i(C_m \Box P_n) = \lceil \frac{cn+\beta}{d} \rceil$ and we keep the word in position i if both i^{th} entries are finite with $X^{20+d}(i) - X^{20}(i) = c$ and also in case $X^{20+d}(i) = X^{20}(i) = +\infty$. However, we remove the word in position i if entries are finite with $X^{20+d}(i) - X^{20}(i) \neq c$ or $X^{20+d}(i), X^{20}(i)$ are non-comparable. This strategy provides a subset $\mathcal{C}$ of correct words that we use in the algorithm, instead of the complete list originally computed in the first step. In cases $6 \leq m \leq 9$, we found the recurrence, as we expected. However, in case $m = 10$, after a first selection of words, recurrence does not appear and we remove a second group of words, with the same criterion. It is shown in Table 3, in rows 10-I and 10-II. After removing the second word group, a recurrence also appears for $m = 10$.

Table 3. Criteria for removing some correct words.

m	d	c	Vector Pair X^{20+d}, X^{20}	Values of Differences	Non-Comparable Pairs	Remove Correct Words in Positions Where Appears	Remaining Words
6	3	4	X^{23}, X^{20}	4, 6	yes	6, non-comparable	69
7	2	3	X^{22}, X^{20}	3, 4	no	4	126
8	5	9	X^{25}, X^{20}	9, 10	yes	10, non-comparable	228
9	2	4	X^{22}, X^{20}	4, 6	no	6	660
10-I	2	4	X^{22}, X^{20}	2, 4, 5, 6	no	2, 5, 6	1077
10-II	2	4	X^{22}, X^{20}	4, 5	no	5	1067

We now apply Algorithm 1, but using the subset $\mathcal{C}$ of remaining words instead of the complete list. We would like to point out that there exists an (m, n)-list with words in $\mathcal{C}$ if and only if the value of $f_m(n) = \min\{\delta_i \colon P_i \in \mathcal{C} \text{ is a final word}\}$ is finite, where $X^n = A^n \boxtimes X^1 = (\delta_1, \ldots, \delta_k)$, X^1 is the initial vector, A is the matrix, both associated to subset $\mathcal{C}$ of correct words of length m, and $|\mathcal{C}| = k$. Clearly, if this auxiliary function $f_m(n)$ is finite, then it provides the minimum number of zeros among all (m, n)-lists with words in $\mathcal{C}$, so it is trivially true that $i_{[1,2]}(C_m \Box P_n) \leq f_m(n)$. Moreover, if we just consider words in $\mathcal{C}$ and the recurrence relationships described in Proposition 5 occurs, then the finitude of boundary values ensures that $f_m(n)$ is finite, for every $n \geq n_0$.

Recurrences and boundary values found by Algorithm 1, when we just use the subset $\mathcal{C}$ of correct words described in Table 3, are shown in Table 4.

Table 4. Recurrence values for reduced sets of correct words.

m	n_0	d	c	Auxiliary Equation	Boundary Values
6	7	3	4	$f_6(n+3) - f_6(n) = 4, n \geq 7$	$f_6(7) = 10, f_6(8) = 12, f_6(9) = 13$
7	6	2	3	$f_7(n+2) - f_7(n) = 3, n \geq 6$	$f_7(6) = 12, f_7(7) = 13$
8	11	5	9	$f_8(n+5) - f_8(n) = 9, n \geq 11$	$f_8(11) = 21, f_8(12) = 24, f_8(13) = 25$ $f_8(14) = 28, f_8(15) = 29$
9	8	2	4	$f_9(n+2) - f_9(n) = 4, n \geq 8$	$f_9(8) = 18, f_9(9) = 20$
10	10	2	4	$f_{10}(n+2) - f_{10}(n) = 4, n \geq 10$	$f_{10}(10) = 24, f_{10}(11) = 26$

The solutions of auxiliary equations are the following:

$$m = 6, n \geq 7: \quad f_6(n) = \begin{cases} \left\lceil \frac{4n}{3} \right\rceil & \text{if } n \equiv 1 \pmod 3, \\ \left\lceil \frac{4n}{3} \right\rceil + 1 & \text{otherwise.} \end{cases}$$

$$m = 7, n \geq 5: \quad f_7(n) = \begin{cases} \left\lceil \frac{3n}{2} \right\rceil + 2 & \text{if } n \equiv 1 \pmod 2, \\ \left\lceil \frac{3n}{2} \right\rceil + 3 & \text{otherwise.} \end{cases}$$

$$m = 8, n \geq 11: \quad f_8(n) = \begin{cases} \left\lceil \frac{9n}{5} \right\rceil + 1 & \text{if } n \equiv 1,3 \pmod 5, \\ \left\lceil \frac{9n}{5} \right\rceil + 2 & \text{if } n \equiv 0,2,4 \pmod 5. \end{cases}$$

$$m = 9, n \geq 8: \quad f_9(n) = 2n + 2.$$

$$m = 10, n \geq 10: f_{10}(n) = 2n + 4.$$

As we mentioned before, $f_m(n)$ is an upper bound of the independent $[1,2]$-number of $C_m \square P_n$, and we now combine these results with the values of $i(C_m \square P_n)$ [14], which is a natural lower bound, that is

$$i(C_m \square P_n) \leq i_{[1,2]}(C_m \square P_n) \leq f_m(n).$$

In cases $m = 6, 7, 9, 10$ and also in case $m = 8, n \equiv 0, 1, 3 \pmod 5$, we obtained $f_m(n) = i(C_m \square P_n)$, so the independent $[1,2]$-number agrees with the independent domination number. In case $m = 8$ and $n \geq 12, n \equiv 2, 4 \pmod 5$, we obtained $f_m(n) = i(C_m \square P_n) + 1$; the bounds do not agree and we can just conclude that $i(C_m \square P_n) \leq i_{[1,2]}(C_m \square P_n) \leq i(C_m \square P_n) + 1$. We computed the independent $[1,2]$-number for $12 \leq n \leq 100, n \equiv 2, 4 \pmod 5$ and it agrees with the upper bound in all cases.

We also computed the independent $[1,2]$-number of small cylinders ($n \leq n_0 - 1$) with the algorithm, by using Equation (2) with the complete list of correct words. We include these values in the final formulas.

$$i_{[1,2]}(C_6 \square P_n) = \begin{cases} 6 & \text{if } n = 3, \\ 9 & \text{if } n = 5, \\ \left\lceil \frac{4n}{3} \right\rceil & \text{if } n \equiv 1 \pmod 3, \\ \left\lceil \frac{4n}{3} \right\rceil + 1 & \text{otherwise.} \end{cases}$$

$$i_{[1,2]}(C_7 \square P_n) = \begin{cases} 2n & \text{if } n = 2,3,4, \\ \lceil \frac{3n}{2} \rceil + 2 & \text{if } n \equiv 1 \pmod 2, \\ \lceil \frac{3n}{2} \rceil + 3 & \text{otherwise.} \end{cases}$$

$$i_{[1,2]}(C_8 \square P_n) = \begin{cases} 2n & \text{if } n = 2,3,4,5,7,9, \\ \lceil \frac{9n}{5} \rceil + 1 & \text{if } n \equiv 1,3 \pmod 5, \\ \lceil \frac{9n}{5} \rceil + 2 & \text{if } n \equiv 0 \pmod 5 \text{ or } n \leq 100, n \equiv 2,4 \pmod 5. \end{cases}$$

$$\lceil \tfrac{9n}{5} \rceil + 1 \leq i_{[1,2]}(C_8 \square P_n) \leq \lceil \tfrac{9n}{5} \rceil + 2, \text{ if } n > 100 \text{ and } n \equiv 2,4 \pmod 5.$$

$$i_{[1,2]}(C_9 \square P_n) = 2n + 2, \text{ if } n \geq 2.$$

$$i_{[1,2]}(C_{10} \square P_n) = \begin{cases} 2n+2 & \text{if } n = 2,4,5, \\ 2n+3 & \text{if } n = 3,6,7,8,9, \\ 2n+4 & \text{otherwise .} \end{cases}$$

4.3. Cases $11 \leq m \leq 15$

In these cases, the independent domination number is not known, so our first task is to compute it. We will use these values as a lower bound of the independent $[1,2]$-number. To this end we have implemented the algorithm described in [16], to compute the independent domination number of the grid $P_m \square P_n$, making the necessary changes to adapt it to the cylinder $C_m \square P_n$. These changes just consist of considering that the first and the last letters in each word are neighbors. Computations can be done following the same steps as in Algorithm 1, but taking into account the rules to define correct words and to compute vector X^1 and matrix A that correspond with the definition of independent domination. For the sake of completeness, we recall these rules from [16]. For an independent dominating set S of $C_m \square P_n$, each vertex $u \in V(C_m \square P_n)$ is identified with an element of the set $\{0,1,2\}$, following these rules:

$u = 0$ if $u \in S$;
$u = 1$ if u has at least one neighbor in S, in its column or in the previous one;
$u = 2$ if u has no neighbors in S, in its column or in the previous one.

Each column in $C_m \square P_n$ can be seen as a word of length m in the alphabet $\{0,1,2\}$, where the first and last letters are consecutive. Correct words are those words not containing the sequences $00, 22, 1111, 1112, 2111, 2112$. For a pair of correct words P, Q, we say that $P = p_1 p_2 \ldots p_m$ can follow $Q = q_1 q_2 \ldots q_m$ if they satisfy the following conditions

(i) if $q_i = 0$, then $p_i = 1$,
(ii) if $q_i = 1$, then either $p_i = 0$ or $\{p_i = 1$ and $\{p_{i-1} \neq 0$ or $p_{i+1} \neq 0,\}\}$ or $p_i = 2$,
(iii) if $q_i = 2$, then $p_i = 0$.

Following the same steps as in Algorithm 1 with the rules described above, recurrence is found for $11 \leq m \leq 15$ and we present the final formulas obtained (we have not included small cases not following the general formula).

$$i(C_{11} \square P_n) = \lceil \tfrac{12n+12}{5} \rceil \quad \text{if } n \geq 4.$$
$$i(C_{12} \square P_n) = \lceil \tfrac{5n+9}{2} \rceil \quad \text{if } n \geq 17.$$
$$i(C_{13} \square P_n) = \lceil \tfrac{20n+20}{7} \rceil \quad \text{if } n \geq 6.$$
$$i(C_{14} \square P_n) = 3n + 4 \quad \text{if } n \geq 12.$$
$$i(C_{15} \square P_n) = 3n + 6 \quad \text{if } n \geq 8.$$

We now follow the same strategy as in the preceding cases and we use vectors X^{50}, X^{50+d} because they provide positive results in all cases. For $m = 11, 12, 13, 15$, the first reduction of the correct words gives positive results and we found the recurrence. However, in case $m = 14$, after a first selection of words, a recurrence does not appear and we remove a second group of words, with the same criterion. After removing the second word group, a recurrence also appears in this case. We show the rules for reducing the correct word set in Table 5. In Table 6, we show recurrences and finite difference equations in each case.

Table 5. Criteria for removing some correct words.

m	d	c	Vector Pair X^{50+d}, X^{50}	Values of Differences	Non-Comparable Pairs	Remove Correct Words in Positions Where Appears	Remaining Words
11	5	12	X^{55}, X^{50}	$11, 12, 13, 14, 15$	no	$11, 13, 14, 15$	2475
12	2	5	X^{52}, X^{50}	$2, 4, 5, 6, 8$	no	$2, 4, 6, 8$	3531
13	7	20	X^{57}, X^{50}	$19, 20, 21, 22, 23, 28$	no	$19, 21, 22, 23, 28$	9438
14-I	2	6	X^{52}, X^{50}	$2, 4, 5, 6, 7, 8$	no	$2, 4, 5, 7, 8$	20792
14-II	2	6	X^{52}, X^{50}	$6, 7$	no	7	19686
15	1	3	X^{51}, X^{50}	$1, 2, 3, 4, 5, 6$	no	$1, 2, 4, 5, 6$	34913

Table 6. Recurrence values for reduced sets of correct words.

m	n_0	d	c	Auxiliary Equation	Boundary Values
11	7	5	12	$f_{11}(n+5) - f_{11}(n) = 12, n \geq 7$	$f_{11}(7) = 20, f_{11}(8) = 22, f_{11}(9) = 24,$ $f_{11}(10) = 27, f_{11}(11) = 29$
12	8	2	5	$f_{12}(n+2) - f_{12}(n) = 5, n \geq 8$	$f_{12}(8) = 25, f_{12}(9) = 27$
13	10	7	20	$f_{13}(n+7) - f_{13}(n) = 20, n \geq 10$	$f_{13}(10) = 32, f_{13}(11) = 35, f_{13}(12) = 38,$ $f_{13}(13) = 40, f_{13}(14) = 44, f_{13}(15) = 46, f_{13}(16) = 50$
14	10	2	6	$f_{14}(n+2) - f_{14}(n) = 6, n \geq 10$	$f_{14}(10) = 34, f_{14}(11) = 37$
15	10	1	3	$f_{15}(n+1) - f_{15}(n) = 3, n \geq 10$	$f_{15}(10) = 36$

Solutions of auxiliary equations are the following

$$m = 11, n \geq 7: \quad f_{11}(n) = \left\lceil \frac{12n+12}{5} \right\rceil \qquad\qquad m = 14, n \geq 10: f_{14}(n) = 3n + 4$$

$$m = 12, n \geq 8: \quad f_{12}(n) = \left\lceil \frac{5n+9}{2} \right\rceil \qquad\qquad m = 15, n \geq 15: f_{15}(n) = 3n + 6$$

$$m = 13, n \geq 10: f_{13}(n) = \begin{cases} \left\lceil \frac{20n+20}{7} \right\rceil + 1 & \text{if } n \equiv 0, 2 \pmod 7, \\ \left\lceil \frac{20n+20}{7} \right\rceil & \text{otherwise.} \end{cases}$$

Final formulas have been obtained by comparing auxiliary equations with the above computed independent domination number and by using inequalities $i(C_m \square P_n) \leq i_{[1,2]}(C_m \square P_n) \leq f_m(n)$. We also include values for small cylinders, with $n \leq n_0 - 1$. In the case $m = 13$, we also include values of $i_{[1,2]}(C_{13} \square P_n)$ for $n \leq 100, n \equiv 0, 2 \pmod 7$.

$$i_{[1,2]}(C_{11} \square P_n) = \begin{cases} 6 & \text{if } n = 2, \\ 9 & \text{if } n = 3, \\ \left\lceil \frac{12n+12}{5} \right\rceil & \text{otherwise.} \end{cases}$$

$$i_{[1,2]}(C_{12} \square P_n) = \begin{cases} 3n & \text{if } n = 2, 3, 4, \\ \left\lceil \frac{5n+5}{2} \right\rceil & \text{if } n = 5, 6, 7, \\ \left\lceil \frac{5n+9}{2} \right\rceil & \text{if } n \geq 8. \end{cases}$$

$$i_{[1,2]}(C_{13}\Box P_n) = \begin{cases} 7 & \text{if } n = 2, \\ 10 & \text{if } n = 3, \\ \left\lceil \frac{20n+11}{7} \right\rceil & \text{if } 4 \leq n \leq 9, \\ \left\lceil \frac{20n+20}{7} \right\rceil & \text{if } n \geq 10, n \not\equiv 0,2 \pmod 7 \text{ or } n = 14,16, \\ \left\lceil \frac{20n+20}{7} \right\rceil + 1 & \text{if } 21 \leq n \leq 100, n \equiv 0,2 \pmod 7. \end{cases}$$

$$\left\lceil \frac{20n+20}{7} \right\rceil \leq i_{[1,2]}(C_{13}\Box P_n) \leq \left\lceil \frac{20n+20}{7} \right\rceil + 1, \text{ if } n > 100 \text{ and } n \equiv 0,2 \pmod 7.$$

$$i_{[1,2]}(C_{14}\Box P_n) = \begin{cases} 8 & \text{if } n = 2, \\ 3n + 3 & \text{if } 3 \leq n \leq 9, \\ 3n + 4 & \text{if otherwise.} \end{cases}$$

$$i_{[1,2]}(C_{15}\Box P_n) = \begin{cases} 8 & \text{if } n = 2, \\ 12 & \text{if } n = 3, \\ 3n + 4 & \text{if } n = 4,5, \\ 3n + 5 & \text{if } n = 6,7, \\ 3n + 6 & \text{otherwise.} \end{cases}$$

5. Conclusions

In this paper, we have deeply studied independent $[1,2]$-sets and their associated parameter, the independent $[1,2]$-number, in cylindrical networks. The main interest of this study lies in the known fact that the cylinder $C_m \Box P_n$ has an efficient dominating set if and only if $m \equiv 0 \pmod 4$ and $n = 2$, so in other cylinders different domination-like sets are needed to dominate them as efficiently as possible. On the other hand, the symmetry of these graphs allows us to focus their study from different points of view.

In Section 2, we have proven that every cylinder $C_m \Box P_n$ with $(m,n) \neq (5,2)$ has an independent $[1,2]$-set. We also provided exact values of $i_{[1,2]}(C_m \Box P_2)$, $m \neq 5$ and $i_{[1,2]}(C_3 \Box P_n)$, $n \geq 2$ and upper bounds for the independent $[1,2]$-number in the rest of the cases.

In Section 3, we presented an adaptation of a known algorithm to compute exact values of $i_{[1,2]}(C_m \Box P_n)$; and we presented the experimental results obtained with the algorithm in Section 4, for $4 \leq m \leq 15$. To this end, we have adapted the CSPARSE library, a fast method for multiplying sparse matrices, to the case of $(min, +)$ multiplication and we have introduced the technique of selecting some correct words when using the algorithm, providing new possibilities of applying this type of recursive computing in cases where the matrix is not irreducible and a recurrence is not found.

Regarding the cases in which we have exactly computed the independent $[1,2]$-number, comparing our results with the values of the domination number [11,12] and leaving aside the small values of n not following the general formula, we may conclude that:

- if $m = 3,4,5,6,9,10,15$, then $i_{[1,2]}(C_m \Box P_n) = \gamma(C_m \Box P_n)$;
- if $m = 7,14$, then $i_{[1,2]}(C_m \Box P_n) = \gamma(C_m \Box P_n) + 1$;
- if $m = 8,12$, then $\gamma(C_m \Box P_n) \leq i_{[1,2]}(C_m \Box P_n) \leq \gamma(C_m \Box P_n) + 1$;
- $i_{[1,2]}(C_{11}\Box P_n) = i(C_{11}\Box P_n)$;
- $i(C_{13}\Box P_n) \leq i_{[1,2]}(C_{13}\Box P_n) \leq i(C_{13}\Box P_n) + 1$.

Summing up, it is known that, in general, the independent $[1,2]$-number does not equal the domination number; however, we have seen that there are some cylinders having this property and some others where both parameters differ by 1. In view of these results, we may conclude that independent $[1,2]$-sets provide an interesting alternative to efficient dominating sets in cylindrical networks.

Supplementary Materials: The following are available online at https://github.com/hpcjmart/cylinders: source code, in programming language C, to perform all the operations described in Algorithm 1, and instructions to generate the executable files and links to the additional libraries necessary for its compilation.

Symmetry **2018**, *10*, 24

Acknowledgments: This project is partially supported by grants MINECO-ERDF TIN2015-66680 and Junta de Andalucía FQM305.

Author Contributions: All authors contributed equally to this work.

Conflicts of Interest: The authors declare no conflict of interest.

References

1. Berge, C. *Theory of Graphs and its Applications*; Collection Universitaire de Mathématiques: Dunod, Paris, France, 1958.
2. Ore, O. *Theory of Graphs*; American Mathematical Society Publication: Providence, RI, USA, 1962; Volume 38.
3. Haynes T.W.; Hedetniemi, S.T.; Slater P.J. *Foundamentals of Domination in Graphs*; Marcel Dekker, Inc.: New York, NY, USA, 1998.
4. Bange, D.W.; Barkauskas, A.E.; Slater, P.J. Efficient dominating sets in graphs. *Appl. Discret. Math. Proc. Third SIAM Conf. Discret. Math.* **1986**, *189*, 189–199.
5. Biggs, N. Perfect codes in graphs. *J. Comb. Theory Ser. B* **1973**, *15*, 288–296.
6. Livingston, M.; Stout, Q.F. Perfect Dominating Sets. *Congr. Numer.* **1990**, *79*, 187–203.
7. Chellali, M.; Favaron, O.; Haynes, T.W.; Hedetniemi, S.T.; McRae, A. Independent $[1, k]$-sets in graphs. *Australas. J. Comb.* **2014**, *59*, 144–156.
8. Vizing, V.G. Some unsolved problems in graph theory. *Uspehi Mat. Nauk.* **1968**, *23*, 117–134.
9. Barbosa, R.; Slater, P. On the efficiency index of a graph. *J. Comb. Optim.* **2016**, *31*, 1134–1141.
10. Gonçalves, D.; Pinlou, A.; Rao, M.; Thomassé, S. The domination number of grids. *SIAM J. Discrete Math.* **2011**, *25*, 1443–1453.
11. Pavlič, P.; Žerovnik, J. A note on the domination number of the Cartesian products of paths and cycles. *Kragujev. J. Math.* **2013**, *37*, 275–285.
12. Crevals, S. Domination of Cylinder Graphs. *Congr. Numer.* **2014**, *219*, 53–63.
13. Chartrand, G.; Lesniak, L.; Zhang, P. *Graphs and Digraphs*, 5th ed.; CRC Press: Boca Raton, FL, USA, 2011.
14. Pavlič, P.; Žerovnik, J. *Formulas for Various Domination Numbers of Products of Paths and Cycles*; IMFM Preprint Series; Institute of Mathematics, Physics and Mechanics: Ljubljana, Slovenia, 2012; Volume 50, ISSN 2232-2094.
15. Spalding, A. Min-Plus Algebra and Graph Domination. Ph.D. Thesis, Department of Applied Mathematics, University of Colorado, Boulder, CO, USA, 1998.
16. Crevals, S.; Östergård, P.R.J. Independent domination of grids. *Discret. Math.* **2015**, *338*, 1379–1384.
17. Aleid, S.A.; Cáceres, J.; Puertas, M.L. Quasi-efficient domination in grids. *arXiv* **2016**, arXiv:1604.08521.
18. Pin, J.-E. *Tropical Semirings*; Publications of the Newton Institute; Jeremy, G., Ed.; Cambridge University Press: Cambridge, UK, 1998; Volume 11, pp. 50–69.
19. Brualdi, R.A.; Ryser H.J. *Combinatorial Matrix Theory. Part of Encyclopedia of Mathematics and its Applications*; Cambridge University Press: Cambridge, UK, 2014.
20. Davis, T. CSPARSE: A Concise Sparse Matrix Package in C. 2006. Available online: http://people.sc.fsu.edu/~jburkardt/c_src/csparse/csparse.html (accessed on 1 July 2017).

symmetry

MDPI

Article

Computing the Metric Dimension of Gear Graphs

Shahid Imran [1,*], Muhammad Kamran Siddiqui [2,3], Muhammad Imran [3,4], Muhammad Hussain [1], Hafiz Muhammad Bilal [1], Imran Zulfiqar Cheema [1], Ali Tabraiz [5] and Zeeshan Saleem [6]

[1] Department of Mathematics, COMSATS University Islamabad, Lahore Campus 54000, Pakistan;
    mhmaths@gmail.com (M.H); hafizbilal331@gmail.com (H.M.B); imran.cheema@hotmail.com (I.Z.C)
[2] Department of Mathematics, COMSATS University Islamabad, Sahiwal Campus 57000, Pakistan;
    kamransiddiqui75@gmail.com
[3] Department of Mathematical Sciences, United Arab Emirates University, P.O. Box 15551, Al Ain,
    United Arab Emirates; imrandhab@gmail.com
[4] Department of Mathematics, School of Natural Sciences (SNS), National University of Sciences and
    Technology (NUST), Sector H-12, Islamabad 44000, Pakistan
[5] Department of Electrical Engineering, University of Central Punjab, Lahore 54000, Pakistan;
    ali.tabraiz@ucp.edu.pk
[6] Department of Mathematics, The University of Lahore, Old Campus Lahore 54000, Pakistan;
    zeeshansaleem009@gmail.com
* Correspondence: shahidimrangondal@gmail.com

Received: 15 May 2018; Accepted: 6 June 2018; Published: 8 June 2018

Abstract: Let $G = (V, E)$ be a connected graph and $d(u, v)$ denote the distance between the vertices u and v in G. A set of vertices W resolves a graph G if every vertex is uniquely determined by its vector of distances to the vertices in W. A metric dimension of G is the minimum cardinality of a resolving set of G and is denoted by $dim(G)$. Let $J_{2n,m}$ be a m-level gear graph obtained by m-level wheel graph $W_{2n,m} \cong mC_{2n} + k_1$ by alternatively deleting n spokes of each copy of C_{2n} and J_{3n} be a generalized gear graph obtained by alternately deleting $2n$ spokes of the wheel graph W_{3n}. In this paper, the metric dimension of certain gear graphs $J_{2n,m}$ and J_{3n} generated by wheel has been computed. Also this study extends the previous result given by Tomescu et al. in 2007.

Keywords: Metric dimension; basis; resolving set; gear graph; generalized gear graph

MSC: 05C12; 05C90; 05C15; 05C62

1. Introduction and Preliminary Results

In a connected graph $G(V, E)$, where V is the set of vertices and E is the set of edges. The distance $d(u, v)$ between two vertices $u, v \in V$ is the length of the shortest path between them and the diameter of G denoted by $diam(G)$ is the maximum distance between pairs of vertices $u, v \in V(G)$. Let $W = \{v_1, v_2, \ldots, v_k\}$ be an order set of vertices of G and u be a vertex of G. The representation $r(u|W)$ of u with respect to W is the $k - $tuple $\{d(u, v_1), d(u, v_2), d(u, v_3), \ldots, d(u, v_k)\}$, where W is called a resolving set or locating set if distinct vertices of G have distinct representations with respect to W. See for more results [1,2].

A resolving set of minimum cardinality is called a metric basis for G and the cardinality of a metric basis is said the metric dimension of G, denoted by $dim(G)$, see [3]. The motivation for this topic stems from chemistry [4]. A common but important problem in the study of chemical structures is to determine ways of representing a set of chemical compounds such that distinct compounds have distinct representations. Moreover the application of this invariant to the navigation of robots in networks are discussed in [5]. The application to problems of pattern recognition and image processing, some of which involve the use of hierarchical data structures are given in [6].

For a given ordered set of vertices $W = \{v_1, v_2, \ldots, v_k\}$ of a graph G, the ith component of $r(u|W)$ is 0 if and only if $u = v_i$. Thus, to show that W is a resolving set it suffices to verify that $r(y|W) \neq r(z|W)$ for each pair of distinct vertices $y, z \in V(G) \backslash W$.

Motivated by the problem of determining uniquely the location of an intruder in a network, the concept of metric dimension was introduced by Slater in [7] and studied independently by Harary and Melter in [8].

Let Ω be a family of connected graphs $F_m : \Omega = (F_m)_{m \geq 1}$ depending on m as follows: $\psi(m) =$ cardinality of the set of vertices of any member F of Ω and $\lim_{m \to \infty} \psi(m) = \infty$. If $\forall m \geq 1, \exists C > 0$ such that $dim(F_m) \leq C$, then we shall say that Ω has bounded metric dimension, otherwise Ω has unbounded metric dimension. If all graphs in Ω have the same metric dimension then F is called a family with constant metric dimension [9].

A connected graph G has $dim(G) = 1$ if and only if G is a path [5], cycle C_n have metric dimension 2 for every $n \geq 3$. Other families of graphs with unbounded metric dimension are regular bipartite graphs [10], wheel graph [11]. The metric dimensions of m-level wheel graphs, convex polytope graphs and antiweb gear graphs are computed in [12]. The metric dimension of honeycomb networks are computed in [13] and t he metric dimension of generators of graphs in [14]. In the following section, some results related to m-level generalized gear graph are given.

2. The Metric Dimension of Double Gear Graph $J_{2n,m}$

Definition 1. *The joining of two graphs G_1 and G_2 is denoted by $G_1 + G_2$ with the following vertex and edge sets:*

$$V(G_1 + G_2) = V(G_1) \cup V(G_2)$$

$$E(G_1 + G_2) = E(G_1) \cup E(G_2) \cup \{uv; u \in V(G_1), v \in V(G_2)\}.$$

Definition 2. *In graph theory, an isomorphism of graphs G_1 and G_2 is a bijection between the vertex sets of G_1 and G_2, $f : V(G_1) \to V(G_2)$ such that any two vertices u and v of G_1 are adjacent in G_1 if and only if $f(u)$ and $f(v)$ are adjacent in G_2. If an isomorphism exists between two graphs, then the graphs are called isomorphic and denoted as $G_1 \cong G_2$.*

Note that the the graph $C_n + K_1$ is isomorphic to wheel graph W_n. In addition, note that $2C_n + K_1$ mean union of two copies of C_n that are joined with K_1.

Definition 3. *A double-wheel graph $W_{n,2}$ can be obtained as join of $2C_n + k_1$ and inductively an m-level wheel graph denoted by $W_{n,m}$ can be constructed as $W_{n,m} \cong mC_n + k_1$.*

Definition 4. *A double gear graph denoted by $J_{2n,2}$ can be obtained from double-wheel $W_{2n,2} = 2C_{2n} + k_1$ by alternatively deleting n spokes of each copy of C_{2n} and inductively an m-level gear graph $J_{2n,m}$ can be constructed from m-level wheel $W_{2n,m} \cong mC_{2n} + k_1$ by alternatively deleting n spokes of each C_{2n} (see [15]). A double gear graph is depicted in Figure 1.*

Construction and Observations

A double gear graph $J_{2n,2}$ (see in Figure 1) is constructed if we consider two even cycles with $n \geq 2$,

$$C_{2n,1} : v_1^1, v_2^1, v_3^1, \ldots, v_{2n}^1, v_1^1 \quad and \quad C_{2n,2} : v_1^2, v_2^2, v_3^2, \ldots, v_{2n}^2, v_1^2$$

Now take a new vertex v adjacent to n vertices of $C_{2n,1} : v_2^1, v_4^1, \ldots, v_{2n}^1$ as well as v is also adjacent to n vertices of $C_{2n,2} : v_2^2, v_4^2, \ldots, v_{2n}^2$. Inductively we can construct an m-level gear graph denoted by $J_{2n,m}$ by taking m even cycles $C_{2n,1}, C_{2n,2}, \ldots, C_{2n,m}$.

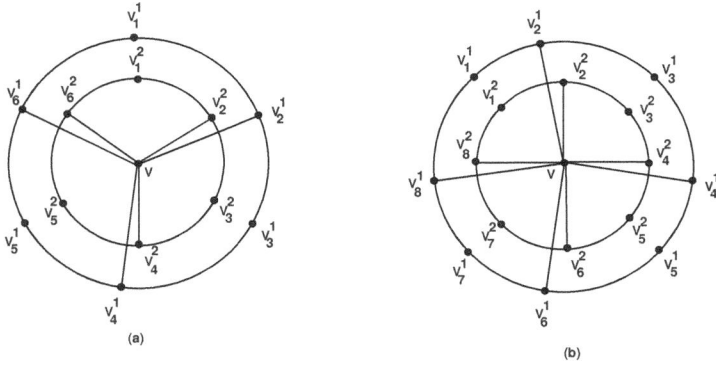

Figure 1. (a) The double gear graph $J_{6,2}$; (b) The double gear graph $J_{8,2}$.

The vertices of $C_{2n,i}$; $1 \leq i \leq 2$, in the graph $J_{2n,2}$ are of two kinds namely the vertices of degree 2 and the vertices of degree 3. Vertices of degree 2 and 3 will be considered as minor and major vertices respectively. One can easily check that:

- When $n = 2$,

  $dim(J_{4,2}) = 3 + 2$, (central vertex v with one major and minor vertex of each $C_{2n,i}$, $1 \leq i \leq 2$ form basis).

- When $n = 3$,

  $dim(J_{6,2}) = 3 + 2$, (central vertex v with two minor vertices of each $C_{2n,i}$, $1 \leq i \leq 2$ form basis).

- When $n = 4$,

  $dim(J_{8,2}) = 2 + 3$, (two minor vertices u^1, w^1 such that $d(u^1, w^1) = 2$ of $C_{2n,1}$ with one minor vertex u^2 and two major vertices w^2, x^2 of $C_{2n,2}$ such that $d(u^2, w^2) = d(u^2, x^2) = 3$ form basis).

- When $n = 5$,

  $dim(J_{10,2}) = 3 + 4$, (three minor vertices u^1, w^1, x^1 satisfying $d(u^1, w^1) = d(w^1, x^1) = 2$, $d(u^1, x^1) = 4$ of $C_{2n,1}$ with three minor vertices u^2, w^2, x^2 and one major vertex z^2 of $C_{2n,2}$ satisfying $d(u^2, w^2) = d(w^2, x^2) = 2$, $d(u^2, x^2) = 4$ and $d(u^2, z^2) = d(w^2, z^2) = d(x^2, z^2) = 3$ form metric basis of $J_{10,2}$).

Consider the gear graph $J_{2n,1}$ in which $C_{2n,1}$ is an outer cycle of length $2n$. If B is a basis of $J_{2n,1}$ then B contains $r \geq 2$ vertices of $C_{2n,1}$ for $n \geq 6$. Suppose $B = \{v_{i_1}, v_{i_2}, \ldots, v_{i_r}\}$ then vertices of B can be ordered as $v_{i_1} < v_{i_2} < \ldots < v_{i_r}$ such that $\{v_{i_t}, v_{i_{t+1}}\}$ for $1 \leq t \leq r - 1$ and $\{v_{i_r}, v_{i_1}\}$ are called neighboring vertices. Vertices of $C_{2n,1}$ lying between any two neighboring vertices of B are called gaps which are denoted by G_{i_t} for $1 \leq t \leq r - 1$ and G_{i_r}, and their cardinalities are said to be the size of gaps. One can easily observe that every vertex of B has two neighboring vertices; gaps generated by these three vertices are called neighboring gaps following a concept already exist in [2] and [17]. A gap determined by neighboring vertices of basis say v_i and v_j will be called an $\alpha - \beta$ with $\alpha \leq \beta$ when $deg(v_i) = \alpha$ and $deg(v_j) = \beta$ or when $deg(v_i) = \beta$ and $deg(v_j) = \alpha$. Hence we have three kinds of gaps namely, $2 - 2$ gap, $2 - 3$ gap and $3 - 3$ gap.

For the graph $J_{2n,2}, n \geq 4$ central vertex v does not belong to any basis. Since $d(v_i^j, v) \leq 2 \forall$, $1 \leq i \leq 2n$, $1 \leq j \leq 2$ and $diam(J_{2n,2}) = 4$, if central vertex v belongs to any metric basis B then there must exists two distinct vertices u_i and u_j for $1 \leq i \neq j \leq 2n$ such that $r(u_i|B) = r(u_j|B)$. Consequently, the basis vertices of $J_{2n,2}$ belong to the cycles induced by $C_{2n,1}$ and $C_{2n,2}$. It is shown in [17] that if B is a basis of $J_{2n,1}$ then B consist only of the vertices of $C_{2n,1}$ that satisfy the following properties.

- If B is a basis of $J_{2n,1}$, $n \geq 6$ then every $2-2$ gap, $2-3$ gap and $3-3$ gap of B contains at most 5, 4 and 3 vertices respectively.
- If B is a basis of $J_{2n,1}$, $n \geq 6$ then it contains at most one major gap.
- If B is a basis of $J_{2n,1}$, $n \geq 6$ then any two neighboring gaps contain together at most six vertices in which one gap is a major gap.
- If B is a basis of $J_{2n,1}$, $n \geq 6$ then any two minor neighboring gaps contain together at most four vertices.

Lemma 1. *Let B be a basis of $J_{2n,2}$, $n \geq 6$, then every $2-2$ gap, $2-3$ gap and $3-3$ gap of B induced by $C_{2n,1}$ and $C_{2n,2}$ contains at most 5, 4 and 3 vertices respectively.*

Proof. Suppose the result is false and there exists a $2-2$ gap of size 7 say $u_1, u_2, u_3, u_4, u_5, u_6, u_7$ consisting of consecutive vertices of $C_{2n,1}$ or $C_{2n,2}$ with $deg(u_1) = deg(u_7) = 3$ then $r(u_3|B) = r(u_5|B)$ which is a contradiction. If there exists a $2-3$ gap of size 6 then we have a path $u_1, u_2, u_3, u_4, u_5, u_6$ consisting of consecutive vertices of $C_{2n,1}$ or $C_{2n,2}$ with $deg(u_1) = 3$ and $deg(u_6) = 2$ then $r(u_3|B) = r(u_5|B)$ which is again a contradiction. The existence of a $3-3$ gap of size 5 say u_1, u_2, u_3, u_4, u_5 induced by $C_{2n,1}$ or $C_{2n,2}$ with $deg(u_1) = deg(u_5) = 2$, would imply $r(u_2|B) = r(u_4|B)$ a contradiction.

The $2-2$ gap $2-3$ gap and $3-3$ gap containing 5, 4 and 3 vertices respectively will be referred to as major gaps and the remaining gaps are called minor gaps. In the proof of Lemmas 2–4, the major vertices will be labeled by a star (*). $\square$

Lemma 2. *Let B be a basis of $J_{2n,2}$, $n \geq 6$ then it contains at most one major gap induced by the vertices of cycles $C_{2n,1}$ and $C_{2n,2}$.*

Proof. Suppose B contains two distinct major gaps induced by the vertices of cycles $C_{2n,1}$ or $C_{2n,2}$.

Case-(i): When both gaps are $3-3$ then we have two distinct paths consisting consecutive vertices u_1, u_2^*, u_3 and w_1, w_2^*, w_3 of $C_{2n,1}$ and $C_{2n,2}$ respectively in this case $r(u_3^*|B) = r(w_3^*|B)$; a contradiction.

Case-(ii): When both gaps are $2-2$ then we have two distinct paths consisting of consecutive vertices $u_1^*, u_2, u_3^*, u_4, u_5^*$ and $w_1^*, w_2, w_3^*, w_4, w_5^*$ of $C_{2n,1}$ and $C_{2n,2}$ respectively but $r(u_3^*|B) = r(w_3^*|B)$; a contradiction.

Case-(iii): When both gaps are $2-3$ then we have two distinct paths consisting of consecutive vertices u_1^*, u_2, u_3^*, u_4 and w_1^*, w_2, w_3^*, w_4 of $C_{2n,1}$ and $C_{2n,2}$ respectively in this case $r(u_3^*|B) = r(w_3^*|B)$; a contradiction.

Case-(iv): When one gap is $3-3$ and other is $2-2$ gap then we have two distinct paths u_1, u_2^*, u_3 and $w_1^*, w_2, w_3^*, w_4, w_5^*$ induced by $C_{2n,1}$ and $C_{2n,2}$ respectively but $r(u_2^*|B) = r(w_3^*|B)$; a contradiction.

Case-(v): When one gap is $3-3$ and other is $2-3$ gap then we have two distinct paths consisting of consecutive vertices u_1, u_2^*, u_3 and w_1^*, w_2, w_3^*, w_4 of $C_{2n,1}$ and $C_{2n,2}$ respectively but $r(u_2^*|B) = r(w_3^*|B)$; a contradiction.

Case-(vi): When one gap is $2-2$ and other is $2-3$ gap then we have two distinct paths consisting of consecutive vertices $u_1^*, u_2, u_3^*, u_4, u_5^*$ and w_1^*, w_2, w_3^*, w_4 of $C_{2n,1}$ and $C_{2n,2}$ respectively in this case $r(u_3^*|B) = r(w_3^*|B)$; a contradiction.

Similarly, if both major gaps are induced by $C_{2n,1}$ then we get a contradiction and a similar contradiction arises if $C_{2n,2}$ induced both major gaps. $\square$

Lemma 3. *Let B be a basis of $J_{2n,2}$, $n \geq 6$, then any two neighboring gaps, one of which being a major gap induced by exactly one of two cycles $C_{2n,1}$ or $C_{2n,2}$ contain together at most six vertices.*

Proof. If the major gap is $3-3$ then there is nothing to prove by Lemma 2. Without loss of any generality we can say that only $C_{2n,1}$ induced a major gap by Lemma 2. If the major gap is a $2-2$ gap having five vertices then its neighboring minor gap contains at most one vertex. If this statement is false and $2-2$ gap, $2-3$ minor gaps having three and two vertices respectively are neighboring gaps of $2-2$ major gap, then we have two paths consisting of consecutive vertices of $C_{2n,1}$:$u_1^*, u_2, u_3^*, u_4, u_5^*, u_6, u_7^*, u_8, u_9^*$ and $w_1^*, w_2, w_3^*, w_4, w_5^*, w_6, w_7^*, w_8$, where $u_4, w_6 \in B$ induced by $2-2$ major, $2-2$ minor gaps and $2-2$ major, $2-3$ minor gaps respectively. In this case $r(u_3^*|B) = r(u_5^*|B)$ and $r(w_5^*|B) = r(w_7^*|B)$; a contradiction. The existence of $2-3$ major gap having four vertices is not possible if its neighboring minor gap is a $2-2$ gap with three vertices. If this case holds then we consider the following path: $u_1^*, u_2, u_3^*, u_4, u_5^*, u_6, u_7^*, u_8$, where $u_4 \in B$ then $r(u_4^*|B) = r(u_5^*|B)$; a contradiction. $\square$

Lemma 4. *Let B be a basis of $J_{2n,2}$, $n \geq 6$, then any two minor neighboring gaps induced by $C_{2n,1}$ or $C_{2n,2}$ contain together at most four vertices.*

Proof. To prove the statement, it is sufficient to prove two cases.

Case-(i): $2-2$ minor gap with three vertices cannot be neighboring gap of $2-2$ minor gap having three vertices, otherwise we have a path consisting of consecutive vertices of $C_{2n,1}$ or $C_{2n,2}$:$u_1^*, u_2, u_3^*, u_4, u_5^*, u_6, u_7^*$, where $u_4 \in B$ in this case $r(u_3^*|B) = r(u_5^*|B)$.

Case-(ii): $2-2$ minor gap with three vertices cannot be neighboring gap of $2-3$ minor gap having two vertices, otherwise we have a path consisting of consecutive vertices of $C_{2n,1}$ or $C_{2n,2}$:$w_1^*, w_2, w_3^*, w_4, w_5^*, w_6$ where $w_4 \in B$ in this case $r(w_3^*|B) = r(w_5^*|B)$; a contradiction. $\square$

Theorem 1. *If $J_{2n,2}$ be a double gear graph for $n \geq 4$, then*

$$dim(J_{2n,2}) = dim(J_{2n,1}) + \left\lceil \frac{2n}{3} \right\rceil$$

Proof. We have seen that $dim(J_{8,2}) = 5 = dim(J_{8,1}) + \lceil \frac{8}{3} \rceil$, $dim(J_{10,2}) = 7 = dim(J_{10,1}) + \lceil \frac{10}{3} \rceil$ and the central vertex v does not belong to any basis B of $J_{2n,2}$. Moreover

$$C_{2n,1} : v_1^1, v_2^1, v_3^1, \ldots, v_{2n}^1, v_1^1$$

and

$$C_{2n,2} : v_1^2, v_2^2, v_3^2, \ldots, v_{2n}^2, v_1^2$$

be the outer cycles of $J_{2n,2}$ at level 1 and 2 respectively. First we prove that $dim(J_{2n,2}) \leq dim(J_{2n,1}) + \lceil \frac{2n}{3} \rceil$ by constructing a resolving set W in $J_{2n,2}$ with $dim(J_{2n,1}) + \lceil \frac{2n}{3} \rceil$ vertices.

We consider three cases according to the residue class modulo 3 to which n belongs.

Case-(i): When $n \equiv 0 \pmod 3$, then we may write $2n = 3k$, where $k \geq 4$, is even and $dim(J_{2n,1}) + \lceil \frac{2n}{3} \rceil = 2k$, in this case W can be considered as:

$$W = \{v_1^j, v_{2n-1}^j; 1 \leq j \leq 2\} \cup \{v_{6i+1}^1, v_{6i+3}^1, v_{6i-1}^2, v_{6i+1}^2; 1 \leq i \leq \frac{k}{2} - 1\}$$

Case-(ii): When $n \equiv 1 \pmod 3$, then we may write $2n = 3k + 2$, where $k \geq 4$ is even and $dim(J_{2n,1}) + \lceil \frac{2n}{3} \rceil = 2k + 1$, in this case W can be considered as:

$$W = \{v_1^1, v_{2n-1}^1, v_1^2\} \cup \{v_{6i+1}^1, v_{6i+3}^1; 1 \leq i \leq \frac{k}{2} - 1\} \cup \{v_{6i-1}^2, v_{6i+1}^2; 1 \leq i \leq \frac{k}{2}\}$$

Case-(iii): When $n \equiv 2 \pmod 3$, then we may write $2n = 3k + 1$, where $k \geq 5$ is even and $dim(J_{2n,1}) + \lceil \frac{2n}{3} \rceil = 2k + 1$, in this case W can be considered as:

$$W = \{v_1^1, v_1^2, v_{2n-1}^2\} \cup \{v_{6i+1}^1, v_{6i+3}^1; 1 \leq i \leq \frac{k-1}{2}\} \cup \{v_{6i-1}^2, v_{6i+1}^2; 1 \leq i \leq \frac{k-1}{2}\}$$

The set W contains a unique $2-2$ major gap having at most five vertices and all other gaps are $2-2$ minor gaps which contain at most three vertices. The set W is a resolving set of $J_{2n,2}$ since any two major or any two minor vertices respectively lying in different gaps or in the same gap are separated by at least one vertex in the set of three vertices of W generating these neighboring gaps. When gaps are not neighboring gaps, then the set of four vertices of W which generate two gaps make the representation unique of each vertex of these two gaps. Representation of central vertex v is $(2,2,2,\ldots,2)$, which is different from the representation of all other vertices of $J_{2n,2}$. Hence,

$$dim(J_{2n,2}) \leq dim(J_{2n,1}) + \left\lceil \frac{2n}{3} \right\rceil \tag{1}$$

Now we show that $dim(J_{2n,2}) \geq dim(J_{2n,1}) + \left\lceil \frac{2n}{3} \right\rceil$. As the central vertex v does not belong to any basis of J_{3n}. Let B be a basis of $J_{2n,2}$ such that $|B| = r$ then we have r gaps. By lemma 2 B contains at most one major gap, without loss of generality we can say major gap lies on $C_{2n,1}$. Hence B induces $\left\lfloor \frac{r}{2} \right\rfloor$ gaps on $C_{2n,1}$ and $\left\lceil \frac{r}{2} \right\rceil$ gaps on $C_{2n,2}$.

We denote the gaps on $C_{2n,1}$ by $G_1^1, G_2^1, G_3^1, \ldots, G_{\lfloor \frac{r}{2} \rfloor}^1$ where G_i^1 and G_{i+1}^1 are called neighboring gaps for $1 \leq i \leq \left\lfloor \frac{r}{2} \right\rfloor - 1$ as well as $G_{\lfloor \frac{r}{2} \rfloor}^1$ is also neighboring gap of G_1^1 and the gaps on $C_{2n,2}$ will be denoted by $G_1^2, G_2^2, G_3^2, \ldots, G_{\lceil \frac{r}{2} \rceil}^2$ where G_i^2 and G_{i+1}^2 are called neighboring gaps for $1 \leq i \leq \left\lceil \frac{r}{2} \right\rceil - 1$ as well as $G_{\lceil \frac{r}{2} \rceil}^2$ is also neighboring gap of G_1^2. By Lemma 2, suppose G_1^1 is a major gap. By Lemmas 3 and 4, we can write

$$|G_1^1 + G_2^1| \leq 6, \quad |G_1^1 + G_{\lfloor \frac{r}{2} \rfloor}^1| \leq 6, \quad \|G_i^1 + G_{i+1}^1| \leq 4, \quad for \ 2 \leq i \leq \left\lfloor \frac{r}{2} \right\rfloor - 1$$

and

$$|G_1^2 + G_2^2| \leq 4, \quad \|G_1^2 + G_{\lceil \frac{r}{2} \rceil}^2| \leq 4, \quad \|G_i^2 + G_{i+1}^2| \leq 4, \quad for \ 2 \leq i \leq \left\lceil \frac{r}{2} \right\rceil - 1$$

We consider two cases according to the residue class modulo 2 to which r belongs.

Case-(i): When $r \equiv 0 (mod2)$: In this case $\left\lfloor \frac{r}{2} \right\rfloor = \left\lceil \frac{r}{2} \right\rceil = \frac{r}{2}$
By summing the above inequality we have

$$2(2n - \frac{r}{2}) = 2 \sum_{i=1}^{\frac{r}{2}} |G_i^1| \leq 2r + 4 \Rightarrow \frac{r}{2} \geq \frac{2n-2}{3} \Rightarrow \frac{r}{2} \geq \left\lfloor \frac{2n}{3} \right\rfloor \tag{2}$$

Again

$$2(2n - \frac{r}{2}) = 2 \sum_{i=1}^{\frac{r}{2}} |G_i^1| \leq 2r \Rightarrow \frac{r}{2} \geq \frac{2n}{3} \Rightarrow \frac{r}{2} \geq \left\lceil \frac{2n}{3} \right\rceil \tag{3}$$

From Equations (2) and (3) we have,

$$r \geq \left\lfloor \frac{2n}{3} \right\rfloor + \left\lceil \frac{2n}{3} \right\rceil \Rightarrow dim(J_{2n,2}) \geq dim(J_{2n,1}) + \left\lceil \frac{2n}{3} \right\rceil$$

Case-(ii): When $r \equiv 1 (mod2)$: In this case $\left\lfloor \frac{r}{2} \right\rfloor = \frac{r-1}{2}$ and $\left\lceil \frac{r}{2} \right\rceil = \frac{r+1}{2}$
By summing the above inequality we have

$$2(2n - \frac{r-1}{2}) = 2 \sum_{i=1}^{\frac{r-1}{2}} |G_i^1| \leq 4 + 4\left(\frac{r-1}{2}\right) \Rightarrow \frac{r-1}{2} \geq \frac{2n-2}{3} \Rightarrow \frac{r-1}{2} \geq \left\lfloor \frac{2n}{3} \right\rfloor \tag{4}$$

and

$$2(2n - \frac{r+1}{2}) = 2 \sum_{i=1}^{\frac{r+}{2}} |G_i^2| \leq 4\left(\frac{r+1}{2}\right) \Rightarrow \frac{r+1}{2} \geq \frac{2n}{3} \Rightarrow \frac{r+1}{2} \geq \left\lceil \frac{2n}{3} \right\rceil \tag{5}$$

From Equations (4) and (5) we have

$$r \geq \left\lfloor \frac{2n}{3} \right\rfloor + \left\lceil \frac{2n}{3} \right\rceil \Rightarrow dim(J_{2n,2}) \geq dim(J_{2n,1}) + \left\lceil \frac{2n}{3} \right\rceil \tag{6}$$

Now from Equations (1) and (6) we conclude that,

$$dim(J_{2n,2}) = dim(J_{2n,1}) + \left\lceil \frac{2n}{3} \right\rceil$$

which complete the proof. □

Theorem 2. *If $J_{2n,m}$ be a double gear graph for $n \geq 4$, $m \geq 3$, then*

$$dim(J_{2n,m}) = dim(J_{2n,1}) + (m-1) \left\lceil \frac{2n}{3} \right\rceil$$

Proof. We will prove this result by induction on levels of gear graph denoted by $J_{2n,m}$.

When $m = 1$, then $dim(J_{2n,1}) = \lfloor \frac{2n}{3} \rfloor$ is obtained in [17].

When $m = 2$, then $dim(J_{2n,2}) = dim(J_{2n,1}) + \lceil \frac{2n}{3} \rceil$ by Theorem 1.

Now we assume that the statement is true for $m = k$, $dim(J_{2n,k}) = dim(J_{2n,1}) + (k-1)\lceil \frac{2n}{3} \rceil$. we will show the result for $m = k+1$, by using concept of Theorem 1 we have $dim(J_{2n,k+1}) = dim(J_{2n,k}) + \lceil \frac{2n}{3} \rceil$.

Now $dim(J_{2n,k+1}) = dim(J_{2n,k}) + \lceil \frac{2n}{3} \rceil = dim(J_{2n,1}) + (k-1)\lceil \frac{2n}{3} \rceil + \lceil \frac{2n}{3} \rceil$. $\Rightarrow dim(J_{2n,k+1}) = dim(J_{2n,1}) + k\lceil \frac{2n}{3} \rceil$. Hence the result is true for all positive integers $m \geq 3$. □

3. The Metric Dimension of Generalized Gear Graph J_{3n}

Definition 5. *To define the generalized gear graph J_{3n}: consider a cycle C_{3n} having vertices $v_1, v_2, v_3, \ldots, v_{3n}, v_1$ with $n \geq 2$, take a new vertex v adjacent to n vertices $v_3, v_6, v_9, \ldots, v_{3n}$ of C_{3n}. The generalized gear graph J_{3n} has order $3n+1$ and size $4n$. It can be obtained from wheel graph W_{3n} by alternately deleting $2n$ spokes.*

Construction and Observations

The vertices of C_{3n} in the graph J_{3n} are two kinds: vertices of degree 2 and 3. Vertices of degree 2 and 3 will be considered as minor and major vertices respectively. The graph J_{3n} is a bipartite graph in which one bipartition class contains minor vertices together with central vertex v and the second bipartition class contain major vertices. In the proof of Lemmas 5–9, major vertices will be represented by a star. One can easily check that:

- When $n = 2$
  $dim(J_6) = 2$, (one minor vertex of C_6 together with central vertex v form basis).

- When $n = 3$
  $dim(J_9) = 2 = dim(J_{12})$, (two minor vertices w_1 and w_2 such that $d(w_1, w_2) = 3$ form basis).

- When $n = 5$
  $dim(J_{15}) = 3$, (three minor vertices w_1, w_2 and w_3 such that $d(w_1, w_2) = d(w_2, w_3) = d(w_3, w_4) = 4$ form basis).

For the graph $J_{3n}, n \geq 4$ central vertex v does not belong to any basis. Since $d(v_i, v) \leq 2\forall$, $1 \leq i \leq 3n$, and $diam(J_{3n}) = 4$ if central vertex v belongs to any metric basis B then there must exist two distinct vertices u_i and u_j for $1 \leq i \neq j \leq 3n$ such that $r(u_i|B) = r(u_j|B)$. If B is a basis of J_{3n} and central vertex v does not belong to B then by using the concept of gap given in Section 2, we have again three kinds of gaps i.e $2-2$ gpa, $2-3$ gap, and $3-3$ gap.

Lemma 5. *If B is a basis of* J_{3n}, $n \geq 6$ *then every* $2-2$ *gap,* $2-3$ *gap and* $3-3$ *gap of B contains at most 8, 7 and 5 points respectively.*

Proof. Suppose the basis set B contains a $2-2$ gap of nine consecutive vertices $u_1, u_2, u_3,$ $u_4, u_5, u_6, u_7, u_8, u_9$ of C_{3n} such that $deg(u_1) = deg(u_9)$ we have $r(u_4|B) = r(u_6|B)$ in this case. If $2-3$ gap contains more than 7 vertices then it contains 9 consecutive vertices $u_1, u_2, u_3, u_4, u_5, u_6, u_7, u_8, u_9$ of C_{3n} such that $deg(u_1) = 3$ and $deg(u_9) = 2$ we have $r(u_4|B) = r(u_7|B)$, a contradiction in this case. If a $3-3$ gap contains more then 5 vertices, then it contains 8 consecutive vertices $u_1, u_2, u_3, u_4, u_5, u_6, u_7, u_8$ such that $deg(u_1) = deg(u_8) = 2$ then $r(u_3|B) = r(u_6|B)$ which is again a contradiction.

The $2-2$ gap, $2-3$ gap and $3-3$ gap containing 8, 7 and 5 vertices respectively will be referred to as major gaps and the remaining gaps are called minor gaps. □

Lemma 6. *If B is a basis of* J_{3n}, $n \geq 6$, *then it contains at most one major gap.*

Proof. Suppose B is basis of J_{3n} and it contains two distinct major gaps.

Case-(i): When both gaps are $3-3$ then we have two distinct paths $u_1, u_2, u_3^*, u_4, u_5$ and $w_1, w_2, w_3^*, w_4, w_5$ but $r(u_3^*|B) = r(w_3^*|B)$.

Case-(ii): When both gaps are $2-2$ then we have two distinct paths $u_1, u_2^*, u_3, u_4, u_5^*, u_6, u_7, u_8^*$ and $w_1, w_2^*, w_3, w_4, w_5^*, w_6, w_7, w_8^*$ but $r(u_5^*|B) = r(w_5^*|B)$.

Case-(iii): When both gaps are $2-3$ then we have two distinct paths $u_1, u_2^*, u_3, u_4, u_5^*, u_6, u_7$ and $w_1, w_2^*, w_3, w_4, w_5^*, w_6, w_7$ but $r(u_5^*|B) = r(w_5^*|B)$.

Case-(iv): When one gap is $3-3$ and other is $2-2$ gap then we have two distinct paths $u_1, u_2, u_3^*, u_4, u_5$ and $w_1, w_2^*, w_3, w_4, w_5^*, w_6, w_7, w_8^*$ but $r(u_3^*|B) = r(w_5^*|B)$.

Case-(v): When one gap is $3-3$ and other is $2-3$ gap then we have two distinct paths $u_1, u_2, u_3^*, u_4, u_5$ and $w_1, w_2^*, w_3, w_4, w_5^*, w_6, w_7$ but $r(u_3^*|B) = r(w_5^*|B)$.

Case-(vi): When one gap is $2-2$ and other is $2-3$ gap then we have two distinct paths $u_1^*, u_2, u_3, u_4^*, u_5, u_6, u_7^*, u_8$ and $w_1, w_2^*, w_3, w_4, w_5^*, w_6, w_7$ but $r(u_4^*|B) = r(w_5^*|B)$. □

Lemma 7. *If B is a basis of* J_{3n}, $n \geq 6$, *containing one major gap either* $2-2$ *gap or* $2-3$ *gap then it does not contain* $2-2$ *gap and* $2-3$ *minor gap having 7 and 6 vertices respectively.*

Proof. *Case-(i)*: When one gap is $2-2$ major gap and the other is $2-2$ minor gap having 7 vertices, then we have two distinct paths $u_1, u_2^*, u_3, u_4, u_5^*, u_6, u_7, u_8^*$ and $w_1^*, w_2, w_3, w_4^*, w_5, w_6, w_7^*$ but $r(u_5^*|B) = r(w_4^*|B)$.

Case-(ii): When one gap is $2-2$ major gap and the other is $2-3$ minor gap having 6 vertices, then we have two distinct paths $u_1, u_2^*, u_3, u_4, u_5^*, u_6, u_7, u_8^*$ and $w_1^*, w_2, w_3, w_4^*, w_5, w_6$ but $r(u_5^*|B) = r(w_4^*|B)$.

Case-(iii): When one gap is $2-3$ major gap and the other is $2-2$ minor gap having 7 vertices, then we have two distinct paths $u_1, u_2^*, u_3, u_4, u_5^*, u_6, u_7$ and $w_1^*, w_2, w_3, w_4^*, w_5, w_6, w_7^*$ but $r(u_5^*|B) = r(w_4^*|B)$.

Case-(iv): When one gap is $2-3$ major gap and the other is $2-3$ minor gap having 6 vertices, then we have two distinct paths $u_1, u_2^*, u_3, u_4, u_5^*, u_6, u_7$ and $w_1^*, w_2, w_3, w_4^*, w_5, w_6$ but $r(u_5^*|B) = r(w_4^*|B)$. □

Lemma 8. *If B is a basis of* J_{3n}, $n \geq 6$ *then any two neighboring gaps contain together at most 13 vertices in which one gap is a major gap.*

Proof. To show the statement, it is sufficient to show that a $2-2$ major gap with 8 vertices has a neighboring $2-2$ minor gap in which 6 vertices cannot occur. If it holds then we have the path $u_1, u_2, u_3^*, u_4, u_5, u_6^*, u_7, u_8, u_9^*, u_{10}, u_{11}, u_{12}^*, u_{13}, u_{14}, u_{15}^*, u_{16}, u_{17}$ with $u_1, u_{10}, u_{17} \in B$ in this case $r(u_7|B) = r(w_{13}|B)$, a contradiction. □

Lemma 9. *If B is a basis of J_{3n}, $n \geq 6$, then any two minor neighboring gaps contain together at most 11 vertices.*

Proof. To show the statement, it is sufficient to show that a $2 - 2$ gap with 6 vertices has a neighboring $2 - 2$ gap with 6 vertices cannot occur. Since gap is $2 - 2$, both base elements must have degree 2. For two consecutive $2 - 2$ gaps having 6 vertices, we have two possible paths. (i) First possible path is $u_1, u_2, u_3^*, u_4, u_5, u_6^*, u_7, u_8, u_9^*, u_{10}, u_{11}, u_{12}^*, u_{13}, u_{14}, u_{15}^*$ with $u_1, u_8, u_{15}^* \in B$ which is not possible as $d(u_1) = 2 = d(u_8)$ but $d(u_{15}^*) = 3 \neq 2$.

(ii) Second possible path is $u_2, u_3^*, u_4, u_5, u_6^*, u_7, u_8, u_9^*, u_{10}, u_{11}, u_{12}^*, u_{13}, u_{14}, u_{15}^*, u_{16}$ with $u_2, u_9^*, u_{16} \in B$ which is not possible as $d(u_2) = 2 = d(u_{16})$ but $d(u_9^*) = 3 \neq 2$. Hence two minor gap contain at most 11 vertices. □

Theorem 3. *If J_{3n} be the generalized gear graph for $n \geq 6$, then $dim(J_{3n}) = \lfloor \frac{n}{2} \rfloor$.*

Proof. First we prove that $dim(J_{3n}) \leq \lfloor \frac{n}{2} \rfloor$ by constructing a resolving set W in J_{3n} with $\lfloor \frac{n}{2} \rfloor$ vertices. We consider two cases according to the residue class modulo 2 to which n belongs.

Case-(i): When $n \equiv 0(mod2)$ then W can be considered as:

$$W = \{v_1, v_{10}, v_{16}\} \cup \{v_{6i+5}; 3 \leq i \leq \frac{n}{2} - 1\}$$

Case-(ii): When $n \equiv 1(mod2)$ then W can be considered as:

$$W = \{v_1, v_{10}, v_{16}\} \cup \{v_{6i+5}; 3 \leq i \leq \frac{n-1}{2} - 1\}$$

□

The set W contains a unique $2 - 2$ major gap and all other gaps are $2 - 2$ minor gap which contain at most five vertices, only one $2 - 2$ minor gap contains six vertices. The set W is a resolving set of J_{3n} since any two major or any two minor vertices lying in different gaps or in the same gap are separated by at least one vertex in the set of three vertices of W generating these neighboring gaps; when gaps are not neighboring gaps then the set of four vertices of W which generate two gaps make the representation of each vertex of these two gaps unique. Representation of central vertex is $(2, 2, 2, \ldots, 2)$, which is different from the representation of all other vertices of J_{3n}. Hence

$$dim(J_{3n}) \leq \left\lfloor \frac{n}{2} \right\rfloor \tag{7}$$

Now we show that $dim(J_{3n}) \geq \lfloor \frac{n}{2} \rfloor$. By Lemma 5 the central vertex v does not belong to any basis of J_{3n}. Let B be a basis of J_{3n} such that $|B| = r$. We have r gaps on C_{3n} generated by elements of B. We denote these gaps by $G_1, G_2, G_3, \ldots, G_r$ where G_i and G_{i+1} are called neighboring gaps for $1 \leq i \leq r - 1$ as well as G_r is also a neighboring gap of G_1. By Lemma 6 at most one of them say G_1 is a major gap. By Lemmas 6 and 7, we have

$$|G_1 + G_2| \leq 13, \quad |G_1 + G_r| \leq 13$$

and by Lemmas 8 and 9, we have,

$$|G_2 + G_3| \leq 11, \quad |G_3 + G_4| \leq 11, \quad |G_i + G_{i+1}| \leq 10, \quad for \ all \ 4 \leq i \leq r - 1$$

Symmetry **2018**, *10*, 209

By summing these inequalities, we get,

$$2(3n - r) = 2\sum_{i=1}^{r}|G_i| \leq 8 + 10r \Rightarrow 6n - 2r \leq 8 + 10r$$

$$\Rightarrow 6n - 8 \leq 12r \Rightarrow r \geq \frac{n}{2} - \frac{2}{3}$$

Hence $r = \left\lfloor \frac{n}{2} \right\rfloor$.

$$\Rightarrow dim(J_{3n}) \geq \left\lfloor \frac{n}{2} \right\rfloor \tag{8}$$

So from Equations (7) and (8), we get

$$dim(J_{3n}) = \left\lfloor \frac{n}{2} \right\rfloor$$

which complete the proof.

4. Conclusions

In the foregoing section, m-level gear graph $J_{2n,m}$ and generalized gear graph J_{3n} are constructed. It is proved that metric dimension of $J_{2n,m}$ is $dim(J_{2n,1}) + (m-1)\left\lceil \frac{2n}{3} \right\rceil$ for every $n \geq 4$ and metric dimension of J_{3n} is $\left\lfloor \frac{n}{2} \right\rfloor$ for every $n \geq 6$. This section is closed by raising the following open problem.

Open Problem. Determine the metric dimension of m-level generalized gear graph $J_{2n,k,m}$.

Author Contributions: S.I. contribute for conceptualization, funding, and analyzed the data. M.K.S. and M.I. contribute for supervision, methodology, and software, validation, designing the experiments and formal analysing. M.H. and H.M.B. contribute for performed experiments, resources, some computations and wrote the initial draft of the paper. A.T. and Z.S. contribute for analyzed the data and investigated this draft and wrote the final draft. All authors read and approved the final version of the paper.

Acknowledgments: The authors are grateful to the anonymous referees for their valuable comments and suggestions that improved this paper. This research is supported by the Start-Up Research Grant 2016 of the United Arab Emirates University (UAEU), Al Ain, United Arab Emirates via Grant No. G00002233 and UPAR Grant of UAEU via Grant No. G00002590. Also This research is supported by The Higher Education Commission of Pakistan Under Research and Development Division, National Research Program for Universities via Grant No.: 5348/Federal/NRPU/R&D/HEC/2016.

Conflicts of Interest: The authors declare no conflict of interest.

References

1. Tomescu, I.; Javaid, I. On the metric dimension of the jahangir graph. *Bull. Math. Soc. Sci. Math. Roum.* **2007**, *50*, 371–376.
2. Chartrand, G.; Eroh, L.; Johnson, M.A.; Oellermann, O.R. Resolvability in graphs and metric dimension of a graph. *Disc. Appl. Math.* **2000**, *105*, 99–113. [CrossRef]
3. Imran, M.; Baig, A.Q.; Bokhary, S.A.; Javaid, I. On the metric dimension of circulant graphs. *Appl. Math. Lett.* **2012**, *25*, 320–325. [CrossRef]
4. Cameron, P.J.; Van Lint, J.H. Designs, Graphs, Codes and their Links. In *London Mathematical Society Student Texts*; Cambridge University Press: Cambridge, UK, 1991; Volume 22.
5. Khuller, S.; Raghavachari, B.; Rosenfeld, A. *Localization in Graphs*; Technical Report CS-TR-3326; University of Maryland at College Park: College Park, MD, USA, 1994.
6. Melter, R.A.; Tomescu, I. Metric bases in digital geometry. *Graph. Image Process.* **1984**, *25*, 113–121. [CrossRef]
7. Slater, P.J. Leaves of trees. *Congress* **1975**, *14*, 549–559.
8. Harary, F.; Melter, R.A. On the metric dimension of a graph. *Ars Combin.* **1976**, *2*, 191–195.
9. Tomescu, I.; Imran, M. metric dimension and R-Sets of connected graph. *Graphs Combin.* **2011**, *27*, 585–591. [CrossRef]
10. Bača, M.; Baskoro, E.T.; Salman, A.N.M.; Saputro, S.W.; Suprijanto, D. On metric dimension of regular bipartite graphs. *Bull. Math. Soc. Sci. Math. Roum.* **2011**, *54*, 15–28.

11. Buczkowski, P.S.; Chartrand, G.; Poisson, C.; Zhang, P. On *k*-dimensional graphs and their bases. *Perioddica Math. Hung.* **2003**, *46*, 9–15. [CrossRef]
12. Siddique, H.M.A.; Imran, M. Computing the metric dimension of wheel related graphs. *Appl. Math. Comput.* **2014**, *242*, 624–632.
13. Manuel, P.; Rajan, B.; Rajasingh, I.; Monica, C. On minimum metric dimension of honeycomb networks. *J. Discret. Algorithms* **2008**, *6*, 20–27. [CrossRef]
14. Sebo, A.; Tannier, E. On metric generators of graphs. *Math. Oper. Res.* **2004**, *29*, 383–393. [CrossRef]
15. Bras, R.L.; Gomes, C.P.; Selman, B. Double-wheel graphs are graceful. In Proceedings of the Twenty-Third International Joint Conference on Artificial Intelligence, Beijing, China, 3–9 August 2013; pp. 587–593.

symmetry

MDPI

Article

Dynamics on Binary Relations over Topological Spaces

Chung-Chuan Chen [1], **J. Alberto Conejero** [2,*] , **Marko Kostić** [3] and **Marina Murillo-Arcila** [4]

[1] Department of Mathematics Education, National Taichung University of Education, Taichung 403, Taiwan; chungchuan@mail.ntcu.edu.tw

[2] Instituto Universitario de Matemática Pura y Aplicada, Universitat Politècnica de València, E-46022 València, Spain

[3] Faculty of Technical Sciences, University of Novi Sad, Trg D. Obradovića 6, 21125 Novi Sad, Serbia; marco.s@verat.net

[4] Departamento de Matemáticas, Universitat Jaume I, Campus de Ríu Sec, E-12071 Castelló de la Plana, Spain; murillom@uji.es

* Correspondence: aconejero@upv.es

Received: 16 April 2018; Accepted: 7 June 2018; Published: 11 June 2018

Abstract: The existence of chaos and the quest of dense orbits have been recently considered for dynamical systems given by multivalued linear operators. We consider the notions of topological transitivity, topologically mixing property, hypercyclicity, periodic points, and Devaney chaos in the general case of binary relations on topological spaces, and we analyze how they can be particularized when they are represented with graphs and digraphs. The relations of these notions with different types of connectivity and with the existence of Hamiltonian paths are also exposed. Special attention is given to the study of dynamics over tournaments. Finally, we also show how disjointness can be introduced in this setting.

Keywords: Devaney chaos; hypercyclicity; topological transitivity; topologically mixing; disjointness; connectivity

1. Introduction

One of the pillars of the study of chaos in dynamical systems is the search of orbits that are dense in the whole space. Typical examples of chaotic maps on the interval, where chaos can be easy visualized, are given by the tent map or by some functions of the logistic family (see, for instance, [1–4]).

The existence of orbits that spread along the whole space has been also studied in the setting of binary relations. When they are considered from one set onto itself, one can consider them from the point of view of graph theory. On the one hand, Hamilton was the first one who started to consider the analysis of graphs containing paths that visit all the nodes, named *Hamiltonian paths*. On the other hand, in terms of connectivity of (directed) graphs, a graph is (strongly) connected if, for every (ordered) pair of nodes, there is a path connecting them.

Roughly speaking, these two areas share the idea of the quest of orbits/paths visiting (nearly) the whole domain. But there are differences in how these problems are addressed. In the case of chaotic maps on the interval, the results usually involve computable conditions on the parameters of the function that defines the system. For instance, this is the case of how it is determined the chaos of the logistic map for $\mu = 4$. However, in the case of graphs, these results are related with the global structure of edges/arcs of the graph and/or on the quantification of the local structure at every node.

In the present work, our goal is to investigate the dynamics on graphs and on the more general frame of binary relations on topological spaces. In this setting, when a relation is composed with another one (or itself), each element of the domain does not need to be necessarily connected with a single element in the range. For this reason, we have set a connection with some recent results of the authors that were introduced to study the dynamics of multivalued continuous linear operators [5]. We will analyze hypercyclicity, topological transitivity, and topologically mixing properties of binary relations. It is worth mentioning that the study of dynamics over finite graphs has been recently considered by Bahi et al. by setting links between Devaney chaos and strong connectivity in order to provide an algorithm for the generation of strongly connected graphs and to construct *Pseudo Random Number Generators* (PRNGs) [6]. This approach has also allowed for the obtainment of PRNGs based on the construction of Hamiltonian cycles over an *N*-cube [7,8]. It is also worth mentioning that such results can also be considered in connection with finite state machines [9] and explained using Turing machines [10].

In the present work, we have also tackled the problem of the link between chaotic properties and connectivity, but analyzing more carefully the implications of using different topologies over the set of nodes. These results will allow us to stretch the connections between graph theory and dynamical systems in order to facilitate the exchange of ideas between both areas.

It is worth mentioning that there have been also recent results concerning the dynamics of continuous linear operators acting on L^p-spaces consisting of functions $V \mapsto \mathbb{K}$, where $\mathbb{K} = \{\mathbb{R}, \mathbb{C}\}$, and V is a multigraph, an infinite or a Cayley graph, a tree, or some similar structure (see, e.g., [11,12] and the references cited therein). We point out that our approach is different to the one of studying associations of directed graphs with finite topologies, as is taken in [13]. It also differs from the one taken by Namayanja; she generalized the dynamics of solution C_0-semigroups of birth-and-death models [14–16] to the case in which the transport equations are defined on the edges of an infinite network [17].

2. Preliminaries

Given two nonempty sets X and Y, a *binary relation* E is a subset of the Cartesian product $X \times Y$. If we consider the following relations $\rho \subseteq X \times Y$ and $\sigma \subseteq Z \times T$ with $Y \cap Z \neq \emptyset$, then we can define the *inverse* of ρ, denoted by $\rho^{-1} \subseteq Y \times X$, as the relation $\rho^{-1} := \{(y,x) \in Y \times X : (x,y) \in \rho\}$, and the composition of relations $\sigma \circ \rho \subseteq X \times T$ by

$$\sigma \circ \rho := \{(x,t) \in X \times T : \exists y \in Y \cap Z \text{ such that } (x,y) \in \rho \text{ and } (y,t) \in \sigma\}.$$

Given $x \in X$, we define its set of *adjacent elements* by $\rho(x) := \{y \in Y : (x,y) \in \rho\}$. The *domain* of a binary relation $\rho \subseteq X \times Y$ is given by $D(\rho) := \{x \in X : \exists y \in Y \text{ such that } (x,y) \in \rho\}$ and the *range* of ρ is defined as $R(\rho) := \{y \in Y : \exists x \in X \text{ such that } (x,y) \in \rho\}$.

We define the *n*-th power of ρ as $\rho^n := \underbrace{\rho \circ \cdots \circ \rho}_{n}$, their inverses $\rho^{-n} := (\rho^n)^{-1}$, and the trivial relation $\rho^0 := \Delta_X := \{(x,x) : x \in X\}$. We also set $D_\infty(\rho) := \bigcap_{n \in \mathbb{N}} D(\rho^n)$ and $\mathbb{N}_n := \{1, \ldots, n\}$, for every $n \in \mathbb{N}$. The definitions of the reflexive, symmetric, anti-symmetric, and transitive properties are assumed to be known, as long as the classes of equivalence and partial order relations.

When $X = Y$, a binary relation can be also understood as the links of a graph. If we distinguish the order in which the elements appear in each pair of the relation, then we speak of a *directed graph* or a *digraph*; if not, we will refer to it as an *undirected graph* or simply as a *graph*. Following the previous notation, a *(di-)graph* $G = (X, \rho)$ is given by a nonempty set X, whose elements are called *nodes* or *vertices*, and a set ρ or (ordered) pairs of elements of X, called *arcs* in the directed case and *edges* in the undirected one. Thus, a binary relation on a graph is just the set of arcs/edges.

If $y \notin \rho(x)$, then y is not adjacent to x. In a graph, we define the *degree* of x as the cardinal of $\rho(x)$, $|\rho(x)|$. In a digraph, the *outer degree* of x, $d^+(x)$, is given by $|\rho(x)|$, and the *inner degree* of x, $d^-(x)$, is given by $|\{y \in Y : (y,x) \in \rho\}|$. The set X of nodes of a (di-)graph G can be considered as a topological space when endowed with certain topology over its elements. In the sequel, we will only deal with non-trivial finite simple (di-)graphs (without multiple edges connecting two nodes, and without any pair (x,x) in ρ).

A *walk* is an ordered sequence of nodes $x_1, \ldots, x_{n+1} \in X$ such that $(x_i, x_{i+1}) \in \rho$, $1 \leq i \leq n$. In this case, we say that its length is n. A *path* is a walk that does not include any node twice, except that its first node can be the same as the last one. Such a (walk) path is called a $x_1 - x_{n+1}$ (walk) path. A path with $x_1 = x_{n+1}$ is called a *closed path* or a *cycle*. Given an element $x \in X$, we define its set of *accessible elements* by $\omega(x) := \{y \in Y : \text{there is an } x - y \text{ path}\}$. We say that a (di-)graph is *(strongly) connected* if, forevery pair of nodes $x, y \in X$, there is an $x - y$ path. A digraph whose underlying non-directed graph, obtained by removing the direction of every arc, is connected is said to be *weakly connected*. Further information on graph theory can be found in [18–20].

Throughout the rest of the paper, we assume that X and Y are two given topological spaces. If it is not explicitly mentioned, we consider that these spaces are endowed with the discrete topology. In this case, the unique dense set of X is the whole set X itself. We will also consider the product of N copies of these spaces, X^N and Y^N with $N \in \mathbb{N}$, equipped with the usual product space topologies.

We introduce several notions of dynamical systems in the setting of binary relations:

Definition 1. *Let $(\rho_n)_{n \in \mathbb{N}}$ be a sequence of binary relations between the spaces X and Y, ρ a binary relation on X, and $x \in X$. Then we say that*

(i) *x is a universal element for the sequence $(\rho_n)_{n \in \mathbb{N}}$ if $x \in \bigcap_{n \in \mathbb{N}} D(\rho_n)$ and for each $n \in \mathbb{N}_0$ there exists an element $y_n \in \rho_n(x)$ such that the set $\{y_n : n \in \mathbb{N}\}$ is dense in Y. As a particular case, if $\rho_n := \rho^n$, then we say x is hypercyclic for ρ.*

(ii) *ρ is topologically transitive if, for every pair of non-empty open sets $U, V \subset X$, there is some $n \in \mathbb{N}$ such that $U \cap \rho^{-n}(V) \neq \varnothing$. If there is some $n_0 \in \mathbb{N}$ such that this last condition holds for all $n \geq n_0$, we say that ρ is topologically mixing.*

Clearly, if x is hypercyclic for $G = (X, \rho)$, and $(z, x) \in \rho^l$ for some $l \in \mathbb{N}$, then z is likewise hypercyclic for G. If ρ is a binary equivalence relation and x is hypercyclic, then the underlying (di-)graph must be (strongly) connected and all the elements of X are hypercyclic for ρ.

Let us consider $(O_n)_{n \in \mathbb{N}}$ a base of non-empty open sets for the topology of X. If we denote by $HC(\rho)$ the set consisting of all hypercyclic elements of ρ, then the following equality holds [21]:

$$HC(\rho) = \bigcap_{n \in \mathbb{N}} \bigcup_{k \in \mathbb{N}} \rho^{-k}(O_n). \tag{1}$$

Remark 1. (i) *Let $G = (X, \rho)$ be a graph with X equipped with the discrete topology. It can be simply proved that the graph G is connected if and only if ρ is topologically transitive or hypercyclic.*

(ii) *In the Definition 1 (ii), it is irrelevant whether we write $\rho^{-n}(V)$ or $\rho^n(V)$. It is also worth noting that ρ does not need to be topologically mixing whenever ρ is topologically transitive: Let us consider a graph $G = (X, \rho)$ that is isomorphic to a square, that is $X = \{x_1, x_2, x_3, x_4\}$ and $\rho = \{(x_1, x_2), (x_2, x_3), (x_3, x_4), (x_4, x_1)\}$, see Figure 1a. Clearly, ρ is topologically transitive, but it does not hold the topologically mixing property since there is no odd number $n \in \mathbb{N}$ such that $x_3 \in \rho^n(x_1)$.*

(iii) *Unlike the linear setting, in our framework, the notion of hypercyclicity cannot be connected to that of topological transitivity in any reasonable way. It is well known that these notions are equivalent for continuous linear operators on Fréchet spaces by Baire's category theorem (see, for instance, [21,22]). However, there exist examples of continuous linear operators on non-metrizable locally convex spaces that are topologically transitive and not hypercyclic [23]. Moreover, for any non-trivial Banach space X there exists a multivalued linear operator $\mathcal{A} = \{0\} \times X$ that is hypercyclic and not topologically transitive (cf. [24]). It is very simple to construct an example of a hypercyclic relation on a finite set that is not topologically transitive, as well: Set the digraph $G = (X, \rho)$ with $X := \{x_1, x_2, x_3\}$ and $\rho := \{(x_1, x_2), (x_2, x_1), (x_1, x_3)\}$, endowing X with the discrete topology, see Figure 1b. Then x_1 is a hypercyclic element for ρ, but ρ is not topologically transitive, since $\{x_2\} \cap \rho^n(\{x_3\}) = \varnothing$ for all $n \in \mathbb{N}$.*

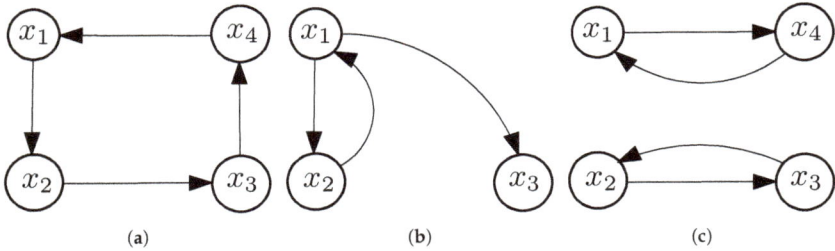

Figure 1. (a) In Remark 1 (ii), we show a graph that, endowed with the discrete topology, it is topologically transitive but not topologically mixing. (b) In Remark 1 (iii), we have a graph that, endowed with the discrete topology, it is hypercyclic but not topologically transitive. (c) In Example 1, different topologies—$\tau_1 = \{\varnothing, \{x_1\}, \{x_2\}, \{x_3\}, \{x_1, x_2\}, \{x_1, x_3\}, \{x_2, x_3\}, \{x_1, x_2, x_3\}, \{x_1, x_2, x_3, x_4\}\}$ and $\tau_2 = \{\varnothing, \{x_1, x_2\}, \{x_1, x_2, x_3, x_4\}\}$—can be defined such that the graph is hypercyclic for τ_1 but not for τ_2.

By considering different topologies on the set of nodes X, the aforementioned dynamical properties can be used to generalize the notion of connectivity. The following examples illustrate these facts:

Example 1. *Let $G = (X, \rho)$ be a graph without isolated nodes, and let τ be the topology on X. Let us denote by $G_1, \ldots, G_k$, the connected components of the graph G. Then ρ is hypercyclic if there exists a number $i \in \mathbb{N}_k$ such that G_i is dense in (G, τ), and any element of G_i is a hypercyclic element of ρ.*

We point out that the topology endowed to X is crucial in order to ensure hypercyclicity, or any other dynamical property. For example, let us take $X = \{x_1, x_2, x_3, x_4\}$, endowed with $\tau_1 = \{\varnothing, \{x_1\}, \{x_2\}, \{x_3\}, \{x_1, x_2\}, \{x_1, x_3\}, \{x_2, x_3\}, \{x_1, x_2, x_3\}, \{x_1, x_2, x_3, x_4\}\}$, and $\tau_2 = \{\varnothing, \{x_1, x_2\}, \{x_1, x_2, x_3, x_4\}\}$, see Figure 1c. We now set the binary relation $\rho := \{(x_1, x_4), (x_4, x_1), (x_2, x_3), (x_3, x_2)\}$, then ρ is not hypercyclic in (X, τ_1) but it is hypercyclic in (X, τ_2).

Finally, observe that ρ is topologically transitive (and ρ is topologically mixing) if, for every pair of non-empty open subsets of $U, V \subset X$, there exists $i \in \mathbb{N}_k$ such that $U \cap G_i \neq \varnothing$ and $V \cap G_i \neq \varnothing$.

The links between connectivity and dynamics will be thoroughly explained in the next section.

3. Hypercyclic and Chaotic Digraphs

We study the relations between different types of connectivity on digraphs and some of the aforementioned dynamical properties. This will enable us to introduce two new important classes of digraphs that are subclasses of the class of weakly connected digraphs and that extend the class of strongly connected ones.

One of the most accepted notions of chaos is the one introduced by Devaney [25] for continuous mappings acting on metric spaces. Three ingredients are considered in this definition: topological transitivity, density of periodic points, and sensitive dependence on the initial conditions (SDIC). Banks et al. [26] proved that SDIC can be deduced from the other two properties. Nevertheless, since topological transitivity does not coincide with hypercyclicity in our setting, we will introduce two different notions of chaos.

Definition 2. *Let ρ be a binary relation on X, and $x \in X$. Then we say that*

1. *x is a periodic point of ρ if $x \in D_\infty(\rho)$ and there exists $n \in \mathbb{N}$ such that $x \in \rho^n(x)$.*
2. *ρ is Devaney-chaotic if it is topologically transitive and it has a dense set of periodic points.*
3. *ρ is chaotic if it is hypercyclic and it has a dense set of periodic points.*

Let us consider $G = (X, \rho)$, where X is endowed with the discrete topology. It immediately follows from our definitions that G is strongly connected if and only if ρ is topologically transitive. Every $x \in X$ is also a periodic element for ρ. Thus, ρ is also Devaney-chaotic. Besides, ρ is also hypercyclic and chaotic, and $x \in HC(\rho)$ for every $x \in X$. Since ρ is hypercyclic, let us pick any $x \in HC(\rho)$. Then, for two different points $y, z \in X$, an $x - y$ path and an $x - z$ path exists in G. Thus, considering the underlying non-directed edges, there exists a walk connecting y and z, and G is thus weakly connected. We summarize all these relations underneath.

$$(G, \rho) \text{ is Devaney-chaotic (equiv., strongly connected)} \Rightarrow (G, \rho) \text{ is chaotic}$$
$$\Rightarrow (G, \rho) \text{ is hypercyclic} \Rightarrow (G, \rho) \text{ is weakly connected.} \tag{2}$$

Therefore, we are able to introduce two new classes of digraphs that are subclasses of the class of weakly connected digraphs and that extend the notion of strong connectivity. Any of these three implications is strict, as the following examples show:

Example 2. *(i) Let $X := \{x_1, x_2, x_3, x_4\}$ be equipped with discrete topology, and let $\rho := \{(x_1, x_2), (x_2, x_1), (x_1, x_3), (x_3, x_4), (x_4, x_3)\}$, see Figure 2a. Then $x_3 \in \rho(x_1)$, $x_4 \in \rho^2(x_1)$, $x_1 \in \rho^3(x_1)$, $x_2 \in \rho^{2n+1}(x_1)$, and $x_1 \in \rho^{2n}(x_1)$ $(n \geq 2)$, which simply yields that $x_1 \in HC(\rho)$. It is also clear that any element of X is periodic for ρ, such that ρ is chaotic. Since there is no path connecting x_3 and x_1 in G, G is not strongly connected and ρ is neither topologically transitive nor Devaney-chaotic. Therefore, the first implication in Equation (2) is strict.*

(ii) Let $C = x_1 \dots x_{n+1}$ be an oriented closed cycle of length n (with $x_1 = x_{n+1}$), and let $\rho := C \cup \{(x_1, x_{n+2})\}$, where $x_{n+2} \neq x_j$ for $1 \leq j \leq n+1$, see Figure 2b. Then it can be easily seen that $G = (X, \rho)$, endowed with the discrete topology, is hypercyclic, since any element lying on the cycle C is hypercyclic for ρ and that G is not chaotic, because the point x_{n+2} cannot be a periodic element for ρ.

(iii) Let $X := \{x_1, x_2, x_3\}$ and $\rho := \{(x_1, x_2), (x_3, x_2)\}$, see Figure 2c. Then $G = (X, \rho)$ is weakly connected but, equipped with the discrete topology, it is not hypercyclic.

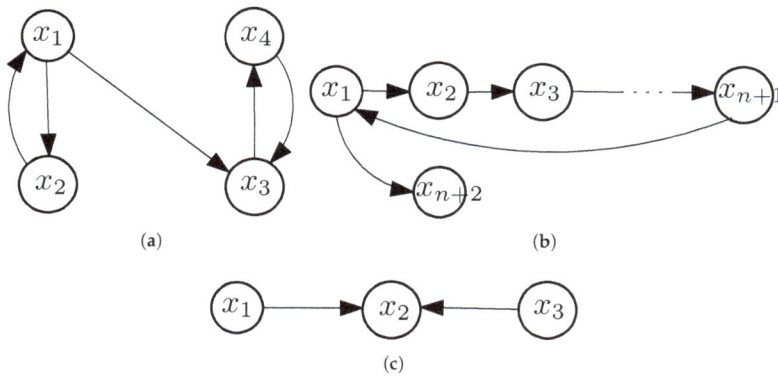

(a)

(b)

(c)

Figure 2. The following graphs are endowing with the discrete topology. (**a**) In Example 2 (i), we show that all in the elements in a graph can be periodic, but this does not imply strong connectivity, Devaney chaos, nor topological transitivity. (**b**) In Example 2 (ii), we show that hypercyclicity does not imply density of periodic points. (**c**) In Example 2 (iii), we show that weak connectivity does not imply hypercyclicity.

A series of recent results in the theory of digraphs are devoted to the study of cyclability: Given a digraph $G = (X, \rho)$, a set $S \subset X$ said to be *cyclable* in G if G contains a cycle through all the nodes of S.

Setting $S = X$, we obtain the classical concept of a Hamiltonian (di-)graph. We refer the interested reader to [27]. It is clear that the cyclability of $S \subset X$ in $G = (X, \rho)$ jointly with the denseness of S in X imply that ρ is (Devaney)-chaotic for ρ. However, this is far from being necessary for ρ to be (Devaney)-chaotic since this condition is automatically satisfied if the points in S lie on a closed path in G not on a circle.

It is worth noting that our conclusions from Example 1 can be reformulated for digraphs only partially; if $G_1, G_2, \ldots, G_k$ denote the strongly connected components of a digraph $G = (X, \rho)$, then the denseness of some G_i in G implies that ρ is (Devaney)-chaotic for ρ, but the converse statement fails to be true even for discrete topology, as the next example shows:

Example 3. *Consider a digraph G consisting of two oriented cycles given by $G_1 := \{x_1, x_2, x_3\}$ and $G_2 := \{x_4, x_5, x_6\}$ joined by an arc (x_1, x_4), see Figure 3a. Then G is Devaney-chaotic but neither G_1 nor G_2 are dense in G.*

4. Dynamics on Tournaments

Without any doubt, tournaments are the best studied class of digraphs (see, for instance, the classical reference of [28]). We recall their definition.

Definition 3. *A tournament $T = (X, \rho)$ is a digraph in which any pair of different nodes $x, y \in X$ are connected by exactly one arc.*

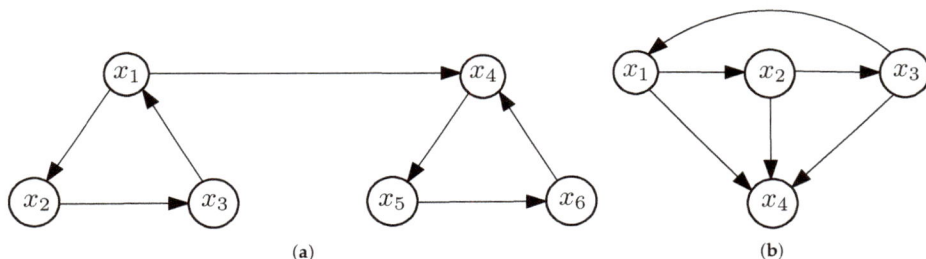

(a)　　　　　　　　　　　(b)

Figure 3. (a) In Example 3, we show a Devaney-chaotic digraph, whose strongly connected components are not dense. (b) In Example 4, we show an example of a tournament that, endowed with the discrete topology, it is hypercyclic but not topologically transitive.

Example 4. *It is straightforward that a tournament $T = (X, \rho)$ with $X = \{x_1, x_2, x_3, x_4\}$ and ρ obtained as the union of a cycle $\{(x_1, x_2), (x_2, x_3), (x_3, x_1)\}$ with the arcs $\{(x_1, x_4), (x_2, x_4), (x_3, x_4)\}$) is hypercyclic but not topologically transitive, see Figure 3b.*

In the following theorem, we completely characterize the class of hypercyclic tournaments.

Theorem 1. *Let $T_n = (X, \rho)$, with $|X| = n$, be a tournament of n nodes equipped with the discrete topology. Then T_n is hypercyclic if and only if $d^+(x) < n - 1$ for every $x \in X$.*

Proof. First, assume that T_n is hypercyclic and $d^+(x) = n - 1$ for some $x \in X$. Then $d^-(x) = 0$, and there is no $k \in \mathbb{N}, z \in \{x_1, x_2, \ldots, x_n\}$ such that $x \in \rho^k(z)$, which contradicts the hypercyclicity of T_n. Now, let us assume that the outdegree of any vertex is strictly less than $n - 1$. As a consequence of the famous theorem attributed to Rédei [29], there exists a non-closed path visiting all nodes of X, that is a Hamiltonian path named $x_1, x_2 \ldots, x_n$. Therefore, $x_1 \in HC(\rho)$. □

It is clear that, in any arbitrary tournament T, we have that $d^+(x) + d^-(x) = n - 1$ for all $x \in X$, and $d^+(x) < n - 1$ if and only if $d^-(x) > 0$. For digraphs, we can prove the following extension of Theorem 1:

Corollary 1. *Let $G = (X, \rho)$ be a digraph and X equipped with discrete topology, with $|X| = n \geq 3$. Suppose that, for any pair of nodes, $x, y \in X$ such that, if $(x, y) \notin \rho$, we have $d^+(x) + d^-(y) \geq n - 1$. Then ρ is hypercyclic if and only if $d^-(x) > 0$ for all $x \in X$.*

Proof. It is clear that the hypercyclicity of ρ implies that the indegree of any node is strictly positive. The converse statement follows from the fact that such assumptions imply that there is a Hamiltonian path in G by Ore's Theorem [20], and the argumentation is in the proof of Theorem 1. □

5. Disjointness on Binary Relations, Graphs, and Digraphs

Disjointness was firstly introduced for dynamical systems by Furstenberg in [30]. For linear operators, it was firstly considered in [31,32] (see also [33]). Further information on the dynamics of linear operators can be found in [34–37]. We introduce analogons of disjointness of the above classes of hypercyclic and (Devaney)-chaotic graphs. Part of these results are inspired by the analogons obtained for the case of multivalued linear operators in [5].

We first introduce the notion of disjointness for hypercyclicity, topological transitivity, and the topologically mixing property.

Definition 4. *Let $N \geq 2$. Let X, Y be two topological spaces. For every $j \in \mathbb{N}_N$, let $(\rho_{j,n})_{n \in \mathbb{N}}$ be a sequence of binary relations between the spaces X and Y, let ρ_j be a binary relation on X, and let $x \in X$. An element $x \in X$ is a* d-universal *element for the sequences $(\rho_{1,n})_{n \in \mathbb{N}}, j \in \mathbb{N}_N$ if, for each $j \in \mathbb{N}_N, n \in \mathbb{N}$, there exist elements $y_{j,n} \in \rho_{j,n}(x)$ such that the set $\{(y_{1,n}, y_{2,n}, \ldots, y_{N,n}) : n \in \mathbb{N}\}$ is dense in Y^N. As a particular case, the binary relations $\rho_1, \ldots, \rho_N$ are called* d-hypercyclic *if there exists a d-universal element x of the sequences $(\rho_1^n)_{n \in \mathbb{N}}, \ldots, (\rho_N^n)_{n \in \mathbb{N}}$. In this case, x is called a* d-hypercyclic *element of the binary relations $\rho_1, \ldots, \rho_N$.*

Definition 5. *Let $N \geq 2$. Let X, Y be two topological spaces. For every $1 \leq j \leq N$, let $(\rho_{j,n})_{n \in \mathbb{N}}$ be a sequence of binary relations between the spaces X and Y and let ρ_j be a binary relation on X.*

1. *The sequences $(\rho_{1,n})_{n \in \mathbb{N}}, \ldots, (\rho_{N,n})_{n \in \mathbb{N}}$ are* d-topologically transitive *if, for every non-empty open subsets $U \subset X$ and $V_1, \ldots, V_N \subset Y$, there exists $n \in \mathbb{N}$ such that*

$$U \cap \rho_{1,n}^{-1}(V_1) \cap \ldots \cap \rho_{N,n}^{-1}(V_N) \neq \emptyset. \tag{3}$$

2. *The sequences $(\rho_{1,n})_{n \in \mathbb{N}}, \ldots, (\rho_{N,n})_{n \in \mathbb{N}}$ are* d-topologically mixing *if, for every non-empty open subsets $U \subset X$ and $V_1, \ldots, V_N \subset Y$, there exists $n_0 \in \mathbb{N}$ such that, for every $n \geq n_0$, we have that Equation (3) holds.*
3. *The binary relations $\rho_1, \cdots, \rho_N$ are* d-topologically transitive (d-topologically mixing) *if the sequences $(\rho_1^n)_{n \in \mathbb{N}}, \ldots, (\rho_N^n)_{n \in \mathbb{N}}$ are d-topologically transitive (d-topologically mixing).*

We also introduce the notion of d-Devaney chaos and d-chaos for binary relations. For this purpose, we define the set of periodic elements

$$\mathcal{P}(\rho_1, \rho_2, \cdots, \rho_N) := \{(x_1, x_2, \ldots, x_N) \in X^N : \exists n \in \mathbb{N} \text{ with } x_j \in \rho_j^n(x_j), j \in \mathbb{N}_N\}. \tag{4}$$

Definition 6. *Given $N \geq 2$, the binary relations $\rho_1, \ldots, \rho_N$ on X are said to be* d-Devaney-chaotic *if they are d-topologically transitive and the set of periodic elements $\mathcal{P}(\rho_1, \rho_2, \cdots, \rho_N)$ is dense in X^N. These relations are* d-chaotic *if they are d-hypercyclic and the set of periodic elements is dense in X^N.*

A formula similar to Equation (1) can be given for d-hypercyclic elements of binary relations [32], and the most important consequences of both formulae can be formulated for continuous mappings acting between topological spaces.

It is well known that two single-valued linear operators acting on a Fréchet space cannot be d-hypercyclic if one of them is a scalar multiple of the other one. This is no longer true for multivalued linear operators since there exists a multivalued linear operator $\mathcal{A}$ on a Banach space such that $\mathcal{A}$ and some arbitrary multiples of it can be d-hypercyclic [5]. Concerning simple graphs, the notion of d-hypercyclicity is much more complicated than the notion of hypercyclicity, and it does not reduce to the connectivity of the graphs, as is described in Example 2 (ii). For the next result, we recall that a graph of n nodes $K_n = (X_n, \rho)$ is *complete* if, for every pair of different nodes $x, y \in X_n$, we have that $(x, y) \in \rho$, i.e., $\rho = X \times X \setminus \Delta_X$.

Theorem 2. *Let $n \geq 2$, and let $K_n = (X_n, \rho)$ denote the complete graph with $X_n = \{x_1, x_2, \ldots, x_n\}$, equipped with discrete topology. Let us consider $N \geq 2$ copies of the graphs K_n. The following affirmations are equivalent:*

i.   *The graphs are d-Devaney-chaotic.*
ii. *The graphs are d-hypercyclic.*
iii. *Each graph contains $n \geq 3$ nodes.*

Proof. Clearly, (i) implies (ii). To prove (ii) implies (iii), we only need to observe that, in the case $n = 2$, we do not have the existence of a natural number $k \in \mathbb{N}$ such that $x_1 \in \rho^k(x_1)$ and $x_2 \in \rho^k(x_1)$. Thus, any N-tuple of the X_2^N containing two different components cannot be an element of the set $\cup_{k \in \mathbb{N}}[(\rho^k(x_1))^N \cup (\rho^k(x_2))^N]$. Hence, neither x_1 nor x_2 are d-hypercyclic elements for the N graphs K_2. Now, let us assume $n \geq 3$ and let us see that the N copies of K_n are d-Devaney-chaotic. For any N-tuple $(x_{i_1}, \ldots, x_{i_N}) \in X_n^N$ and $i \in \mathbb{N}_N$, we have that $(x_{i_1}, \ldots, x_{i_N}) \in (\rho^2(x_i))^N$ so that x_i is a d-hypercyclic element for the N-tuple of graphs K_n. Finally, the set of periodic points $\mathcal{P}(\rho, \ldots, \rho)$ coincides with X_n^N, even in the case $n = 2$. $\square$

In a similar way, it can be seen the equivalence of d-chaos and d-transitivity with the condition $n \geq 3$.

Another remarkable class of graphs are bipartite graphs. A graph $G = (X, \rho)$ is said to be *bipartite* if $X = X_1 \cup X_2$ with $X_1 \cap X_2 = \emptyset$ and for every $(x, y) \in \rho$, either $x \in X_1$ and $y \in X_2$ or $x \in X_2$ and $y \in X_1$. In this case, the d-hypercyclicity can never hold. If we consider a family of bipartite graphs $G_1, \ldots, G_N$, sharing the partition of the set of nodes X into two sets X_1 and X_2, then for any $x \in X$ and every $k \in \mathbb{N}$ either $\rho_1^k(x) \times \ldots \times \rho_N^k(x) \subset X_1^N$ or $\rho_1^k(x) \times \ldots \times \rho_N^k(x) \subset X_2^N$, depending on where is x and if k is even or odd.

Before proceeding further, we would like to observe that the notion of (d-topological transitivity) d-hypercyclicity is equivalent for simple graphs equipped with the discrete topology to the notion of d-Devaney chaos, since for any simple graph $G = (X, \rho)$ and for any $x \in X$ we have that $x \in \rho^{2k}(x)$.

In the remaining part of this section, we will only consider digraphs.

For any digraph $G = (X, \rho)$, where $X = \{x_1, \ldots, x_n\}$, we introduce the *adjacency matrix* of A, $[A] = [a_{i,j}]$ shortly, by $a_{i,j} := 1$ if $(x_i, x_j) \in \rho$ and $a_{i,j} := 0$ if $(x_i, x_j) \notin \rho$, for every $1 \leq i, j \leq n$.

Let $G_1, \ldots, G_N$ be digraphs with, $G_i = (X_i, \rho_i), i \in \mathbb{N}_N$, each of which is equipped with discrete topology, and let $A_1, \ldots, A_N$ be their corresponding adjacency matrices. Denote, for every $l \in \mathbb{N}_N$ and $k \in \mathbb{N}$, the k-th power of the adjacency matrix A_l as $A_l^k = [a_{i,j}^{l,k}]_{1 \leq i,j \leq n}$. As is well known, the element $a_{i,j}^{l,k}$ of matrix A_l^k represents the exact number of $x_i - x_j$ walks of length k in G_l. Using this result and our definition of d-hypercyclic elements, we can simply clarify the following necessary and sufficient conditions for an element to be a d-hypercyclic element.

Theorem 3. *Let $G_1 \ldots G_N$ be digraphs over $X = \{x_1, \ldots, x_n\}$, equipped with the discrete topology, with adjacency matrices $A_1, \ldots, A_N$. An element $x_i \in X$ is d-hypercyclic for $G_1, \ldots, G_N$ if and only if, for every $(j_1, j_2, \ldots, j_N) \in \mathbb{N}_n^N$, there exists $k \in \mathbb{N}$ such that, for all $l \in \mathbb{N}_N, a_{i,j_l}^{l,k} \geq 1$.*

By taking $V_l = \{x_{j_l}\}$, $l \leq \mathbb{N}_N$ and $U = \{x_i\}$ in Condition 3, we obtain the following characterization of d-topological transitivity of digraphs $G_1, \ldots, G_N$ equipped with discrete topologies, and we can also characterize d-topologically mixing property in a similar way:

Theorem 4. *Let $G_1 \ldots G_N$ be digraphs over $X = \{x_1, \ldots, x_n\}$, equipped with the discrete topology, with adjacency matrices $A_1, \ldots, A_N$. The digraphs $G_1, \ldots, G_N$ are d-topologically transitive if and only if, for every $(j_1, j_2, \ldots, j_N) \in \mathbb{N}_n^N$ and for every $i \in \mathbb{N}$, there exists $k \in \mathbb{N}$ such that, for all $l \in \mathbb{N}_N$, $a_{i,j_l}^{l,k} \geq 1$.*

Thus, d-topological transitivity of digraphs $G_1, \ldots, G_N$ equipped with discrete topologies immediately implies that these digraphs are d-hypercyclic with any $x \in G$ being a d-hypercyclic element. It is clear that d-Devaney chaos of $G_1, G_2, \ldots, G_N$ implies d-chaos, which also implies d-hypercyclicity. The converse implications do not hold in general, as can be illustrated with the following examples.

Example 5. (i) *Let G be the graph appearing in Example 2 (i), and let K_4 be the complete graph of these 4 nodes, and both are equipped with discrete topologies, see Figure 4a. It can be easily seen that G and H are d-chaotic but not d-Devaney-chaotic.*

(ii) *Let $X := \{x_1, x_2, x_3, x_4\}$, and let $\rho := \bigcup_{1 \leq i,j \leq 3}(x_i, x_j) \cup \{(x_1, x_4), (x_2, x_4)\}$, see Figure 4b. If $G = (X, \rho)$ is equipped with the discrete topology, then the pair G, G is d-hypercyclic, since, for any even number $k \in \mathbb{N}$ and for every $x \in X$, we have $x \in \rho^k(x_1)$, but it is not d-chaotic since x_4 is not periodic in G.*

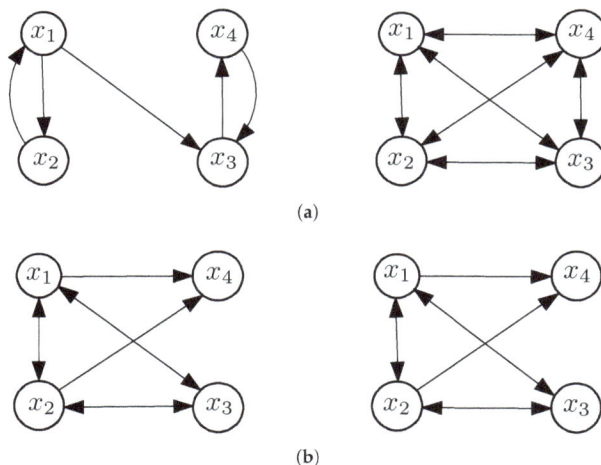

(a)

(b)

Figure 4. The following graphs are endowed with the discrete topology. (**a**) In Example 5 (i), we show that the graph of Example 2 (i) and K_4 are d-chaotic but not d-Devaney-chaotic. (**b**) In Example 5 (ii), these two copies of the same graph are d-hypercyclic but not d-chaotic.

Disjointness of tournaments depends on the size of the set of nodes.

Theorem 5. *Let $n \leq 3$, and let $T_1, \ldots, T_N$ be N tournaments equipped with discrete topologies. Then $T_1, \ldots, T_N$ cannot be d-hypercyclic nor d-topologically transitive.*

Proof. Without loss of generality, we may assume that $N = 2$. If $n = 2$, then it can be easily seen that T_1 and T_2 must be isomorphic to an oriented segment, which is neither hypercyclic nor topologically transitive.

If $n = 3$, there exist only two non-isomorphic tournaments of order 3. Since T_1 and T_2 should be hypercyclic (topologically transitive), and only the Hamiltonian circles are hypercyclic (topologically transitive), it follows that T_1 and T_2 have to be a Hamiltonian circle—either x_1, x_2, x_3, x_1 or x_1, x_3, x_2, x_1. A careful inspection of all possible cases leads us to the impossibility of satisfying any of these dynamical properties. □

By Theorems 2 and 5, looking at the case $n = 3$, we easily observe the existence of digraphs that are not d-chaotic, but whose underlying simple graphs are d-chaotic. In the case of $n \geq 4$, we have the following important result.

Theorem 6. *Let $n \geq 4$, and let $T_1, \ldots, T_N$ be tournaments over a set $X = \{x_1, \ldots, x_n\}$ equipped with discrete topologies. The following are equivalent:*

1. $T_1, \ldots, T_N$ *are d-Devaney-chaotic.*
2. $T_1, \ldots, T_N$ *are d-topologically transitive.*
3. T_l *is strongly connected for all $l \in \mathbb{N}_N$.*
4. T_l *is a Hamiltonian tournament for all $l \in \mathbb{N}_N$.*

Proof. Let $T_1, \ldots, T_N$ be d-chaotic. Then these tournaments are d-topologically transitive such that any $T_l, l \in \mathbb{N}_N$, is topologically transitive and therefore strongly connected. By Camion's theorem (see, for instance, [28]), T_l is strongly connected if and only if each T_l is Hamiltonian.

Therefore, it remains to be proven that the strong connectivity of all T_l values implies that $T_1, \ldots, T_N$ are d-chaotic. To see this, let us recall that the strong connectivity of any directed graph is equivalent to its irreducibility [38]. Applying Th. 1 of [38], we have that the adjacency matrix A_l of $T_l, l \in \mathbb{N}_N$, is primitive, i.e., there exists a natural number q_l such that $A_l^{q_l}$ is strictly positive. In fact, Wielandt estimates that $q_l \leq (n-1)^2 + 1$ [39]. The result now follows from an application of Theorem 4 and from the observation that, for a given tuple $(j_1, j_2, \ldots, j_N) \in \mathbb{N}_n^N$, $(x_{j_1}, x_{j_2}, \ldots, x_{j_N})$ is a periodic point for $T_1, \ldots, T_N$ if and only if there exists $k \in \mathbb{N}$ such that, for each $l \in \mathbb{N}_N$, one has $a_{j_l, j_l}^{l, k} \geq 1$. □

Any of the above equivalencies implies that $T_1, \ldots, T_N$ are also d-hypercyclic. However, the situation is not so simple because the strong connectivity of T_l values for all $l \in \mathbb{N}_N$ is not equivalent to the fact that $T_1, \ldots, T_N$ were d-hypercyclic:

Example 6. *It is well known that there exist only four non-isomorphic tournaments of order four. Only two of them, T_1 and T_2, defined as follows, are hypercyclic: T_1 is the union of the Hamiltonian cycle x_2, x_3, x_4, x_2 and oriented segments (x_2, x_1), (x_3, x_1), and (x_4, x_1), while T_2 is the union of the Hamiltonian cycle x_1, x_2, x_3, x_4, x_1 and oriented segments (x_1, x_3) and (x_2, x_4). Furthermore, T_1 is hypercyclic but not topologically transitive, whereas T_2 is. It can be verified that the pair T_1, T_1 is not d-hypercyclic. Hence, the hypercyclicity of components does not imply d-hypercyclicity of a tuple. We also point out that the pair T_1, T_2 is d-hypercyclic, but T_1 is not strongly connected.*

We close this work by providing one more example.

Example 7. *Let $n \geq 5$, and $T_1, \ldots T_N$ be tournaments over a set of nodes $X = \{x_1, \ldots, x_n\}$. Suppose that T_1 is a hypercyclic, non-Hamiltonian tournament and that, for $l = 2, \ldots, N$, T_l is a Hamiltonian tournament. Let x be a hypercyclic element for T_1. Then the proof of Theorem 6 shows that $T_1, \ldots T_N$ are d-hypercyclic, where x is their d-hypercyclic element. However, it is clear that $T_1, \ldots T_N$ cannot be d-topologically transitive.*

Author Contributions: All the authors have equally participated in the work.

Acknowledgments: The second and fourth authors were supported by MEC Project MTM2016-75963-P.

Symmetry **2018**, *10*, 211

Conflicts of Interest: The authors declare no conflict of interest. The founding sponsors had no role in the design of the study; in the collection, analyses, or interpretation of data; in the writing of the manuscript, and in the decision to publish the results.

References

1. Strogatz, S.H. Nonlinear Dynamics and Chaos: With Applications to Physics, Biology, Chemistry, and Engineering (Studies in Nonlinearity). In *Nonlinear Dynamics and Chaos: With Applications to Physics, Biology, Chemistry, and Engineering (Studies in Nonlinearity)*, 2nd ed.; Addison-Wesley Publishing Company: Boston, MA, USA, 1994.
2. Hilborn, R.C. *Chaos and Nonlinear Dynamics: An Introduction for Scientists and Engineers*; Oxford University Press: Oxford, UK, 2000.
3. Perc, M. Visualizing the attraction of strange attractors. *Eur. J. Phys.* **2005**, *26*, 579. [CrossRef]
4. Kodba, S.; Perc, M.; Marhl, M. Detecting chaos from a time series. *Eur. J. Phys.* **2004**, *26*, 205. [CrossRef]
5. Chen, C.-C.; Conejero, J.A.; Kostić, M.; Murillo-Arcila, M. Dynamics of multivalued linear operators. *Open Math.* **2017**, *15*, 948–958. [CrossRef]
6. Bahi, J.M.; Couchot, J.-F.; Guyeux, C.; Richard, A. On the link between strongly connected iteration graphs and chaotic Boolean discrete-time dynamical systems. In *Fundamentals of Computation Theory*; Lecture Notes in Computer Science; Springer: Heidelberg, Germany, 2011; Volume 6914, pp. 126–137.
7. Contassot-Vivier, S.; Couchot, J.-F.; Guyeux, C.; Heam, P.-C. Random walk in a *N*-cube without Hamiltonian cycle to chaotic pseudorandom number generation: Theoretical and practical considerations. *Int. J. Bifurc. Chaos* **2017**, *27*, 1750014. [CrossRef]
8. Bakiri, M.; Couchot, J.-F.; Guyeux, C. One random jump and one permutation: Sufficient conditions to chaotic, statistically faultless, and large throughput prng for fpga. *arXiv* **2017**, arXiv:1706.08093.
9. Bahi, J.M.; Guyeux, C. *Discrete Dynamical Systems and Chaotic Machines: Theory and Applications*; CRC Press: Boca Raton, FL, USA, 2013.
10. Guyeux, C.; Wang, Q.; Fang, X.; Bahi, J. Introducing the truly chaotic finite state machines and their applications in security field. *arXiv* **2017**, arXiv:1708.04963.
11. Chen, C.-C. Hypercyclic and chaotic operators on l^p spaces of Cayley graphs. Preprint, 2017.
12. Martínez-Avendano, R.A. Hypercyclicity of shifts on weighted L^p spaces of directed trees. *J. Math. Anal. Appl.* **2017**, *446*, 823–842. [CrossRef]
13. Marijuán, C. Finite topologies and digraphs. *Proyecciones* **2010**, *29*, 291–307. [CrossRef]
14. Banasiak, J. Chaos in Kolmogorov systems with proliferation—General criteria and applications. *J. Math. Anal. Appl.* **2011**, *378*, 89–97. [CrossRef]
15. Banasiak, J.; Moszyński, M. Dynamics of birth-and-death processes with proliferation—Stability and chaos. *Discret. Contin. Dyn. Syst.* **2011**, *29*, 67–79.
16. Aroza, J.; Peris, A. Chaotic behaviour of birth-and-death models with proliferation. *J. Differ. Equ. Appl.* **2012**, *18*, 647–655. [CrossRef]
17. Namayanja, P. Chaotic phenomena in a transport equation on a network. *Discret. Contin. Dyn. Syst. Ser. B* **2018**, in press.
18. Bondy, J.A.; Murty, U.S.R. *Graph Theory with Applications*; American Elsevier Publishing Co., Inc.: New York, NY, USA, 1976.
19. Chartrand, G.; Lesniak, L.; Zhang, P. *Graphs & Digraphs*, 6th ed.; Textbooks in Mathematics; CRC Press: Boca Raton, FL, USA, 2016.
20. Petrović, V. *Graph Theory*; University of Novi Sad: Novi Sad, Serbia, 1998.
21. Grosse-Erdmann, K.-G. Universal families and hypercyclic operators. *Bull. Am. Math. Soc. (N.S.)* **1999**, *36*, 345–381. [CrossRef]
22. Godefroy, G.; Shapiro, J.H. Operators with dense, invariant, cyclic vector manifolds. *J. Funct. Anal.* **1991**, *98*, 229–269. [CrossRef]
23. Bonet, J.; Frerick, L.; Peris, A.; Wengenroth, J. Transitive and hypercyclic operators on locally convex spaces. *Bull. Lond. Math. Soc.* **2005**, *37*, 254–264. [CrossRef]
24. Cross, R. *Multivalued Linear Operators*; Monographs and Textbooks in Pure and Applied Mathematics; Marcel Dekker, Inc.: New York, NY, USA, 1998; Volume 213.

25. Devaney, R.L. *An Introduction to Chaotic Dynamical Systems*, 2nd ed.; Addison-Wesley Studies in Nonlinearity; Addison-Wesley Publishing Company, Advanced Book Program: Redwood City, CA, USA, 1989.
26. Banks, J.; Brooks, J.; Cairns, G.; Davis, G.; Stacey, P. On Devaney's definition of chaos. *Am. Math. Mon.* **1992**, *99*, 332–334. [CrossRef]
27. Bermond, J.-C.; Thomassen, C. Cycles in digraphs—A survey. *J. Graph Theory* **1981**, *5*, 1–43. [CrossRef]
28. Moon, J.W. *Topics on Tournaments*; Holt, Rinehart and Winston: New York, NY, USA; Montreal, QC, Canada; London, UK, 1968.
29. Gross, J.L.; Yellen, J. *Graph Theory and Its Applications*, 2nd ed.; Discrete Mathematics and Its Applications (Boca Raton); Chapman & Hall/CRC: Boca Raton, FL, USA, 2006.
30. Furstenberg, H. Disjointness in ergodic theory, minimal sets, and a problem in Diophantine approximation. *Math. Syst. Theory* **1967**, *1*, 1–49. [CrossRef]
31. Bernal-González, L. Disjoint hypercyclic operators. *Stud. Math.* **2007**, *182*, 113–131. [CrossRef]
32. Bès, J.; Peris, A. Disjointness in hypercyclicity. *J. Math. Anal. Appl.* **2007**, *336*, 297–315. [CrossRef]
33. Bès, J.; Martin, Ö.; Peris, A.; hkarin, S. Disjoint mixing operators. *J. Funct. Anal.* **2012**, *263*, 1283–1322. [CrossRef]
34. Bayart, F.; Matheron, É. *Dynamics of Linear Operators*; Cambridge Tracts in Mathematics; Cambridge University Press: Cambridge, UK, 2009; Volume 179.
35. Grosse-Erdmann, K.-G.; Peris, A. *Linear Chaos*; Universitext; Springer: London, UK, 2011.
36. Bernal-González, L.; Pellegrino, D.; Seoane-Sepúlveda, J.B. Linear subsets of nonlinear sets in topological vector spaces. *Bull. Am. Math. Soc. (N.S.)* **2014**, *51*, 71–130. [CrossRef]
37. Aron, R.M.; Bernal-González, L.; Pellegrino, D.; Seoane-Sepúlveda, J.B. *Lineability: The Search for Linearity in Mathematics, Monographs and Research Notes in Mathematics*; CRC Press: Boca Raton, FL, USA, 2016.
38. Moon, J.W.; Pullman, N.J. On the powers of tournament matrices. *J. Comb. Theory* **1967**, *3*, 1–9. [CrossRef]
39. Wielandt, H. Unzerlegbare, nichtnegative Matrizen. *Math. Z.* **1950**, *52*, 642–648. [CrossRef]

symmetry

MDPI

Article

Topological Properties of Crystallographic Structure of Molecules

Jia-Bao Liu [1] , Muhammad Kamran Siddiqui [2,]* , Manzoor Ahmad Zahid [2],
Muhammad Naeem [3] and Abdul Qudair Baig [3]

[1] School of Mathematics and Physics, Anhui Jianzhu University, Hefei 230601, China; liujiabaoad@163.com
[2] Department of Mathematics, COMSATS University Islamabad, Sahiwal Campus, Sahiwal 57000, Pakistan;
    manzoor@cuisahiwal.edu.pk
[3] Department of Mathematics, The University of Lahore, Pakpattan Campus, Pakpattan 57400, Pakistan;
    naeempkn@gmail.com (M.N.); aqbaig1@gmail.com (A.Q.B.)
* Correspondence: kamransiddiqui75@gmail.com

Received: 17 June 2018; Accepted: 3 July 2018; Published: 5 July 2018

Abstract: Chemical graph theory plays an important role in modeling and designing any chemical structure. The molecular topological descriptors are the numerical invariants of a molecular graph and are very useful for predicting their bioactivity. In this paper, we study the chemical graph of the crystal structure of titanium difluoride TiF_2 and the crystallographic structure of cuprite Cu_2O. Furthermore, we compute degree-based topological indices, mainly ABC, GA, ABC_4, GA_5 and general Randić indices. Furthermore, we also give exact results of these indices for the crystal structure of titanium difluoride TiF_2 and the crystallographic structure of cuprite Cu_2O.

Keywords: topological indices; cuprite; atom bond connectivity index; Zagreb indices; geometric arithmetic index; general Randić index; titanium difluoride

1. Introduction

Graph theory is one of the most special and unique branches of mathematics by which the demonstration of any structure is made conceivable. Recently, it has attained much attention among researchers because of its wide range of applications in computer science, electrical networks, interconnected networks, biological networks, chemistry, etc. The chemical graph theory CGT is a fast growing area among researchers. It helps in understanding the structural properties of a molecular graph. There are many chemical compounds that possess a variety of applications in the fields of commercial, industrial, pharmaceutical chemistry and daily life and in the laboratory.

A relationship exists between chemical compounds and their molecular structures. The manipulation and examination of chemical structural information is made conceivable using molecular descriptors. Chemical graph theory is a branch of mathematical chemistry in which the tools of graph theory are applied to model the chemical phenomenon mathematically. Furthermore, it relates to the nontrivial applications of graph theory for solving molecular problems. This theory contributes to a prominent role in the field of chemical sciences; see for details [1–3].

Chem-informatics is a new subject, which is a combination of chemistry, mathematics and information science. It examines the quantitative structure–activity relationship $(QSAR)$ and quantitative structure-property relationship $(QSPR)$, which are utilized to predict the bioactivity and physicochemical properties of chemical compounds [4]. The field of chemical graph theory has attained much attention and consideration among researchers [5,6].

In solid state physics, the electrons of a single, isolated atom occupy atomic orbitals, each of which has a discrete energy level. When two atoms join together to form a molecule, their atomic orbitals overlap [7]. The Pauli exclusion principle dictates that no two electrons can have the same quantum numbers in a molecule. Therefore, if two identical atoms combine to form a diatomic molecule, each atomic orbital splits into two molecular orbitals of different energy, allowing the electrons in the former atomic orbitals to occupy the new orbital structure without any having the same energy. Similarly if a large number N of identical atoms come together to form a solid, such as a crystal lattice, the atoms' atomic orbitals overlap [8]. Since the Pauli exclusion principle dictates that no two electrons in the solid have the same quantum numbers, each atomic orbital splits into N discrete molecular orbitals, each with a different energy.

In chemical graph, the vertices represent atoms, and edges refer to the chemical bonds in the underlying chemical structure. A topological index is a numerical value that is computed mathematically from the molecular graph. It is associated with the chemical constitution indicating the correlation of the chemical structure with many physical, chemical properties and biological activities. The exact formulas of topological indices of certain chemical graphs have been computed and plotted in [9,10].

Let $G = (V, E)$ be a graph where V is the vertex set and E is the edge set of G. The degree $\deg(t)$ (or d_t) of v is the number of edges of G incident with t. The length of the shortest path in a graph G is a distance $d(s, t)$ between s and t. A graph can be represented by a polynomial, a numerical value or by matrix form. There are certain types of topological indices, mainly eccentric-based, degree-based, distance-based indices, etc. In this paper, we deal with degree-based topological indices.

The first and oldest degree-based index was introduced by Milan Randić [11] in 1975 and is defined in the following equation.

$$R_{-\frac{1}{2}}(G) = \sum_{st \in E(G)} \frac{1}{\sqrt{d_s d_t}}$$

In 1988, Bollobás et al. [12] and Amic et al. [13] proposed the general Randić index independently. For more details about the Randić index, its properties and important results, see [14,15]. The general Randić index is defined as:

$$R_\alpha(G) = \sum_{st \in E(G)} (d_s d_t)^\alpha$$

The atom bond connectivity index is of vital importance and was introduced by Estrada et al. [16]. It is defined as:

$$ABC(G) = \sum_{st \in E(G)} \sqrt{\frac{d_s + d_t - 2}{d_s d_t}}$$

The geometric arithmetic index GA of a graph G was introduced by Vukičević et al. [17]. It is defined as:

$$GA(G) = \sum_{st \in E(G)} \frac{2\sqrt{d_s d_t}}{d_s + d_t}$$

The first Zagreb index was introduced in 1972 by [18]. Later on, the second Zagreb index was introduced by [19]. The first and second Zagreb indices are formulated as:

$$M_1(G) = \sum_{st \in E(G)} (d_s + d_t)$$

$$M_2(G) = \sum_{st \in E(G)} (d_s d_t)$$

The fourth version of the atom bond connectivity index ABC_4 of a graph G was introduced by Ghorbhani et al. [20]. It is defined as:

$$ABC_4(G) = \sum_{st \in E(G)} \sqrt{\frac{S_s + S_t - 2}{S_s S_t}}$$

where $S_s = \sum\limits_{st \in E(G)} d_t$ and $S_t = \sum\limits_{st \in E(G)} d_s$.

Another molecular descriptor was the fifth version of the geometric arithmetic index GA_5 of a graph G introduced by Graovac et al. [21]. It is defined as:

$$GA_5(G) = \sum_{st \in E(G)} \frac{2\sqrt{S_s S_t}}{S_s + S_t}$$

2. Research Aim

Our aim in this article is to compute the additive topological indices, mainly the atom bond connectivity index, geometric arithmetic index, fourth atom bond connectivity index ABC_4, fifth geometric arithmetic index GA_5 and general Randić index R_α, for $\alpha = \{-1, 1, \frac{1}{2}, -\frac{1}{2}\}$ for $Cu_2O[m, n, t]$ and $TiF_2[m, n, t]$. Moreover, the graphical representation of these exact result is depicted for further explanation of the behavior of these topological indices.

3. Applications of Topological Indices

The atom-bond connectivity (ABC) index provides a very good correlation for the stability of linear alkanes, as well as the branched alkanes and for computing the strain energy of cyclo alkanes [22]. The Randi/'c index is a topological descriptor that has been correlated with many chemical characteristics of molecules and has been found to the parallel to computing the boiling point and Kovats constants of the molecules. To correlate with certain physicochemical properties, the GA index has much better predictive power than the predictive power of the Randić connectivity index [23,24]. The first and second Zagreb index were found to occur for the computation of the total π-electron energy of the molecules within specific approximate expressions [25]. These are among the graph invariants, which were proposed for the measurement of the skeleton of the branching of the carbon-atom [26].

4. Crystallographic Structure of the Molecule Cu_2O

Among various transition metal oxides, Cu_2O has attracted much attention in recent years owing to its distinguished properties and non-toxic nature, low-cost, abundance and simple fabrication process [27]. Nowadays, the promising applications of Cu_2O mainly focus on chemical sensors, solar cells, photocatalysis, lithium-ion batteries and catalysis [28]. The chemical graph of the crystallographic structure of Cu_2O is described in Figures 1 and 2; see details in [29]. Let $G \cong Cu_2O[m, n, t]$ be the chemical graph of Cu_2O with $m \times n$ unit cells in the plane and t layers. We construct this graph first by taking $m \times n$ units in the mn-plane and then storing it up in t layers. The number of vertices and edges of $Cu_2O[m, n, t]$ is $(m + 1)(n + 1)(t + 1) + 5mnt$ and $8mnt$, respectively. In $Cu_2O[m, n, t]$, the number of vertices of degree zero is four; the number of vertices of degree one is $4m + 4n + 4t - 8$; the number of vertices of degree two is $4mnt + 2mn + 2mt + 2nt - 4n - 4m - 4t + 6$; and the number of vertices of degree four is $2nmt - nm - nt - mt + n + m + t - 1$. Furthermore, the edge partition of $Cu_2O[m, n, t]$ based on the degrees of end vertices of each edge is depicted in Table 1.

Table 1. Edge partition of $Cu_2O[m, n, t]$ based on the degrees of end vertices of each edge.

(d_s, d_t)	Frequency	Set of Edges
$(1, 2)$	$4n + 4m + 4t - 8$	E_1
$(2, 2)$	$4nm + 4nt + 4mt - 8n - 8m - 8t + 12$	E_2
$(2, 4)$	$4(2nmt - nm - nt - mt + n + m + t - 1)$	E_3

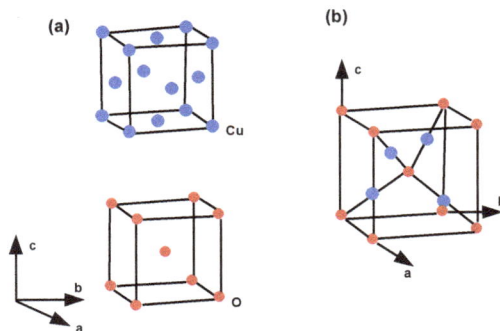

Figure 1. Crystallographic structure of the molecule Cu_2O. (**a**) Structural characteristics of Cu and O atoms in the Cu_2O lattice. The Cu_2O lattice is formed by interpenetrating the Cu and O lattices with each other. (**b**) Unit cell of Cu_2O. Copper atoms are shown as small blue spheres, and oxygen atoms are shown as large red spheres. In the Cu_2O lattice, each Cu atom is coordinated with two O atoms, and each O atom is coordinated with four Cu atoms.

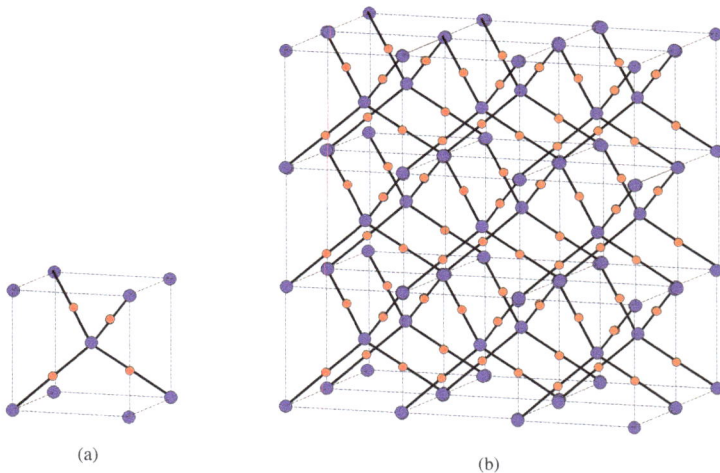

(a)

(b)

Figure 2. (**a**) Unit cell of $Cu_2O[1,1,1]$ (**b**) Crystallographic structure of $Cu_2O[3,2,3]$.

Theorem 1. *Consider the graph of $G \cong Cu_2O[m,n,t]$ with $m,n,t \geq 1$, then its general Randić index is equal to,*

$$R_\alpha G = \begin{cases} 8[8mnt - 2(mn + mt + nt) + m + n + t], & \text{if } \alpha = 1, \\[2mm] \frac{1}{2}(2mnt + mn + mt + nt + m + n + t - 3), & \text{if } \alpha = -1, \\[2mm] 4(4\sqrt{2}mnt + 2(1 - \sqrt{2})(mn + mt + nt) \\ + (3\sqrt{2} - 4)(m + n + t) - 4\sqrt{2} + 6), & \text{if } \alpha = \frac{1}{2}, \\[2mm] 2\sqrt{2}mnt + (2 - \sqrt{2})(mn + mt + nt) \\ + (3\sqrt{2} - 4)(m + n + t) - 5\sqrt{2} + 6, & \text{if } \alpha = -\frac{1}{2}. \end{cases}$$

Proof. Let G be the crystallographic structure of $Cu_2O[m, n, t]$. The general Randić index,

For $\alpha = 1$.

$$
\begin{aligned}
R_1(G) &= \sum_{st \in E(G)} (d_s \times d_t) \\
&= \sum_{st \in E_1(G)} (d_s \times d_t) + \sum_{st \in E_2(G)} (d_s \times d_t) + \sum_{st \in E_3(G)} (d_s \times d_t) \\
&= (4m + 4n + 4t - 8)(1 \times 2) + (4mn + 4mt + 4nt - 8m - 8n - 8t + 12)(2 \times 2) \\
&+ (8mnt - 4mn - 4mt - 4nt + 4m + 4n + 4t - 4)(2 \times 4) \\
&= 8\big[8mnt - 2(mn + mt + nt) + m + n + t\big]
\end{aligned}
$$

For $\alpha = -1$,

$$
\begin{aligned}
R_{-1}(G) &= \sum_{st \in E(G)} \frac{1}{(d_s \times d_t)} \\
&= \sum_{st \in E_1(G)} \frac{1}{(d_s \times d_t)} + \sum_{st \in E_2(G)} \frac{1}{(d_s \times d_t)} + \sum_{st \in E_3(G)} \frac{1}{(d_s \times d_t)} \\
&= (4m + 4n + 4t - 8)\frac{1}{(1 \times 2)} + (4mn + 4mt + 4nt - 8m - 8n - 8t + 12)\frac{1}{(2 \times 2)} \\
&+ (8mnt - 4mn - 4mt - 4nt + 4m + 4n + 4t - 4)\frac{1}{(2 \times 4)} \\
&= \frac{1}{2}(2mnt + mn + mt + nt + m + n + t - 3)
\end{aligned}
$$

For $\alpha = \frac{1}{2}$,

$$
\begin{aligned}
R_{\frac{1}{2}}(G) &= \sum_{st \in E(G)} \sqrt{(d_s \times d_t)} \\
&= \sum_{st \in E_1(G)} \sqrt{(d_s \times d_t)} + \sum_{st \in E_2(G)} \sqrt{(d_s \times d_t)} + \sum_{st \in E_3(G)} \sqrt{(d_s \times d_t)} \\
&= (4m + 4n + 4t - 8)\sqrt{(1 \times 2)} + (4mn + 4mt + 4nt - 8m - 8n - 8t + 12)\sqrt{(2 \times 2)} \\
&+ (8mnt - 4mn - 4mt - 4nt + 4m + 4n + 4t - 4)\sqrt{(2 \times 4)} \\
&= 4\big(4\sqrt{2}mnt + 2(1 - \sqrt{2})(mn + mt + nt) + (3\sqrt{2} - 4)(m + n + t) - 4\sqrt{2} + 6\big)
\end{aligned}
$$

For $\alpha = -\frac{1}{2}$,

$$
\begin{aligned}
R_{-\frac{1}{2}}(G) &= \sum_{st \in E(G)} \frac{1}{\sqrt{(d_s \times d_t)}} \\
&= \sum_{st \in E_1(G)} \frac{1}{\sqrt{(d_s \times d_t)}} + \sum_{st \in E_2(G)} \frac{1}{\sqrt{(d_s \times d_t)}} + \sum_{st \in E_3(G)} \frac{1}{\sqrt{(d_s \times d_t)}} \\
&= (4m + 4n + 4t - 8) \frac{1}{\sqrt{(1 \times 2)}} + (4mn + 4mt + 4nt - 8m - 8n - 8t + 12) \frac{1}{\sqrt{(2 \times 2)}} \\
&\quad + (8mnt - 4mn - 4mt - 4nt + 4m + 4n + 4t - 4) \frac{1}{\sqrt{(2 \times 4)}} \\
&= 2\sqrt{2}mnt + (2 - \sqrt{2})(mn + mt + nt) + (3\sqrt{2} - 4)(m + n + t) - 5\sqrt{2} + 6
\end{aligned}
$$

□

Theorem 2. *Consider the graph of $G \cong Cu_2O[m, n, t]$ with $m, n, t \geq 1$, then its atom bond connectivity index is equal to,*

$$ABC(G) = 4\sqrt{2}mnt.$$

Proof. Let G be the crystallographic structure of $Cu_2O[m, n, t]$. The result for the atom bond connectivity index is as follows:

$$
\begin{aligned}
ABC(G) &= \sum_{st \in E(G)} \sqrt{\frac{d_s + d_t - 2}{d_s d_t}} \\
&= \sum_{st \in E_1(G)} \sqrt{\frac{d_s + d_t - 2}{d_s d_t}} + \sum_{st \in E_2(G)} \sqrt{\frac{d_s + d_t - 2}{d_s d_t}} + \sum_{st \in E_3(G)} \sqrt{\frac{d_s + d_t - 2}{d_s d_t}} \\
&= (4m + 4n + 4t - 8) \sqrt{\frac{1 + 2 - 2}{1 \times 2}} + (4mn + 4mt + 4nt - 8m - 8n - 8t + 12) \sqrt{\frac{2 + 2 - 2}{2 \times 2}} \\
&\quad + (8mnt - 4mn - 4mt - 4nt + 4m + 4n + 4t - 4) \sqrt{\frac{2 + 4 - 2}{2 \times 4}} \\
&= 4\sqrt{2}mnt.
\end{aligned}
$$

□

Theorem 3. *Consider the graph of $G \cong Cu_2O[m, n, t]$ with $m, n, t \geq 1$, then its geometric arithmetic index is equal to,*

$$GA(G) = 4\left[\frac{4\sqrt{2}mnt}{3} - \left(\frac{2\sqrt{2} - 3}{3} \right)(mn + mt + nt - 2m - 2n - 2t) - 2\sqrt{2} + 3 \right].$$

Proof. Let G be the crystallographic structure of $Cu_2O[m, n, t]$. The geometric arithmetic index is computed as below:

$$GA(G) = \sum_{st \in E(G)} \frac{2\sqrt{d_s d_t}}{d_s + d_t}$$

$$= \sum_{st \in E_1(G)} \frac{2\sqrt{d_s d_t}}{d_s + d_t} + \sum_{st \in E_2(G)} \frac{2\sqrt{d_s d_t}}{d_s + d_t} + \sum_{st \in E_3(G)} \frac{2\sqrt{d_s d_t}}{d_s + d_t}$$

$$= (4m + 4n + 4t - 8)\frac{2\sqrt{1 \times 2}}{1 + 2} + (4mn + 4mt + 4nt - 8m - 8n - 8t + 12)\frac{2\sqrt{2 \times 2}}{2 + 2}$$

$$+ (8mnt - 4mn - 4mt - 4nt + 4m + 4n + 4t - 4)\frac{2\sqrt{2 \times 4}}{2 + 4}$$

$$= 4\left[\frac{4\sqrt{2}mnt}{3} - \left(\frac{2\sqrt{2} - 3}{3}\right)(mn + mt + nt - 2m - 2n - 2t) - 2\sqrt{2} + 3\right]$$

□

Theorem 4. *Consider the graph of $G \cong Cu_2O[m, n, t]$ with $m, n, t \geq 1$, then its first and second Zagreb indices are equal to,*

$$M_1(G) = 4(12mnt - 2(mn + mt + nt) + m + n + t)$$

$$M_2(G) = 8(8mnt - 2(mn + mt + nt) + m + n + t).$$

Proof. Let G be the crystallographic structure of $Cu_2O[m, n, t]$. The first Zagreb index is computed as below:

$$M_1(G) = \sum_{st \in E(G)} (d_s + d_t)$$

$$= \sum_{st \in E_1(G)} (d_s + d_t) + \sum_{st \in E_2(G)} (d_s + d_t) + \sum_{st \in E_3(G)} (d_s + d_t)$$

$$= (4m + 4n + 4t - 8)(1 + 2) + (4mn + 4mt + 4nt - 8m - 8n - 8t + 12)(2 + 2)$$

$$+ (8mnt - 4mn - 4mt - 4nt + 4m + 4n + 4t - 4)(2 + 4)$$

$$= 4(12mnt - 2(mn + mt + nt) + m + n + t)$$

The second Zagreb index is computed as below:

$$M_2(G) = \sum_{st \in E(G)} (d_s d_t)$$

$$= \sum_{st \in E_1(G)} (d_s d_t) + \sum_{st \in E_2(G)} (d_s d_t) + \sum_{st \in E_3(G)} (d_s d_t)$$

$$= (4m + 4n + 4t - 8)(1 \times 2) + (4mn + 4mt + 4nt - 8m - 8n - 8t + 12)(2 \times 2)$$

$$+ (8mnt - 4mn - 4mt - 4nt + 4m + 4n + 4t - 4)(2 \times 4)$$

$$= 8(8mnt - 2(mn + mt + nt) + m + n + t)$$

□

Table 2 shows the edge partition of the chemical graph $Cu_2O[m, n, t]$ based on the degree sum of end vertices of each edge.

Table 2. Edge partition of $Cu2O[m, n, t]$ with $m, n, t \geq 2$ based on the degree sum of end vertices of each edge.

(S_s, S_t)	Frequency	Set of Edges
(2, 4)	$4m + 4n + 4t - 8$	E_1
(4, 6)	$4mn + 4mt + 4nt - 8m - 8n - 8t + 12$	E_2
(5, 8)	$4n + 4m + 4t - 8$	E_3
(6, 8)	$4mn + 4mt + 4nt - 8m - 8n - 8t + 12$	E_4
(8, 8)	$8mnt - 8mn - 8mt - 8nt + 8m + 8n + 8t - 8$	E_5

Theorem 5. *Consider the graph $G \cong Cu_2O[m, n, t]$ with $m, n, t \geq 2$, then its fourth atom bond connectivity index is equal to,*

$$ABC_4(G) = \sqrt{14}mnt + \left(\frac{4}{\sqrt{3}} - \sqrt{14} + 2\right)(mn + mt + nt) - 4\sqrt{2} + 4\sqrt{3} - \sqrt{14} - \frac{2\sqrt{110}}{5} + 6$$

$$+ \left(2\sqrt{2} - \frac{8}{\sqrt{3}} + \sqrt{14} + \frac{\sqrt{110}}{5} - 4\right)(m + n + t).$$

Proof. Let G be the crystallographic structure of $Cu_2O[m, n, t]$. The fourth atom bond connectivity index is computed by using Table 2 in the following equation.

$$ABC_4(G) = \sum_{st \in E(G)} \sqrt{\frac{S_s + S_t - 2}{S_s S_t}}$$

$$= \sum_{st \in E_1(G)} \sqrt{\frac{S_s + S_t - 2}{S_s S_t}} + \sum_{st \in E_2(G)} \sqrt{\frac{S_s + S_t - 2}{S_s S_t}} + \sum_{st \in E_3(G)} \sqrt{\frac{S_s + S_t - 2}{S_s S_t}}$$

$$+ \sum_{st \in E_4(G)} \sqrt{\frac{S_s + S_t - 2}{S_s S_t}} + \sum_{st \in E_5(G)} \sqrt{\frac{S_s + S_t - 2}{S_s S_t}}$$

$$ABC_4(G) = (4m + 4n + 4t - 8)\sqrt{\frac{2 + 4 - 2}{2 \times 4}} + (4mn + 4mt + 4nt - 8m - 8n - 8t + 12)\sqrt{\frac{4 + 6 - 2}{4 \times 6}}$$

$$+ (4m + 4n + 4t - 8)\sqrt{\frac{5 + 8 - 2}{5 \times 8}} + (4mn + 4mt + 4nt - 8m - 8n - 8t + 12)\sqrt{\frac{6 + 8 - 2}{6 \times 8}}$$

$$+ (8mnt - 8mn - 8mt - 8nt + 8m + 8n + 8t - 8)\sqrt{\frac{8 + 8 - 2}{8 \times 8}}$$

$$= \sqrt{14}mnt + \left(\frac{4}{\sqrt{3}} - \sqrt{14} + 2\right)(mn + mt + nt) - 4\sqrt{2} + 4\sqrt{3} - \sqrt{14} - \frac{2\sqrt{110}}{5} + 6$$

$$+ \left(2\sqrt{2} - \frac{8}{\sqrt{3}} + \sqrt{14} + \frac{\sqrt{110}}{5} - 4\right)(m + n + t)$$

□

Theorem 6. *Consider the graph $G \cong Cu_2O[m, n, t]$ with $m, n, t \geq 2$, then its fifth geometric arithmetic index is equal to,*

$$GA_5(G) = 8mnt + \left(\frac{16\sqrt{3}}{7} + \frac{8\sqrt{6}}{5} - 8\right)(mn + mt + nt) - \frac{16\sqrt{2}}{3} + \frac{48\sqrt{3}}{7} + \frac{24\sqrt{6}}{5} - \frac{32\sqrt{10}}{13} - 8$$

$$+ \left(\frac{8\sqrt{2}}{3} - \frac{32\sqrt{3}}{7} - \frac{16\sqrt{6}}{5} + \frac{16\sqrt{10}}{13} + 8\right)(m + n + t)$$

Proof. Let G be the crystallographic structure of $Cu_2O[m, n, t]$. The fifth geometric arithmetic index is computed as below:

$$GA_5(G) = \sum_{st \in E(G)} \frac{2\sqrt{S_s S_t}}{S_s + S_t}$$

$$= \sum_{st \in E_1(G)} \frac{2\sqrt{S_s S_t}}{S_s + S_t} + \sum_{st \in E_2(G)} \frac{2\sqrt{S_s S_t}}{S_s + S_t} + \sum_{st \in E_3(G)} \frac{2\sqrt{S_s S_t}}{S_s + S_t}$$

$$+ \sum_{st \in E_4(G)} \frac{2\sqrt{S_s S_t}}{S_s + S_t} + \sum_{st \in E_5(G)} \frac{2\sqrt{S_s S_t}}{S_s + S_t}$$

$$= (4m + 4n + 4t - 8)\frac{2\sqrt{2 \times 4}}{2 + 4} + (4mn + 4mt + 4nt - 8m - 8n - 8t + 12)\frac{2\sqrt{4 \times 6}}{4 + 6}$$

$$+ (4m + 4n + 4t - 8)\frac{2\sqrt{5 \times 8}}{5 + 8} + (4mn + 4mt + 4nt - 8m - 8n - 8t + 12)\frac{2\sqrt{6 \times 8}}{6 + 8}$$

$$+ (8mnt - 8mn - 8mt - 8nt + 8m + 8n + 8t - 8)\frac{2\sqrt{8 \times 8}}{8 + 8}$$

$$= 8mnt + \left(\frac{16\sqrt{3}}{7} + \frac{8\sqrt{6}}{5} - 8\right)(mn + mt + nt) - \frac{16\sqrt{2}}{3} + \frac{48\sqrt{3}}{7} + \frac{24\sqrt{6}}{5} - \frac{32\sqrt{10}}{13} - 8$$

$$+ \left(\frac{8\sqrt{2}}{3} - \frac{32\sqrt{3}}{7} - \frac{16\sqrt{6}}{5} + \frac{16\sqrt{10}}{13} + 8\right)(m + n + t)$$

□

5. Crystal Structure of Titanium Difluoride

Titanium difluoride is a water-insoluble titanium source for use in oxygen-sensitive applications, such as metal production. Fluoride compounds have diverse applications in current technologies and science, from oil refining and etching to synthetic organic chemistry and the manufacture of pharmaceuticals. The chemical graph of the crystal structure of titanium difluoride $TiF_2[m, n, t]$ is described in Figure 3; for more details, see [30]. Let $G \cong TiF_2[m, n, t]$ be the chemical graph of TiF_2 with $m \times n$ unit cells in the plane and t layers. We construct this graph first by taking $m \times n$ units in the mn-plane and then storing it up in t layers. The number of vertices and edges of $TiF_2[m, n, t]$ is $12mnt + 2mn + 2mt + 2nt + m + n + t + 1$ and $32mnt$, respectively. In $TiF_2[m, n, t]$, the number of vertices of degree one is eight; the number of vertices of degree two is $4m + 4n + 4t - 12$; the number of vertices of degree four is $8mnt + 4mn + 4mt + 4nt - 4n - 4m - 4t + 6$; and the number of vertices of degree eight is $4mnt - 2(mn + mt + nt) + m + n + t - 1$. The edge partition of $TiF_2[m, n, t]$ based on the degrees of end vertices of each edge is depicted in Table 3.

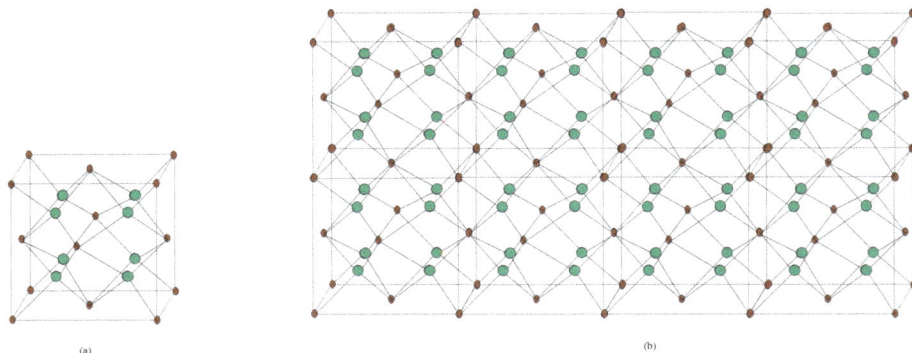

Figure 3. (a)The unit cell of of $TiF_2[m, n, t]$ with Ti atoms in red and F atoms in green; (b) the crystal structure of $TiF_2[4, 1, 2]$.

Table 3. Edge partition of $TiF_2[m, n, t]$ based on the degrees of end vertices of each edge.

(d_s, d_t)	Frequency	Set of Edges
$(1, 4)$	8	E_1
$(2, 4)$	$8(m + n + t - 3)$	E_2
$(4, 4)$	$16(mn + mt + nt) - 16(m + n + t) + 24$	E_3
$(4, 8)$	$32mnt - 16(mt + mn + nt) + 8(m + n + t) - 8$	E_4

Theorem 7. *Consider the graph $G \cong TiF_2[m, n, t]$ with $m, n, t \geq 1$, then its general Randić index is equal to,*

$$R_\alpha G = \begin{cases} 32\left[32mnt - 8(mn + mt + nt) + 2(m + n + t) - 1\right], & \text{if } \alpha = 1, \\[2mm] \frac{1}{4}(4mnt + 2(mn + mt + nt) + m + n + t + 1), & \text{if } \alpha = -1, \\[2mm] 16(8\sqrt{2}mnt + 4(1 - \sqrt{2})(mn + mt + nt) \\ +(3\sqrt{2} - 4)(m + n + t) - 5\sqrt{2} + 7), & \text{if } \alpha = \frac{1}{2}, \\[2mm] 4\sqrt{2}mnt + 2(2 - \sqrt{2})(mn + mt + nt) \\ +(3\sqrt{2} - 4)(m + n + t) - 7\sqrt{2} + 10, & \text{if } \alpha = -\frac{1}{2}. \end{cases}$$

Proof. Let $G \cong TiF_2[m, n, t]$ be the crystal structure of titanium difluoride. The general Randić index,

For $\alpha = 1$.

$$\begin{aligned} R_1(G) &= \sum_{st \in E(G)} (d_s \times d_t) \\ &= \sum_{st \in E_1(G)} (d_s \times d_t) + \sum_{st \in E_2(G)} (d_s \times d_t) + \sum_{st \in E_3(G)} (d_s \times d_t) + \sum_{st \in E_4(G)} (d_s \times d_t) \\ &= (8)(1 \times 4) + (8m + 8n + 8t - 24)(2 \times 4) \\ &\quad + (16(mn + mt + nt) - 16(m + n + t) + 24)(4 \times 4) \\ &\quad + (32mnt - 16(mn + mt + nt) + 8(m + n + t) - 8)(4 \times 8) \\ &= 32\left[32mnt - 8(mn + mt + nt) + 2(m + n + t) - 1\right] \end{aligned}$$

For $\alpha = -1$,

$$
\begin{aligned}
R_{-1}(G) &= \sum_{st \in E(G)} \frac{1}{(d_s \times d_t)} \\
&= \sum_{st \in E_1(G)} \frac{1}{(d_s \times d_t)} + \sum_{st \in E_2(G)} \frac{1}{(d_s \times d_t)} + \sum_{st \in E_3(G)} \frac{1}{(d_s \times d_t)} + \sum_{st \in E_4(G)} \frac{1}{(d_s \times d_t)}
\end{aligned}
$$

$$
\begin{aligned}
R_{-1}(G) &= (8)\frac{1}{(1 \times 4)} + (8m + 8n + 8t - 24)\frac{1}{(2 \times 4)} \\
&\quad + \big(16(mn + mt + nt) - 16(m + n + t) + 24\big)\frac{1}{(4 \times 4)} \\
&\quad + \big(32mnt - 16(mn + mt + nt) + 8(m + n + t) - 8\big)\frac{1}{(4 \times 8)} \\
&= \frac{1}{4}(4mnt + 2(mn + mt + nt) + m + n + t + 1)
\end{aligned}
$$

For $\alpha = \frac{1}{2}$,

$$
\begin{aligned}
R_{\frac{1}{2}}(G) &= \sum_{st \in E(G)} \sqrt{(d_s \times d_t)} \\
&= \sum_{st \in E_1(G)} \sqrt{(d_s \times d_t)} + \sum_{st \in E_2(G)} \sqrt{(d_s \times d_t)} + \sum_{st \in E_3(G)} \sqrt{(d_s \times d_t)} + \sum_{st \in E_4(G)} \sqrt{(d_s \times d_t)} \\
&= (8)\sqrt{(1 \times 4)} + (8m + 8n + 8t - 24)\sqrt{(2 \times 4)} \\
&\quad + \big(16(mn + mt + nt) - 16(m + n + t) + 24\big)\sqrt{(4 \times 4)} \\
&\quad + \big(32mnt - 16(mn + mt + nt) + 8(m + n + t) - 8\big)\sqrt{(4 \times 8)} \\
&= 16(8\sqrt{2}mnt + 4(1 - \sqrt{2})(mn + mt + nt) + (3\sqrt{2} - 4)(m + n + t) - 5\sqrt{2} + 7)
\end{aligned}
$$

For $\alpha = -\frac{1}{2}$,

$$
\begin{aligned}
R_{-\frac{1}{2}}(G) &= \sum_{st \in E(G)} \frac{1}{\sqrt{(d_s \times d_t)}} \\
&= \sum_{st \in E_1(G)} \frac{1}{\sqrt{(d_s \times d_t)}} + \sum_{st \in E_2(G)} \frac{1}{\sqrt{(d_s \times d_t)}} + \sum_{st \in E_3(G)} \frac{1}{\sqrt{(d_s \times d_t)}} + \sum_{st \in E_4(G)} \frac{1}{\sqrt{(d_s \times d_t)}} \\
&= (8)\frac{1}{\sqrt{(1 \times 4)}} + (8m + 8n + 8t - 24)\frac{1}{\sqrt{(2 \times 4)}} \\
&\quad + \big(16(mn + mt + nt) - 16(m + n + t) + 24\big)\frac{1}{\sqrt{(4 \times 4)}} \\
&\quad + \big(32mnt - 16(mn + mt + nt) + 8(m + n + t) - 8\big)\frac{1}{\sqrt{(4 \times 8)}} \\
&= 4\sqrt{2}mnt + 2(2 - \sqrt{2})(mn + mt + nt) + (3\sqrt{2} - 4)(m + n + t) - 7\sqrt{2} + 10
\end{aligned}
$$

□

Theorem 8. *Consider the graph* $G \cong TiF_2[m, n, t]$ *with* $m, n, t \geq 1$, *then its atom bond connectivity index is equal to,*

$$
\begin{aligned}
ABC(G) &= 2\left[4\sqrt{5}mnt - 2(\sqrt{5} - \sqrt{6})(mn + mt + nt) + (2\sqrt{2} + \sqrt{5} - 2\sqrt{6})(m + n + t)\right] \\
&+ 2\left[-6\sqrt{2} + 2\sqrt{3} - \sqrt{5} + 3\sqrt{6}\right]
\end{aligned}
$$

Proof. Let $G \cong TiF_2[m, n, t]$ be the crystal structure of titanium difluoride. The atom bond connectivity index can be calculated by using Table 3 in the following equation.

$$
\begin{aligned}
ABC(G) &= \sum_{st \in E(G)} \sqrt{\frac{d_s + d_t - 2}{d_s d_t}} \\
&= \sum_{st \in E_1(G)} \sqrt{\frac{d_s + d_t - 2}{d_s d_t}} + \sum_{st \in E_2(G)} \sqrt{\frac{d_s + d_t - 2}{d_s d_t}} + \sum_{st \in E_3(G)} \sqrt{\frac{d_s + d_t - 2}{d_s d_t}} + \sum_{st \in E_4(G)} \sqrt{\frac{d_s + d_t - 2}{d_s d_t}} \\
&= (8)\sqrt{\frac{1 + 4 - 2}{1 \times 4}} + (8m + 8n + 8t - 24)\sqrt{\frac{2 + 4 - 2}{2 \times 4}} \\
&+ (16(mn + mt + nt) - 16(m + n + t) + 24)\sqrt{\frac{4 + 4 - 2}{4 \times 4}} \\
&+ (32mnt - 16(mn + mt + nt) + 8(m + n + t) - 8)\sqrt{\frac{4 + 8 - 2}{4 \times 8}} \\
&= 2\left[4\sqrt{5}mnt - 2(\sqrt{5} - \sqrt{6})(mn + mt + nt) + (2\sqrt{2} + \sqrt{5} - 2\sqrt{6})(m + n + t)\right] \\
&+ 2\left[-6\sqrt{2} + 2\sqrt{3} - \sqrt{5} + 3\sqrt{6}\right]
\end{aligned}
$$

□

Theorem 9. *Consider the graph* $G \cong TiF_2[m, n, t]$ *with* $m, n, t \geq 1$, *then its geometric arithmetic index is equal to,*

$$
GA(G) = 8\left[\frac{8\sqrt{2}(mnt - 1)}{3} - \left(\frac{4\sqrt{2}}{3} - 2\right)(mn + mt + nt - m - n - t) + \frac{19}{5}\right]
$$

Proof. Let $G \cong TiF_2[m, n, t]$ be the crystal structure of titanium difluoride. The geometric arithmetic index is computed as below:

$$
\begin{aligned}
GA(G) &= \sum_{st \in E(G)} \frac{2\sqrt{d_s d_t}}{d_s + d_t} \\
&= \sum_{st \in E_1(G)} \frac{2\sqrt{d_s d_t}}{d_s + d_t} + \sum_{st \in E_2(G)} \frac{2\sqrt{d_s d_t}}{d_s + d_t} + \sum_{st \in E_3(G)} \frac{2\sqrt{d_s d_t}}{d_s + d_t} + \sum_{st \in E_4(G)} \frac{2\sqrt{d_s d_t}}{d_s + d_t} \\
&= (8)\frac{2\sqrt{1 \times 4}}{1 + 4} + (8m + 8n + 8t - 24)\frac{2\sqrt{2 \times 4}}{2 + 4} \\
&+ (16(mn + mt + nt) - 16(m + n + t) + 24)\frac{2\sqrt{4 \times 4}}{4 + 4} \\
&+ (32mnt - 16(mn + mt + nt) + 8(m + n + t) - 8)\frac{2\sqrt{4 \times 8}}{4 + 8} \\
&= 8\left[\frac{8\sqrt{2}(mnt - 1)}{3} - \left(\frac{4\sqrt{2}}{3} - 2\right)(mn + mt + nt - m - n - t) + \frac{19}{5}\right]
\end{aligned}
$$

□

Theorem 10. *Consider the graph* $G \cong TiF_2[m, n, t]$ *with* $m, n, t \geq 1$, *then its first and second Zagreb indices are equal to,*

$$M_1(G) = 8\big[48mnt - 8(mn + mt + nt) + 2(m + n + t) - 1\big],$$

$$M_2(G) = 32\big[32mnt - 8(mn + mt + nt) + 2(m + n + t) - 1\big].$$

Proof. Let $G \cong TiF_2[m, n, t]$ be the crystal structure of titanium difluoride. The first and second Zagreb indices are computed as below:

$$
\begin{aligned}
M_1(G) &= \sum_{st \in E(G)} (d_s + d_t) \\
&= \sum_{st \in E_1(G)} (d_s + d_t) + \sum_{st \in E_2(G)} (d_s + d_t) + \sum_{st \in E_3(G)} (d_s + d_t) + \sum_{st \in E_4(G)} (d_s + d_t) \\
&= (8)(1 + 4) + (8m + 8n + 8t - 24)(2 + 4) + (16(mn + mt + nt) - 16(m + n + t) + 24)(4 + 4) \\
&\quad + (32mnt - 16(mn + mt + nt) + 8(m + n + t) - 8)(4 + 8) \\
&= 8\big(48mnt - 8(mn + mt + nt) + 2(m + n + t) - 1\big)
\end{aligned}
$$

$$
\begin{aligned}
M_2(G) &= \sum_{st \in E(G)} (d_s d_t) \\
&= \sum_{st \in E_1(G)} (d_s d_t) + \sum_{st \in E_2(G)} (d_s d_t) + \sum_{st \in E_3(G)} (d_s d_t) + \sum_{st \in E_4(G)} (d_s d_t) \\
&= (8)(1 \times 4) + (8m + 8n + 8t - 24)(2 \times 4) + (16(mn + mt + nt) - 16(m + n + t) + 24)(4 \times 4) \\
&\quad + (32mnt - 16(mn + mt + nt) + 8(m + n + t) - 8)(4 \times 8) \\
&= 32\big[32mnt - 8(mn + mt + nt) + 2(m + n + t) - 1\big]
\end{aligned}
$$

□

Table 4 shows the edge partition of the chemical graph $TiF_2[m, n, t]$ based on the degree sum of the end vertices of each edge.

Table 4. Edge partition of $TiF_2[m, n, t]$, $m, n, s \geq 2$ based on the degree sum of the end vertices of each edge.

(S_s, S_t)	Frequency	Set of Edges
(4, 13)	8	E_1
(8, 18)	$8(m + n + t - 3)$	E_2
(13, 16)	16	E_3
(16, 18)	$16(mn + mt + nt) - 16(m + n + t) + 8$	E_4
(16, 24)	$32mnt - 16(mn + mt + nt) + 8$	E_5
(18, 32)	$8(m + n + t - 2)$	E_6

Theorem 11. *Consider the graph* $G \cong TiF_2[m, n, t]$ *with* $m, n, t > 1$, *then its fourth atom bond connectivity index is equal to,*

$$
\begin{aligned}
ABC_4(G) &= \frac{4\sqrt{57}mnt}{3} - \left(\frac{2\sqrt{57}}{3} - \frac{16}{3}\right)(mn + mt + nt) + \left(\frac{4}{\sqrt{3}} + \frac{4\sqrt{6}}{3} - \frac{16}{3}\right)(m + n + t) \\
&\quad - 4\sqrt{6} - \frac{8}{\sqrt{3}} + \frac{12\sqrt{39}}{13} + \frac{\sqrt{57}}{3} + \frac{4\sqrt{195}}{13} + \frac{8}{3}
\end{aligned}
$$

Proof. Let $G \cong TiF_2[m, n, t]$ be the crystal structure of titanium difluoride. The fourth atom bond connectivity index is computed as below:

$$ABC_4(G) = \sum_{st \in E(G)} \sqrt{\frac{S_s + S_t - 2}{S_s S_t}}$$

$$= \sum_{st \in E_1(G)} \sqrt{\frac{S_s + S_t - 2}{S_s S_t}} + \sum_{st \in E_2(G)} \sqrt{\frac{S_s + S_t - 2}{S_s S_t}} + \sum_{st \in E_3(G)} \sqrt{\frac{S_s + S_t - 2}{S_s S_t}}$$

$$+ \sum_{st \in E_4(G)} \sqrt{\frac{S_s + S_t - 2}{S_s S_t}} + \sum_{st \in E_5(G)} \sqrt{\frac{S_s + S_t - 2}{S_s S_t}} + \sum_{st \in E_6(G)} \sqrt{\frac{S_s + S_t - 2}{S_s S_t}}$$

$$ABC_4(G) = (8)\sqrt{\frac{4 + 13 - 2}{4 \times 13}} + (8m + 8n + 8t - 24)\sqrt{\frac{8 + 18 - 2}{8 \times 18}} + (16)\sqrt{\frac{13 + 16 - 2}{13 \times 16}}$$

$$+ (16(mn + mt + nt) - 16(m + n + t) + 8)\sqrt{\frac{16 + 18 - 2}{16 \times 18}}$$

$$+ (32mnt - 16(mn + mt + nt) + 8)\sqrt{\frac{16 + 24 - 2}{16 \times 24}} + (8m + 8n + 8t - 16)\sqrt{\frac{18 + 32 - 2}{18 \times 32}}$$

$$= \frac{4\sqrt{57}mnt}{3} - \left(\frac{2\sqrt{57}}{3} - \frac{16}{3}\right)(mn + mt + nt) + \left(\frac{4}{\sqrt{3}} + \frac{4\sqrt{6}}{3} - \frac{16}{3}\right)(m + n + t)$$

$$- 4\sqrt{6} - \frac{8}{\sqrt{3}} + \frac{12\sqrt{39}}{13} + \frac{\sqrt{57}}{3} + \frac{4\sqrt{195}}{13} + \frac{8}{3}$$

$\square$

Theorem 12. *Consider the graph $G \cong TiF_2[m, n, t]$ with $m, n, t \geq 2$, then its fifth geometric arithmetic index is equal to,*

$$GA_5(G) = \frac{64\sqrt{6}mnt}{5} + \left(\frac{192\sqrt{2}}{17} - \frac{32\sqrt{6}}{5}\right)(mn + mt + nt) - \left(\frac{192\sqrt{2}}{17} - \frac{4896}{325}\right)(m + n + t)$$

$$+ \frac{96\sqrt{2}}{17} + \frac{16\sqrt{6}}{5} + \frac{3104\sqrt{13}}{493} - \frac{12192}{325}$$

Proof. Let $G \cong TiF_2[m, n, t]$ be the crystal structure of titanium difluoride. The fifth geometric arithmetic index is computed as below:

$$GA_5(G) = \sum_{st \in E(G)} \frac{2\sqrt{S_s S_t}}{S_s + S_t}$$

$$= \sum_{st \in E_1(G)} \frac{2\sqrt{S_s S_t}}{S_s + S_t} + \sum_{st \in E_2(G)} \frac{2\sqrt{S_s S_t}}{S_s + S_t} + \sum_{st \in E_3(G)} \frac{2\sqrt{S_s S_t}}{S_s + S_t}$$

$$+ \sum_{st \in E_4(G)} \frac{2\sqrt{S_s S_t}}{S_s + S_t} + \sum_{st \in E_5(G)} \frac{2\sqrt{S_s S_t}}{S_s + S_t} + \sum_{st \in E_6(G)} \frac{2\sqrt{S_s S_t}}{S_s + S_t}$$

$$= (8)\frac{2\sqrt{4 \times 13}}{4 + 13} + (8m + 8n + 8t - 24)\frac{2\sqrt{8 \times 18}}{8 + 18} + (16)\frac{2\sqrt{13 \times 16}}{13 + 16}$$

$$+ (16(mn + mt + nt) - 16(m + n + t) + 8)\frac{2\sqrt{16 \times 18}}{16 + 18}$$

$$+ (32mnt - 16(mn + mt + nt) + 8)\frac{2\sqrt{16 \times 24}}{16 + 24} + (8m + 8n + 8t - 16)\frac{2\sqrt{18 \times 32}}{18 + 32}$$

$$= \frac{64\sqrt{6}mnt}{5} + \left(\frac{192\sqrt{2}}{17} - \frac{32\sqrt{6}}{5}\right)(mn + mt + nt) - \left(\frac{192\sqrt{2}}{17} - \frac{4896}{325}\right)(m + n + t)$$

$$+ \frac{96\sqrt{2}}{17} + \frac{16\sqrt{6}}{5} + \frac{3104\sqrt{13}}{493} - \frac{12192}{325}$$

$\square$

6. Discussion

Since the topological indices have many applications in different branches of science, namely pharmaceutical, chemistry and biological drugs, the graphical representation of these calculated results is helpful to scientists. The graphical representations of topological indices for $Cu_2O[m, n, t]$ are depicted for Randić indices in Figures 4 and 5. The atomic bond connectivity index and geometric arithmetic index for $Cu_2O[m, n, t]$ are depicted in Figure 6. The first and second Zagreb indices for $Cu_2O[m, n, t]$ are depicted in Figure 7. The fourth atomic bond connectivity index and the fifth geometric arithmetic index for $Cu_2O[m, n, t]$ are depicted in Figure 8.

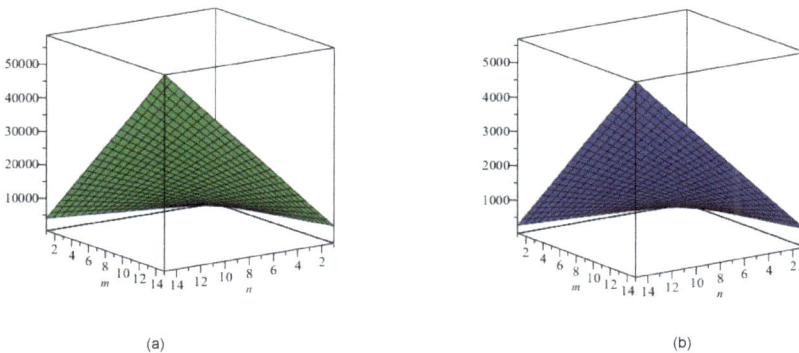

(a)   (b)

Figure 4. The graphical representation of the Randić index for (a) $\alpha = 1$ and (b) for $\alpha = -1$.

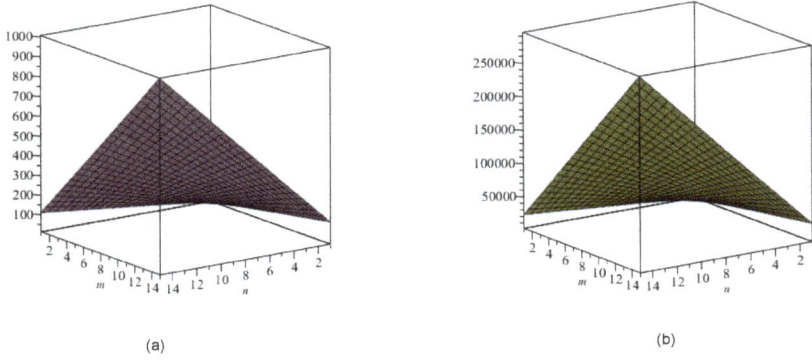

(a)

(b)

Figure 5. The graphical representation of the Randić index for (**a**) $\alpha = \frac{1}{2}$ and (**b**) for $\alpha = \frac{-1}{2}$.

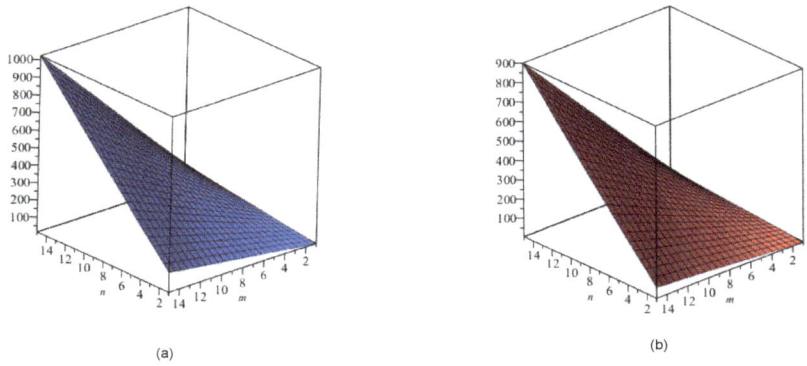

(a)

(b)

Figure 6. The graphical representation of the (**a**) ABC index and (**b**) GA index.

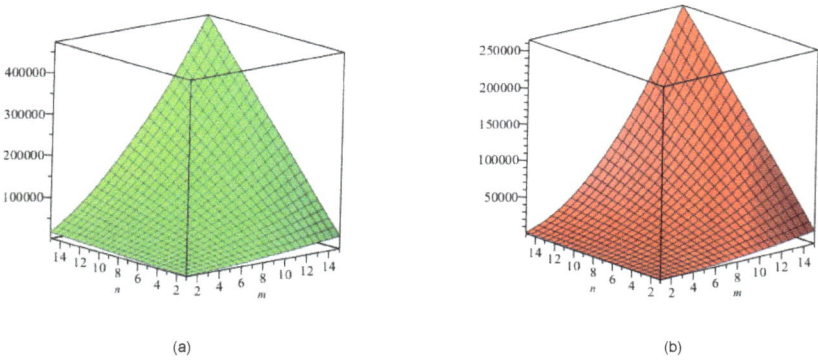

(a)

(b)

Figure 7. The graphical representation of the (**a**) first Zagreb index and (**b**) second Zagreb index.

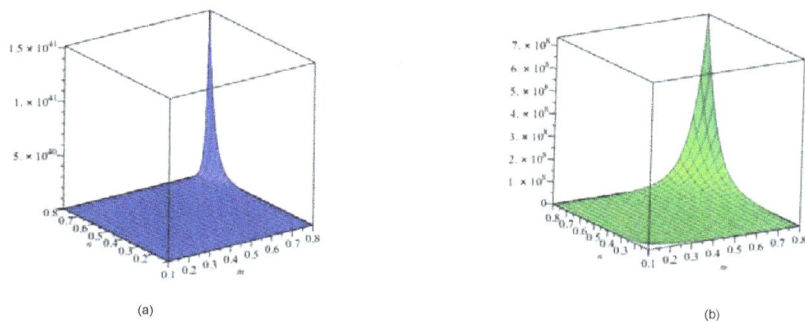

(a)   (b)

Figure 8. The graphical representation of the (a) ABC_4 index and (b) GA_5 index.

The graphical representations of topological indices for titanium difluoride TiF_2 are depicted for Randić indices in Figures 9 and 10. The atomic bond connectivity index and geometric arithmetic index for titanium difluoride TiF_2 are depicted in Figure 11. The first and second Zagreb indices for titanium difluoride TiF_2 are depicted in Figure 12. The fourth atomic bond connectivity index and the fifth geometric arithmetic index for titanium difluoride TiF_2 are depicted in Figure 13.

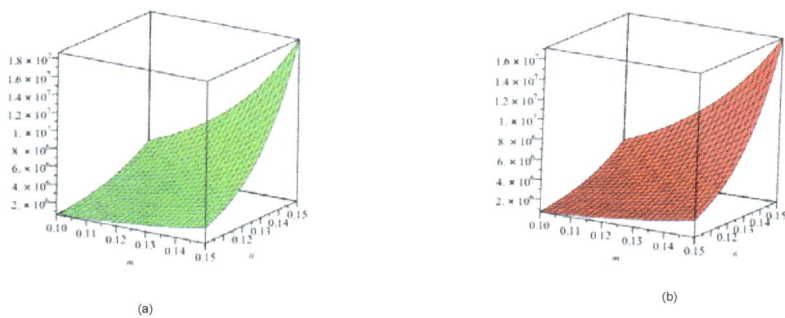

(a)   (b)

Figure 9. The graphical representation of the Randić index for (a) $\alpha = 1$ and (b) for $\alpha = -1$.

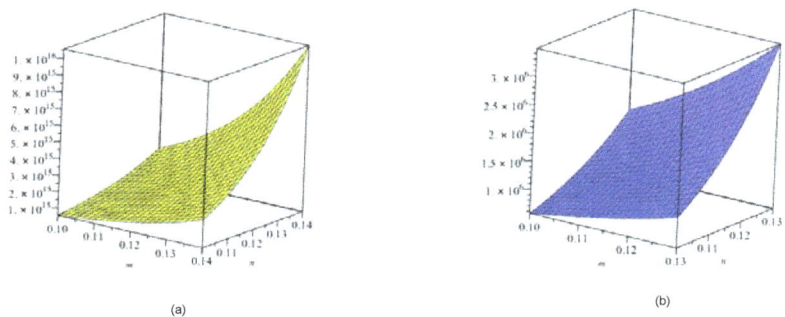

(a)   (b)

Figure 10. The graphical representation of the Randić index for (a) $\alpha = \frac{1}{2}$ and (b) for $\alpha = \frac{-1}{2}$.

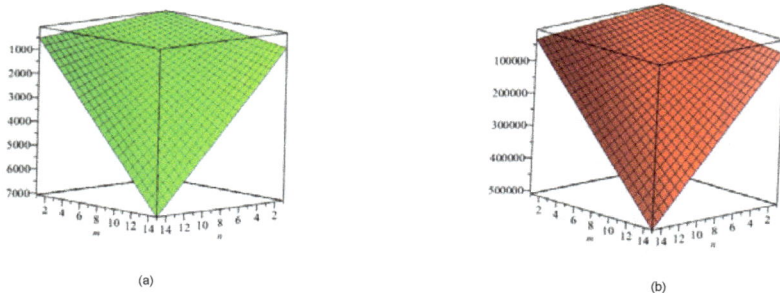

(a)

(b)

Figure 11. The graphical representation of the (**a**) ABC index and (**b**) GA index.

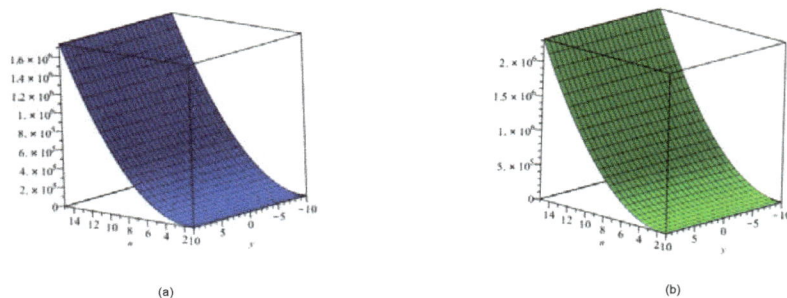

(a)

(b)

Figure 12. The graphical representation of the (**a**) first Zagreb index and (**b**) second Zagreb index.

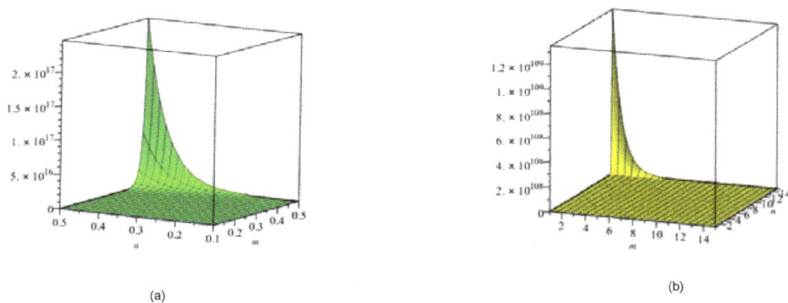

(a)

(b)

Figure 13. The graphical representation of the (**a**) ABC_4 index and (**b**) GA_5 index.

7. Conclusions

In this paper, we have computed some degree-based topological indices, namely the atom bond connectivity index ABC, the geometric arithmetic index GA, the general Randić index, the GA_5 index, the ABC_4 index and the first and second Zagreb indices for the chemical graph of the crystal structure of titanium difluoride TiF_2 and crystallographic structure of cuprite Cu_2O.

In the future, we are interested in computing the distance-based and counting-related topological indices for these structures.

Author Contributions: J.-B.L. contribute for conceptualization, designing the experiments, funding and analyzed the data curation. M.K.S. and A.Q.B. contribute for supervision, methodology, validation, project administration and formal analysing. M.A.Z. and M.N. contribute for performed experiments, resources, software, some computations and wrote the initial draft of the paper which were investigated and approved by M.K.S. and A.Q.B. and wrote the final draft. All authors read and approved the final version of the paper.

Funding: This research was partially supported by Doctoral Science Foundation of Anhui Jianzhu University under grant No. 2016QD116.

Acknowledgments: The authors are grateful to the anonymous referees for their valuable comments and suggestions that improved this paper.

Conflicts of Interest: The authors declare no conflict of interest.

References

1. Farahani, M.R. Computing fourth atom bond connectivity index of V-phenylenic nanotubes and nanotori. *Acta Chim. Slov.* **2013**, *60*, 429–432. [PubMed]
2. Imran, M.; Ali, M.A.; Ahmad, S.; Siddiqui, M.K.; Baig, A.Q. Topological characterization of the symmetrical structure of bismuth tri-iodide. *Symmetry* **2018**, *10*, 201. [CrossRef]
3. Gao, W.; Siddiqui, M.K. Molecular descriptors of nanotube, oxide, silicate, and triangulene networks. *J. Chem.* **2017**, *2017*, 6540754. [CrossRef]
4. Wu, W.; Zhang, C.; Lin, W.; Chen, Q.; Guo, X.; Qian, Y. Quantitative Structure-Property Relationship (QSPR) Modeling of Drug-Loaded Polymeric Micelles via Genetic Function Approximation. *PLoS ONE* **2015**, *10*, 0119575. [CrossRef] [PubMed]
5. Hayat, S.; Malik, M.A.; Imran, M. Computing topological indices of honeycomb derived networks. *Roman. J. Inf. Sci. Technol.* **2015**, *18*, 144–165.
6. Bie, R.J.; Siddiqui, M.K.; Razavi, R.; Taherkhani, M.; Najaf, M. Possibility of C_{38} and $Si_{19}Ge_{19}$ nanocages in anode of metal ion batteries: computational examination. *Acta Chim. Slov.* **2018**, *65*, 303–311. [CrossRef]
7. Holgate, S.A. *Understanding Solid State Physics*; CRC Press: Boca Raton, FL, USA, 2009; pp. 177–178, ISBN 1420012320.
8. Van, Z.B. Section 2.3: Energy Bands. Principles of Semiconductor Devices. Ph.D Thesis, Electrical, Computer, Energy Engineering Department, University of Colorado, Boulder, CO, USA, 13 March 2017.
9. Siddiqui, M.K.; Imran, M.; Ahmad, A. On Zagreb indices, Zagreb polynomials of some nanostar dendrimers. *Appl. Math. Comput.* **2016**, *280*, 132–139. [CrossRef]
10. Siddiqui, M.K.; Naeem, M.; Rahman, N.A.; Imran, M. Computing topological indicesof certain networks. *J. Optoelectron. Adv. Mater.* **2016**, *18*, 884–892.
11. Randić, M. On characterization of molecular branching. *J. Am. Chem. Soc.* **1975**, *97*, 6609–6615. [CrossRef]
12. Bollobás, B.; Erdös, P. Graphs of extremal weights. *Ars Comb.* **1998**, *50*, 225–233. [CrossRef]
13. Amic, D.; Beslo, D.; Lucic, B.; Nikolic, S.; Trinajstić, N. The vertex-connectivity index revisited. *J. Chem. Inf. Comput. Sci.* **1998**, *38*, 819–822. [CrossRef]
14. Shao, Z.; Siddiqui, M.K.; Muhammad, M.H. Computing zagreb indices and zagreb polynomials for symmetrical nanotubes. *Symmetry* **2018**, *10*, 244. [CrossRef]
15. Li, X.; Gutman, I. *Mathematical Aspects of Randić type Molecular Structure Descriptors*; Mathematical Chemistry Monographs No. 1: Kragujevac, Serbia, 2006; Volume 330.
16. Estrada, E.; Torres, L.; Rodríguez, L.; Gutman, I. An atom-bond connectivity inde. Modelling the enthalpy of formation of alkanes. *Indian J. Chem.* **1998**, *37A*, 849–855.
17. Vukičević, D.; Furtula, B. Topological index based on the ratios of geometrical and arithmetical means of end-vertex degrees of edges. *J. Math. Chem.* **2009**, *46*, 1369–1376.
18. Gutman, I.; Trinajstć, N. Graph theory and molecular orbitals, Total π-electron energy of alternant hydrocarbons. *Chem. Phys. Lett.* **1972**, *17*, 535–538. [CrossRef]
19. Gutman, I.; Das, K.C. The first Zagreb index 30 years after. *MATCH Commun. Math. Comput. Chem.* **2004**, *50*, 83–92.
20. Ghorbani, A.; Hosseinzadeh, M.A. Computing ABC_4 index of nanostar dendrimers. *Optoelectron. Adv. Mater.-Rapid Commun.* **2010**, *4*, 1419–1422.
21. Graovac, A.; Ghorbani, M.; Hosseinzadeh, M.A. Computing fifth geometric–arithmetic index for nanostar dendrimers. *J. Math. Nanosci.* **2011**, *1*, 33–42.

22. Gao, W.; Siddiqui, M.K.; Naeem, M.; Rehman, N.A. Topological Characterization of Carbon Graphite and Crystal Cubic Carbon Structures. *Molecules* **2017**, *22*, 1496. [CrossRef] [PubMed]
23. Shao, Z.; Wu, P.; Zhang, X.; Dimitrov, D.; Liu, J. On the maximum ABC index of graphs with prescribed size and without pendent vertices. *IEEE Access* **2018**, *6*, 27604–27616. [CrossRef]
24. Shao, Z.; Wu, P.; Gao, Y.; Gutman, I.; Zhang, X. On the maximum ABC index of graphs without pendent vertices. *Appl. Math. Comput.* **2017**, *315*, 298–312. [CrossRef]
25. Imran, M.; Siddiqui, M.K.; Naeem, M.; Iqbal, M.A. On Topological Properties of Symmetric Chemical Structures. *Symmetry* **2018**, *10*, 173. [CrossRef]
26. Gao, W.; Siddiqui, M.K.; Imran, M.; Jamil, M.K.; Farahani, M.R. Forgotten Topological Index of Chemical Structure in Drugs. *Saudi Pharm. J.* **2016**, *24*, 258–267. [CrossRef] [PubMed]
27. Chen, K.; Sun, C.; Song, S.; Xue, D. Polymorphic crystallization of Cu_2O compound. *CrystEngComm* **2014**, *16*, 52–57. [CrossRef]
28. Yuhas, B.D.; Yang, P. Nanowire-Based All-Oxide Solar Cells. *J. Am. Chem. Soc.* **2009**, *131*, 3756–3761. [CrossRef] [PubMed]
29. Zhang, J.; Liu, J.; Peng, Q.; Wang, X.; Li, Y. Nearly Monodisperse Cu_2O and CuO Nanospheres: Preparation and Applications for Sensitive Gas Sensors. *Chem. Mater.* **2006**, *18*, 867–871. [CrossRef]
30. Cotton, F.A.; Wilkinson, G.; Murillo, C.A.; Bochmann, M. *Advanced Inorganic Chemistry*; John Wiley and Sons: Hoboken, NJ, USA, 1999.

symmetry

MDPI

Article

Hyperbolicity of Direct Products of Graphs

Walter Carballosa [1,2], **Amauris de la Cruz** [3], **Alvaro Martínez-Pérez** [4] (iD)
and **José M. Rodríguez** [3,*] (iD)

[1] Department of Mathematics and Statistics, Florida International University, 11200 SW 8th Street,
    Miami, FL 33199, USA; waltercarb@gmail.com
[2] Department of Mathematics, Miami Dade College, 300 NE Second Ave. Miami, FL 33132, USA
[3] Departamento de Matemáticas, Universidad Carlos III de Madrid, Avenida de la Universidad 30,
    Leganés, 28911 Madrid, Spain; alcruz@math.uc3m.es
[4] Facultad CC. Sociales de Talavera, Universidad de Castilla La Mancha, Avda. Real Fábrica de Seda, s.n.
    Talavera de la Reina, 45600 Toledo, Spain; alvaro.martinezperez@uclm.es
* Correspondence: jomaro@math.uc3m.es; Tel.: +34-91-624-9098

Received: 11 June 2018; Accepted: 9 July 2018; 12 July 2018

(check for updates)

Abstract: It is well-known that the different products of graphs are some of the more symmetric classes of graphs. Since we are interested in hyperbolicity, it is interesting to study this property in products of graphs. Some previous works characterize the hyperbolicity of several types of product graphs (Cartesian, strong, join, corona and lexicographic products). However, the problem with the direct product is more complicated. The symmetry of this product allows us to prove that, if the direct product $G_1 \times G_2$ is hyperbolic, then one factor is bounded and the other one is hyperbolic. Besides, we prove that this necessary condition is also sufficient in many cases. In other cases, we find (not so simple) characterizations of hyperbolic direct products. Furthermore, we obtain good bounds, and even formulas in many cases, for the hyperbolicity constant of the direct product of some important graphs (as products of path, cycle and even general bipartite graphs).

Keywords: direct product of graphs; geodesics; Gromov hyperbolicity; bipartite graphs

---

1. Introduction

An interesting topic in graph theory is the study of the different types of products of graphs [1]. In particular, given two graphs G_1, G_2, the *direct product* $G_1 \times G_2$ is defined as the graph with vertices the (Cartesian) product of $V(G_1)$ and $V(G_2)$, and two vertices $(u_1, v_1), (u_2, v_2) \in V(G_1 \times G_2)$ are connected by an edge if and only if $[u_1, u_2] \in E(G_1)$ and $[v_1, v_2] \in E(G_2)$. The direct product is associative and commutative. Direct product was introduced in Principia Mathematica by Russell and Whitehead.

Weichsel observed that $G_1 \times G_2$ is connected if and only if the graphs G_1 and G_2 are connected and G_1 or G_2 is not a bipartite graph [2], i.e., there exists an odd cycle. The direct product is known with different names: tensor product, conjunction, categorical product, Kronecker product and cardinal product. There are many works studying several properties of direct products. These works include structural results [3–8], hamiltonian properties [9,10], and above all the well-known Hedetniemi's conjecture (see [11,12]). Imrich has an algorithm in [13] which can recognize in polynomial time if a graph is a direct product; furthermore, the algorithm provides a factorization if the graph is a direct product. This fact facilitates the computational use of the direct product of graphs.

Hyperbolic spaces are an important tool in geometry and group theory [14–16]. Gromov hyperbolicity is a meeting point for different spaces: some of them continuous (hyperbolic plane and many Riemannian manifolds with negative curvature) and some of them discrete (trees and many graphs) [14–16].

Gromov hyperbolicity was introduced in the context of finitely generated groups [16], and it was applied, in the science of computation, to the study of automatic groups [17,18]. Gromov hyperbolicity is useful in networking, algorithms and discrete mathematics [19–24]; also, many real networks are hyperbolic [25–29]. Besides, there are several important applications of hyperbolic spaces to the Internet [30–34] and to random graphs [35–37]. It has recently been pointed out that also some aspects of biological systems require hyperbolicity for proper functioning [38]. In [39], it was proven that, for a large class of Riemannian surfaces endowed with a metric of negative curvature, there is a very simple graph related with the surface such that the surface is hyperbolic if and only if the graph is hyperbolic; therefore, it is interesting to study hyperbolic graphs to understand hyperbolic surfaces.

All these facts show the increasing interest of hyperbolic graphs (see, e.g., [19,24–27,32,33,35–37,39–47] and the references therein).

In this paper, let us denote by $G = (V, E) = (V(G), E(G))$ a connected graph with $V(G) \neq \emptyset$. We consider that the length of each edge is 1. In addition, we assume that the graph does not have either multiple edges or loops.

Trees are the graphs with hyperbolicity constant zero. Thus, we can view the hyperbolicity constant as a measure of how "tree-like" the space is. This is an important subject (see, e.g., [48,49]).

From a computational viewpoint, we can obtain $\delta(G)$ in time $O(n^{3.69})$ for graphs with n vertices [50]. In addition, there is an algorithm which decides if a Cayley graph is hyperbolic [51]. In [52], this algorithm is improved, allowing to obtain $\delta(G)$ in time $O(n^2)$, but only if the graph is given in terms of its distance-matrix. However, it is usually very difficult to decide if an infinite graph is hyperbolic. Therefore, it is useful to study hyperbolicity for particular classes of graphs. There are many works dealing with the hyperbolicity of different types of graphs: median graphs [53], line graphs [54–56], cubic graphs [57], complement graphs [58], regular graphs [59], chordal graphs [25,42,45,60], planar graphs [61,62], bipartite and intersection graphs [63], vertex-symmetric graphs [64], periodic graphs [65,66], expanders [34], bridged graphs [67], short graphs [68], graph minors [69], graphs with small hyperbolicity constant [70], Mycielskian graphs [71], geometric graphs [56,72], and some types of products of graphs: Cartesian product and sum [46,73], strong product [74], lexicographic product [75], and corona and join product [76].

Some of these works give results about the hyperbolicity of some unary operations in graphs:

A line graph is hyperbolic if and only if the original graph does [54–56].

For a large class of minor graphs, the minor graph is hyperbolic if and only if the original graph does [69].

Mycielskian graphs are always hyperbolic [71].

Now, we summarize the known results about the hyperbolicity of the main class of binary operations in graphs: products of graphs.

The Cartesian product is hyperbolic if and only if one factor graph is bounded and the other one is hyperbolic [46].

The same holds for the strong product [74].

The corona product $G_1 \diamond G_2$ is hyperbolic if and only if the first factor G_1 is hyperbolic, and the join $G_1 \uplus G_2$ is always hyperbolic [76].

The Cartesian sum $G_1 \oplus G_2$ is always hyperbolic, if the factors have at least two vertices [73].

The lexicographic product graph $G_1 \circ G_2$ is hyperbolic if and only if G_1 does, if the first factor has at least two vertices [75].

The goal of this paper is the characterization in many cases of the direct product of graphs which are hyperbolic. Here, the situation is more complicated than with other products of graphs. This is partly because the direct product of two bipartite graphs (i.e., graphs without odd cycles) is already disconnected and the formula for the distance in $G_1 \times G_2$ is more complicated that in the case of other products of graphs. The symmetry of this product allows us to show that, if $G_1 \times G_2$ is hyperbolic, then one factor is hyperbolic and the other one is bounded (see Theorem 10). Besides, we prove that this necessary condition is also sufficient in many cases. If G_1 is a hyperbolic graph and G_2 is a bounded graph, then we prove that $G_1 \times G_2$ is hyperbolic when G_2 has some odd cycle (Theorem 3) or G_1 and G_2 do not have odd cycles (Theorem 4). One could think that otherwise (if G_1 has some odd cycle and G_2 does not have odd cycles) this necessary condition is also sufficient; however, Theorem 15 allows constructing in an easy way examples G_1, G_2 (with G_1 hyperbolic and G_2 bounded) such that $G_1 \times G_2$ is not hyperbolic. This shows that the characterization of hyperbolic direct products is a more difficult task when G_1 has some odd cycle and G_2 does not have odd cycles. Theorems 11 and 12 provide sufficient conditions for non-hyperbolicity and hyperbolicity, respectively. Besides, Theorems 15 and Corollary 5 characterize the hyperbolicity of $G_1 \times G_2$ under some additional conditions. Furthermore, we obtain good bounds, and even formulas in many cases, for the hyperbolicity constant of the direct product of some important graphs; in particular, Theorem 18 provides the hyperbolicity constant of many direct products of bipartite graphs, and Theorems 17 and 19 give the hyperbolicity constant of many direct products of path and cycle graphs.

We want to remark that, in a general context, the hypothesis on the existence (or non-existence) of odd cycles is artificial in the context of Gromov hyperbolicity. However, it is an essential hypothesis in the works on direct products (see Theorem 1). Throughout the development of this work, we have verified that the existence of odd cycles is also essential in the study of hyperbolic product graphs.

2. Definitions and Background

Let (X, d) be a metric space, and denote by L the length associated to the distance d. A *geodesic* is a curve $g : [a, b] \to X$ satisfying $L(g|_{[t,s]}) = d(g(t), g(s)) = |t - s|$ for every $s, t \in [a, b]$ (here, $g|_{[t,s]}$ is the restriction of g to $[t, s]$). We say that the metric space X is a *geodesic metric space* if for each $p, q \in X$ there is a geodesic connecting them; we denote by $[pq]$ any geodesic form p to q. Hence, a geodesic metric space is a connected space. When X is a graph and $p, q \in V(X)$, $[p, q]$ denotes the edge connecting p and q if they are adjacent.

Along this paper, we consider the graphs as geodesic metric spaces. To do that, we identify any edge $[p, q] \in E(G)$ with the real interval $[0, 1]$; therefore, the points in a graph are the vertices and also the points in the interior of the edges. Hence, we can define a natural distance on the points of a connected graph G by taking shortest paths in G, and so, we consider G as a metric graph. If p and q are points in different connected components of the graph, we define $d(p, q) = \infty$.

Some authors do not consider the internal points of edges in the study. Although this approach has some advantages, we prefer to consider the internal points since these graphs are geodesic metric spaces. We use this approach since to work with geodesic metric spaces provides an interesting geometric viewpoint (for instance, Theorem 2 holds for geodesic metric spaces).

Given a geodesic metric space X and three points $x_1, x_2, x_3 \in X$, the *geodesic triangle* $T = \{x_1, x_2, x_3\}$ is the union of three geodesics $[x_1 x_2]$, $[x_2 x_3]$ and $[x_3 x_1]$. The points x_1, x_2, x_3 are the vertices of the triangle T. The geodesic triangle T is *δ-thin* if any side of T is contained in the δ-neighborhood of the union of the two other sides. We define the thin constant of the triangle T by $\delta(T) := \inf\{\delta \geq 0 : T \text{ is } \delta\text{-thin}\}$, and the *hyperbolicity constant* of the space X as $\delta(X) := \sup\{\delta(T) : T \text{ is a geodesic triangle in } X\}$. The space X is *hyperbolic* if $\delta(X) < \infty$, and it is *δ-hyperbolic* if X is hyperbolic and the constant δ satisfies $\delta \geq \delta(X)$. We say that a triangle with two identical vertices is a "bigon". Of course, each bigon in a space (which is δ-hyperbolic) is δ-thin. If $\{X_i\}_{i \in I}$ are the connected components of X, then we can define $\delta(X) := \sup_{i \in I} \delta(X_i)$, and X is hyperbolic if and only if $\delta(X) < \infty$.

We want to remark that in the classical references on hyperbolicity [14,15,77] appear many different definitions of Gromov hyperbolicity. However, the definitions are equivalent: if X is δ_1-hyperbolic for a definition, then it is δ_2-hyperbolic for every definition, where the constant δ_2 can be obtained from δ_1.

We refer to the classical book [1] for definitions and background about direct product graphs.

We need bounds for the distance between points in the direct product. We use the definition given in [1].

Definition 1. *Let* $G_1 = (V(G_1), E(G_1))$ *and* $G_2 = (V(G_2), E(G_2))$ *be two graphs. The* direct product $G_1 \times G_2$ *of* G_1 *and* G_2 *has* $V(G_1) \times V(G_2)$ *as vertex set, so that two distinct vertices* (u_1, v_1) *and* (u_2, v_2) *of* $G_1 \times G_2$ *are adjacent if* $[u_1, u_2] \in E(G_1)$ *and* $[v_1, v_2] \in E(G_2)$.

If G_1 and G_2 are isomorphic, we write $G_1 \simeq G_2$. It is clear that, if $G_1 \simeq G_2$, then $\delta(G_1) = \delta(G_2)$.

It is clear that the direct product of two graphs is commutative, i.e., $G_1 \times G_2 \simeq G_2 \times G_1$. Therefore, the conclusion of every result in this paper with some "non-symmetric" hypothesis also holds if we change the roles of G_1 and G_2 (see, e.g., Theorems 3, 4, 11, 12 and 15 and Corollary 5).

Denote by π_i the projection map $\pi_i : V(G_1 \times G_2) \to V(G_i)$ for $i \in \{1, 2\}$. In fact, this projection is well defined as a map $\pi_i : G_1 \times G_2 \to G_i$ for $i \in \{1, 2\}$.

We need some previous results of [1]. If $u, u' \in V(G)$, then by a u, u'-*walk* in G we mean a path joining u and u' where repeating vertices is allowed.

Proposition 1. *([1], Proposition 5.7) Suppose* (u, v) *and* (u', v') *are vertices of the direct product* $G_1 \times G_2$, *and* n *is an integer for which* G_1 *has a* u, u'-*walk of length* n *and* G_2 *has a* v, v'-*walk of length* n. *Then,* $G_1 \times G_2$ *has a walk of length* n *from* (u, v) *to* (u', v'). *The smallest such* n *(if it exists) equals* $d_{G_1 \times G_2}((u, v), (u', v'))$. *If no such* n *exists, then* $d_{G_1 \times G_2}((u, v), (u', v')) = \infty$.

Proposition 2. *([1], Proposition 5.8) Suppose* x *and* y *are vertices of* $G_1 \times G_2$. *Then,*

$$d_{G_1 \times G_2}(x, y) = \min \left\{ n \in \mathbb{N} \mid \text{each factor } G_i \text{ has a } \pi_i(x), \pi_i(y)\text{-walk of length } n \text{ for } i = 1, 2 \right\},$$

where it is understood that $d_{G_1 \times G_2}(x, y) = \infty$ *if no such* n *exists.*

Definition 2. *If* G *is a connected graph, the* diameter *of its vertices is*

$$\operatorname{diam} V(G) := \sup \{ d_G(u, v) : u, v \in V(G) \},$$

and the diameter *of* G *is*

$$\operatorname{diam} G := \sup \{ d_G(x, y) : x, y \in G \}.$$

Corollary 1. *We have for every* $(u, v), (u', v') \in V(G_1 \times G_2)$

$$d_{G_1 \times G_2}((u, v), (u', v')) \geq \max \left\{ d_{G_1}(u, u'), d_{G_2}(v, v') \right\}$$

and, consequently,

$$\operatorname{diam} V(G_1 \times G_2) \geq \max \left\{ \operatorname{diam} V(G_1), \operatorname{diam} V(G_2) \right\}.$$

Furthermore, if $d_{G_1}(u, u')$ *and* $d_{G_2}(v, v')$ *have the same parity, then*

$$d_{G_1 \times G_2}((u, v), (u', v')) = \max \left\{ d_{G_1}(u, u'), d_{G_2}(v, v') \right\}$$

and, consequently,

$$\operatorname{diam} V(G_1 \times G_2) = \max \left\{ \operatorname{diam} V(G_1), \operatorname{diam} V(G_2) \right\}.$$

By *trivial graph*, we mean a graph which has only a vertex.

The following result characterizes when a direct product is connected. By *cycle*, we mean a simple closed curve, i.e., a path with different vertices, unless the last one, which is equal to the first vertex.

Theorem 1. *([1], Theorem 5.9) Suppose G_1 and G_2 are connected non-trivial graphs. If at least one of G_1 or G_2 has an odd cycle, then $G_1 \times G_2$ is connected. If both G_1 and G_2 are bipartite, then $G_1 \times G_2$ has exactly two connected components.*

Corollary 2. *([1], Corollary 5.10) A direct product of connected non-trivial graphs is connected if and only if at most one of the factors is bipartite. In fact, the product has $2^{\max\{k,1\}-1}$ connected components, where k is the number of bipartite factors.*

Consider the metric spaces (X, d_X) and (Y, d_Y). Given constants $\alpha \geq 1$, $\beta \geq 0$, a map $f : X \longrightarrow Y$ is an (α, β)-*quasi-isometric embedding* if

$$\alpha^{-1} d_X(x, y) - \beta \leq d_Y(f(x), f(y)) \leq \alpha d_X(x, y) + \beta,$$

for $x, y \in X$. We say that f is ε-*full* if for each $y \in Y$ there is $x \in X$ with $d_Y(f(x), y) \leq \varepsilon$.

We say that f is a *quasi-isometry* if there exist constants $\alpha, \beta, \varepsilon$, such that f is an ε-full (α, β)-quasi-isometric embedding.

Two metric spaces X and Y are *quasi-isometric* if there exists a quasi-isometry $f : X \longrightarrow Y$. One can check that to be quasi-isometric is an equivalence relation. An (α, β)-*quasi-geodesic* in X is an (α, β)-quasi-isometric embedding between an interval of $\mathbb{R}$ and X.

We need the following result ([15], p. 88).

Theorem 2 (Invariance of hyperbolicity). *Let $f : X \longrightarrow Y$ be an (α, β)-quasi-isometric embedding between the geodesic metric spaces X and Y. If Y is δ_Y-hyperbolic, then X is δ_X-hyperbolic, where δ_X is a constant which just depends on α, β, δ_Y.*

Besides, if f is ε-full for some $\varepsilon \geq 0$ (a quasi-isometry) and X is δ_X-hyperbolic, then Y is δ_Y-hyperbolic, where δ_Y is a constant which just depends on $\alpha, \beta, \delta_X, \varepsilon$.

There are several explicit expressions for $\delta_X = \delta_X(\alpha, \beta, \delta_Y)$, some of them very complicated. In [78] appears the best possible formula for δ_X:

$$\delta_X(\alpha, \beta, \delta_Y) = 8\alpha(2\alpha^2(A_1 b + A_2\delta_Y) + 4\delta_Y + \beta).$$

for some explicit constants A_1, A_2.

3. Hyperbolic Direct Products

Let us start with a necessary condition for hyperbolicity.

Proposition 3. *Let G_1 and G_2 be two unbounded connected graphs. Then, $G_1 \times G_2$ is not hyperbolic.*

Proof. Since G_1 and G_2 are unbounded graphs, for each positive integer n there exist two geodesic paths $P_1 := [w_1, w_2] \cup [w_2, w_3] \cup \cdots \cup [w_{n-1}, w_n]$ in G_1 and $P_2 := [v_1, v_2] \cup [v_2, v_3] \cup \cdots \cup [v_{n-1}, v_n]$ in G_2. If n is odd, then we can consider the geodesic triangle T in $G_1 \times G_2$ (see Figure 1) defined by the following geodesics:

$$\gamma_1 := [(w_1, v_2), (w_2, v_1)] \cup [(w_2, v_1), (w_3, v_2)] \cup [(w_3, v_2), (w_4, v_1)] \cup \cdots \cup [(w_{n-1}, v_1), (w_n, v_2)],$$

$$\gamma_2 := [(w_1, v_2), (w_2, v_3)] \cup [(w_2, v_3), (w_1, v_4)] \cup [(w_1, v_4), (w_2, v_5)] \cup \cdots \cup [(w_1, v_{n-1}), (w_2, v_n)],$$

$$\gamma_3 := [(w_2, v_n), (w_3, v_{n-1})] \cup [(w_3, v_{n-1}), (w_4, v_{n-2})] \cup [(w_4, v_{n-2}), (w_5, v_{n-3})] \cup \cdots \cup [(w_{n-1}, v_3), (w_n, v_2)],$$

Corollary 1 gives that $\gamma_1, \gamma_2, \gamma_3$ are geodesics.

Let $m := \frac{n+1}{2}$ and consider the vertex (w_m, v_{m+1}) in γ_3. For every vertex (w_i, v_j) in $\gamma_1, j \in \{1, 2\}$, we have $d_{G_1 \times G_2}((w_m, v_{m+1}), (w_i, v_j)) \geq d_{G_2}(v_{m+1}, v_j) \geq m + 1 - 2 = \frac{n-1}{2}$ by Corollary 1. We have for every vertex (w_i, v_j) in γ_2, $i \in \{1, 2\}$, by Corollary 1, $d_{G_1 \times G_2}((w_m, v_{m+1}), (w_i, v_j)) \geq d_{G_1}(w_m, w_i) \geq m - 2 = \frac{n-3}{2}$. Hence, $d_{G_1 \times G_2}((w_m, v_{m+1}), \gamma_1 \cup \gamma_2) \geq \frac{n-3}{2}$ and $\delta(G_1 \times G_2) \geq \delta(T) \geq \frac{n-3}{2}$. Since n is arbitrarily large, $G_1 \times G_2$ is not hyperbolic. $\square$

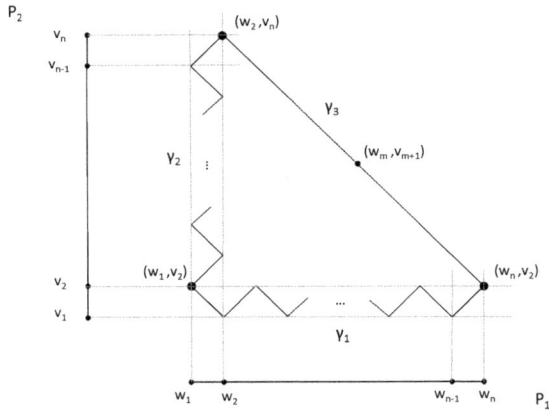

Figure 1. If G_1 and G_2 are unbounded, for any odd n, there is a geodesic triangle $T \subset G_1 \times G_2$ with $\delta(T) \geq \frac{n-3}{2}$.

Lemma 1. *Consider two connected graphs G_1 and G_2. If $f : V(G_1) \longrightarrow V(G_2)$ is an (α, β)-quasi-isometric embedding, then there exists an $(\alpha, \alpha + \beta)$-quasi-isometric embedding $g : G_1 \longrightarrow G_2$ with $g = f$ on $V(G_1)$. In addition, if f is ε-full, then g is $(\varepsilon + \frac{1}{2})$-full.*

Proof. For each $x \in G_1$, let us choose a closest point $v_x \in V(G_1)$ from x, and define $g(x) := f(v_x)$. Note that $v_x = x$ if $x \in V(G_1)$ and so $g = f$ on $V(G_1)$. Given $x, y \in G_1$, we have

$$d_{G_2}(g(x), g(y)) = d_{G_2}(f(v_x), f(v_y)) \leq \alpha d_{G_1}(v_x, v_y) + \beta \leq \alpha(d_{G_1}(x, y) + 1) + \beta,$$
$$d_{G_2}(g(x), g(y)) = d_{G_2}(f(v_x), f(v_y)) \geq \alpha^{-1} d_{G_1}(v_x, v_y) - \beta \geq \alpha^{-1}(d_{G_1}(x, y) - 1) - \beta,$$

and g is an $(\alpha, \alpha + \beta)$-quasi-isometric embedding, since $\alpha \geq 1 \geq \alpha^{-1}$.

In addition, if f is ε-full, then g is $(\varepsilon + \frac{1}{2})$-full since $g(G_1) = f(V(G_1))$. $\square$

Given a graph G, let $g_I(G)$ denote the *odd girth* of G, that is, the length of the shortest odd cycle in G.

Theorem 3. *Let G_1 be a connected graph and G_2 be a non-trivial bounded connected graph with some odd cycle. Then, $G_1 \times G_2$ is hyperbolic if and only if G_1 is hyperbolic.*

Proof. Fix $v_0 \in V(G_2)$ with v_0 contained in an odd cycle C with $L(C) = g_I(G_2)$. Consider the map $i : V(G_1) \to V(G_1 \times G_2)$ such that $i(w) := (w, v_0)$ for every $w \in V(G_1)$.

By Corollary 1, for every $w_1, w_2 \in V(G_1)$, $d_{G_1}(w_1, w_2) \leq d_{G_1 \times G_2}((w_1, v_0), (w_2, v_0))$. In addition, Proposition 2 gives the following.

If a geodesic joining w_1 and w_2 has even length, then

$$d_{G_1 \times G_2}((w_1, v_0), (w_2, v_0)) = d_{G_1}(w_1, w_2).$$

If a geodesic joining w_1 and w_2 has odd length, then C defines a v_0, v_0-walk with odd length and

$$d_{G_1 \times G_2}((w_1, v_0), (w_2, v_0)) \leq \max\{d_{G_1}(w_1, w_2), g_I(G_2)\} \leq d_{G_1}(w_1, w_2) + g_I(G_2).$$

Thus, i is a $(1, g_I(G_2))$ quasi-isometric embedding.

Consider any $(w, v) \in V(G_1 \times G_2)$. Then, if the geodesic joining v and v_0 has even length,

$$d_{G_1 \times G_2}((w, v), (w, v_0)) = d_{G_2}(v, v_0).$$

If a geodesic joining v and v_0 has odd length, $[vv_0] \cup C$ defines a v, v_0-walk with even length. Therefore,

$$d_{G_1 \times G_2}((w, v), (w, v_0)) \leq d_{G_2}(v, v_0) + g_I(G_2).$$

Thus, i is $(\operatorname{diam}(V(G_2)) + g_I(G_2))$-full.

Hence, by Lemma 1, there is a $(\operatorname{diam}(V(G_2)) + g_I(G_2) + \frac{1}{2})$-full $(1, g_I(G_2) + 1)$-quasi-isometry, $j : G_1 \to G_1 \times G_2$, and $G_1 \times G_2$ is hyperbolic if and only if G_1 is hyperbolic by Theorem 2. □

Theorem 4. *Let G_1 be a connected graph without odd cycles and G_2 be a non-trivial bounded connected graph without odd cycles. Then, $G_1 \times G_2$ is hyperbolic if and only if G_1 is hyperbolic.*

Proof. Fix some vertex $w_0 \in V(G_1)$ and some edge $[v_1, v_2] \in E(G_2)$.

By Theorem 1, there are exactly two components in $G_1 \times G_2$. Since there are no odd cycles, there is no $(w_0, v_1), (w_0, v_2)$-walk in $G_1 \times G_2$. Thus, let us denote by $(G_1 \times G_2)^1$ the component containing the vertex (w_0, v_1) and by $(G_1 \times G_2)^2$ the component containing the vertex (w_0, v_2).

Consider $i : V(G_1) \to V(G_1 \times G_2)^1$ defined as $i(w) := (w, v_1)$ for every $w \in V(G_1)$ such that every w_0, w-walk has even length and $i(w) := (w, v_2)$ for every $w \in V(G_1)$ such that every w_0, w-walk has odd length.

By Proposition 2, $d_{G_1 \times G_2}(i(w_1), i(w_2)) = d_{G_1}(w_1, w_2)$ for every $w_1, w_2 \in V(G_1)$ and i is a $(1, 0)$-quasi-isometric embedding.

Let $(w, v) \in V(G_1 \times G_2)^1$. Let v_j with $j \in \{1, 2\}$ such that every v, v_j-walk has even length. Then, by Proposition 2, $d_{G_1 \times G_2}((w, v), (w, v_j)) = d_{G_2}(v, v_j) \leq \operatorname{diam}(G_2)$. Therefore, i is $\operatorname{diam}(G_2)$-full.

Hence, by Lemma 1, there is a $(\operatorname{diam}(G_2) + \frac{1}{2})$-full $(1, 1)$-quasi-isometry, $j : G_1 \to (G_1 \times G_2)^1$, and $(G_1 \times G_2)^1$ is hyperbolic if and only if G_1 is hyperbolic by Theorem 2.

The same argument proves that $(G_1 \times G_2)^2$ is hyperbolic. □

Denote by P_2 the path graph with two vertices and an edge.

Lemma 2. *Let G_1 be a connected graph with some odd cycle and G_2 a non-trivial bounded graph without odd cycles. Then, $G_1 \times G_2$ and $G_1 \times P_2$ are quasi-isometric and $\delta(G_1 \times P_2) \leq \delta(G_1 \times G_2)$.*

Proof. By Theorem 1, we know that $G_1 \times G_2$ and $G_1 \times P_2$ are connected graphs.

Denote by v_1 and v_2 the vertices of P_2 and fix $[w_1, w_2] \in E(G_2)$. The map $f : V(G_1 \times P_2) \longrightarrow V(G_1 \times [w_1, w_2])$ defined as $f(u, v_j) := (u, w_j)$ for every $u \in V(G_1)$ and $j = 1, 2$, is an isomorphism of graphs; hence, it suffices to prove that $G_1 \times G_2$ and $G_1 \times [w_1, w_2]$ are quasi-isometric.

Consider the inclusion map $i : V(G_1 \times [w_1, w_2]) \longrightarrow V(G_1 \times G_2)$. Since $G_1 \times [w_1, w_2]$ is a subgraph of $G_1 \times G_2$, we have $d_{G_1 \times G_2}(x, y) \leq d_{G_1 \times [w_1, w_2]}(x, y)$ for every $x, y \in V(G_1 \times [w_1, w_2])$.

Since G_2 is a graph without odd cycles, every w_1, w_2-walk has odd length and every w_j, w_j-walk has even length for $j = 1, 2$. Thus, Proposition 2 gives, for every $x = (u, w_1), y = (v, w_2) \in V(G_1 \times [w_1, w_2])$,

$$d_{G_1 \times [w_1, w_2]}(x, y) = d_{G_1 \times G_2}(x, y) = \min\{L(g) \mid g \text{ is a } u, v\text{-walk of odd length}\}.$$

Furthermore, for every $x = (u, w_j), y = (v, w_j) \in V(G_1 \times [w_1, w_2])$ and $j = 1, 2$,

$$d_{G_1 \times [w_1, w_2]}(x, y) = d_{G_1 \times G_2}(x, y) = \min \{L(g) \mid g \text{ is a } u, v\text{-walk of even length}\}.$$

Hence, $d_{G_1 \times [w_1, w_2]}(x, y) = d_{G_1 \times G_2}(x, y)$ for every $x, y \in V(G_1 \times [w_1, w_2])$, and the inclusion map i is an $(1, 0)$-quasi-isometric embedding. Therefore, $\delta(G_1 \times P_2) = \delta(G_1 \times [w_1, w_2]) \leq \delta(G_1 \times G_2)$.

Since G_2 is a graph without odd cycles, given any $w \in V(G_2)$, we have either that every w, w_1-walk has even length and every w, w_2-walk has odd length or that every w, w_2-walk has even length and every w, w_1-walk has odd length. In addition, since G_1 is connected, for each $u \in V(G_1)$ there is some $u' \in V(G_1)$ such that $[u, u'] \in E(G_1)$. Therefore, by Proposition 2, for every $(u, w) \in V(G_1 \times G_2)$, if $\min \{d_{G_2}(w, w_1), d_{G_2}(w, w_2)\}$ is even, then

$$d_{G_1 \times G_2}\big((u, w), V(G_1 \times [w_1, w_2])\big) = d_{G_1 \times G_2}\big((u, w), V(u \times [w_1, w_2])\big) = \min \{d_{G_2}(w, w_1), d_{G_2}(w, w_2)\},$$

and if $\min \{d_{G_2}(w, w_1), d_{G_2}(w, w_2)\}$ is odd, then

$$d_{G_1 \times G_2}\big((u, w), V(G_1 \times [w_1, w_2])\big) = d_{G_1 \times G_2}\big((u, w), V(u' \times [w_1, w_2])\big) = \min \{d_{G_2}(w, w_1), d_{G_2}(w, w_2)\}.$$

In both cases,

$$d_{G_1 \times G_2}\big((u, w), V(G_1 \times [w_1, w_2])\big) \leq \operatorname{diam} V(G_2),$$

and i is $(\operatorname{diam} V(G_2))$-full. By Lemma 1, there exists a $\big(\operatorname{diam} V(G_2) + \frac{1}{2}\big)$-full $(1, 1)$-quasi-isometry $g : G_1 \times [w_1, w_2] \longrightarrow G_1 \times G_2$. $\square$

A subgraph Γ of G is said *isometric* if $d_\Gamma(x, y) = d_G(x, y)$ for any $x, y \in \Gamma$. One can check that Γ is isometric if and only if $d_\Gamma(u, v) = d_G(u, v)$ for any $u, v \in V(\Gamma)$.

Lemma 3. *([47], Lemma 5) If Γ is an isometric subgraph of G, then $\delta(\Gamma) \leq \delta(G)$.*

A u, v-walk g in G is a *shortcut* of a cycle C if $g \cap C = \{u, v\}$ and $L(g) < d_C(u, v)$ where d_C denotes the length metric on C.

A cycle C' is a *reduction* of the cycle C if both have odd length and C' is the union of a subarc η of C and a shortcut of C joining the endpoints of η. Note that $L(C') \leq L(C) - 2$. We say that a cycle is *minimal* if it has odd length and it does not have a reduction.

Lemma 4. *If C is a minimal cycle of G, then $L(C) \leq 4\delta(G)$.*

Proof. We prove first that C is an isometric subgraph of G. Assume that C is not an isometric subgraph. Thus, there exists a shortcut g of C with endpoints u, v. There are two subarcs η_1, η_2 of C joining u and v; since C has odd length, we can assume that η_1 has even length and η_2 has odd length. If g has even length, then $C' := g \cup \eta_2$ is a reduction of C. If g has odd length, then $C'' := g \cup \eta_1$ is a reduction of C. Hence, C is not minimal, a contradiction, and so C is an isometric subgraph of G.

It is easy to show that any isometric cycle C has length $4\delta(C)$. This fact and Lemma 3 give $L(C) = 4\delta(C) \leq 4\delta(G)$. $\square$

Given any w_0, w_k-walk $g = [w_0, w_1] \cup [w_1, w_2] \cup \cdots \cup [w_{k-1}, w_k]$ in G_1 and $P_2 = [v_1, v_2]$, if $L(g)$ is either odd or even, then we define the $(w_0, v_1), (w_k, v_i)$-walk for $i \in 1, 2$,

$$\Gamma_1 g := [(w_0, v_1), (w_1, v_2)] \cup [(w_1, v_2), (w_2, v_1)] \cup [(w_2, v_1), (w_3, v_2)] \cup \cdots \cup [(w_{k-1}, v_1), (w_k, v_2)],$$
$$\Gamma_1 g := [(w_0, v_1), (w_1, v_2)] \cup [(w_1, v_2), (w_2, v_1)] \cup [(w_2, v_1), (w_3, v_2)] \cup \cdots \cup [(w_{k-1}, v_2), (w_k, v_1)],$$

respectively.

Remark 1. *By Proposition 2, if g is a geodesic path in G_1, then $\Gamma_1 g$ is a geodesic path in $G_1 \times P_2$.*

Let us define the map $R : V(G_1 \times P_2) \to V(G_1 \times P_2)$ as $R(w, v_1) = (w, v_2)$ and $R(w, v_2) = (w, v_1)$ for every $w \in V(G_1)$, and the path $\Gamma_2 g$ as $\Gamma_2 g = R(\Gamma_1 g)$.

Let us define the map $(\Gamma_1 g)' : g \to \Gamma_1 g$ which is an isometry on the edges and such that $(\Gamma_1 g)'(w_j) = (w_j, v_1)$ if j is even and $(\Gamma_1 g)'(w_j) = (w_j, v_2)$ if j is odd. In addition, let $(\Gamma_2 g)' : g \to \Gamma_2 g$ be the map defined by $(\Gamma_2 g)' := R \circ (\Gamma_1 g)'$.

Given a graph G, denote by $\mathcal{C}$ the set of minimal cycles of G.

Lemma 5. *Let G_1 be a connected graph with some odd cycle and $P_2 = [v_1, v_2]$. Consider a geodesic $g = [w_0 w_k] = [w_0, w_1] \cup [w_1, w_2] \cup \cdots \cup [w_{k-1}, w_k]$ in G_1. Let us define $w_0' := (\Gamma_1 g)'(w_0) = (w_0, v_1)$ and $w_k' := (\Gamma_2 g)'(w_k)$, i.e., $w_k' := (w_k, v_1)$ or $w_k' := (w_k, v_2)$ if k is odd or even, respectively. Then, $d_{G_1 \times P_2}(w_0', w_k') > \sqrt{d_{G_1}(w_j, \mathcal{C}(G_1))}$ for every $0 \le j \le k$.*

Proof. Fix $0 \le j \le k$. Define

$$\mathcal{P} := \{\sigma \mid \sigma \text{ is a } w_0, w_k\text{-walk such that } L(\sigma) \text{ has a parity different from that of } k\}.$$

Proposition 2 gives

$$d_{G_1 \times P_2}(w_0', w_k') = \min\{L(\sigma) \mid \sigma \in \mathcal{P}\}.$$

Choose $\sigma_0 \in \mathcal{P}$ such that $L(\sigma_0) = d_{G_1 \times P_2}(w_0', w_k')$. Since $L(g) + L(\sigma_0)$ is odd, we have $L(g) + L(\sigma_0) = 2t + 1$ for some positive integer t. Thus, $d_{G_1 \times P_2}(w_0', w_k') = L(\sigma_0) > \frac{1}{2}(2t + 1)$.

If $g \cup \sigma_0$ is a cycle, then let us define $C_0 := g \cup \sigma_0$. Thus, $L(C_0) = 2t + 1$ and $d_{G_1}(w_j, C_0) = 0$ for every $0 \le j \le k$. Otherwise, we may assume that $g \cap \sigma_0 = [w_0 w_{i_1}] \cup [w_{i_2} w_k]$ for some $0 \le i_1 < i_2 \le k$. If $\sigma_1 = \sigma_0 \setminus g$, then let us define $C_0 := [w_{i_1} w_{i_2}] \cup \sigma_1$ (where $[w_{i_1} w_{i_2}] \subset g$). Hence, C_0 is a cycle, $L(C_0) \le 2t - 1$ and $d_{G_1}(w_j, C_0) < \frac{1}{2}(2t + 1)$.

If C_0 is not minimal, then consider a reduction C_1 of C_0. Let us repeat the process until we obtain a minimal cycle C_s. Note that $L(C_1) \le L(C_0) - 2$ and for every point $p_1 \in C_0$, $d_{G_1}(p_1, C_1) < \frac{1}{2} L(C_0)$. Now, repeating the argument, for every $1 < i \le s$, $L(C_i) \le L(C_{i-1}) - 2$ and for every point $p_i \in C_{i-1}$, $d_{G_1}(p_i, C_i) < \frac{1}{2} L(C_{i-1})$. Therefore,

$$d_{G_1}(w_j, \mathcal{C}(G_1)) \le d_{G_1}(w_j, C_s) \le d_{G_1}(w_j, C_0) + \frac{1}{2} L(C_0) + \frac{1}{2} L(C_1) + \cdots + \frac{1}{2} L(C_s)$$
$$< \frac{1}{2}(2t + 1) + \frac{1}{2}(2t - 1) + \cdots + \frac{5}{2} + \frac{3}{2}.$$

Hence,

$$d_{G_1}(w_j, \mathcal{C}(G_1)) < \frac{1}{2} \sum_{i=1}^{t} (2i + 1) = \frac{1}{2} t^2 + t < \left(\frac{1}{2}(2t + 1)\right)^2 < \left(d_{G_1 \times P_2}(w_0', w_k')\right)^2.$$

$\square$

Corollary 3. *Let G_1 be a hyperbolic connected graph with some odd cycle and $P_2 = [v_1, v_2]$. Consider a geodesic $g = [w_0 w_k] = [w_0, w_1] \cup [w_1, w_2] \cup \cdots \cup [w_{k-1}, w_k]$ in G_1. Let us define $w_0' := (\Gamma_1 g)'(w_0) = (w_0, v_1)$ and $w_k' := (\Gamma_2 g)'(w_k)$. Then, we have for every $0 \le j \le k$,*

$$\frac{1}{2}\left(k + \sqrt{d_{G_1}(w_j, \mathcal{C}(G_1))}\right) \le d_{G_1 \times P_2}(w_0', w_k') \le k + 2d_{G_1}(w_j, \mathcal{C}(G_1)) + 4\delta(G_1).$$

Proof. Corollary 1 and Lemma 5 give $d_{G_1 \times P_2}(w_0', w_k') \ge k$ and $d_{G_1 \times P_2}(w_0', w_k') \ge \sqrt{d_{G_1}(w_j, \mathcal{C}(G_1))}$, and these inequalities provide the lower bound of $d_{G_1 \times P_2}(w_0', w_k')$.

Consider a geodesic γ joining w_j and $C \in \mathcal{C}(G_1)$ with $L(\gamma) = d_{G_1}(w_j, C) = d_{G_1}(w_j, \mathcal{C}(G_1))$ and the w_0, w_k-walk

$$g' := [w_0 w_j] \cup \gamma \cup C \cup \gamma \cup [w_j w_k].$$

One can check that $\Gamma_1 g'$ is a w_0', w_k'-walk in $G_1 \times P_2$, and so Lemma 4 gives

$$d_{G_1 \times P_2}(w_0', w_k') \leq L(\Gamma_1 g') = L(g') = k + 2d_{G_1}(w_j, \mathcal{C}(G_1)) + L(C) \leq k + 2d_{G_1}(w_j, \mathcal{C}(G_1)) + 4\delta(G_1).$$

$\square$

If $[v_1, v_2]$ is an edge of G, then the point $x \in [v_1, v_2]$ with $d_G(x, v_1) = d_G(x, v_2) = 1/2$ is the *midpoint* of the edge $[v_1, v_2]$. Denote by $J(G)$ the set of vertices and midpoints of edges in G. Consider the set $\mathbb{T}_1(G)$ of geodesic triangles T in G which are cycles and such that the vertices of T are in $J(G)$. We denote by $\delta_1(G)$ the infimum of the constants μ such that any triangle in $\mathbb{T}_1(G)$ is μ-thin.

The following three results are used throughout the paper.

Theorem 5. *([40], Theorem 2.5) For every connected graph G, we have $\delta_1(G) = \delta(G)$.*

Theorem 6. *([40], Theorem 2.6) Let G be any connected graph. Then, $\delta(G)$ is always a multiple of $1/4$.*

Theorem 7. *([40], Theorem 2.7) For any hyperbolic connected graph G, there exists a geodesic triangle $T \in \mathbb{T}_1(G)$ such that $\delta(T) = \delta(G)$.*

Consider the set $\mathbb{T}_v(G)$ of geodesic triangles T in G that are cycles and such that the three vertices of the triangle T are also vertices of G. $\delta_v(G)$ denotes the infimum of the constants μ such that every triangle in $\mathbb{T}_v(G)$ is μ-thin.

Theorem 8. *For every connected graph G, we have $\delta_v(G) \leq \delta(G) \leq 4\delta_v(G) + 1/2$. Hence, G is hyperbolic if and only if $\delta_v(G) < \infty$. Furthermore, if G is hyperbolic, then there are a geodesic triangle $T = \{a, b, c\} \in \mathbb{T}_v(G)$ and $q \in [ab] \cap J(G)$ such that $d(p, [ac] \cup [cb]) = \delta(T) = \delta_v(G)$. In addition, $\delta_v(G)$ is an integer multiple of $1/2$.*

Proof. The inequality $\delta_v(G) \leq \delta(G)$ is direct.

Consider the set $\mathbb{T}_v'(G)$ of geodesic triangles T in G such that the three vertices of the triangle T belong to $V(G)$, and denote by $\delta_v'(G)$ the infimum of the constants μ such that every triangle in $\mathbb{T}_v'(G)$ is μ-thin. The argument in the proof of (ref. [79], Lemma 2.1) gives that $\delta_v'(G) = \delta_v(G)$.

Let us prove now $\delta(G) \leq 4\delta_v(G) + 1/2$. Let us assume that G is hyperbolic. If $\delta_v'(G) = \infty$, then the inequality is trivial. Thus, it suffices to consider the case $\delta_v'(G) < \infty$. By Theorem 7, there is a triangle $T = \{a, b, c\}$ that is a cycle with $a, b, c \in J(G)$ and $q \in [ab]$ such that $d(q, [ac] \cup [cb]) = \delta(T) = \delta(G)$. Assume that $a, b, c \in J(G) \setminus V(G)$ (otherwise, the argument is simpler). Let $a_1, a_2, b_1, b_2, c_1, c_2 \in T \cap V(G)$ such that $a \in [a_1, a_2], b \in [b_1, b_2], c \in [c_1, c_2]$ and $a_2, b_1 \in [ab], c_2, d_1 \in [cd], d_2, a_1 \in [ac]$. Since $H := \{a_2, b_1, b_2, c_1, c_2, a_1\}$ is a geodesic hexagon with vertices in $V(G)$, it is $4\delta_v'(G)$-thin and every point $w \in [b_1, b_2] \cup [b_2 c_1] \cup [c_1, c_2] \cup [c_2 a_1] \cup [a_1, a_2]$ verifies $d(w, [ac] \cup [cb]) \leq 1/2$, we have

$$\delta(G) = d(q, [ac] \cup [cb]) \leq d(q, [b_1, b_2] \cup [b_2 c_1] \cup [c_1, c_2] \cup [c_2 a_1] \cup [a_1, a_2]) + 1/2$$
$$\leq 4\delta_v'(G) + 1/2 = 4\delta_v(G) + 1/2.$$

Assume that G is not hyperbolic. Therefore, for each $M > 0$ there is a triangle $T = \{a, b, c\}$ which is a cycle with $a, b, c \in J(G)$ and $q \in [ab]$ with $d(q, [ac] \cup [cb]) \geq M$. The previous argument gives $M \leq 4\delta_v(G) + 1/2$ and, since M is arbitrary, we conclude $\delta_v(G) = \infty = \delta(G)$.

Finally, consider any geodesic triangle $T = \{a, b, c\}$ in $\mathbb{T}_v(G)$. Since $d(q, [ac] \cup [cb]) = d(q, ([ac] \cup [cb]) \cap V(G))$, $d(q, [ac] \cup [cb])$ attains its maximum value when $q \in J(G)$. Hence, $\delta(T)$ is a multiple of $1/2$ for any triangle $T \in \mathbb{T}_v(G)$. Since the set of non-negative numbers that are multiple of $1/2$ is a discrete set, $\delta(G)$ is an integer multiple of $1/2$ if G is hyperbolic, and there is a triangle $T = \{a, b, c\} \in \mathbb{T}_v(G)$ and $q \in [ab] \cap J(G)$ with $d(q, [ac] \cup [cb]) = \delta(T) = \delta_v(G)$. This finishes the proof. $\square$

Theorem 9. *If G_1 is a non-hyperbolic connected graph, then $G_1 \times P_2$ is not hyperbolic.*

Proof. Since G_1 is not hyperbolic, by Theorem 8, given any $R > 0$ there exists a triangle $T = \{x, y, z\}$ wich is a cycle, with $x, y, z \in V(G_1)$ and such that T is not R-thin. Therefore, there exists some point $m \in T$, let us assume that $m \in [xy]$, such that $d_{G_1}(m, [yz] \cup [zx]) > R$.

Seeking for a contradiction let us assume that $G_1 \times P_2$ is δ-hyperbolic.

Suppose that for some $R > \delta$, there is a geodesic triangle $T = \{x, y, z\}$ that is an even cycle in G_1, with $x, y, z \in V(G_1)$ and such that T is not R-thin. Consider the (closed) path $\Lambda = [xy] \cup [yz] \cup [zx]$. Then, since T has even length, the path $\Gamma_1 \Lambda$ defines a cycle in $G_1 \times P_2$. Let $\gamma_1, \gamma_2, \gamma_3$ be the paths in $\Gamma_1 \Lambda$ corresponding to $[xy], [yz], [zx]$, respectively. By Corollary 1, the curves γ_1, γ_2 and γ_3 are geodesics, and $d_{G_1 \times P_2}((\Gamma_1 \Lambda)'(m), \gamma_2 \cup \gamma_3) > \delta$, leading to contradiction.

Suppose that, for every $R > 0$, there is a geodesic triangle $T = \{x, y, z\}$ which is an odd cycle, with $x, y, z \in V(G_1)$ and such that T is not R-thin.

Let $T_1 = \{x, y, z\}$ be a geodesic triangle as above and let us assume that $\operatorname{diam}(T_1) = D > 8\delta$.

Let $T_2 = \{x', y', z'\}$ be another triangle as above such that T_2 is not $3(D + 8\delta)$-thin, this is, there is a point m in one of the sides, let us call it σ, of T_2 such that $d_{G_1}(m, T_2 \setminus \sigma) > 3(D + 8\delta)$.

Let $g = [w_0 w_k]$ with $w_0 \in T_1$ and $w_k \in T_2$ be a shortest geodesic in G_1 joining T_1 and T_2 (if T_1 and T_2 intersect, just assume that g is a single vertex, $w_0 = w_k$, in the intersection). See Figure 2.

Let us assume that $w_0 \in [xz]$ and $w_k \in [x'z']$. Then, let us consider the closed path C in G_1 given by the union of the geodesics in T_1, g, the geodesics in T_2 and the inverse of g from w_k to w_0, this is,

$$C := [w_0 x] \cup [xy] \cup [yz] \cup [zw_0] \cup [w_0 w_k] \cup [w_k x'] \cup [x'y'] \cup [y'z'] \cup [z'w_k] \cup [w_k w_0].$$

Since T_1, T_2 are odd cycles, C is an even closed cycle. Therefore, $\Gamma_1 C$ defines a cycle in $G_1 \times P_2$. Moreover, by Remark 1, $\Gamma_1 C$ is a geodesic decagon in $G_1 \times P_2$ with sides $\gamma_1 = (\Gamma_1 C)'([w_0 x])$, $\gamma_2 = (\Gamma_1 C)'([xy])$, $\gamma_3 = (\Gamma_1 C)'([yz])$, $\gamma_4 = (\Gamma_1 C)'([zw_0])$, $\gamma_5 = (\Gamma_1 C)'([w_0 w_k])$, $\gamma_6 = (\Gamma_1 C)'([w_k x'])$, $\gamma_7 = (\Gamma_1 C)'([x'y'])$, $\gamma_8 = (\Gamma_1 C)'([y'z'])$, $\gamma_9 = (\Gamma_1 C)'([z'w_k])$ and $\gamma_{10} = (\Gamma_1 C)'([w_k w_0])$.

Since we are assuming that $G_1 \times P_2$ is δ-hyperbolic, then for every $1 \leq i \leq 10$ and every point $p \in \gamma_i$, $d_{G_1 \times P_2}(p, C \setminus \gamma_i) \leq 8\delta$.

Let $p := (\Gamma_1 C)'(m)$.

Case 1. Suppose that $d_{G_1}(m, T_1 \cup g) > 8\delta$. See Figure 2.

By assumption, $d_{G_1}(m, T_2 \setminus \sigma) > 8\delta$. If $\sigma = [x'y']$ (resp. $\sigma = [y'z']$), then $p \in \gamma_7$ (resp. $p \in \gamma_8$) and, by Corollary 1, $d_{G_1 \times P_2}(p, C \setminus \gamma_7) > 8\delta$ (resp. $d_{G_1 \times P_2}(p, C \setminus \gamma_8) > 8\delta$) leading to contradiction. If $\sigma = [x'z']$, since $[x'z'] = [x'w_k] \cup [w_k z']$, let us assume $m \in [x'w_k]$. Then, since $d_{G_1}(m, w_k) > 8\delta$, it follows that $d_{G_1}(m, [w_k z']) > 8\delta$. Thus, $p \in \gamma_6$ and, by Corollary 1, $d_{G_1 \times P_2}(p, C \setminus \gamma_6) > 8\delta$ leading to contradiction.

Case 2. Suppose that $d_{G_1}(m, T_1 \cup g) \leq 8\delta$ and $L(g) \leq 8\delta$. See the left side of Figure 3. Then, for every point q in $T_1 \cup g$, $d_{G_1}(m, q) \leq 8\delta + D + 8\delta$. In particular, $d_{G_1}(m, w_k) \leq 8\delta + D + 8\delta$. Therefore, $m \in [x'z']$ and let us assume that $m \in [x'w_k]$. Since $d_{G_1}(m, x') \geq d_{G_1}(m, [x'y'] \cup [y'z']) > 3(D + 8\delta)$, there is a point $m' \in [x'm] \subset [x'w_k]$ such that $d_{G_1}(m, m') = 2(D + 8\delta)$. Then, $d_{G_1}(m', T_1 \cup g) \geq 2(D + 8\delta) - D - 8\delta - 8\delta = D > 8\delta$. In addition, it is trivial to check that $d_{G_1}(m', [x'y'] \cup [y'z']) > 3(D + 8\delta) - 2(D + 8\delta) > 8\delta$ and since $[x'z']$ is a geodesic, $d_{G_1}(m', [z'w_k]) > 8\delta$. Thus, if $p' := (\Gamma_1 C)'(m')$, then $p' \in \gamma_6$ and, by Corollary 1, $d_{G_1 \times P_2}(p', C \setminus \gamma_6) > 8\delta$ leading to contradiction.

Case 3. Suppose that $d_{G_1}(m, T_1 \cup g) \leq 8\delta$ and $L(g) > 8\delta$. See the right side of Figure 3. Since g is a shortest geodesic in G_1 joining T_1 and T_2, this implies that $d_{G_1}(T_1, T_2) > 8\delta$ and $d_{G_1}(m, [w_0 w_k]) \leq 8\delta$.

Moreover, $d_{G_1}(m, w_k) \leq 16\delta$. Otherwise, there is a point $q \in [w_0 w_k]$ such that $d_{G_1}(m, q) \leq 8\delta$ and $d_{G_1}(q, w_k) > 8\delta$ which means that $d_{G_1}(q, w_0) < d_{G_1}(w_0, w_k) - 8\delta$ and $d_{G_1}(m, w_0) < d_{G_1}(w_0, w_k)$ leading to contradiction.

Since $d_{G_1}(m, w_k) \leq 16\delta$, $m \in [x'z']$. Let us assume that $m \in [x'w_k]$. Since $d_{G_1}(m, [x'y'] \cup [y'z']) > 3(D + 8\delta)$, there is a point $m' \in [x'm] \subset [x'w_k]$ such that $d_{G_1}(m, m') = 2(D + 8\delta)$. Let us see that $d_{G_1}(m', [w_0 w_k]) > 8\delta$. Suppose there is some $q \in [w_0 w_k]$ such that $d_{G_1}(m', q) \leq 8\delta$. Since $m' \in T_2$ and g is a shortest geodesic joining T_1 and T_2, $d_{G_1}(q, w_k) \leq 8\delta$. However, $32\delta < 2(D + 8\delta) = d_{G_1}(m', m) \leq d_{G_1}(m', q) + d_{G_1}(q, w_k) + d_{G_1}(w_k, m) \leq 8\delta + 8\delta + 16\delta$ which is a contradiction. Hence, $d_{G_1}(m', [w_0 w_k]) > 8\delta$. In addition, it is trivial to check that $d_{G_1}(m', [x'y'] \cup [y'z']) > 3(D + 8\delta) - 2(D + 8\delta) > 8\delta$ and since $[x'z']$ is a geodesic, $d_{G_1}(m', [z'w_k]) > 8\delta$. Thus, if $p' := (\Gamma_1 C)'(m')$, then $p' \in \gamma_6$ and, by Corollary 1, $d_{G_1 \times P_2}(p', C \backslash \gamma_6) > 8\delta$ leading to contradiction. $\square$

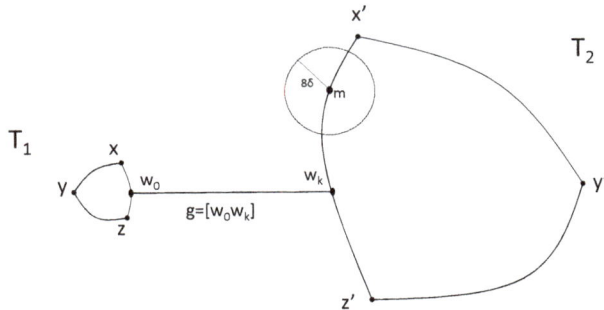

Figure 2. Two geodesic triangles, T_1, T_2, which are odd cycles and a geodesic g joining them define an even closed path.

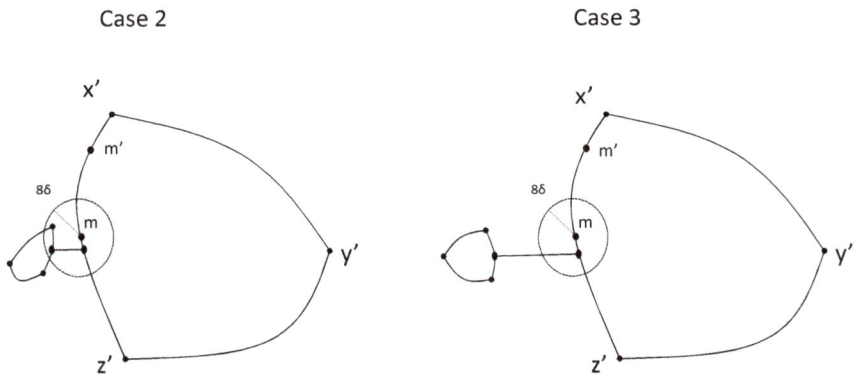

Figure 3. If $d_{G_1}(m, T_1 \cup g) \leq 8\delta$, then $m \in [x'z']$ and there is a point $m' \in [x'm] \subset [x'w_k]$ such that $d_{G_1}(m, m') = 2(D + 8\delta)$.

Proposition 3, Lemma 2 and Theorems 3, 4 and 9 have the following consequence.

Corollary 4. *If G_1 is a non-hyperbolic connected graph and G_2 is some non-trivial connected graph, then $G_1 \times G_2$ is not hyperbolic.*

Proposition 3 and Corollary 4 provide a necessary condition for the hyperbolicity of $G_1 \times G_2$.

Theorem 10. *Let G_1, G_2 be non-trivial connected graphs. If $G_1 \times G_2$ is hyperbolic, then one factor graph is hyperbolic and the other one is bounded.*

Theorems 3 and 4 show that this necessary condition is also sufficient if either G_2 has some odd cycle or G_1 and G_2 do not have odd cycles (when G_1 is a hyperbolic graph and G_2 is a bounded graph). We deal now with the other case, when G_1 has some odd cycle and G_2 does not have odd cycles.

Theorem 11. *Let G_1 be a connected graph with some odd cycle and G_2 a non-trivial bounded connected graph without odd cycles. Assume that G_1 satisfies the following property: for each $M > 0$ there exist a geodesic g joining two minimal cycles of G_1 and a vertex $u \in g \cap V(G_1)$ with $d_{G_1}(u, \mathcal{C}(G_1)) \geq M$. Then, $G_1 \times G_2$ is not hyperbolic.*

Proof. If G_1 is not hyperbolic, then Corollary 4 gives that $G_1 \times G_2$ is not hyperbolic. Assume now that G_1 is hyperbolic. By Theorem 2 and Lemma 2, we can assume that $G_2 = P_2$ and $V(P_2) = \{v_1, v_2\}$.

Fix $M > 0$ and choose a geodesic $g = [w_0 w_k] = [w_0, w_1] \cup [w_1, w_2] \cup \cdots \cup [w_{k-1}, w_k]$ joining two minimal cycles in G_1 and $0 < r < k$ with $d_{G_1}(w_r, \mathcal{C}(G_1)) \geq M$.

Define the paths g_1 and g_2 in $G_1 \times P_2$ as $g_1 := \Gamma_1 g$ and $g_2 := \Gamma_2 g$. Since $L(g_1) = L(g_2) = L(g) = d_{G_1}(w_0, w_k)$, we have

$$d_{G_1 \times P_2}(g_1(w_0), g_1(w_k)) \leq L(g_1) = d_{G_1}(w_0, w_k), \qquad d_{G_1 \times P_2}(g_2(w_0), g_2(w_k)) \leq L(g_2) = d_{G_1}(w_0, w_k).$$

Corollary 1 gives that

$$d_{G_1 \times P_2}(g_1(w_0), g_1(w_k)) \geq d_{G_1}(w_0, w_k), \qquad d_{G_1 \times P_2}(g_2(w_0), g_2(w_k)) \geq d_{G_1}(w_0, w_k).$$

Hence, g_1 and g_2 are geodesics in $G_1 \times P_2$. Choose geodesics $g_3 = [g_1(w_0)g_2(w_0)]$ and $g_4 = [g_1(w_k)g_2(w_k)]$ in $G_1 \times P_2$. Since $d_{P_2}(v_1, v_2) = 1$ is odd, Proposition 2 gives

$$d_{G_1 \times P_2}(g_1(w_0), g_2(w_0)) = \min\{L(\sigma) \mid \sigma \text{ is a } w_0, w_0\text{-walk}\}$$
$$= \min\{L(\sigma) \mid \sigma \text{ cycle of odd length containing } w_0\}.$$

Since w_0 belongs to a minimal cycle, $L(g_3) \leq 4\delta(G_1)$ by Lemma 4. In a similar way, we obtain $L(g_4) \leq 4\delta(G_1)$.

Consider the geodesic quadrilateral $Q := \{g_1, g_2, g_3, g_4\}$ in $G_1 \times P_2$. Thus, $d_{G_1 \times P_2}(g_1(w_r), g_2 \cup g_3 \cup g_4) \leq 2\delta(G_1 \times P_2)$. Since $\max\{L(g_3), L(g_4)\} \leq 4\delta(G_1)$, we deduce $d_{G_1 \times P_2}(g_1(w_r), g_2) \leq 2\delta(G_1 \times P_2) + 4\delta(G_1)$.

Let $0 \leq j \leq k$ with $d_{G_1 \times P_2}(g_1(w_r), g_2) = d_{G_1 \times P_2}(g_1(w_r), g_2(w_j))$. Let us define $w'_r := g_1(w_r)$ and $w'_j := g_2(w_j)$. Thus, Lemma 5 gives

$$\sqrt{M} \leq \sqrt{d_{G_1}(w_r, \mathcal{C}(G_1))} \leq d_{G_1 \times P_2}(w'_r, w'_j) = d_{G_1 \times P_2}(w'_r, g_2) \leq 2\delta(G_1 \times P_2) + 4\delta(G_1),$$

and since M is arbitrarily large, we deduce that $G_1 \times P_2$ is not hyperbolic. □

Lemma 6. *Let G_1 be a hyperbolic connected graph and suppose there is some constant $K > 0$ such that for every vertex $w \in G_1$, $d_{G_1}(w, \mathcal{C}(G_1)) \leq K$. Then, $G_1 \times P_2$ is hyperbolic.*

Proof. Denote by v_1 and v_2 the vertices of P_2. Let $i : V(G_1) \to V(G_1 \times P_2)$ defined as $i(w) := (w, v_1)$ for every $w \in G_1$.

For every $x, y \in V(G_1)$, by Corollary 1, $d_{G_1}(x, y) \leq d_{G_1 \times P_2}(i(x), i(y))$. By Corollary 3,

$$d_{G_1 \times P_2}(i(x), i(y)) \leq d_{G_1}(x, y) + 2d_{G_1}(x, \mathcal{C}(G_1)) + 4\delta(G_1) \leq d_{G_1}(x, y) + 2K + 4\delta(G_1).$$

Therefore, $i : V(G_1) \to V(G_1 \times P_2)$ is a $(1, 2K + 4\delta(G_1))$-quasi-isometric embedding.

Notice that for every $(w, v_1) \in V(G_1 \times P_2)$, $(w, v_1) = i(w)$. In addition, for any $(w, v_2) \in V(G_1 \times P_2)$, since G_1 is connected, there is some edge $[w, w'] \in E(G_1)$ and we have $[(w, v_2), (w', v_1)] \in E(G_1 \times P_2)$. Therefore, $i : V(G_1) \to V(G_1 \times P_2)$ is 1-full.

Thus, by Lemma 1, G_1 and $G_1 \times P_2$ are quasi-isometric and, by Theorem 2, $G_1 \times P_2$ is hyperbolic. $\square$

Theorem 3 and Lemmas 2 and 6 give the following result.

Theorem 12. *Let G_1 be a hyperbolic connected graph and G_2 some non-trivial bounded connected graph. If there is some constant $K > 0$ such that for every vertex $w \in G_1$, $d_{G_1}(w, C(G_1)) \leq K$, then $G_1 \times G_2$ is hyperbolic.*

We finish this section with a characterization of the hyperbolicity of $G_1 \times G_2$, under an additional hypothesis. We present first some lemmas.

Let J be a finite or infinite index set. Now, given a graph G_1, we define some graphs related to G_1 which will be useful in the following results. Let $B_j := B_{G_1}(w_j, K_j)$ with $w_j \in V(G_1)$ and $K_j \in \mathbb{Z}^+$, for any $j \in J$, such that $\sup_j K_j = K < \infty$, $\overline{B}_{j_1} \cap \overline{B}_{j_2} = \varnothing$ if $j_1 \neq j_2$, and every odd cycle C in G_1 satisfies $C \cap B_j \neq \varnothing$ for some $j \in J$. Denote by G_1' the subgraph of G_1 induced by $V(G_1) \setminus (\cup_j B_j)$. Let $N_j := \partial B_j = \{w \in V(G_1) : d_{G_1}(w, w_j) = K_j\}$. Denote by G_1^* the graph with $V(G_1^*) = V(G_1') \cup (\cup_j \{w_j^*\})$, where w_j^* are additional vertices, and $E(G_1^*) = E(G_1') \cup (\cup_j \{[w, w_j^*] : w \in N_j\})$. We have $G_1' = G_1 \cap G_1^*$.

Lemma 7. *Let G_1 be a connected graph as above. Then, there is a quasi-isometry $g : G_1 \to G_1^*$ such that $g(w_j) = w_j^*$ for every $j \in J$.*

Proof. Let $f : V(G_1) \to V(G_1^*)$ defined as $f(u) = u$ for every $u \in V(G_1')$, and $f(u) = w_i^*$ for every $u \in V(B_i)$. It is clear that $f : V(G_1) \to V(G_1^*)$ is 0-full.

Now, we focus on proving that $f : V(G_1) \to V(G_1^*)$ is a $(K, 2K)$-quasi-isometric embedding. For every $u, v \in V(G_1)$, it is clear that $d_{G_1^*}(f(u), f(v)) \leq d_{G_1}(u, v)$.

Let us prove the other inequality. Fix $u, v \in V(G_1)$ and consider an oriented geodesic γ in G_1^* from $f(u)$ to $f(v)$.

Assume that $u, v \in V(G_1')$. If $L(\gamma) = d_{G_1}(u, v)$, then $d_{G_1}(u, v) = d_{G_1^*}(f(u), f(v))$. If $L(\gamma) < d_{G_1}(u, v)$, then γ meets some w_j^*. Since γ is a compact set, it intersects only a finite number of w_j^*'s, which we denote by $w_{j_1}^*, \dots w_{j_r}^*$. Since γ is an oriented curve from $f(u)$ to $f(v)$, we can assume that γ meets $w_{j_1}^*, \dots w_{j_r}^*$ in this order.

Let us define the following vertices in γ

$$ w_i^1 = [f(u) w_{j_i}^*] \cap N_{j_i}, \qquad w_i^2 = [w_{j_i}^* f(v)] \cap N_{j_i}, $$

for every $1 \leq i \leq r$. Note that $[w_i^2 w_{i+1}^1] \subset G_1'$ for every $1 \leq i < r$ (it is possible to have $w_i^2 = w_{i+1}^1$).

Since $d_{G_1^*}(w_i^1, w_i^2) = 2$ and $d_{G_1}(w_i^1, w_i^2) \leq 2K$, we have $d_{G_1^*}(w_i^1, w_i^2) \geq \frac{1}{K} d_{G_1}(w_i^1, w_i^2)$ for every $1 \leq i \leq r$. Thus,

$$ d_{G_1^*}(f(u), f(v)) = d_{G_1^*}(f(u), w_1^1) + \sum_{i=1}^{r} d_{G_1^*}(w_i^1, w_i^2) + \sum_{i=1}^{r-1} d_{G_1^*}(w_i^2, w_{i+1}^1) + d_{G_1^*}(w_r^2, f(v)) $$

$$ \geq d_{G_1}(u, w_1^1) + \frac{1}{K} \sum_{i=1}^{r} d_{G_1}(w_i^1, w_i^2) + \sum_{i=1}^{r-1} d_{G_1}(w_i^2, w_{i+1}^1) + d_{G_1}(w_r^2, v) $$

$$ \geq \frac{1}{K} \left(d_{G_1}(u, w_1^1) + \sum_{i=1}^{r} d_{G_1}(w_i^1, w_i^2) + \sum_{i=1}^{r-1} d_{G_1}(w_i^2, w_{i+1}^1) + d_{G_1}(w_r^2, v) \right) $$

$$ \geq \frac{1}{K} d_{G_1}(u, v). $$

Assume that $f(u) = f(v)$. Therefore, there exists j with $u, v \in B_j$ and

$$d_{G_1^*}(f(u), f(v)) = 0 > d_{G_1}(u, v) - 2K.$$

Assume now that u and/or v does not belong to $V(G_1')$ and $f(u) \neq f(v)$. Let u_0, v_0 be the closest vertices in $V(G_1') \cap \gamma$ to $f(u), f(v)$, respectively (it is possible to have $u_0 = f(u)$ or $v_0 = f(v)$). Since $u_0, v_0 \in V(G_1')$, $u_0 = f(u_0)$, $v_0 = f(v_0)$, we have $d_{G_1}(u, u_0) < 2K$ and $d_{G_1}(v, v_0) < 2K$. Hence,

$$
\begin{aligned}
d_{G_1^*}(f(u), f(v)) &= d_{G_1^*}(f(u), u_0) + d_{G_1^*}(u_0, v_0) + d_{G_1^*}(v_0, f(v)) \\
&\geq d_{G_1^*}(f(u_0), f(v_0)) \\
&\geq \frac{1}{K} d_{G_1}(u_0, v_0) \\
&\geq \frac{1}{K}\left(d_{G_1}(u, v) - d_{G_1}(u, u_0) - d_{G_1}(v, v_0)\right) \\
&> \frac{1}{K} d_{G_1}(u, v) - 4.
\end{aligned}
$$

If $K \geq 2$, then $d_{G_1^*}(f(u), f(v)) > \frac{1}{K} d_{G_1}(u, v) - 2K$. If $K = 1$, then $d_{G_1}(u, u_0) \leq 1, d_{G_1}(v, v_0) \leq 1$, and $d_{G_1^*}(f(u), f(v)) \geq d_{G_1}(u, v) - 2$.

Finally, we conclude that $f : V(G_1) \to V(G_1^*)$ is a $(K, 2K)$-quasi-isometric embedding. Thus, Lemma 1 provides a quasi-isometry $g : G_1 \to G_1^*$ with the required property. $\square$

Definition 3. *Given a connected graph G_1 and some index set J, let $\mathcal{B}_J = \{B_j\}_{j \in J}$ be a family of balls where $B_j := B_{G_1}(w_j, K_j)$ with $w_j \in V(G_1)$, $K_j \in \mathbb{Z}^+$ for any $j \in J$, $\sup_j K_j = K < \infty$ and $\overline{B}_{j_1} \cap \overline{B}_{j_2} = \emptyset$ if $j_1 \neq j_2$. Suppose that every odd cycle C in G_1 satisfies that $C \cap B_j \neq \emptyset$ for some $j \in J$. If there is some constant $M > 0$ such that for every $j \in J$, there is an odd cycle C_j such that $C_j \cap B_j \neq \emptyset$ with $L(C_j) < M$, then we say that $\mathcal{B}_J$ is M-regular.*

Remark 2. *If J is finite, then there exists $M > 0$ such that $\{B_j\}_{j \in J}$ is M-regular.*

Denote by G^* the graph with $V(G^*) = V(G_1' \times P_2) \cup (\cup_j \{w_j^*\})$, where G_1' is a graph as above and w_j^* are additional vertices, and $E(G^*) = E(G_1' \times P_2) \cup (\cup_j \{[w, w_j^*] : \pi_1(w) \in N_j\})$.

Lemma 8. *Let G_1 be a connected graph as above and P_2 with $V(P_2) = \{v_1, v_2\}$. If G_1 is hyperbolic and $\mathcal{B}_J$ as above is M-regular, then there exists a quasi-isometry $f : G_1 \times P_2 \to G^*$ with $f(w_j, v_i) = w_j^*$ for every $j \in J$ and $i \in \{1, 2\}$.*

Proof. Let $F : V(G_1 \times P_2) \to V(G^*)$ defined as $F(v, v_i) = (v, v_i)$ for every $v \in V(G_1')$, and $F(v, v_i) = w_j^*$ for every $v \in V(B_j)$. It is clear that $F : V(G_1 \times P_2) \to V(G^*)$ is 0-full. Recall that we denote by $\pi_1 : G_1 \times P_2 \to G_1$ the projection map. Define $\pi^* : G^* \to G_1$ as $\pi^* = \pi_1$ on $G_1' \times P_2$ and $\pi^*(x) = w_j$ for every x with $d_{G^*}(x, w_j^*) < 1$ for some $j \in J$.

Now, we focus on proving that $F : V(G_1 \times P_2) \to V(G^*)$ is a quasi-isometric embedding. For every $(w, v_i), (w', v_{i'}) \in V(G_1 \times P_2)$, one can check

$$d_{G^*}(F(w, v_i), F(w', v_{i'})) \leq d_{G_1 \times P_2}((w, v_i), (w', v_{i'})).$$

To prove the other inequality, let us fix $(w, v_i), (w', v_{i'}) \in V(G_1' \times P_2)$ (the inequalities in other cases can be obtained from the one in this case, as in the proof of Lemma 7). Consider a geodesic $\gamma := [F(w, v_i)F(w', v_{i'})]$ in G^*. If $L(\gamma) = d_{G_1 \times P_2}((w, v_i), (w', v_{i'}))$, then

$$d_{G^*}(F(w, v_i), F(w', v_{i'})) = d_{G_1 \times P_2}((w, v_i), (w', v_{i'})).$$

If $L(\gamma) < d_{G_1 \times P_2}((w, v_i), (w', v_{i'}))$, then $\pi^*(\gamma)$ meets some B_j. Since γ is a compact set, $\pi^*(\gamma)$ intersects just a finite number of B_j's, which we denote by $B_{j_1}, \ldots B_{j_r}$. We consider γ as an oriented curve from $F(w, v_i)$ to $F(w', v_{i'})$; thus we can assume that $\pi^*(\gamma)$ meets $B_{j_1}, \ldots, B_{j_r}$ in this order.

Let us define the following set of vertices in γ

$$\{w_i^1, w_i^2\} := \gamma \cap (N_{j_i} \times P_2),$$

for every $1 \leq i \leq r$, such that $d_{G_1 \times P_2}((w, v_i), w_i^1) < d_{G_1 \times P_2}((w, v_i), w_i^2)$. Note that $[w_i^2 w_{i+1}^1] \subset G_1' \times P_2$ for every $1 \leq i < r$ and $d_{G_1 \times P_2}(w_i^2, w_{i+1}^1) \geq 1$ since $\overline{B}_{j_i} \cap \overline{B}_{j_{i+1}} = \varnothing$.

If $d_{G_1}(\pi(w_i^1), \pi(w_i^2)) = d_{G_1 \times P_2}(w_i^1, w_i^2)$ for some $1 \leq i \leq r$, then $d_{G_1 \times P_2}(w_i^1, w_i^2) \leq 2K$. Since $d_{G_1 \times P_2}(w_i^2, w_{i+1}^1) \geq 1$ for $1 \leq i < r$, we have that $d_{G_1 \times P_2}(w_i^1, w_i^2) \leq 2K \, d_{G_1 \times P_2}(w_i^2, w_{i+1}^1)$ in this case.

If $d_{G_1}(\pi_1(w_i^1), \pi_1(w_i^2)) < d_{G_1 \times P_2}(w_i^1, w_i^2)$ for some $1 \leq i \leq r$, then $d_{G_1}(\pi_1(w_i^1), \pi_1(w_i^2)) + d_{G_1 \times P_2}(w_i^1, w_i^2)$ is odd.

Since $\mathcal{B}_J$ is M-regular, consider an odd cycle C with $C \cap B_{j_i} \neq \varnothing$ and $L(C) < M$, and let $b_i \in C \cap B_{j_i}$ and $[\pi_1(w_i^1)b_i]$, $[b_i \pi_1(w_i^2)]$ geodesics in G_1. Thus, $[\pi_1(w_i^1)b_i] \cup [b_i \pi_1(w_i^2)]$ and $[\pi_1(w_i^1)b_i] \cup C \cup [b_i \pi_1(w_i^2)]$ have different parity which means that one of them has different parity from $[\pi_1(w_i^1)\pi_1(w_i^2)]$. Then, $d_{G_1 \times P_2}(w_i^1, w_i^2) \leq L([\pi_1(w_i^1)b_i] \cup C \cup [b_i \pi_1(w_i^2)]) \leq 4K + M$. Since $d_{G_1 \times P_2}(w_i^2, w_{i+1}^1) \geq 1$ for $1 \leq i < r$, we have that $d_{G_1 \times P_2}(w_i^1, w_i^2) \leq (4K + M) d_{G_1 \times P_2}(w_i^2, w_{i+1}^1)$ in this case.

Thus, we have that $d_{G_1 \times P_2}(w_i^1, w_i^2) \leq 4K + M$ for every $1 \leq i \leq r$ and $d_{G_1 \times P_2}(w_i^1, w_i^2) \leq (4K + M) d_{G_1 \times P_2}(w_i^2, w_{i+1}^1)$ for every $1 \leq i < r$. Therefore,

$$d_{G_1 \times P_2}((w, v_i), (w', v_{i'})) \leq d_{G_1 \times P_2}((w, v_i), w_1^1) + \sum_{i=1}^{r} d_{G_1 \times P_2}(w_i^1, w_i^2) + \sum_{i=1}^{r-1} d_{G_1 \times P_2}(w_i^2, w_{i+1}^1)$$
$$+ d_{G_1 \times P_2}(w_r^2, (w', v_{i'}))$$

$$\leq d_{G_1 \times P_2}((w, v_i), w_1^1) + d_{G_1 \times P_2}(w_r^2, (w', v_{i'})) + (4K + M + 1) \sum_{i=1}^{r-1} d_{G_1 \times P_2}(w_i^2, w_{i+1}^1)$$
$$+ d_{G_1 \times P_2}(w_r^1, w_r^2)$$

$$= d_{G^*}(F(w, v_i), F(w_r^1)) + d_{G^*}(F(w_r^2), F(w', v_{i'})) + (4K + M + 1) \sum_{i=1}^{r-1} d_{G^*}(F(w_i^2), F(w_{i+1}^1))$$
$$+ d_{G_1 \times P_2}(w_r^1, w_r^2)$$

$$\leq (4K + M + 1) \left(d_{G^*}(F(w, v_i), F(w_r^1)) + d_{G^*}(F(w_r^2), F(w', v_{i'})) + \sum_{i=1}^{r-1} d_{G^*}(F(w_i^2), F(w_{i+1}^1)) \right) + 4K + M$$

$$\leq (4K + M + 1) d_{G^*}(F(w, v_i), F(w', v_{i'})) + 4K + M.$$

We conclude that $F : V(G_1 \times P_2) \to V(G^*)$ is a quasi-isometric embedding. Thus, Lemma 1 provides a quasi-isometry $f : G_1 \times P_2 \to G^*$ with the required property. $\square$

Definition 4. *Given a geodesic metric space X and closed connected pairwise disjoint subsets $\{\eta_j\}_{j \in J}$ of X, we consider another copy X' of X. The double DX of X is the union of X and X' obtained by identifying the corresponding points in each η_j and η_j'.*

Definition 5. *Let us consider $H > 0$, a metric space X, and subsets $Y, Z \subseteq X$. The set $V_H(Y) := \{x \in X : d(x, Y) \leq H\}$ is called the H-neighborhood of Y in X. The Hausdorff distance of Y to Z is defined by $\mathcal{H}(Y, Z) := \inf\{H > 0 : Y \subseteq V_H(Z), Z \subseteq V_H(Y)\}$.*

The following results in [15,80] will be useful.

Theorem 13. ([*80*], *Theorem 3.2*) *Let us consider a geodesic metric space X and closed connected pairwise disjoint subsets* $\{\eta_j\}_{j \in J}$ *of X, such that the double DX is a geodesic metric space. Then, the following conditions are equivalent:*

(1)   *DX is hyperbolic.*
(2)   *X is hyperbolic and there exists a constant c_1 such that for every $k, l \in J$ and $a \in \eta_k, b \in \eta_l$ we have*
       $d_X(x, \cup_{j \in J} \eta_j) \leq c_1$ *for every $x \in [ab] \subset X$.*
(3)   *X is hyperbolic and there exist constants c_2, α, β such that for every $k, l \in J$ and $a \in \eta_k, b \in \eta_l$ we have*
       $d_X(x, \cup_{j \in J} \eta_j) \leq c_2$ *for every x in some (α, β)-quasi-geodesic joining a with b in X.*

Theorem 14. ([*15*], *p. 87*) *For each $\delta \geq 0$, $a \geq 1$ and $b \geq 0$, there exists a constant $H = H(\delta, a, b)$ with the following property:*

Let us consider a δ-hyperbolic geodesic metric space X and an (a, b)-quasigeodesic g starting in x and finishing in y. If γ is a geodesic joining x and y, then $\mathcal{H}(g, \gamma) \leq H$.

This property is called geodesic stability. It is well-known that hyperbolicity is, in fact, equivalent to geodesic stability [*81*].

Theorem 15. *Let G_1 be a connected graph and $B_j := B_{G_1}(w_j, K_j)$ with $w_j \in V(G_1)$ and $K_j \in \mathbb{Z}^+$, for any $j \in J$, such that $\sup_j K_j = K < \infty$, $\overline{B}_{j_1} \cap \overline{B}_{j_2} = \emptyset$ if $j_1 \neq j_2$, and every odd cycle C in G_1 satisfies $C \cap B_j \neq \emptyset$ for some $j \in J$. Suppose $\{B_j\}_{j \in J}$ is M-regular for some $M > 0$. Let G_2 be a non-trivial bounded connected graph without odd cycles. Then, the following statements are equivalent:*

(1) $G_1 \times G_2$ *is hyperbolic.*
(2) G_1 *is hyperbolic and there exists a constant c_1, such that for every $k, l \in J$ and $w_k \in B_k$, $w_l \in B_l$ there exists a geodesic $[w_k w_l]$ in G_1 with $d_{G_1}(x, \cup_{j \in J} w_j) \leq c_1$ for every $x \in [w_k w_l]$.*
(3) G_1 *is hyperbolic and there exist constants c_2, α, β, such that for every $k, l \in J$ we have $d_{G_1}(x, \cup_{j \in J} w_j) \leq c_2$ for every x in some (α, β)-quasi-geodesic joining w_k with w_l in G_1.*

Proof. Items (2) and (3) are equivalent by geodesic stability in G_1 (see Theorem *14*).

Assume that (2) holds. By Lemma *7*, there exists an (α, β)-quasi-isometry $f : G_1 \rightarrow G_1^*$ with $f(w_j) = w_j^*$ for every $j \in J$. Given $k, l \in J$, $f([w_k w_l])$ is an (α, β)-quasi-geodesic with endpoints w_k^* and w_l^* in G_1^*. Given $x \in f([w_k w_l])$, we have $x = f(x_0)$ with $x_0 \in [w_k w_l]$ and $d_{G_1^*}(x, \cup_{j \in J} w_j^*) \leq \alpha d_{G_1}(x_0, \cup_{j \in J} w_j) + \beta \leq \alpha c_1 + \beta$. Taking $X = G_1^*, DX = G^*$ and $\eta_j = w_j^*$ for every $j \in J$, Theorem *13* gives that G^* is hyperbolic. Now, Lemma *8* gives that $G_1 \times P_2$ is hyperbolic and we conclude that $G_1 \times G_2$ is hyperbolic by Lemma *2*.

Now, suppose (1) holds. By Lemma *2*, $G_1 \times P_2$ is hyperbolic and, by Theorem *9*, G_1 is hyperbolic. Then, Lemma *8* gives that G^* is hyperbolic and taking $X = G_1^*, DX = G^*$ and $\eta_j = w_j^*$ for every $j \in J$, by Theorem *13*, (2) holds. $\square$

Theorem *15* and Remark *2* have the following consequence.

Corollary 5. *Let G_1 be a connected graph and suppose that there are a positive integer K and a vertex $w \in G_1$, such that every odd cycle in G_1 intersects the open ball $B := B_{G_1}(w, K)$. Let G_2 be a non-trivial bounded connected graph without odd cycles. Then, $G_1 \times G_2$ is hyperbolic if and only if G_1 is hyperbolic.*

4. Bounds for the Hyperbolicity Constant of Some Direct Products

The following well-known result will be useful (see a proof, e.g., in ([*47*], Theorem 8)).

Theorem 16. *In any connected graph G the inequality $\delta(G) \leq (\text{diam } G)/2$ holds.*

Remark 3. *Note that, if G_1 is a bipartite connected graph, then $\text{diam } G_1 = \text{diam } V(G_1)$. Furthermore, if G_2 is a bipartite connected graph, then the product $G_1 \times G_2$ has exactly two connected components, which are denoted* by $(G_1 \times G_2)^1$ and $(G_1 \times G_2)^2$, where each one is a bipartite graph and, consequently, $\mathrm{diam}(G_1 \times G_2)^i = \mathrm{diam}\, V((G_1 \times G_2)^i)$ for $i \in \{1, 2\}$.

Remark 4. *Let P_m, P_n be two path graphs with $m \geq n \geq 2$. The product $P_m \times P_n$ has exactly two connected components, which will be denoted by $(P_m \times P_n)^1$ and $(P_m \times P_n)^2$. If $u, v \in V((P_m \times P_n)^i)$ for $i \in \{1, 2\}$, then $d_{(P_m \times P_n)^i}(u, v) = \max\{d_{P_m}(\pi_1(u), \pi_1(v)), d_{P_n}(\pi_2(u), \pi_2(v))\}$ and $\mathrm{diam}(P_m \times P_n)^i = \mathrm{diam}\, V((P_m \times P_n)^i) = m - 1$.*

Furthermore, if $m_1 \leq m$ and $n_1 \leq n$, then $\delta(P_m \times P_n) \geq \delta(P_{m_1} \times P_{n_1})$.

Lemma 9. *Let P_m, P_n be two path graphs with $m \geq n \geq 3$, and let γ be a geodesic in $P_m \times P_n$ such that there are two different vertices u, v in γ, with $\pi_1(u) = \pi_1(v)$. Then, $L(\gamma) \leq n - 1$.*

Remark 5. *Note that the conclusion of Lemma 9 does not hold for $n = 2$, since we always have $L(\gamma) \geq 2$.*

Proof. Let $\gamma := [xy]$, and let $V(P_m) = \{v_1, \ldots, v_m\}, V(P_n) = \{w_1, \ldots, w_n\}$ be the sets of vertices in P_m, P_n, respectively, such that $[v_j, v_{j+1}] \in E(P_m)$ and $[w_i, w_{i+1}] \in E(P_n)$ for $1 \leq j < m, 1 \leq i < n$. Seeking for a contradiction, assume that $L(\gamma) > n - 1$. Notice that if $[uv]$ denotes the geodesic contained in γ joining u and v, then π_2 restricted to $[uv]$ is injective. Consider two vertices $u', v' \in \gamma$ such that $[uv] \subseteq [u'v'] \subseteq \gamma$, π_2 is injective in $[u'v']$ and $\pi_2(u') = w_{i_1}, \pi_2(v') = w_{i_2}$ with $i_2 - i_1$ maximal under these conditions. See Figure 4.

Since $L(\gamma) > n - 1 \geq i_2 - i_1$, either there is an edge $[v', w]$ in $G_1 \times G_2$ such that $[v', w] \cap (\gamma \setminus [u'v']) \neq \varnothing$ or there is an edge $[u', w']$ in $G_1 \times G_2$ such that $[u', w'] \cap (\gamma \setminus [u'v']) \neq \varnothing$. In addition, since $L(\gamma) > n - 1$, notice that π_2 is not injective in γ. Moreover, since $i_2 - i_1$ is maximal, if $\pi_2(w) = w_{i_2+1}$, then $w \notin \gamma$, and since $L(\gamma) > n - 1, u' \notin \{x, y\}$ and $\pi_2(w') = w_{i_1+1}$. Thus, either $\pi_2(w) = w_{i_2-1}$ or $\pi_2(w') = w_{i_1+1}$.

Hence, let us assume that there is an edge $[v', w]$ in $G_1 \times G_2$ such that $[v', w] \cap (\gamma \setminus [u'v']) \neq \varnothing$ with $\pi_2(w) = w_{i_2-1}$ (otherwise, if there is an edge $[u', w']$ in $G_1 \times G_2$ such that $[u', w'] \cap (\gamma \setminus [u'v']) \neq \varnothing$ with $\pi_2(w') = w_{i_1+1}$, the proof is similar).

Suppose $\pi_1(v') = v_j$. Let v'' be the vertex in $[u'v']$ such that $\pi_2(v'') = w_{i_2-1}$. Then, by construction of $G_1 \times G_2$, since $v'' \neq w$, it follows that $\{\pi_1(v''), \pi_1(w)\} = \{v_{j-1}, v_{j+1}\}$. Therefore, in particular, $1 < j < m$.

Assume that $v'' = (v_{j-1}, w_{i_2-1})$ (if $v'' = (v_{j+1}, w_{i_2-1})$, then the argument is similar). Therefore, $w = (v_{j+1}, w_{i_2-1})$.

Consider the geodesic

$$\sigma = [(v_{j+1}, w_{i_2-1}), (v_j, w_{i_2-2})] \cup [(v_j, w_{i_2-2}), (v_{j-1}, w_{i_2-3})] \cup [(v_{j-1}, w_{i_2-3}), (v_{j-2}, w_{i_2-4})] \cup \ldots$$

Since $\pi_1(u) = \pi_1(v)$, there is a vertex ξ of $V(P_m \times P_n)$ in $[u'v'] \cap \sigma$. Let $s \in [v', w] \cap \gamma$ with $s \neq v'$. Let σ_0 be the geodesic contained in σ joining ξ and w. Let γ_0 be the geodesic contained in γ joining ξ and s. Hence, $L(\sigma_0 \cup [ws]) < L(\sigma_0) + 1 < L(\gamma_0)$ leading to contradiction. $\square$

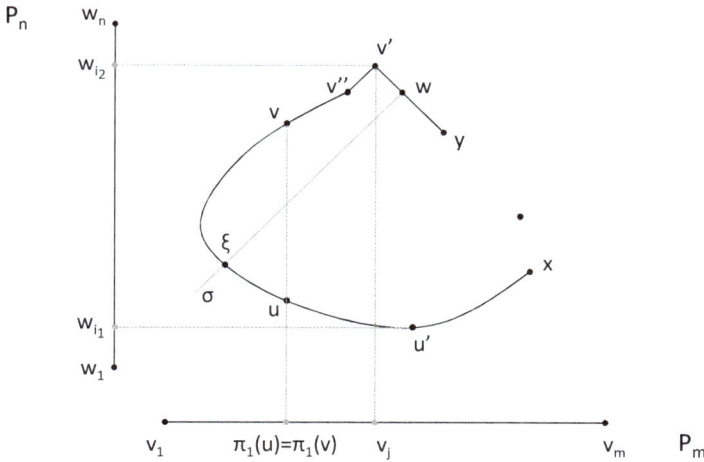

Figure 4. For any geodesic γ in $P_m \times P_n$ with $\pi_1(u) = \pi_1(v)$ for some different vertices u, v in γ, then $L(\gamma) \leq n - 1$.

Theorem 17. *Let P_m, P_n be two path graphs with $m \geq n \geq 2$. If $n = 2$, then $\delta(P_m \times P_2) = 0$. If $n \geq 3$, then*

$$\min\left\{\frac{m}{2}, n-1\right\} - 1 \leq \delta(P_m \times P_n) \leq \min\left\{\frac{m}{2}, n\right\} - \frac{1}{2}.$$

Furthermore, if $m \leq 2n - 3$ and m is odd, then $\delta(P_m \times P_n) = (m-1)/2$.

Proof. If $m \geq 2$, then $P_m \times P_2$ has two connected components isomorphic to P_m, and $\delta(P_m \times P_2) = 0$.

Assume that $n \geq 3$. By symmetry, it suffices to prove the inequalities for $\delta((P_m \times P_n)^1)$. Hence, Theorem 16 and Remark 4 give $\delta((P_m \times P_n)^1) \leq \frac{m-1}{2}$. By Theorem 7, there exists a geodesic triangle $T = \{x, y, z\} \in \mathbb{T}_1(P_m \times P_n)$ with $p \in \gamma_1 := [xy], \gamma_2 := [xz], \gamma_3 := [yz]$, and $\delta((P_m \times P_n)^1) = \delta(T) = d_{(P_m \times P_n)^1}(p, \gamma_2 \cup \gamma_3)$. Let $u \in V(\gamma_1)$ such that $d_{(P_m \times P_n)^1}(p, u) \leq 1/2$.

To prove $\delta((P_m \times P_n)^1) \leq n - 1/2$, we consider two cases.

Assume first that there is at least a vertex $v \in V((P_m \times P_n)^1) \cap T \setminus \{u\}$ such that $\pi_1(u) = \pi_1(v)$. If $v \notin \gamma_1$, then $v \in \gamma_2 \cup \gamma_3$ and

$$\delta(T) = d_{(P_m \times P_n)^1}(p, \gamma_2 \cup \gamma_3) \leq 1/2 + d_{(P_m \times P_n)^1}(u, v) \leq n - 1/2.$$

If $v \in \gamma_1$, then $L(\gamma_1) \leq n - 1$ by Lemma 9, and

$$\delta(T) = d_{(P_m \times P_n)^1}(p, \gamma_2 \cup \gamma_3) \leq d_{(P_m \times P_n)^1}(p, \{x, y\}) \leq (n-1)/2 < n - 1/2.$$

Assume now that there is not a vertex $v \in V((P_m \times P_n)^1) \cap T \setminus \{u\}$ such that $\pi_1(u) = \pi_1(v)$. Then, there exist two different vertices v_1, v_2 in $T \setminus \{u\}$ such that $d_{(P_m \times P_n)^1}(u, v_1) = d_{(P_m \times P_n)^1}(u, v_2) = 1$, and $\pi_1(v_1) = \pi_1(v_2)$. If v_1 or v_2 belongs to $\gamma_2 \cup \gamma_3$, then $\delta(T) = d_{(P_m \times P_n)^1}(p, \gamma_2 \cup \gamma_3) \leq 3/2 \leq n - 1/2$. Otherwise, $v_1, v_2 \in \gamma_1 \setminus \{u\}$. Lemma 9 gives $L(\gamma_1) \leq n - 1$, and we have that

$$\delta(T) = d_{(P_m \times P_n)^1}(p, \gamma_2 \cup \gamma_3) \leq d_{(P_m \times P_n)^1}(p, \{x, y\}) \leq (n-1)/2 < n - 1/2.$$

To prove the lower bound, denote the vertices of P_m and P_n by $V(P_m) = \{w_1, w_2, w_3, \ldots, w_m\}$ and $V(P_n) = \{v_1, v_2, v_3, \ldots, v_n\}$, with $[w_i, w_{i+1}] \in E(P_m)$ for $1 \le i < m$ and $[v_i, v_{i+1}] \in E(P_n)$ for $1 \le i < n$.

Let $(P_m \times P_n)^1$ be the connected component of $P_m \times P_n$ containing (w_1, v_{n-1}).

Assume first that $m \ge 2n - 3$. Consider the following curves in $(P_m \times P_n)^1$:

$$\gamma_1 := [(w_1, v_{n-1}), (w_2, v_n)] \cup [(w_2, v_n), (w_3, v_{n-1})] \cup [(w_3, v_{n-1}), (w_4, v_n)] \cup \cdots \cup [(w_{2n-4}, v_n), (w_{2n-3}, v_{n-1})],$$

$$\gamma_2 := [(w_1, v_{n-1}), (w_2, v_{n-2})] \cup [(w_2, v_{n-2}), (w_3, v_{n-3})] \cup \cdots \cup [(w_{n-2}, v_2), (w_{n-1}, v_1)] \cup [(w_{n-1}, v_1), (w_n, v_2)]$$
$$\cup \cdots \cup [(w_{2n-4}, v_{n-2}), (w_{2n-3}, v_{n-1})].$$

Corollary 1 gives that γ_1, γ_2 are geodesics. If B is the geodesic bigon $B = \{\gamma_1, \gamma_2\}$, then Remark 4 gives that

$$\delta(P_m \times P_n) \ge \delta(B) \ge d_{(P_m \times P_n)^1}((w_{n-1}, v_1), \gamma_1) = n - 2.$$

If m is odd with $m \le 2n - 3$, then $n - (m+1)/2 \ge 1$ and we can consider the curves in $(P_m \times P_n)^1$:

$$\gamma_1 := [(w_1, v_{n-1}), (w_2, v_n)] \cup [(w_2, v_n), (w_3, v_{n-1})] \cup [(w_3, v_{n-1}), (w_4, v_n)] \cup \cdots \cup [(w_{m-1}, v_n), (w_m, v_{n-1})],$$

$$\gamma_2 := [(w_1, v_{n-1}), (w_2, v_{n-2})] \cup [(w_2, v_{n-2}), (w_3, v_{n-3})] \cup \cdots \cup [(w_{(m+1)/2-1}, v_{n-(m+1)/2+1}), (w_{(m+1)/2}, v_{n-(m+1)/2})]$$
$$\cup [(w_{(m+1)/2}, v_{n-(m+1)/2}), (w_{(m+1)/2+1}, v_{n-(m+1)/2+1})] \cup \cdots \cup [(w_{m-1}, v_{n-2}), (w_m, v_{n-1})].$$

Corollary 1 gives that γ_1, γ_2 are geodesics. If $B = \{\gamma_1, \gamma_2\}$, then Remark 4 gives that

$$\delta(P_m \times P_n) \ge \delta(B) \ge d_{(P_m \times P_n)^1}((w_{(m+1)/2}, v_{n-(m+1)/2}), \gamma_1) = (m-1)/2.$$

By Remark 4, if m is even with $m - 1 \le 2n - 3$, then we have that

$$\delta(P_m \times P_n) \ge \delta(P_{m-1} \times P_n) \ge (m-2)/2.$$

Hence,

$$\delta(P_m \times P_n) \ge \begin{cases} n - 2, & \text{if } m \ge 2n - 3 \\ (m-2)/2, & \text{if } m \le 2n - 2 \end{cases} = \min\left\{n - 2, \frac{m-2}{2}\right\} = \min\left\{\frac{m}{2}, n - 1\right\} - 1.$$

Furthermore, if $m \le 2n - 3$ and m is odd, then we have proven $(m - 1)/2 \le \delta(P_m \times P_n) \le (m - 1)/2$. $\square$

Theorem 18. *If G_1 and G_2 are bipartite connected graphs with $k_1 := \operatorname{diam} V(G_1)$ and $k_2 := \operatorname{diam} V(G_2)$ such that $k_1 \ge k_2 \ge 1$, then*

$$\max\left\{\min\left\{\frac{k_1 - 1}{2}, k_2 - 1\right\}, \delta(G_1), \delta(G_2)\right\} \le \delta(G_1 \times G_2) \le \frac{k_1}{2}.$$

Furthermore, if $k_1 \le 2k_2 - 2$ and k_1 is even, then $\delta(G_1 \times G_2) = k_1/2$.

Proof. Corollary 1, Theorem 16 and Remark 3 give us the upper bound.

To prove the lower bound, we can see that there exist two path graphs P_{k_1+1}, P_{k_2+1} which are isometric subgraphs of G_1 and G_2, respectively. It is easy to check that $P_{k_1+1} \times P_{k_2+1}$ is an isometric subgraph of $G_1 \times G_2$. By Lemma 3 and Theorem 17, we have

$$\min\left\{\frac{k_1 - 1}{2}, k_2 - 1\right\} \le \delta(P_{k_1+1} \times P_{k_2+1}) \le \delta(G_1 \times G_2).$$

Using a similar argument as above, we have $\delta(P_2 \times G_2) \le \delta(G_1 \times G_2)$ and $\delta(G_1 \times P_2) \le \delta(G_1 \times G_2)$. Thus, since $(G_1 \times P_2)^i \simeq G_1$ and $(P_2 \times G_2)^i \simeq G_2$ for $i \in \{1, 2\}$, we obtain the first statement.

Furthermore, if $k_1 + 1 \leq 2(k_2 + 1) - 3$ and $k_1 + 1$ is odd, then Theorem 17 gives $\delta(P_{k_1+1} \times P_{k_2+1}) = k_1/2$, and we conclude $\delta(G_1 \times G_2) = k_1/2$. $\square$

The following result deals just with odd cycles since otherwise we can apply Theorem 18.

Theorem 19. *For every odd number $m \geq 3$ and every $n \geq 2$,*

$$\delta(C_m \times P_n) = \begin{cases} m/2, & \text{if } n - 1 \leq m, \\ (n-1)/2, & \text{if } m < n - 1 < 2m, \\ m - 1/2, & \text{if } n - 1 \geq 2m. \end{cases}$$

Proof. Let $V(C_m) = \{w_1, \ldots, w_m\}$ and $V(P_n) = \{v_1, \ldots, v_n\}$ be the sets of vertices in C_m and P_n, respectively, such that $[w_1, w_m], [w_j, w_{j+1}] \in E(C_m)$ and $[v_i, v_{i+1}] \in E(P_n)$ for $j \in \{1, \ldots, m-1\}$, $i \in \{1, \ldots, n-1\}$. Note that for $1 \leq j, r \leq m$ and $1 \leq i, s \leq n$, we have $d_{C_m \times P_n}((w_j, v_i), (w_r, v_s)) = \max\{|i - s|, |j - r|\}$, if $|i - s| \equiv |j - r| \pmod 2$, or $d_{C_m \times P_n}((w_j, v_i), (w_r, v_s)) = \max\{|i - s|, m - |j - r|\}$, if $|i - s| \not\equiv |j - r| \pmod 2$. Besides, we have $\operatorname{diam}(C_m \times P_n) = \operatorname{diam}(V(C_m \times P_n))$, i.e., $\operatorname{diam}(C_m \times P_n) = m$ if $n - 1 \leq m$, and $\operatorname{diam}(C_m \times P_n) = n - 1$ if $n - 1 > m$. Thus, by Theorem 16, we have

$$\delta(C_m \times P_n) \leq \begin{cases} m/2, & \text{if } n - 1 \leq m, \\ (n-1)/2, & \text{if } n - 1 > m. \end{cases}$$

Assume first that $n - 1 \leq m$. Note that $C_m \times P_2 \simeq C_{2m}$ and $C_m \times P_{n'}$ is an isometric subgraph of $C_m \times P_n$, if $n' \leq n$. By Lemma 3, we have $\delta(C_m \times P_n) \geq \delta(C_{2m}) = m/2$, and we obtain the result in this case.

Assume now that $n - 1 > m$. Consider the geodesic triangle T in $C_m \times P_n$ defined by the following geodesics

$$\gamma_1 := [(w_1, v_n), (w_2, v_{n-1})] \cup [(w_2, v_{n-1}), (w_3, v_n)] \cup [(w_3, v_n), (w_4, v_{n-1})] \cup \ldots \cup [(w_{m-1}, v_{n-1}), (w_m, v_n)],$$

$$\gamma_2 := [(w_{(m+1)/2}, v_1), (w_{(m-1)/2}, v_2)] \cup [(w_{(m-1)/2}, v_2), (w_{(m-3)/2}, v_3)] \cup \ldots \cup [(w_2, v_{(m-1)/2}), (w_1, v_{(m+1)/2})] \cup$$
$$[(w_1, v_{(m+1)/2}), (w_m, v_{(m+3)/2})] \cup [(w_m, v_{(m+3)/2}), (w_1, v_{(m+5)/2})] \cup [(w_1, v_{(m+5)/2}), (w_m, v_{(m+7)/2})] \cup \ldots,$$

$$\gamma_3 := [(w_{(m+1)/2}, v_1), (w_{(m+3)/2}, v_2)] \cup [(w_{(m+3)/2}, v_2), (w_{(m+5)/2}, v_3)] \cup \ldots \cup [(w_{m-1}, v_{(m-1)/2}), (w_m, v_{(m+1)/2})] \cup$$
$$[(w_m, v_{(m+1)/2}), (w_1, v_{(m+3)/2})] \cup [(w_1, v_{(m+3)/2}), (w_m, v_{(m+5)/2})] \cup [(w_m, v_{(m+5)/2}), (w_1, v_{(m+7)/2})] \cup \ldots,$$

where (w_1, v_n) (respectively, (w_m, v_n)) is an endpoint of either γ_2 or γ_3, depending of the parity of n. Since T is a geodesic triangle in $C_m \times P_n$, we have $\delta(C_m \times P_n) \geq \delta(T)$. If $n - 1 < 2m$ and M is the midpoint of the geodesic γ_3, then $\delta(C_m \times P_n) \geq \delta(T) = d_{C_m \times P_n}(M, \gamma_1 \cup \gamma_2) = L(\gamma_3)/2 = (n-1)/2$. Therefore, the result for $m < n - 1 < 2m$ follows.

Finally, assume that $n - 1 \geq 2m$. Let us consider $N \in \gamma_3$ such that $d_{C_m \times P_n}(N, (w_{(m+1)/2}, v_1)) = m - 1/2$. Thus, $\delta(C_m \times P_n) \geq \delta(T) \geq d_{C_m \times P_n}(N, \gamma_1 \cup \gamma_2) = d_{C_m \times P_n}(N, (w_{(m+1)/2}, v_1)) = m - 1/2$. To finish the proof, it suffices to prove that $\delta(C_m \times P_n) \leq m - 1/2$. Seeking for a contradiction, assume that $\delta(C_m \times P_n) > m - 1/2$. By Theorems 6 and 7, there is a geodesic triangle $\triangle = \{x, y, z\} \in \mathbb{T}_1(C_m \times P_n)$ and $p \in [xy]$ with $d_{C_m \times P_n}(p, [yz] \cup [zx]) = \delta(C_m \times P_n) \geq m - 1/4$. Then, $L([xy]) = d_{C_m \times P_n}(x, p) + d_{C_m \times P_n}(p, y) \geq 2m - 1/2$. Let V_x (respectively, V_y) be the closest vertex to x (respectively, y) in $[xy]$, and consider a vertex V_p in $[xy]$ such that $d_{C_m \times P_n}(p, V(C_m \times P_n)) = d_{C_m \times P_n}(p, V_p)$. Note that $d_{C_m \times P_n}(p, [yz] \cup [zx]) \geq m - 1/4$ implies that $d_{C_m \times P_n}(p, V_p) \leq 1/2$. Since $x, y, z \in J(C_m \times P_n)$, we have $d_{C_m \times P_n}(V_x, V_y) \geq 2m - 1 > m$ and, consequently, $\pi_2([xy])$ is a geodesic in P_n. Since $\pi_2([yz] \cup [zx])$ is a path in P_n joining $\pi_2(x)$ and $\pi_2(y)$, there exists a vertex $(u, v) \in [xz] \cup [zy]$ such that $\pi_2(V_p) = v$ and $u \neq \pi_1(V_p)$. Therefore, $d_{C_m \times P_n}(V_p, (u, v)) \leq m - 1$ and, consequently, $d_{C_m \times P_n}(p, [xz] \cup [zy]) \leq d_{C_m \times P_n}(p, V_p) + d_{C_m \times P_n}(V_p, [xz] \cup [zy]) \leq 1/2 + m - 1$, leading to contradiction. $\square$

5. Conclusions

In this paper, we characterize in many cases the hyperbolic direct product of graphs. Here, the situation is more complex than with other graph products, partly because the direct product of two bipartite graphs is already disconnected and the formula for the distance in $G_1 \times G_2$ is more complicated than in the case of other products of graphs. Although in the study of hyperbolicity in a general context the hypothesis on the existence (or non-existence) of odd cycles is artificial, in the study of hyperbolic direct products, it is an essential hypothesis. We have proven that, if $G_1 \times G_2$ is hyperbolic, then one factor is hyperbolic and the other one is bounded. Besides, we prove that this necessary condition is also sufficient in many cases. If G_1 is a hyperbolic graph and G_2 is a bounded graph, then we prove that $G_1 \times G_2$ is hyperbolic when G_2 has some odd cycle or G_1 and G_2 do not have odd cycles. Otherwise, the characterization of hyperbolic direct products is a more difficult task. If G_1 has some odd cycle and G_2 does not have odd cycles, we provide sufficient conditions for non-hyperbolicity and hyperbolicity, respectively. Besides, we characterize the hyperbolicity of $G_1 \times G_2$ under some additional conditions.

A natural open problem is the complete characterization of hyperbolic direct products.

A second open problem is to compute the precise value of the hyperbolicity constant of the graphs appearing in Theorems 17 and 18 with unknown hyperbolicity constant.

Direct product of graphs is a subject closely related to lift of graphs, which have been intensively studied (see, e.g., [82] and the references therein). Another interesting problem is to study the hyperbolicity of lift of graphs. We think that it is possible to obtain some similar results in this context, although the odd cycles may not play an important role in the study of hyperbolic lifts of graphs.

Author Contributions: The authors contributed equally to this work.

Funding: This work was supported in part by four grants from Ministerio de Economía y Competititvidad (MTM2012-30719, MTM2013-46374-P, MTM2016-78227-C2-1-P and MTM2015-69323-REDT), Spain.

Acknowledgments: We thank the referees for their suggestions and helpful remarks.

References

1. Hammack, R.; Imrich, W.; Klavžar, S. *Handbook of Product Graphs*, 2rd ed.; Discrete Mathematics and Its Applications Series; CRC Press: Boca Raton, FL, USA, 2011.
2. Weichsel, P.M. The Kronecker product of graphs. *Proc. Am. Math. Soc.* **1962**, *13*, 47–52. [CrossRef]
3. Bendall, S.; Hammack, R. Centers of *n*-fold tensor products of graphs. *Discuss. Math. Graph Theory* **2004**, *24*, 491–501. [CrossRef]
4. Brešar, B.; Imrich, W.; Klavžar, S.; Zmazek, B. Hypercubes as direct products. *SIAM J. Discret. Math.* **2005**, *18*, 778–786. [CrossRef]
5. Hammack, R. Minimum cycle bases of direct products of bipartite graphs. *Australas. J. Combin.* **2006**, *36*, 213–222.
6. Imrich, W.; Rall, D.F. Finite and infinite hypercubes as direct products. *Australas. J. Combin.* **2006**, *36*, 83–90.
7. Imrich, W.; Stadler, P. A prime factor theorem for a generalized direct product. *Discuss. Math. Graph Theory* **2006**, *26*, 135–140. [CrossRef]
8. Jha, P.K.; Klavžar, S. Independence in direct-product graphs. *Ars Combin.* **1998**, *50*, 53–63.
9. Balakrishnan, R.; Paulraja, P. Hamilton cycles in tensor product of graphs. *Discret. Math.* **1998**, *186*, 1–13. [CrossRef]
10. Kheddouci, H.; Kouider, M. Hamiltonian cycle decomposition of Kronecker product of some cubic graphs by cycles. *J. Combin. Math. Combin. Comput.* **2000**, *32*, 3–22.

11. Imrich, W.; Klavžar, S. *Product Graphs: Structure and Recognition*; John Wiley & Sons: New York, NY, USA, 2000.
12. Zhu, X. A survey on Hedetniemi's conjecture. *Taiwanese J. Math.* **1998**, *2*, 1–24. [CrossRef]
13. Imrich, W. Factoring cardinal product graphs in polynomial time. *Discr. Math.* **1998**, *192*, 119–144. [CrossRef]
14. Alonso, J.; Brady, T.; Cooper, D.; Delzant, T.; Ferlini, V.; Lustig, M.; Mihalik, M.; Shapiro, M.; Short, H. Notes on word hyperbolic groups. In *Group Theory from a Geometrical Viewpoint*; Ghys, E., Haefliger, A., Verjovsky, A., Eds.; World Scientific: Singapore, 1992.
15. Ghys, E.; de la Harpe, P. *Sur les Groupes Hyperboliques d'après Mikhael Gromov*; Progress in Mathematics 83; Birkhäuser Boston Inc.: Boston, MA, USA, 1990.
16. Gromov, M. Hyperbolic groups. In *Essays in Group Theory*; Gersten, S.M., Ed.; Mathematical Sciences Research Institute Publications; Springer: Berlin, Germany, 1987; Volume 8, pp. 75–263.
17. Oshika, K. *Discrete Groups*; AMS Bookstore: Providence, RI, USA, 2002.
18. Charney, R. Artin groups of finite type are biautomatic. *Math. Ann.* **1992**, *292*, 671–683. [CrossRef]
19. Chepoi, V.; Estellon, B. Packing and covering δ-hyperbolic spaces by balls. In *Approximation, Randomization, and Combinatorial Optimization. Algorithms and Techniques*; Springer: Berlin/Heidelberg, Germany, 2007; pp. 59–73.
20. Eppstein, D. Squarepants in a tree: Sum of subtree clustering and hyperbolic pants decomposition. In Proceedings of the eighteenth annual ACM-SIAM symposium on Discrete algorithms (SODA'2007), Barcelona, Spain, 16–19 January 2017; pp. 29–38.
21. Gavoille, C.; Ly, O. Distance labeling in hyperbolic graphs. In *Proceedings of the International Symposium on Algorithms and Computation (ISAAC)*; Springer: Berlin/Heidelberg, Germany, 2005; pp. 171–179.
22. Krauthgamer, R.; Lee, J.R. Algorithms on negatively curved spaces. In Proceedings of the 47th Annual IEEE Symposium on Foundations of Computer Science (FOCS'06), Berkeley, CA, USA, 21–24 October 2006; pp. 1–11.
23. Shavitt, Y.; Tankel, T. On Internet embedding in hyperbolic spaces for overlay construction and distance estimation. *IEEE/ACM Trans. Netw.* **2008**, *16*, 25–36. [CrossRef]
24. Verbeek, K.; Suri, S. Metric embeddings, hyperbolic space and social networks. In Proceedings of the 30th Annual Symposium on Computational Geometry, Kyoto, Japan, 8–11 June 2014; pp. 501–510.
25. Wu, Y.; Zhang, C. Chordality and hyperbolicity of a graph. *Electr. J. Comb.* **2011**, *18*, P43.
26. Abu-Ata, M.; Dragan, F.F. Metric tree-like structures in real-life networks: An empirical study. *Networks* **2016**, *67*, 49–68. [CrossRef]
27. Adcock, A.B.; Sullivan, B.D.; Mahoney, M.W. Tree-like structure in large social and information networks. In Proceedings of the 13th Int Conference Data Mining (ICDM), Dallas, TX, USA, 7–10 December 2013; pp. 1–10.
28. Krioukov, D.; Papadopoulos, F.; Kitsak, M.; Vahdat, A.; Boguná, M. Hyperbolic geometry of complex networks. *Phys. Rev. E* **2010**, *82*, 036106. [CrossRef] [PubMed]
29. Montgolfier, F.; Soto, M.; Viennot, L. Treewidth and Hyperbolicity of the Internet. In Proceedings of the 10th IEEE International Symposium on Network Computing and Applications (NCA), Cambridge, MA, USA, 25–27 August 2011; pp. 25–32.
30. Chepoi, V.; Dragan, F.F.; Vaxès, Y. Core congestion is inherent in hyperbolic networks. In Proceedings of the Twenty-Eighth Annual ACM-SIAM Symposium on Discrete Algorithms, Barcelona, Spain, 16–19 January 2017; pp. 2264–2279.
31. Grippo, E.; Jonckheere, E.A. Effective resistance criterion for negative curvature: application to congestion control. In Proceedings of the 2016 IEEE Conference on Control Applications (CCA), Buenos Aires, Argentina, 19–22 September 2016.
32. Jonckheere, E.A. Contrôle du traffic sur les réseaux à géométrie hyperbolique—Vers une théorie géométrique de la sécurité l'acheminement de l'information. *J. Eur. Syst. Autom.* **2002**, *8*, 45–60.
33. Jonckheere, E.A.; Lohsoonthorn, P. Geometry of network security. In Proceedings of the 2004 American Control Conference, Boston, MA, USA, 30 June–2 July 2004; pp. 111–151.
34. Li, S.; Tucci, G.H. Traffic Congestion in Expanders, (p, δ)-Hyperbolic Spaces and Product of Trees. *Int. Math.* **2015**, *11*, 134–142.
35. Shang, Y. Lack of Gromov-hyperbolicity in colored random networks. *Pan-Am. Math. J.* **2011**, *21*, 27–36.

36. Shang, Y. Lack of Gromov-hyperbolicity in small-world networks. *Cent. Eur. J. Math.* **2012**, *10*, 1152–1158. [CrossRef]

37. Shang, Y. Non-hyperbolicity of random graphs with given expected degrees. *Stoch. Models* **2013**, *29*, 451–462. [CrossRef]

38. Gosak, M.; Markovič, R.; Dolenšek, J.; Slak Rupnik, M.; Marhl, M.; Stožer, A.; Perc, M. Network science of biological systems at different scales: A review. *Phys. Life Rev.* **2018**, *24*, 118–135. [CrossRef] [PubMed]

39. Tourís, E. Graphs and Gromov hyperbolicity of non-constant negatively curved surfaces. *J. Math. Anal. Appl.* **2011**, *380*, 865–881. [CrossRef]

40. Bermudo, S.; Rodríguez, J.M.; Sigarreta, J.M. Computing the hyperbolicity constant. *Comput. Math. Appl.* **2011**, *62*, 4592–4595. [CrossRef]

41. Bermudo, S.; Rodríguez, J.M.; Sigarreta, J.M.; Vilaire, J.-M. Gromov hyperbolic graphs. *Discr. Math.* **2013**, *313*, 1575–1585. [CrossRef]

42. Brinkmann, G.; Koolen J.; Moulton, V. On the hyperbolicity of chordal graphs. *Ann. Comb.* **2001**, *5*, 61–69. [CrossRef]

43. Chepoi, V.; Dragan, F.F.; Estellon, B.; Habib, M.; Vaxes, Y. Notes on diameters, centers, and approximating trees of δ-hyperbolic geodesic spaces and graphs. *Electr. Notes Discr. Math.* **2008**, *31*, 231–234. [CrossRef]

44. Dragan, F.; Mohammed, A. Slimness of graphs. *arXiv* **2017**, arXiv:1705.09797.

45. Martínez-Pérez, A. Chordality properties and hyperbolicity on graphs. *Electr. J. Comb.* **2016**, *23*, P3.51.

46. Michel, J.; Rodríguez, J.M.; Sigarreta, J.M.; Villeta, M. Gromov hyperbolicity in Cartesian product graphs. *Proc. Indian Acad. Sci. Math. Sci.* **2010**, *120*, 1–17. [CrossRef]

47. Rodríguez, J.M.; Sigarreta, J.M.; Vilaire, J.-M.; Villeta, M. On the hyperbolicity constant in graphs. *Discr. Math.* **2011**, *311*, 211–219. [CrossRef]

48. Chen, B.; Yau, S.-T.; Yeh, Y.-N. Graph homotopy and Graham homotopy. *Discret. Math.* **2001**, *241*, 153–170. [CrossRef]

49. Shang, Y. On the likelihood of forests. *Phys. A Stat. Mech. Appl.* **2016**, *456*, 157–166. [CrossRef]

50. Fournier, H.; Ismail, A.; Vigneron, A. Computing the Gromov hyperbolicity of a discrete metric space. *Inf. Process. Lett.* **2015**, *115*, 576–579. [CrossRef]

51. Papasoglu, P. An algorithm detecting hyperbolicity. In *Geometric and Computational Perspectives on Infinite Groups, DIMACS*; Series in Discrete Mathematics and Theoretical Computer Science; AMS: Hong Kong, China, 1996; Volume 25, pp.193–200.

52. Chalopin, J.; Chepoi, V.; Papasoglu, P.; Pecatte, T. Cop and robber game and hyperbolicity. *SIAM J. Discr. Math.* **2015**, *28*, 1987–2007. [CrossRef]

53. Sigarreta, J.M. Hyperbolicity in median graphs. *Proc. Indian Acad. Sci. Math. Sci.* **2013**, *123*, 455–467. [CrossRef]

54. Carballosa, W.; Rodríguez, J.M.; Sigarreta, J.M. New inequalities on the hyperbolicity constant of line graphs. *Ars Combin.* **2016**, *129*, 367–386.

55. Carballosa, W.; Rodríguez, J.M.; Sigarreta, J.M.; Villeta, M. Gromov hyperbolicity of line graphs. *Electr. J. Comb.* **2011**, *18*, P210.

56. Cohen, N.; Coudert, D.; Ducoffe, G.; Lancin, A. Applying clique-decomposition for computing Gromov hyperbolicity. *Theor. Comp. Sci.* **2017**./j.tcs.2017.06.001. [CrossRef]

57. Pestana, D.; Rodríguez, J.M.; Sigarreta, J.M.; Villeta, M. Gromov hyperbolic cubic graphs. *Central Eur. J. Math.* **2012**, *10*, 1141–1151. [CrossRef]

58. Bermudo, S.; Rodríguez, J.M.; Sigarreta, J.M.; Tourís, E. Hyperbolicity and complement of graphs. *Appl. Math. Lett.* **2011**, *24*, 1882–1887. [CrossRef]

59. Hernández, J.C.; Rodríguez, J.M.; Sigarreta, J.M.; Torres-Nuñez, Y.; Villeta, M. Gromov hyperbolicity of regular graphs. *Ars Combin.* **2017**, *130*, 395–416.

60. Bermudo, S.; Carballosa, W.; Rodríguez, J.M.; Sigarreta, J.M. On the hyperbolicity of edge-chordal and path-chordal graphs. *Filomat* **2016**, *30*, 2599–2607. [CrossRef]

61. Carballosa, W.; Portilla, A.; Rodríguez, J.M.; Sigarreta, J.M. Planarity and hyperbolicity in graphs. *Graphs Combin.* **2015**, *31*, 1311–1324. [CrossRef]

62. Portilla, A.; Rodríguez, J.M.; Sigarreta, J.M.; Vilaire, J.-M. Gromov hyperbolic tessellation graphs. *Utilitas Math.* **2015**, *97*, 193–212.

63. Coudert, D.; Ducoffe, G. On the hyperbolicity of bipartite graphs and intersection graphs. *Discr. Appl. Math.* **2016**, *214*, 187–195. [CrossRef]

64. Calegari, D.; Fujiwara, K. Counting subgraphs in hyperbolic graphs with symmetry. *J. Math. Soc. Jpn.* **2015**, *67*, 1213–1226. [CrossRef]

65. Cantón, A.; Granados, A.; Pestana, D.; Rodríguez, J.M. Gromov hyperbolicity of periodic graphs. *Bull. Malays. Math. Sci. Soc.* **2016**, *39*, S89–S116. [CrossRef]

66. Cantón, A.; Granados, A.; Pestana, D.; Rodríguez, J.M. Gromov hyperbolicity of periodic planar graphs. *Acta Math. Sin.* **2014**, *30*, 79–90. [CrossRef]

67. Koolen, J.H.; Moulton, V. Hyperbolic Bridged Graphs. *Eur. J. Comb.* **2002**, *23*, 683–699. [CrossRef]

68. Rodríguez, J.M. Characterization of Gromov hyperbolic short graphs. *Acta Math. Sin.* **2014**, *30*, 197–212. [CrossRef]

69. Carballosa, W.; Rodríguez, J.M.; Rosario, O.; Sigarreta, J.M. Gromov hyperbolicity of graph minors. *Bull. Iran. Math. Soc.* **2018**, *44*, 481–503. [CrossRef]

70. Bermudo, S.; Rodríguez, J.M.; Rosario, O.; Sigarreta, J.M. Small values of the hyperbolicity constant in graphs. *Discr. Math.* **2016**, *339*, 3073–3084. [CrossRef]

71. Granados, A.; Pestana, D.; Portilla, A.; Rodríguez, J.M. Gromov hyperbolicity in Mycielskian Graphs. *Symmetry* **2017**, *9*, 131. [CrossRef]

72. Hernández, J.C.; Reyes, R.; Rodríguez, J.M.; Sigarreta, J.M. Mathematical properties on the hyperbolicity of interval graphs. *Symmetry* **2017**, *9*, 255. [CrossRef]

73. Carballosa, W.; de la Cruz, A.; Rodríguez, J.M. Gromov hyperbolicity in the Cartesian sum of graphs. *Bull. Iran. Math. Soc.* **2017**, 1–20. [CrossRef]

74. Carballosa, W.; Casablanca, R.M.; de la Cruz, A.; Rodríguez, J.M. Gromov hyperbolicity in strong product graphs. *Electr. J. Comb.* **2013**, *20*, 2.

75. Carballosa, W.; de la Cruz, A.; Rodríguez, J.M. Gromov hyperbolicity in lexicographic product graphs. *arXiv* **2015**, arXiv 1506.06034

76. Carballosa, W.; Rodríguez, J.M.; Sigarreta, J.M. Hyperbolicity in the corona and join of graphs. *Aequ. Math.* **2015**, *89*, 1311–1327. [CrossRef]

77. Bowditch, B.H. Notes on Gromov's hyperbolicity criterion for path-metric spaces. In *Group Theory from a Geometrical Viewpoint*; Ghys, E., Haefliger, A., Verjovsky, A., Eds.; World Scientific: River Edge, NJ, USA, 1991; pp. 64–167.

78. Shchur, V. A quantitative version of the Morse lemma and quasi-isometries fixing the ideal boundary. *J. Funct. Anal.* **2013**, *264*, 815–836. [CrossRef]

79. Rodríguez, J.M.; Touris, E. Gromov hyperbolicity through decomposition of metric spaces. *Acta Math. Hung.* **2004**, *103*, 53–84. [CrossRef]

80. Alvarez, V.; Portilla, A.; Rodríguez, J.M.; Touris, E. Gromov hyperbolicity of Denjoy domains. *Geom. Dedic.* **2006**, *121*, 221–245. [CrossRef]

81. Bonk, M. Quasi-geodesics segments and Gromov hyperbolic spaces. *Geom. Dedicata* **1996**, *62*, 281–298. [CrossRef]

82. Shang, Y. Random Lifts Of Graphs: Network Robustness Based On The Estrada Index. *Appl. Math. E-Notes* **2012**, *12*, 53–61.

![symmetry logo] *symmetry*

MDPI

Article

Computing Metric Dimension and Metric Basis of 2D Lattice of Alpha-Boron Nanotubes

Zafar Hussain [1], Mobeen Munir [2,*] [iD], Maqbool Chaudhary [1] and Shin Min Kang [3,4,*]

[1] Department of Mathematics and Statistics, the University of Lahore, Lahore 54000, Pakistan; hussainzafar888@gmail.com (Z.H.); maqboolchaudhri@gmail.com (M.C.)
[2] Division of Science and Technology, University of Education, Lahore 54000, Pakistan
[3] Department of Mathematics and RINS, Gyeongsang National University, Jinju 52828, Korea
[4] Center for General Education, China Medical University, Taichung 40402, Taiwan
[*] Correspondence: mmunir@ue.edu.pk (M.M.), smkang@gnu.ac.kr (S.M.K.)

Received: 16 June 2018; Accepted: 19 July 2018; Published: 25 July 2018

Abstract: Concepts of resolving set and metric basis has enjoyed a lot of success because of multi-purpose applications both in computer and mathematical sciences. For a connected graph G(V,E) a subset W of V(G) is a resolving set for G if every two vertices of G have distinct representations with respect to W. A resolving set of minimum cardinality is called a metric basis for graph G and this minimum cardinality is known as metric dimension of G. Boron nanotubes with different lattice structures, radii and chirality's have attracted attention due to their transport properties, electronic structure and structural stability. In the present article, we compute the metric dimension and metric basis of 2D lattices of alpha-boron nanotubes.

Keywords: alpha-boron nanotube; resolving set; metric basis; metric dimension

---

1. Introduction

In a complex network, one is always interested to uniquely identify the location of nodes by assigning an address with reference to a particular set. Such a particular set with minimum possible nodes is known as the metric basis and its cardinality is known as the metric dimension.

These facts have been efficiently utilized in drug design to attack on particular nodes. Some computational aspects of carbon and boron nanotubes have been summed up in [1]. Similarly, a moving point in a graph may be located by finding the distance from the point to the collection of sonar stations, which have been properly positioned in the graph [2]. Thus, finding a minimal sufficiently large set of labeled vertices, such that a robot can find its position, is a problem known as robot navigation, already well-studied in [3]. This sufficiently large set of labeled vertices is a resolving set of the graph space and the cardinality of such a set with minimum possible elements is the metric dimension. Similarly, on another node, a real-world problem is the study of networks whose structure has not been imposed by a central authority but is also brought into light from local and distributed processes. Obtaining a map of all nodes and the links between them is difficult as well as expensive. To have a good approximation of the real network, a frequently used technique is to attain a local view of network from multiple dimensions and join them. The metric dimension also has some applications in this respect as well.

In nanomaterials, nanowires, nanocrystals, and nanotubes formulate three main classes. Boron nanotubes are becoming highly attractive due to their extraordinary features, including work function, transport properties, electronic structure, and structural stability, [3,4]. Triangular boron and α-boron are deduced from a triangular sheet and an α-sheet as shown in Figure 1 below.

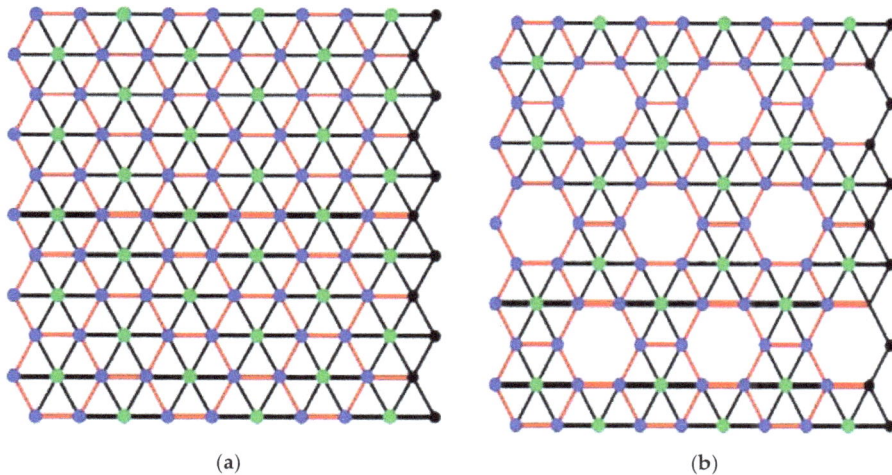

Figure 1. (a) 2D lattice of Triangular Boron Tubes; (b) 2D-lattice of alpha-boron Tubes.

The first boron nanotubes were made, in 2004, from a buckled triangular latticework [4]. The other famous type, alpha-boron, is constructed from an α-sheet. Both types are more conductive than carbon nanotubes regardless of their structure and chiralities. Due to an additional atom at the center of some of the hexagons, alpha-boron nanotubes have a more complex structure than triangular boron nanotubes [4]. This structure is the most stable known theoretical structure for boron nanotubes. With this specimen, boron nanotubes should have variable electrical properties, where wider ones should be metallic conductors, but narrower ones should be semiconductors. These tubes will replace carbon nanotubes in Nano devices like diodes and transistors. The following figure represents alpha-boron nanotubes.

The subject matter of the present article is the metric dimension of the 2D-lattices of alpha-boron nanotubes. An elementary problem in chemistry is to provide a distinct mathematical representation for the set of atoms, molecules, or compounds in a big structure. Consequently, the huge structure of a chemical compound under discussion can be represented by a labeled graph whose vertex and edge labels specify the atom and bond types, respectively. So, a graph-theoretic interpretation of this problem is to provide unique mathematical representations for the vertices of a graph in such a way that distinct vertices have distinct representations [5]. Going with a similar idea, we associate a 2D planar graph corresponding to the structure where nodes or vertices are represented for atoms, and where edges are actually the bonds between them. For the basics of graph theory, we refer to [6].

Let G be a connected graph and u, w be any two vertices of G. The length of the shortest path between u and w is called the distance between u and w and the number of edges between u and v in this shortest path is denoted by $d(u, v)$. Let $W = \{w_1, w_2, w_3, \ldots, w_n\}$ be an ordered set of vertices of G and v ∈ V(G). The k-vector $(d(v, w_1), d(v, w_2), d(v, w_3), \ldots, d(v, w_n))$ is called the representation $r(v|w)$ of v with respect to w. If the distinct vertices of G have a distinct representation with respect to w, then w is called a resolving set of G (see [7–10]). A resolving set of minimum cardinality is called a basis of G and this minimum cardinality is the metric dimension of G, denoted by dim(G).

The concept of the metric dimension was first crafted for metric spaces of a continuous nature but later on, these concepts were used for graphs. In fact, Slater initiated the concepts of metric dimension and resolving sets and these concepts were also studied by Melter and Harary independently in [11,12]. Resolving sets have been analyzed a lot since then. The resolving sets have applications in many fields including network discovery and verification [13], connected joins in graphs, strategies for the mastermind games [14], applications to problems of pattern recognition, image processing, combinatorial optimization, pharmaceutical chemistry, and game theory. In [8,10], authors computed metric dimension of some graphs and proved that it is 1 if and only if graph is the path P_n. The metric dimension of complete graph K_n is $n - 1$ for $n > 1$ and the metric dimension of cycle graph C_n is 2 for $n > 1$ [8]. In [7], authors computed the metric dimension of the Cartesian products of some graphs. In [9], Imran et al. computed the metric dimension of the generalized Peterson graph. Also, generalized Petersen graphs $p(n,2)$, antiprisms A_n, and circulant graphs $C_n(1,2)$ are families of graphs with a constant metric dimension [15]. In [16], Imran et. al. discussed some families of graphs with a constant metric dimension. Ali et al. computed partial results about the metrics dimension of classical Mobius Ladders in [17], whereas Munir et al. computed the exact and full results for this family in [18]. In [19], authors computed the metric dimension of a generalized wheel graph and ant-web gear graph. Authors also gave a new family of convex polytopes with an unbounded metric dimension [19]. Recently authors in [20] computed metric dimension of some families of Gear graphs. Manuel et. al. computed the metric dimension of a honey-comb network in [21]. In [22], the authors computed the metric dimension of circulant graphs. In [23], authors computed explicit formula for the metric dimension of a regular bipartite graph. Imran et al. computed the metric dimension of a Jahangir graph in [24]. Authors discussed the metric dimension of the circulant and Harary graph in [25].

In the present article, we intend to compute the metric dimension of 2D lattices of an α-boron Nanotube. For the rest of this article, we reserve the symbol $T_{m,n}$ for the 2D lattice of the α-boron Nanotube of dimensions m and n. We use the term lattice only to denote the 2D sheet of the alpha-boron tubes. The vertices of the alpha-boron sheet in the first row are $u_{11}, u_{12}, u_{13}, \ldots, u_{1n}$, in second row $u_{21}, u_{22}, \ldots, u_{2n}$, in third row $u_{31}, u_{32}, u_{34}, u_{35}, u_{37}, u_{38} \ldots, u_{3n}$, etc. The representation of vertices is u_{ij}, where i is the row number and j is the column number. For the sake of simplicity, we label the vertices in Figure 2 as 11, 12, 13 etc. instead of u_{11}, u_{12}, u_{13} etc. Please see below, Figure 2.

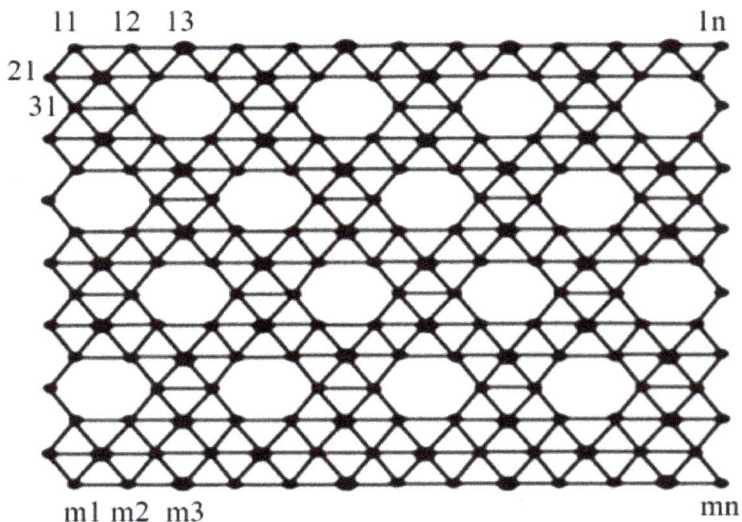

Figure 2. Alpha boron sheet.

In the tubes, m1 and 1n are connected with each other, whereas in the 2D-lattice these vertices are at $n - 1$ distance apart, see Figure 3.

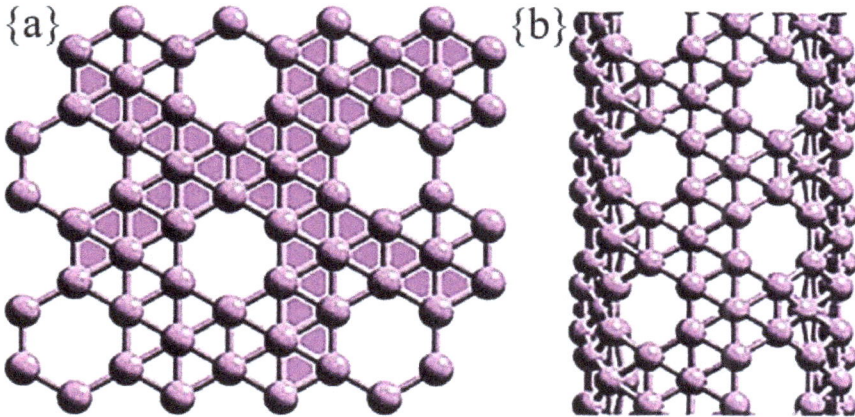

Figure 3. Alpha-boron nanotubes. (**a**) Sheet view and (**b**) tube view.

2. Main Results

Theorem 1. *For all $m, n \in Z^+$ and $m < n$, we have $dim(T_{m,n}) = 2$.*

Proof. We consider the following labelling of vertices of 2D-lattice of alpha-boron tubes as depicted in the above figure. Consider $m \times n$ 2D-lattice of α − boron nanotubes. The vertex set of G is partitioned as

$$\{u_{11}, u_{12}, u_{13}, \ldots, u_{1n}, u_{21}, u_{22}, \ldots, u_{2n}, u_{41}, u_{42}, \ldots, u_{4n}, u_{51}, u_{52}, \ldots, u_{5n}, u_{71}, u_{72}, \ldots, u_{7n}, \ldots, u_{m1}, u_{m2},$$
$$\ldots, u_{mn}\} \cup \{u_{31}, u_{32}, u_{34}, u_{35}, u_{37}, u_{38} \ldots, u_{3n}, u_{61}, u_{63}, u_{64}, u_{66}, u_{67}, u_{69}, \ldots, u_{6n}, u_{91}, u_{92}, u_{94}, u_{95}, u_{97}, u_{98}, \ldots \ldots, u_{9n}\}$$

If m < n

Let $W = \{u_{11}, u_{1n}\}$. We prove that W is a resolving set for $T_{m,n}$. The representations of different vertices of $T_{m,n}$ are

$$For\ i = 1,\ \ r(u_{ij}|W) = (j-1, n-j);\quad 1 \leq j \leq n$$

$$For\ i = 2,\ \ r(u_{ij}|W) = \begin{cases} (1, n);\ j = 1 \\ (j-1, n+1-j);\ 2 \leq j \leq n \end{cases}$$

In general, for $3 < i \leq m$ where i is odd and $i \neq 3k$

$$r(u_{ij}|W) = \begin{cases} (i-1, n+\frac{i-1}{2}-j); & 1 \leq j \leq \frac{i-1}{2} \\ (j+\frac{i-3}{2}, n+\frac{i-1}{2}-j); & \frac{i+1}{2} \leq j \leq n-\frac{i-1}{2} \\ (j+\frac{i-3}{2}, i-1); & n-\frac{i-3}{2} \leq j \leq n \end{cases}$$

In general, for $4 \leq i \leq m$ where i is even and $i \neq 3k$

$$r(u_{ij}|W) = \begin{cases} (i-1, n+\frac{i}{2}-j); & 1 \leq j \leq \frac{i}{2} \\ (j+\frac{i-4}{2}, n+\frac{i}{2}-j); & \frac{i+2}{2} \leq j \leq n-\frac{i-2}{2} \\ (j+\frac{i-4}{2}, i-1); & n-\frac{i-4}{2} \leq j \leq n \end{cases}$$

For $i = 3k$, $3k$ is odd and $j \neq 3p$

$$r(u_{ij}|W) = \begin{cases} (i-1, n+\frac{i-1}{2}-j); & 1 \leq j \leq \frac{i-1}{2} \\ (j+\frac{i-3}{2}, n+\frac{i-1}{2}-j); & \frac{i+1}{2} \leq j \leq n-\frac{i-1}{2} \\ (j+\frac{i-3}{2}, i-1); & n-\frac{i-3}{2} \leq j \leq n \end{cases}$$

For $i = 3k$, $3k$ is even and $j \neq 3p-1$

$$r(u_{ij}|W) = \begin{cases} (i-1, n+\frac{i}{2}-j); & 1 \leq j \leq \frac{i}{2} \\ (j+\frac{i-4}{2}, n+\frac{i}{2}-j); & \frac{i+2}{2} \leq j \leq n-\frac{i-2}{2} \\ (j+\frac{i-4}{2}, i-1); & n-\frac{i-4}{2} \leq j \leq n \end{cases}$$

These representations are distinct. So W is a resolving set for $T_{m,n}$ and the $\dim(T_{m,n}) \leq 2$. Since $T_{m,n}$ is not a path so $\dim(T_{m,n}) \geq 2$. Hence the $\dim(T_{m,n}) = 2$ in this case. $\square$

Theorem 2. *For all $m, n \in Z^+$ and $m \geq n$, we have $\dim(T_{m,n}) \leq 3$,*

Proof. Let $W = \{u_{11}, u_{1n}, u_{m1}\}$. We prove that W is a resolving set. The representations of vertices u_{ij} with respect to W are

Case I: m is odd with $n \leq m < 2n$ and $m \neq 6k+1$

$$\text{For } i = 1, r(u_{ij}|W) = \begin{cases} (j-1, n-j, m-1); & 1 \leq j \leq \frac{m+1}{2} \\ (j-1, n-j, j+\frac{m-3}{2}); & \frac{m+3}{2} \leq j \leq n \end{cases}$$

$$\text{For } i = 2, r(u_{ij}|W) = \begin{cases} (1, n, m-2); & j = 1 \\ (j-1, n+1-j, m-2); & 2 \leq j \leq \frac{m+1}{2} \\ (j-1, n+1-j, j+\frac{m-5}{2}); & \frac{m+3}{2} \leq j \leq n \end{cases}$$

If i is odd and $i \neq 3k$ then
for $3 < i \leq \lceil \frac{m}{2} \rceil$

$$r(u_{ij}|W) = \begin{cases} (i-1, n+\lfloor \frac{i}{2} \rfloor - j, m-i); & 1 \leq j \leq \frac{i-1}{2}, \\ (j+\lfloor \frac{i-3}{2} \rfloor, n+\lfloor \frac{i}{2} \rfloor - j, m-i); & \frac{i+1}{2} \leq j \leq \frac{m-(i-2)}{2}, \\ (j+\lfloor \frac{i-3}{2} \rfloor, n+\lfloor \frac{i}{2} \rfloor - j, j+\frac{m-(i+2)}{2}); & \frac{m-(i-4)}{2} \leq j \leq n-\frac{i-1}{2} \\ (j+\lfloor \frac{i-3}{2} \rfloor, i-1, j+\frac{m-(i+2)}{2}); & n-\frac{i-3}{2} \leq j \leq n \end{cases}$$

for $\lceil \frac{m}{2} \rceil + 1 \leq i \leq m$

$$r(u_{ij}|W) = \begin{cases} (i-1, n+\lfloor \frac{i}{2} \rfloor - j, m-i); & 1 \leq j \leq \frac{m-(i-2)}{2}, \\ (i-1, n+\lfloor \frac{i}{2} \rfloor - j, j+\frac{m-(i+2)}{2}); & \frac{m-(i-4)}{2} \leq j \leq \frac{i-1}{2}, \\ (j+\lfloor \frac{i-3}{2} \rfloor, n+\lfloor \frac{i}{2} \rfloor - j, j+\frac{m-(i+2)}{2}); & \frac{i+1}{2} \leq j \leq n-\frac{i-1}{2} \\ (j+\lfloor \frac{i-3}{2} \rfloor, i-1, j+\frac{m-(i+2)}{2}); & n-\frac{i-3}{2} \leq j \leq n \end{cases}$$

If i is even and $i \neq 3k$ then

for $3 < i \le \lceil \frac{m}{2} \rceil$

$$
r(u_{ij}|W) = \begin{cases}
(i-1, n + \lfloor \frac{i}{2} \rfloor - j, m - i); & 1 \le j \le \frac{i}{2}, \\
(j + \lfloor \frac{i-3}{2} \rfloor, n + \lfloor \frac{i}{2} \rfloor - j, m - i); & \frac{i+2}{2} \le j \le \frac{m-(i-3)}{2}, \\
(j + \lfloor \frac{i-3}{2} \rfloor, n + \lfloor \frac{i}{2} \rfloor - j, j + \frac{m-(i+3)}{2}); & \frac{m-(i-5)}{2} \le j \le n - \frac{i-2}{2} \\
(j + \lfloor \frac{i-3}{2} \rfloor, i - 1, j + \frac{m-(i+3)}{2}); & n - \frac{i-4}{2} \le j \le n
\end{cases}
$$

for $\lceil \frac{m}{2} \rceil + 1 \le i \le m$

$$
r(u_{ij}|W) = \begin{cases}
(i-1, n + \lfloor \frac{i}{2} \rfloor - j, m - i); & 1 \le j \le \frac{m-(i-3)}{2}, \\
(i-1, n + \lfloor \frac{i}{2} \rfloor - j, j + \frac{m-(i+3)}{2}); & \frac{m-(i-5)}{2} \le j \le \frac{i}{2}, \\
(j + \lfloor \frac{i-3}{2} \rfloor, n + \lfloor \frac{i}{2} \rfloor - j, j + \frac{m-(i+3)}{2}); & \frac{i+2}{2} \le j \le n - \frac{i-2}{2} \\
(j + \lfloor \frac{i-3}{2} \rfloor, i - 1, j + \frac{m-(i+3)}{2}); & n - \frac{i-4}{2} \le j \le n
\end{cases}
$$

If $m = 6k + 1$ with $i \ne 3k$ and i is odd
for $3 < i \le \lceil \frac{m}{2} \rceil$

$$
r(u_{ij}|W) = \begin{cases}
(i-1, n + \lfloor \frac{i}{2} \rfloor - j, m - i); & 1 \le j \le \frac{i-1}{2}, \\
(j + \lfloor \frac{i-3}{2} \rfloor, n + \lfloor \frac{i}{2} \rfloor - j, m - i); & \frac{i+1}{2} \le j \le \frac{m-i}{2}, \\
(j + \lfloor \frac{i-3}{2} \rfloor, n + \lfloor \frac{i}{2} \rfloor - j, m - i + 1); & j = \frac{m-(i-2)}{2} \\
(j + \lfloor \frac{i-3}{2} \rfloor, n + \lfloor \frac{i}{2} \rfloor - j, j + \frac{m-(i+2)}{2}); & \frac{m-(i-4)}{2} \le j \le n - \frac{i-1}{2}, \\
(j + \lfloor \frac{i-3}{2} \rfloor, i - 1, j + \frac{m-(i+2)}{2}); & n - \frac{i-3}{2} \le j \le n,
\end{cases}
$$

for $\lceil \frac{m}{2} \rceil + 1 \le i \le m$

$$
r(u_{ij}|W) = \begin{cases}
(i-1, n + \lfloor \frac{i}{2} \rfloor - j, m - i); & 1 \le j \le \frac{m-i}{2}, \\
(i-1, n + \lfloor \frac{i}{2} \rfloor - j, m - i + 1); & j = \frac{m-(i-2)}{2}, \\
(i-1, n + \lfloor \frac{i}{2} \rfloor - j, j + \frac{m-(i+2)}{2}); & \frac{m-(i-4)}{2} \le j \le \frac{i-1}{2} \\
(j + \lfloor \frac{i-3}{2} \rfloor, n + \lfloor \frac{i}{2} \rfloor - j, j + \frac{m-(i+2)}{2}); & \frac{i+1}{2} \le j \le n - \frac{i-1}{2}, \\
(j + \lfloor \frac{i-3}{2} \rfloor, i - 1, j + \frac{m-(i+2)}{2}); & n - \frac{i-3}{2} \le j \le n,
\end{cases}
$$

If $m = 6k + 1$ with $i \ne 3k$ and i is even
for $3 < i \le \lceil \frac{m}{2} \rceil$

$$
r(u_{ij}|W) = \begin{cases}
(i-1, n + \lfloor \frac{i}{2} \rfloor - j, m - i); & 1 \le j \le \frac{i}{2}, \\
(j + \lfloor \frac{i-3}{2} \rfloor, n + \lfloor \frac{i}{2} \rfloor - j, m - i); & \frac{i+2}{2} \le j \le \frac{m-(i-1)}{2}, \\
(j + \lfloor \frac{i-3}{2} \rfloor, n + \lfloor \frac{i}{2} \rfloor - j, m - i + 1); & j = \frac{m-(i-3)}{2} \\
(j + \lfloor \frac{i-3}{2} \rfloor, n + \lfloor \frac{i}{2} \rfloor - j, j + \frac{m-(i+2)}{2}); & \frac{m-(i-5)}{2} \le j \le n - \frac{i-2}{2}, \\
(j + \lfloor \frac{i-3}{2} \rfloor, i - 1, j + \frac{m-(i+2)}{2}); & n - \frac{i-4}{2} \le j \le n,
\end{cases}
$$

for $\lceil \frac{m}{2} \rceil + 1 \le i \le m$

$$r(u_{ij}|W) = \begin{cases} (i-1, n+\lfloor \frac{i}{2} \rfloor - j, m-i); & 1 \le j \le \frac{m-(i-1)}{2}, \\ (i-1, n+\lfloor \frac{i}{2} \rfloor - j, m-i+1); & j = \frac{m-(i-3)}{2}, \\ (i-1, n+\lfloor \frac{i}{2} \rfloor - j, j+\frac{m-(i+2)}{2}); & \frac{m-(i-5)}{2} \le j \le \frac{i}{2}, \\ (j+\lfloor \frac{i-3}{2} \rfloor, n+\lfloor \frac{i}{2} \rfloor - j, j+\frac{m-(i+2)}{2}); & \frac{i+2}{2} \le j \le n-\frac{i-2}{2}, \\ (j+\lfloor \frac{i-3}{2} \rfloor, i-1, j+\frac{m-(i+2)}{2}); & n-\frac{i-4}{2} \le j \le n, \end{cases}$$

If $i = 3k$ and $3k$ is odd
for $3 < i \le \lceil \frac{m}{2} \rceil$

$$r(u_{ij}|W) = \begin{cases} (i-1, n+\lfloor \frac{i}{2} \rfloor - j, m-i); & 1 \le j \le \frac{i-1}{2}, j \ne 3l \\ (j+\lfloor \frac{i-3}{2} \rfloor, n+\lfloor \frac{i}{2} \rfloor - j, m-i); & \frac{i+1}{2} \le j \le \frac{m-(i-2)}{2}, j \ne 3l \\ (j+\lfloor \frac{i-3}{2} \rfloor, n+\lfloor \frac{i}{2} \rfloor - j, j+\frac{m-(i+2)}{2}); & \frac{m-(i-4)}{2} \le j \le n-\frac{i-1}{2}, j \ne 3l \\ (j+\lfloor \frac{i-3}{2} \rfloor, i-1, j+\frac{m-(i+2)}{2}); & n-\frac{i-3}{2} \le j \le n, j \ne 3l \end{cases}$$

for $\lceil \frac{m}{2} \rceil + 1 \le i \le m$

$$r(u_{ij}|W) = \begin{cases} (i-1, n+\lfloor \frac{i}{2} \rfloor - j, m-i); & 1 \le j \le \frac{m-(i-2)}{2}, j \ne 3l \\ (i-1, n+\lfloor \frac{i}{2} \rfloor - j, j+\frac{m-(i+2)}{2}); & \frac{m-(i-4)}{2} \le j \le \frac{i-1}{2}, j \ne 3l \\ (j+\lfloor \frac{i-3}{2} \rfloor, n+\lfloor \frac{i}{2} \rfloor - j, j+\frac{m-(i+2)}{2}); & \frac{i+1}{2} \le j \le n-\frac{i-1}{2}, j \ne 3l \\ (j+\lfloor \frac{i-3}{2} \rfloor, i-1, j+\frac{m-(i+2)}{2}); & n-\frac{i-3}{2} \le j \le n, j \ne 3l \end{cases}$$

If $i = 3k$ and $3k$ is even
for $3 < i \le \lceil \frac{m}{2} \rceil$

$$r(u_{ij}|W) = \begin{cases} (i-1, n+\lfloor \frac{i}{2} \rfloor - j, m-i); & 1 \le j \le \frac{i}{2}, j \ne 3l-1 \\ (j+\lfloor \frac{i-3}{2} \rfloor, n+\lfloor \frac{i}{2} \rfloor - j, m-i); & \frac{i+2}{2} \le j \le \frac{m-(i-3)}{2}, j \ne 3l-1 \\ (j+\lfloor \frac{i-3}{2} \rfloor, n+\lfloor \frac{i}{2} \rfloor - j, j+\frac{m-(i+3)}{2}); & \frac{m-(i-5)}{2} \le j \le n-\frac{i-2}{2}, j \ne 3l-1 \\ (j+\lfloor \frac{i-3}{2} \rfloor, i-1, j+\frac{m-(i+3)}{2}); & n-\frac{i-4}{2} \le j \le n, j \ne 3l-1 \end{cases}$$

for $\lceil \frac{m}{2} \rceil + 1 \le i \le m$

$$r(u_{ij}|W) = \begin{cases} (i-1, n+\lfloor \frac{i}{2} \rfloor - j, m-i); & 1 \le j \le \frac{m-(i-3)}{2}, j \ne 3l-1 \\ (i-1, n+\lfloor \frac{i}{2} \rfloor - j, j+\frac{m-(i+3)}{2}); & \frac{m-(i-5)}{2} \le j \le \frac{i}{2}, j \ne 3l-1 \\ (j+\lfloor \frac{i-3}{2} \rfloor, n+\lfloor \frac{i}{2} \rfloor - j, j+\frac{m-(i+3)}{2}); & \frac{i+2}{2} \le j \le n-\frac{i-2}{2}, j \ne 3l-1 \\ (j+\lfloor \frac{i-3}{2} \rfloor, i-1, j+\frac{m-(i+3)}{2}); & n-\frac{i-4}{2} \le j \le n, j \ne 3l-1 \end{cases}$$

Case II: If m is even and $m \ne 6k+2$
For i is odd and $i \ne 3k$

for $3 < i \leq \lceil \frac{m}{2} \rceil$

$$
r(u_{ij}|W) = \begin{cases}
(i-1, n + \lfloor \frac{i}{2} \rfloor - j, m - i); & 1 \leq j \leq \frac{i-1}{2}, \\
(j + \lfloor \frac{i-3}{2} \rfloor, n + \lfloor \frac{i}{2} \rfloor - j, m - i); & \frac{i+1}{2} \leq j \leq \frac{m-(i-1)}{2}, \\
(j + \lfloor \frac{i-3}{2} \rfloor, n + \lfloor \frac{i}{2} \rfloor - j, j + \frac{m-(i+1)}{2}); & \frac{m-(i-3)}{2} \leq j \leq n - \frac{i-1}{2}, \\
(j + \lfloor \frac{i-3}{2} \rfloor, i - 1, j + \frac{m-(i+1)}{2}); & n - \frac{i-3}{2} \leq j \leq n
\end{cases}
$$

for $\lceil \frac{m}{2} \rceil + 1 \leq i \leq m$

$$
r(u_{ij}|W) = \begin{cases}
(i-1, n + \lfloor \frac{i}{2} \rfloor - j, m - i); & 1 \leq j \leq \frac{m-(i-1)}{2}, \\
(i-1, n + \lfloor \frac{i}{2} \rfloor - j, j + \frac{m-(i+1)}{2}); & \frac{m-(i-3)}{2} \leq j \leq \frac{i-1}{2}, \\
(j + \lfloor \frac{i-3}{2} \rfloor, n + \lfloor \frac{i}{2} \rfloor - j, j + \frac{m-(i+1)}{2}); & \frac{i+1}{2} \leq j \leq n - \frac{i-1}{2}, \\
(j + \lfloor \frac{i-3}{2} \rfloor, i - 1, j + \frac{m-(i+1)}{2}); & n - \frac{i-3}{2} \leq j \leq n
\end{cases}
$$

If i is even $i \neq 3k$
for $3 < i \leq \lceil \frac{m}{2} \rceil$

$$
r(u_{ij}|W) = \begin{cases}
(i-1, n + \lfloor \frac{i}{2} \rfloor - j, m - i); & 1 \leq j \leq \frac{i}{2}, \\
(j + \lfloor \frac{i-3}{2} \rfloor, n + \lfloor \frac{i}{2} \rfloor - j, m - i); & \frac{i+2}{2} \leq j \leq \frac{m-(i-2)}{2}, \\
(j + \lfloor \frac{i-3}{2} \rfloor, n + \lfloor \frac{i}{2} \rfloor - j, j + \frac{m-(i+2)}{2}); & \frac{m-(i-4)}{2} \leq j \leq n - \frac{i-2}{2}, \\
(j + \lfloor \frac{i-3}{2} \rfloor, i - 1, j + \frac{m-(i+2)}{2}); & n - \frac{i-4}{2} \leq j \leq n
\end{cases}
$$

for $\lceil \frac{m}{2} \rceil + 1 \leq i \leq m$

$$
r(u_{ij}|W) = \begin{cases}
(i-1, n + \lfloor \frac{i}{2} \rfloor - j, m - i); & 1 \leq j \leq \frac{m-(i-2)}{2}, \\
(i-1, n + \lfloor \frac{i}{2} \rfloor - j, j + \frac{m-(i+2)}{2}); & \frac{m-(i-4)}{2} \leq j \leq \frac{i}{2}, \\
(j + \lfloor \frac{i-3}{2} \rfloor, n + \lfloor \frac{i}{2} \rfloor - j, j + \frac{m-(i+2)}{2}); & \frac{i+2}{2} \leq j \leq n - \frac{i-2}{2}, \\
(j + \lfloor \frac{i-3}{2} \rfloor, i - 1, j + \frac{m-(i+2)}{2}); & n - \frac{i-4}{2} \leq j \leq n
\end{cases}
$$

If $m = 6k + 2$ with $i \neq 3k$ and i is odd
for $3 < i \leq \lceil \frac{m}{2} \rceil$

$$
r(u_{ij}|W) = \begin{cases}
(i-1, n + \lfloor \frac{i}{2} \rfloor - j, m - i); & 1 \leq j \leq \frac{i-3}{2}, \\
(j + \lfloor \frac{i-3}{2} \rfloor, n + \lfloor \frac{i}{2} \rfloor - j, m - i); & \frac{i-1}{2} \leq j \leq \frac{m-(i+1)}{2}, \\
(j + \lfloor \frac{i-3}{2} \rfloor, n + \lfloor \frac{i}{2} \rfloor - j, m - i + 1); & j = \frac{m-(i-1)}{2} \\
(j + \lfloor \frac{i-3}{2} \rfloor, n + \lfloor \frac{i}{2} \rfloor - j, j + \frac{m-(i+2)}{2}); & \frac{m-(i-3)}{2} \leq j \leq n - \frac{i-1}{2}, \\
(j + \lfloor \frac{i-3}{2} \rfloor, i - 1, j + \frac{m-(i+2)}{2}); & n - \frac{i-3}{2} \leq j \leq n,
\end{cases}
$$

for $\lceil \frac{m}{2} \rceil + 1 \leq i \leq m$

$$r(u_{ij}|W) = \begin{cases} (i-1, n + \lfloor \frac{i}{2} \rfloor - j, m - i); & 1 \leq j \leq \frac{m-(i+1)}{2}, \\ (i-1, n + \lfloor \frac{i}{2} \rfloor - j, m - i + 1)); & j = \frac{m-(i-1)}{2}, \\ (i-1, n + \lfloor \frac{i}{2} \rfloor - j, j + \frac{m-(i+2)}{2}); & \frac{m-(i-3)}{2} \leq j \leq \frac{i-3}{2}, \\ (j + \lfloor \frac{i-3}{2} \rfloor, n + \lfloor \frac{i}{2} \rfloor - j, j + \frac{m-(i+2)}{2}); & \frac{i-1}{2} \leq j \leq n - \frac{i-1}{2}, \\ (j + \lfloor \frac{i-3}{2} \rfloor, i - 1, j + \frac{m-(i+2)}{2}); & n - \frac{i-3}{2} \leq j \leq n, \end{cases}$$

If $m = 6k + 2$ with $i \neq 3k$ and i is even
for $3 < i \leq \lceil \frac{m}{2} \rceil$

$$r(u_{ij}|W) = \begin{cases} (i-1, n + \lfloor \frac{i}{2} \rfloor - j, m - i); & 1 \leq j \leq \frac{i}{2}, \\ (j + \lfloor \frac{i-3}{2} \rfloor, n + \lfloor \frac{i}{2} \rfloor - j, m - i); & \frac{i+2}{2} \leq j \leq \frac{m-i}{2}, \\ (j + \lfloor \frac{i-3}{2} \rfloor, n + \lfloor \frac{i}{2} \rfloor - j, m - i + 1); & j = \frac{m-(i-2)}{2} \\ (j + \lfloor \frac{i-3}{2} \rfloor, n + \lfloor \frac{i}{2} \rfloor - j, j + \frac{m-(i+2)}{2}); & \frac{m-(i-4)}{2} \leq j \leq n - \frac{i-2}{2}, \\ (j + \lfloor \frac{i-3}{2} \rfloor, i - 1, j + \frac{m-(i+2)}{2}); & n - \frac{i-4}{2} \leq j \leq n, \end{cases}$$

for $\lceil \frac{m}{2} \rceil + 1 \leq i \leq m$

$$r(u_{ij}|W) = \begin{cases} (i-1, n + \lfloor \frac{i}{2} \rfloor - j, m - i); & 1 \leq j \leq \frac{m-i}{2}, \\ (i-1, n + \lfloor \frac{i}{2} \rfloor - j, m - i + 1); & j = \frac{m-(i-2)}{2}, \\ (i-1, n + \lfloor \frac{i}{2} \rfloor - j, j + \frac{m-(i+2)}{2}); & \frac{m-(i-4)}{2} \leq j \leq \frac{i-2}{2}, \\ (j + \lfloor \frac{i-3}{2} \rfloor, n + \lfloor \frac{i}{2} \rfloor - j, j + \frac{m-(i+2)}{2}); & \frac{i}{2} \leq j \leq n - \frac{i-2}{2}, \\ (j + \lfloor \frac{i-3}{2} \rfloor, i - 1, j + \frac{m-(i+2)}{2}); & n - \frac{i-4}{2} \leq j \leq n, \end{cases}$$

If $i = 3k$ and $3k$ is odd
for $3 < i \leq \lceil \frac{m}{2} \rceil$

$$r(u_{ij}|W) = \begin{cases} (i-1, n + \lfloor \frac{i}{2} \rfloor - j, m - i); & 1 \leq j \leq \frac{i-1}{2}, j \neq 3k \\ (j + \lfloor \frac{i-3}{2} \rfloor, n + \lfloor \frac{i}{2} \rfloor - j, m - i); & \frac{i+1}{2} \leq j \leq \frac{m-(i-2)}{2}, j \neq 3k \\ (j + \lfloor \frac{i-3}{2} \rfloor, n + \lfloor \frac{i}{2} \rfloor - j, j + \frac{m-(i+2)}{2}); & \frac{m-(i-4)}{2} \leq j \leq n - \frac{i-1}{2}, j \neq 3k \\ (j + \lfloor \frac{i-3}{2} \rfloor, i - 1, j + \frac{m-(i+2)}{2}); & n - \frac{i-3}{2} \leq j \leq n, j \neq 3k \end{cases}$$

for $\lceil \frac{m}{2} \rceil + 1 \leq i \leq m$

$$r(u_{ij}|W) = \begin{cases} (i-1, n + \lfloor \frac{i}{2} \rfloor - j, m - i); & 1 \leq j \leq \frac{m-(i-2)}{2}, j \neq 3k \\ (i-1, n + \lfloor \frac{i}{2} \rfloor - j, j + \frac{m-(i+2)}{2}); & \frac{m-(i-4)}{2} \leq j \leq \frac{i-1}{2}, j \neq 3k \\ (j + \lfloor \frac{i-3}{2} \rfloor, n + \lfloor \frac{i}{2} \rfloor - j, j + \frac{m-(i+2)}{2}); & \frac{i+1}{2} \leq j \leq n - \frac{i-1}{2}, j \neq 3k \\ (j + \lfloor \frac{i-3}{2} \rfloor, i - 1, j + \frac{m-(i+2)}{2}); & n - \frac{i-3}{2} \leq j \leq n, j \neq 3k \end{cases}$$

If $i = 3k$ and $3k$ is odd

for $3 < i \leq \lceil \frac{m}{2} \rceil$

$$r(u_{ij}|W) = \begin{cases} (i-1, n + \lfloor \frac{i}{2} \rfloor - j, m-i); & 1 \leq j \leq \frac{i}{2}, j \neq 3k-1 \\ (j + \lfloor \frac{i-3}{2} \rfloor, n + \lfloor \frac{i}{2} \rfloor - j, m-i); & \frac{i+2}{2} \leq j \leq \frac{m-(i-3)}{2}, j \neq 3k-1 \\ (j + \lfloor \frac{i-3}{2} \rfloor, n + \lfloor \frac{i}{2} \rfloor - j, j + \frac{m-(i+3)}{2}); & \frac{m-(i-5)}{2} \leq j \leq n - \frac{i-2}{2}, j \neq 3k-1 \\ (j + \lfloor \frac{i-3}{2} \rfloor, i-1, j + \frac{m-(i+3)}{2}); & n - \frac{i-4}{2} \leq j \leq n, j \neq 3k-1 \end{cases}$$

for $\lceil \frac{m}{2} \rceil + 1 \leq i \leq m$

$$r(u_{ij}|W) = \begin{cases} (i-1, n + \lfloor \frac{i}{2} \rfloor - j, m-i); & 1 \leq j \leq \frac{m-(i-3)}{2}, j \neq 3k-1 \\ (i-1, n + \lfloor \frac{i}{2} \rfloor - j, j + \frac{m-(i+3)}{2}); & \frac{m-(i-5)}{2} \leq j \leq \frac{i}{2}, j \neq 3k-1 \\ (j + \lfloor \frac{i-3}{2} \rfloor, n + \lfloor \frac{i}{2} \rfloor - j, j + \frac{m-(i+3)}{2}); & \frac{i+2}{2} \leq j \leq n - \frac{i-2}{2}, j \neq 3k-1 \\ (j + \lfloor \frac{i-3}{2} \rfloor, i-1, j + \frac{m-(i+3)}{2}); & n - \frac{i-4}{2} \leq j \leq n, j \neq 3k-1 \end{cases}$$

For $i = 2n - 1$, $r(u_{ij}|W) = (2n-2, 2n-2, j-1)$; $1 \leq j \leq n$

For $i = 2n - 2$, $r(u_{ij}|W) = \begin{cases} (2n-3, 2n-2, 1); & j = 1 \\ (2n-3, 2n-3, j-1); & 2 \leq j \leq n \end{cases}$

If m is odd, $r \geq 2$ and $rn \leq m < (r+1)n$

For $i = rn + k$ where $0 \leq k \leq n - 1$

If i is odd,

$$r(u_{ij}|W) = \begin{cases} (rn+k-1, rn+k-1, m-i); & 1 \leq j \leq \frac{m-(rn+k-2)}{2} \\ \left(rn+k-1, rn+k-1, j + \frac{m-(rn+k+2)}{2}\right); & \frac{m-(rn+k-4)}{2} \leq j \leq n \end{cases}$$

If i is even,

$$r(u_{ij}|W) = \begin{cases} (rn+k-1, rn+k-1, m-i); & 1 \leq j \leq \frac{m-(rn+k-3)}{2} \\ \left(rn+k-1, rn+k-1, j + \frac{m-(rn+k+3)}{2}\right); & \frac{m-(rn+k-5)}{2} \leq j \leq n \end{cases}$$

If $i = 3k$, i is odd and $j \neq 3p$ then

$$r(u_{ij}|W) = \begin{cases} (rn+k-1, rn+k-1, m-i); & 1 \leq j \leq \frac{m-(rn+k-2)}{2} \\ \left(rn+k-1, rn+k-1, j + \frac{m-(rn+k+2)}{2}\right); & \frac{m-(rn+k-4)}{2} \leq j \leq n \end{cases}$$

If $i = 3k$, i is even and $j \neq 3p - 1$ then

$$r(u_{ij}|W) = \begin{cases} (rn+k-1, rn+k-1, m-i); & 1 \leq j \leq \frac{m-(rn+k-3)}{2} \\ \left(rn+k-1, rn+k-1, j + \frac{m-(rn+k+3)}{2}\right); & \frac{m-(rn+k-5)}{2} \leq j \leq n \end{cases}$$

If m is even, $r \geq 2$ and $rn \leq m < (r+1)n$

For $i = rn + k$ where $0 \leq k \leq n - 1$ If i is odd,

$$r(u_{ij}|W) = \begin{cases} (rn + k - 1, rn + k - 1, m - i); & 1 \leq j \leq \frac{m-(rn+k-1)}{2} \\ \left(rn + k - 1, rn + k - 1, j + \frac{m-(rn+k+1)}{2}\right); & \frac{m-(rn+k-3)}{2} \leq j \leq n \end{cases}$$

If i is even,

$$r(u_{ij}|W) = \begin{cases} (rn + k - 1, rn + k - 1, m - i); & 1 \leq j \leq \frac{m-(rn+k-2)}{2} \\ \left(rn + k - 1, rn + k - 1, j + \frac{m-(rn+k+2)}{2}\right); & \frac{m-(rn+k-4)}{2} \leq j \leq n \end{cases}$$

If $i = 3k$, i is odd and $j \neq 3p$ then

$$r(u_{ij}|W) = \begin{cases} (rn + k - 1, rn + k - 1, m - i); & 1 \leq j \leq \frac{m-(rn+k-1)}{2} \\ \left(rn + k - 1, rn + k - 1, j + \frac{m-(rn+k+1)}{2}\right); & \frac{m-(rn+k-3)}{2} \leq j \leq n \end{cases}$$

If $i = 3k$, i is even and $j \neq 3p - 1$ then

$$r(u_{ij}|W) = \begin{cases} (rn + k - 1, rn + k - 1, m - i); & 1 \leq j \leq \frac{m-(rn+k-2)}{2} \\ \left(rn + k - 1, rn + k - 1, j + \frac{m-(rn+k+2)}{2}\right); & \frac{m-(rn+k-4)}{2} \leq j \leq n \end{cases}$$

These representations are distinct. So, W is a resolving set for $T_{m,n}$. Therefore, the metric dimension of $T_{m,n}$ is ≤ 3. Now we prove that the metric dimension of $T_{m,n}$ is greater than 2. For this we shall prove that any set of cardinality two does not resolve. $\square$

Theorem 3. *For all $m, n \in Z^+$ and $m \geq n$, we have $dim(T_{m,n}) \geq 3$,*

Proof. Let $W = \{u_{ij}, u_{pq}\}$ be a resolving set for $T_{m,n}$. We consider all possibilities and come up with a contradistinction in each case. The following three possibilities arise

Possibility 1: If u_{ij}, u_{pq} lie on the same row then $i = p$.

If $W = \{u_{11}, u_{1n}\}$ then $r(u_{n+1,\frac{n}{2}}|W) = r(u_{n+1,\frac{n+2}{2}}|W)$ if n is even and $r(u_{n+1,\frac{n+1}{2}}|W) = r(u_{n+1,\frac{n+3}{2}}|W)$, if n is odd so both cases result in contradiction. In all remaining possibilities, we denote u_x instead of $r(u_x|W)$ where no confusion arises.

(i)   If $W = \{u_{ij}, u_{iq}\}$ and $1 \leq i < m$ and $1 \leq j < q < n$ then $u_{i,q+1} = u_{i+1,q+1}$, a contradiction.

(ii)   If $W = \{u_{ij}, u_{iq}\}$ and $1 \leq i < m$, $i = 3k$ and $1 \leq j < q < n$ then $u_{i-1,j} = u_{i+1,j}$ or $u_{i-1,j-1} = u_{i+1,j-1}$, a contradiction.

(iii)   If $W = \{u_{ij}, u_{iq}\}$ and $1 \leq i < m$ and $1 < j < q = n$ then $u_{i,j-1} = u_{i+1,j}$, a contradiction.

(iv)   If $W = \{u_{mj}, u_{mq}\}$ and $1 \leq j < q < n$ then $u_{m,q+1} = u_{m-1,q}$ or $u_{m,q+1} = u_{m-1,q+1}$, a contradiction.

(v)   If $W = \{u_{mj}, u_{mq}\}$ and $1 < j < q = n$ then $u_{m,j-1} = u_{m-1,j}$, a contradiction.

(vi)   If $W = \{u_{i1}, u_{in}\}$ and $1 < i < m$ then $u_{i-1,1} = u_{i+1,1}$, a contradiction.

(vii)   If $W = \{u_{m1}, u_{mn}\}$ and then $u_{m-n,3} = u_{m-n,4}$, a contradiction.

Possibility 2: If u_{ij}, u_{pq} lie on the same column then $j = q$.

(i)   If $W = \{u_{11}, u_{m1}\}$ then $u_{21} = u_{22}$, a contradiction.

(ii)   If $W = \{u_{i1}, u_{j1}\}$ and $1 \leq i < j < m$ then $u_{j3} = u_{j+1,3}$ or $u_{j3} = u_{j+1,2}$, a contradiction.

(iii)   If $W = \{u_{i1}, u_{j1}\}$ and $1 < i < j = m$ then $u_{i3} = u_{i-1,2}$ or $u_{i3} = u_{i-1,3}$, a contradiction.

(iv) If $W = \{u_{iq}, u_{jq}\}$, $1 \leq i < j < m$ and $1 < q < n$ then $u_{j+1,q} = u_{j+1,q+1}$ or $u_{j+1,q} = u_{j+1,q-1}$, a contradiction.

(v) If $W = \{u_{iq}, u_{jq}\}$, $1 < i < j \leq m$ and $1 < q < n$ then $u_{i-1,j-1} = u_{i-1,j}$ or $u_{i-1,j} = u_{i-1,j+1}$, a contradiction.

(vi) If $W = \{u_{iq}, u_{jq}\}$, $i = 1, j = m$ and $1 < q < n$ then $u_{2,j} = u_{2,j+1}$, a contradiction.

(vii) If $W = \{u_{i,n-1}, u_{j,n-1}\}$ and $i < j < m$ then $u_{j,n-2} = u_{j+1,n-2}$ or $u_{j,n-3} = u_{j+1,n-2}$, a contradiction.

(viii) If $W = \{u_{1,n}, u_{m,n}\}$ then either $u_{m-1,n-1} = u_{m-1,n}$ or $u_{m-2,n-1} = u_{m-2,n}$, a contradiction.

Possibility 3: If u_{ij}, u_{pq} lie neither on the same row nor the same column so $i \neq p$ and $j \neq q$. Let $W = \{u_{i,j}, u_{p,q}\}$. Since $i \neq p$ so let $i < p$.

(i) If $j < q$ then $u_{p,q+1} = u_{p+1,q+1}$ or $u_{p,q+1} = u_{p+1,q}$, a contradiction

(ii) If $j > q$ then $u_{p-1,q} = u_{p,q+1}$ or $u_{p-1,q+1} = u_{p,q+1}$, a contradiction

(iii) If $i = 1$ and $p = m$ then $u_{2,j} = u_{2,j+1}$ or $u_{1,j+1} = u_{2,j+1}$ or $u_{m-1,q+1} = u_{m,q-1}$, a contradiction.

So any set with cardinality 2 does not resolve $T_{m,n}$. So, the metric dimension of $T_{m,n}$ is greater than 2. Hence metric dimension of $T_{m,n}$ is 3 if $m > n$. □

3. Conclusions and Discussion

In the present article, we computed the metric dimension of a 2D-lattice of alpha-boron nanotubes, $T_{m,n}$, and have come up with the following summarized result:

$$\dim(T_{m,n}) = \begin{cases} 2 \text{ if } m < n \\ 3 \text{ if } m \geq n \end{cases}$$

It is evident that the dimension depends upon the size of the 2D sheet. These results have applications in drug design, networking communication, robot navigations, designing new nano-devices and nano-engineering. Actually, these results are useful for engineer and hardware-designers that use alpha-boron sheets in different industries. It is overwhelming that they can capture the whole sheet uniquely if they know the resolving set and metric dimension. It can save time and cost if they choose only two or three vertices depending upon the size of the sheet using our results. We can conclude that every atom in the 2D sheet of alpha-boron nanotubes can be uniquely accessed and controlled by its metric basis whose cardinality is 2 or 3 depending upon the dimensions of the sheet.

Author Contributions: Z.H. did all computation, M.M. and S.M.K. conceived the idea and M.C. reviewed it and made corrections.

Funding: This research received no external funding.

Acknowledgments: Authors are extremely thankful to reviewers' comments and suggestions for improving this article. Authors are also thankful to University of Education for providing a chance to present this article at ICE2018.

Conflicts of Interest: The authors declare no conflicts of interest.

References

1. Manuel, P. Computational Aspects of Carbon and Boron Nanotubes. *Molecules* **2010**, *15*, 8709–8722. [CrossRef] [PubMed]

2. Khuller, S.; Raghavachari, B.; Rosenfeld, A. Landmarks in graphs. *Discret. Appl. Math.* **1996**, *70*, 217–229. [CrossRef]

3. Chartrand, G.; Poisson, C.; Zhang, P. Resolvability and the upper dimension of graphs. *Comput. Math. Appl.* **2000**, *39*, 19–28. [CrossRef]

4.   Bezugly, V.; Kunstmann, J.; Grundkötter-Stock, B.; Frauenheim, T.; Niehaus, T.; Cuniberti, G. Highly Conductive Boron Nanotubes: Transport Properties, Work Functions, and Structural Stabilities. *ACS Nano* **2011**, *5*, 4997–5005. [CrossRef] [PubMed]

5.   Cameron, P.; Lint, J. *Designs, Graphs, Codes and Their Links, London Mathematical Society Student Texts*; Cambridge University Press: Cambridge, UK, 1991; Volume 22.

6.   Deistal, R. *Graph Theory*; GTM 173; Springer: Berlin, Germany, 2017; ISBN 978-3-662-53621-6.

7.   Caceres, J.; Hernando, C.; Mora, M.; Pelayo, I.M.; Puertas, M.J.; Seara, C.; Wood, D.R. On the metric dimension of cartesian product of graphs. *SIAM J. Discret. Math.* **2007**, *2*, 423–441. [CrossRef]

8.   Buczkowski, P.S.; Chartrand, G.; Poisson, C.; Zhang, P. On *k*-dimensional graphs and their bases. *Period. Math. Hung.* **2003**, *46*, 9–15. [CrossRef]

9.   Imran, M.; Baig, A.Q.; Shafiq, M.K.; Tomescu, I. On the metric dimension of generalized Petersen graphs P(n, 3). *Ars Comb.* **2014**, *117*, 113–130.

10.   Chartrand, G.; Eroh, L.; Johnson, M.A.; Oellermann, O.R. Resolvability in graphs and the metric dimension of a graph. *Discret. Appl. Math.* **2000**, *105*, 99–113. [CrossRef]

11.   Harary, F.; Melter, R.A. On the metric dimension of a graph. *Ars Comb.* **1976**, *2*, 191–195.

12.   Slater, P.J. Leaves of trees. *Congr. Numer.* **1975**, *14*, 549–559.

13.   Sebő, A.; Tannier, E. On metric generators of graphs. *Math. Oper. Res.* **2004**, *29*, 383–393. [CrossRef]

14.   Chvatal, V. Mastermind. *Combinatorica* **1983**, *3*, 325–329. [CrossRef]

15.   Hernando, C.; Mora, M.; Pelayo, I.M.; Seara, C.; Caceres, J.; Puertas, M.L. On the metric dimension of some families of graphs. *Electron. Notes Discret. Math.* **2005**, *22*, 129–133. [CrossRef]

16.   Javaid, I.; Rahim, M.T.; Ali, K. Families of regular graphs with constant metric dimension. *Util. Math.* **2008**, *75*, 21–34.

17.   Ali, M.; Ali, G.; Imran, M.; Baig, A.Q.; Shafiq, M.K. On the metric dimension of Mobius ladders. *Ars Comb.* **2012**, *105*, 403–410.

18.   Munir, M.; Nizami, A.R.; Saeed, H. On the metric dimension of Möbius Ladder. *Ars Comb.* **2017**, *135*, 249–256.

19.   Siddique, H.M.A.; Imran, M. Computing the metric dimension of wheel related graphs. *Appl. Math. Comput.* **2014**, *242*, 624–632.

20.   Imran, S.; Siddiqui, M.K.; Imran, M.; Hussain, M.; Bilal, H.M.; Cheema, I.Z.; Tabraiz, A.; Saleem, Z. Computing the Metric Dimension of Gear Graphs. *Symmetry* **2018**, *10*, 209. [CrossRef]

21.   Manuel, P.; Rajan, B.; Rajasingh, I.; Monica, C. On minimum metric dimension of honeycomb networks. *J. Discret. Algorithms* **2008**, *6*, 20–27. [CrossRef]

22.   Imran, M.; Baig, A.Q.; Bokhary, S.A.; Javaid, I. On the metric dimension of circulant graphs. *Appl. Math. Lett.* **2012**, *25*, 320–325. [CrossRef]

23.   Baca, M.; Baskoro, E.T.; Salman, A.N.M.; Saputro, S.W.; Suprijanto, D. On metric dimension of regular bipartite graphs. *Bull. Math. Soc. Sci. Math. Roum.* **2011**, *54*, 15–28.

24.   Tomescu, I.; Javaid, I. On the metric dimension of the Jahangir graph. *Bull. Math. Soc. Sci. Math. Roum.* **2007**, *50*, 371–376.

25.   Grigorious, C.; Manuel, P.; Miller, M.; Rajan, B.; Stephen, S. On the metric dimension of circulant and Harary graphs. *Appl. Math. Comput.* **2014**, *248*, 47–54. [CrossRef]

symmetry

MDPI

Article

On the Distinguishing Number of Functigraphs

Muhammad Fazil [1], Muhammad Murtaza [1], Zafar Ullah [2], Usman Ali [1] and Imran Javaid [1,*]

[1]   Centre for Advanced Studies in Pure and Applied Mathematics, Bahauddin Zakariya University Multan,
       Multan 60800, Pakistan; mfazil@bzu.edu.pk (M.F.); mahru830@gmail.com (M.M.); uali@bzu.edu.pk (U.A.)
[2]   Department of Mathematics, University of Education Lahore, Campus Dera Ghazi Khan 32200, Pakistan;
       zafarbhatti73@gmail.com
*   Correspondence: imran.javaid@bzu.edu.pk; Tel.: +92-332-459-8377

Received: 3 July 2018; Accepted: 4 August 2018; Published: 9 August 2018

Abstract: Let G_1 and G_2 be disjoint copies of a graph G and $g : V(G_1) \to V(G_2)$ be a function. A functigraph F_G consists of the vertex set $V(G_1) \cup V(G_2)$ and the edge set $E(G_1) \cup E(G_2) \cup \{uv : g(u) = v\}$. In this paper, we extend the study of distinguishing numbers of a graph to its functigraph. We discuss the behavior of distinguishing number in passing from G to F_G and find its sharp lower and upper bounds. We also discuss the distinguishing number of functigraphs of complete graphs and join graphs.

Keywords: distinguishing number; functigraph; complete graph

1. Introduction

Given a key ring of apparently identical keys to open different doors, how many colors are needed to identify them? This puzzle was given by Rubin [1] for the first time. In this puzzle, there is no need for coloring to be a proper one. Indeed, one cannot find a reason why adjacent keys must be assigned different colors, whereas in other problems, like storing chemicals and scheduling meetings, a proper coloring is needed with a small number of colors required.

Inspired by this puzzle, Albertson and Collins [2] introduced the concept of the distinguishing number of a graph as follows: a labeling $f : V(G) \to \{1, 2, 3, ..., t\}$ is called *t-distinguishing* if no non-trivial automorphism of a graph G preserves the vertex labels. The least integer t such that a graph G has a labeling which is *t*-distinguishing for the graph G, is called the *distinguishing number* of G and it is denoted by $Dist(G)$. For example, the distinguishing number of a complete graph K_n is n, the distinguishing number of a path graph P_n is 2 and the distinguishing number of a cyclic graph C_n, $n \geq 6$ is 2. For a graph G of order n, $1 \leq Dist(G) \leq n$ [2].

Harary [3] gave different methods (orienting some of the edges, coloring some of the vertices with one or more colors and same for the edges, labeling vertices or edges, adding or deleting vertices or edges) of destroying the symmetries of a graph. Collins and Trenk [4] defined the distinguishing chromatic number where the authors used proper *t*-distinguishing for vertex labeling. The authors have given a comparison between the distinguishing number, the distinguishing chromatic number and the chromatic number of families like complete graphs, path graphs, cyclic graphs, Petersen graph and trees, etc. Kalinowski and Pilsniak [5] defined similar graph parameters, the distinguishing index and the distinguishing chromatic index where the authors labeled edges instead of vertices. The authors also gave a comparison between the distinguishing number and the distinguishing index of a connected graph G of order $n \geq 3$. Boutin [6] introduced the concept of determining sets. Albertson and Boutin [7] proved that a graph has a $(t - 1)$-distinguishable determining set if and only if the graph is *t*-distinguishable. The authors also proved that every Kneser graph $K_{n:k}$ with $n \geq 6$ and $k \geq 2$ is 2-distinguishable. A considerable amount of literature has been developed in this area—for example, see [8–12].

Unless otherwise specified, all graphs considered in this paper are simple, non-trivial and connected. The set of all vertices that are adjacent to a vertex $u \in V(G)$ is called the *open neighborhood* of u and it is denoted by $N(u)$. The set of vertices $\{u\} \cup N(u)$ is called the *closed neighborhood* of u and it is denoted by $N[u]$. If two distinct vertices u, v of a graph G have the same open neighborhood, then these are called *non-adjacent twins*. If the two vertices have the same closed neighborhood, then these are called *adjacent twins*. In the both cases, u and v are called *twins*. A vertex v of a graph G is called *saturated*, if it is adjacent to all other vertices of G. A graph H whose vertex set $V(H)$ and edge set $E(H)$ are subsets of $V(G)$ and $E(G)$, respectively, then H is called a *subgraph* of graph G. Let $S \subset V(G)$ be any subset of vertices of G. The *induced subgraph*, denoted by $< S >$, is the graph whose vertex set is S and whose edge set is the set of all those edges in $E(G)$ which have both end vertices in S. A *spanning subgraph* H of a graph G is a subgraph such that $V(H) = V(G)$ and $E(H) \subseteq E(G)$. An *automorphism* α of G, $\alpha : V(G) \to V(G)$, is a bijective mapping such that $\alpha(u)\alpha(v) \in E(G)$ if and only if $uv \in E(G)$. Thus, each automorphism α of G is a permutation of the vertex set $V(G)$ which preserves adjacencies and non-adjacencies. The *automorphism group* of a graph G, denoted by $\Gamma(G)$, is the set of all automorphisms of a graph G. A graph with a trivial automorphism group is called a *rigid (or asymmetric)* graph. The minimum number of vertices in a rigid graph is 6 [13]. The distinguishing number of a rigid graph is 1.

The idea of a permutation graph was introduced by Chartrand and Harary [14] for the first time. The authors defined a permutation graph as follows: a permutation graph consists of two identical disjoint copies of a graph G, say G_1 and G_2, along with $|V(G)|$ additional edges joining $V(G_1)$ and $V(G_2)$ according to a given permutation on $\{1, 2, ..., |V(G)|\}$. Dorfler [15] defined a mapping graph as follows: a *mapping graph* of a graph G on n vertices consists of two disjoint identical copies of graph G with n additional edges between the vertices of two copies, where the additional edges are defined by a function. The mapping graph was rediscovered and studied by Chen et al. [16], where it was called the functigraph. A functigraph is an extension of a permutation graph. Formally, the functigraph is defined as follows: let G_1 and G_2 be disjoint copies of a connected graph G and let $g : V(G_1) \to V(G_2)$ be a function. A *functigraph* F_G of a graph G consists of the vertex set $V(G_1) \cup V(G_2)$ and the edge set $E(G_1) \cup E(G_2) \cup \{uv : g(u) = v\}$. Linda et al. [17,18] and Kang et al. [19] studied the functigraphs for some graph invariants like metric dimension, domination and zero forcing number. In [20], we have studied the fixing number of some functigraphs. The aim of this paper is to study the distinguishing number of functigraphs.

Network science and graph theory are two interconnected research fields that have synonymous structures, problems and their solutions. The notions 'network' and 'graph' are identical and these can be used interchangeably subject to the nature of application. The roads network, railway network, social networks, scholarly networks, etc are among the examples of networks. In the recent past, the network science has imparted to a functional understanding and the analysis of the complex real world networks. The basic premise in these fields is to relate metabolic networks, proteomic and genomic with disease networks [21] and information cascades in complex networks [22]. Real systems of quite a different nature can have the same network representation. Even though these real systems have different nature, appearance or scope, they can be represented as the same network. Since a functigraph consists of two copies of the same graph (network) with the additional edges described by a function, a mathematical model involving two systems with the same network representation and additional links (edges) between nodes (vertices) of two systems can be represented by a functigraph. The present study is useful in distinguishing the nodes of such pair of the same networks (systems) that can be represented by a functigraph.

Throughout the paper, we denote the functigraph of G by F_G, $V(G_1) = A$, $V(G_2) = B$, g denotes a function $g : A \to B$, f denotes the distinguishing labeling, and $g(V(G_1)) = I$, $|g(V(G_1))| = |I| = s$.

In order to understand the concept of functigraphs, we consider an example of a functigraph of $G = K_9$. Take $V(G_1) = A = \{u_1, ..., u_n\}$ and $V(G_2) = B = \{v_1, ..., v_n\}$ and function g is defined as follows:

$$g(u_i) = \begin{cases} v_1 & if & 1 \le i \le 3, \\ v_2 & if & 4 \le i \le 5, \\ v_{i-3} & if & 6 \le i \le 9. \end{cases} \tag{1}$$

The corresponding functigraph is shown in Figure 1.

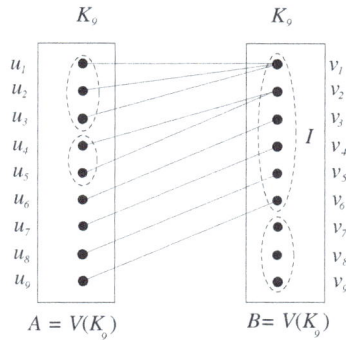

Figure 1. The functigraph F_G when $G = K_9$ and function g is as defined below.

This paper is organized as follows: in Section 2, we give sharp lower and upper bounds for the distinguishing number of functigraphs. This section also establishes connections between the distinguishing number of graphs and their corresponding functigraphs in the form of realizable results. In Section 3, we compute the distinguishing number of functigraphs of complete graphs and joining of path graphs. Some useful results related to these families have also been presented in this section.

2. Bounds and Realizable Results

The sharp lower and upper bounds on the distinguishing number of functigraphs are given in the following result.

Proposition 1. *Let G be a connected graph of order $n \ge 2$; then,*

$$1 \le Dist(F_G) \le Dist(G) + 1.$$

Both bounds are sharp.

Proof. Obviously, $1 \le Dist(F_G)$ by definition. Let $Dist(G) = t$ and f be a t-distinguishing labeling for the graph G. In addition, let $u_i \in A$ and $v_i \in B$, $1 \le i \le n$. We extend labeling f to F_G as: $f(u_i) = f(v_i)$ for all $1 \le i \le n$. We have the following two cases for g:

1. If g is not bijective, then f as defined earlier is a t-distinguishing labeling for F_G. Hence, $Dist(F_G) \le t$.
2. If g is bijective, then f as defined earlier destroys all non-trivial automorphisms of F_G except possible flipping of G_1 and G_2 in F_G. Let F'_G and F_G be the functigraph of G when g is an identity function, i.e., $g(u_i) = v_i$ for all i, $1 \le i \le n$ and when g is not identity function, respectively. The flipping of G_1 and G_2 is possible in the cases when either g is an identity function or when g is not the identity function but the corresponding functigraph F'_G is isomorphic to F_G. In order to break this automorphism (flipping), only one vertex of either G_1 or G_2 must be labeled with an extra color, and hence $Dist(F_G) \le t + 1$.

For the sharpness of bounds, we consider a rigid graph G on $n \geq 6$ vertices. For the sharpness of the lower bound, take a functigraph F_G in which g is a constant function. For the sharpness of the upper bound, take functigraph F_G in which g is an identity function. $\square$

We discuss an example for Proposition 1, where we consider a rigid graph G with the smallest number of vertices i.e., $|V(G)| = 6$ as shown in Figure 2a. Since $Dist(G) = 1$, we label its vertices with a red color. Figure 2b shows F_G, when g is a constant function. In this case, F_G is a rigid graph and hence $Dist(F_G) = 1$. Figure 2c shows F_G, when g is the identity function. In this case, F_G has a non-trivial automorphism i.e., horizontal flipping of F_G. We label vertex v_6 of copy G_2 with blue color to break the non-trivial automorphism. Hence, $Dist(F_G) = 2 = Dist(G) + 1$.

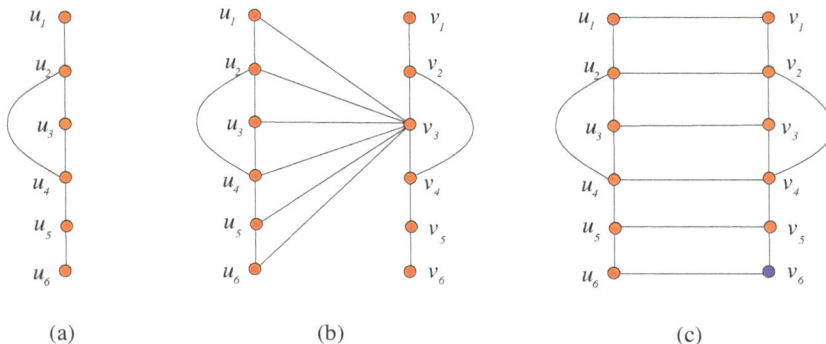

Figure 2. (a) a rigid graph G with six vertices; (b) a functigraph F_G when g is a constant function; (c) a functigraph F_G when g is the identity function i.e., $g(u_i) = v_i$ for all i $(1 \leq i \leq 6)$.

Since at least m colors are required to break all automorphisms of a twin-set of cardinality m, we have the following proposition.

Proposition 2. *Let* $U_1, U_2, ..., U_t$ *be disjoint twin-sets in a connected graph* G *of order* $n \geq 3$ *and* $m = max\{|U_i| : 1 \leq i \leq t\}$,

(i) $Dist(G) \geq m$,

(ii)   *If* $Dist(G) = m$, *then* $Dist(F_G) \leq m$.

Two vertices in a graph G are said to be *similar vertices*, if both can be mapped on each other under some automorphism of graph G.

Lemma 1. *Let* G *be a connected graph of order* $n \geq 2$ *and* g *be a constant function, then* $Dist(F_G) = Dist(G)$.

Proof. Let $I = \{v\} \subset B$. In the functigraph F_G, we label the vertices in copy G_1 of G with $Dist(G)$ colors. Now, v is the only vertex of F_G with the largest degree (as we can see in Figure 2b $I = \{v_3\}$ and v_3 is the vertex of F_G with the largest degree); therefore, it is not similar to any other vertex of F_G and hence it can also be labeled with one of $Dist(G)$ colors. Thus, vertices in $A \cup \{v\}$ are labeled by $Dist(G)$ colors. Since g is a constant function, all vertices in $V(F_G) \setminus \{A \cup \{v\}\}$ are not similar to any vertex in $A \cup \{v\}$ in functigraph F_G. If two disjoint subsets of vertices of a graph are such that every vertex of one set is not similar to any vertex of the other set, then the vertices of both sets can be labeled by the same set of colors; therefore, the vertices in $V(F_G) \setminus \{A \cup \{v\}\}$ and $A \cup \{v\}$ can be labeled by $Dist(G)$ colors. Hence, $Dist(F_G) = Dist(G)$. $\square$

Remark 1. *Let G be a connected graph and $Dist(F_G) = m_1$, if g is constant and $Dist(F_G) = m_2$, if g is not constant; then, $m_1 \geq m_2$.*

Now, we discuss a special type of connected subgraph H of a connected graph G such that $Dist(H) \leq Dist(G)$. We define a set $S(H) = \{u \in V(H) : u \text{ is similar to } v(\neq u) \text{ for some } v \in V(H)\}$. If the graph G has a connected subgraph H in which all vertices in $S(H)$ are either adjacent to all vertices in $V(G) - V(H)$ or non-adjacent to all vertices in $V(G) - V(H)$, then we discuss in Remark 2 that $Dist(G) \geq Dist(H)$.

Lemma 2. *Let H be a connected subgraph of a connected graph G such that all vertices in $S(H)$ are either adjacent to all vertices in $V(G) - V(H)$ or non-adjacent to all vertices in $V(G) - V(H)$, then every automorphism of H can be extended to an automorphism of G.*

Proof. Let $\alpha \in \Gamma(H)$ be an arbitrary automorphism. We define an extension α' of α on $V(G)$ as:

$$\alpha'(w) = \begin{cases} \alpha(w) & if \quad w \in V(H), \\ w & if \quad w \in V(G) - V(H). \end{cases}$$

Since $\alpha'(w) = w$ for all $w \in V(H) - S(H)$, α' being an identity function preserves the relation of adjacency among the vertices in $V(G) - S(H)$. In addition, $\alpha' = \alpha$ being an automorphism of the subgraph H preserves the relation of adjacency among the vertices in $V(H)$. Next, we will prove that α' also preserves the relation of adjacency among the vertices in $\{V(G) - V(H)\} \cup S(H)$. Suppose $u \in S(H)$ and $y \in V(G) - V(H)$, where both y and u are arbitrary vertices of their sets. Since $\alpha \in \Gamma(H)$, $\alpha(u) \in V(H)$. We discuss two cases for the subgraph H in graph G:

1. All vertices in $S(H)$ are adjacent to all vertices in $V(G) - V(H)$; then, u is adjacent to y in G. In addition, $\alpha'(u) = \alpha(u)$ being a vertex of H is adjacent to $\alpha'(y) = y$. Hence, α' preserves the relation of adjacency among the vertices in $\{V(G) - V(H)\} \cup S(H)$.
2. All vertices in $S(H)$ are non-adjacent to all vertices in $V(G) - V(H)$; then, u is non-adjacent to y in G. In addition, $\alpha'(u) = \alpha(u)$ being a vertex of H is non-adjacent to $\alpha'(y) = y$. Hence, α' preserves the relation of adjacency among the vertices in $\{V(G) - V(H)\} \cup S(H)$.

Thus, α' preserves the relation of adjacency among the vertices of $V(G)$.   □

Let H be a connected subgraph of a graph G such that H satisfies the hypothesis of Lemma 2, then every distinguishing labeling of G requires at least $Dist(H)$ colors to break the extended automorphism g' of G, therefore $Dist(G) \geq Dist(H)$ for the subgraph H. It can be seen in Figure 3 that subgraph H of graph G satisfies the hypothesis of Lemma 2 and $Dist(G) = 2 = Dist(H)$. We label the vertices of the graph with red and white colors.

Remark 2. *Let H be a connected subgraph of a connected graph G such that all vertices in $S(H)$ are either adjacent to all vertices in $V(G) - V(H)$ or non-adjacent to all vertices in $V(G) - V(H)$, then $Dist(G) \geq Dist(H)$.*

A vertex v of degree at least three in a connected graph G is called a *major vertex*. Two paths rooted from the same major vertex and having the same length are called the *twin stems*.

We define a function $\psi : \mathbb{N} \setminus \{1\} \to \mathbb{N} \setminus \{1\}$ as $\psi(m) = k$, where k is the least number such that $m \leq 2\binom{k}{2} + k$. For example, $\psi(19) = 5$. Note that ψ is well-defined.
In the following lemma, we find a lower bound of the distinguishing number of a graph having twin stems of length 2 rooted at the same major vertex, in terms of the function ψ.

Lemma 3. *If a graph G has $t \geq 2$ twin stems of length 2 rooted at the same major vertex, then $Dist(G) \geq \psi(t)$.*

Proof. Let $x \in V(G)$ be a major vertex and $xu_i u_i'$ where $1 \leq i \leq t$ are the twin stems of length 2 attached with x. Let $H = <\{x, u_i, u_i'\}>$ and $k = \psi(t)$. Since $xu_i u_i'$ where $1 \leq i \leq t$ are twin stems in the graph G, the subgraph H satisfies the hypothesis of Lemma 2. We define a labeling $f : V(H) \rightarrow \{1, 2, ..., k\}$ as:

$$f(x) = k,$$

$$f(u_i) = \begin{cases} 1 & if & 1 \leq i \leq k, \\ 2 & if & k+1 \leq i \leq 2k, \\ 3 & if & 2k+1 \leq i \leq 3k, \\ \vdots & & \vdots \\ k & if & (k-1)k+1 \leq i \leq k^2. \end{cases} \tag{2}$$

$$f(u_i') = \begin{cases} i \bmod(k) & if & 1 \leq i \bmod(k) \leq k-1, \\ k & if & i \bmod(k) = 0. \end{cases} \tag{3}$$

Using this labeling, one can see that f is a t-distinguishing labeling for H. With permutations with a repetition of k colors, when two of them are taken at a time equal to $2\binom{k}{2} + k$, at least k colors are needed to label the vertices in t-stems. Thus, k is the least integer for which subgraph H has k-distinguishing labeling, and hence $Dist(H) = k$. Thus, $Dist(G) \geq \psi(t)$ by Remark 2. □

It can be seen that the graph G as shown in Figure 3a has four twin stems of length 2 rooted at the same major vertex; therefore, by Lemma 3, $Dist(G) \geq \psi(4) = 2$. The following result gives the existence of a graph G and its functigaph F_G, such that both have the same distinguishing number.

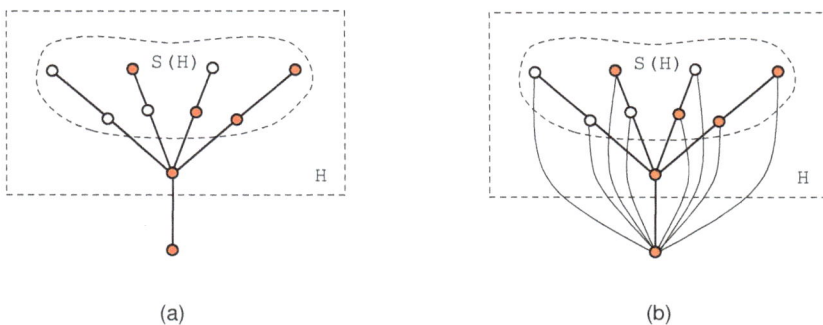

(a)   (b)

Figure 3. (**a**) a graph G and its subgraph H such that all vertices of $S(H)$ are non-adjacent to all vertices of $V(G) - V(H)$; (**b**) a graph G and its subgraph H such that all vertices of $S(H)$ are adjacent to all vertices of $V(G) - V(H)$.

Lemma 4. *For any integer* $t \geq 2$, *there exists a connected graph* G *and a function* g *such that* $Dist(G) = t = Dist(F_G)$.

Proof. Construct a graph G as follows: let $P_{(t-1)^2+1} : x_1 x_2 x_3 ... x_{(t-1)^2+1}$ be a path graph. Join $(t-1)^2 + 1$ twin stems $x_1 u_i u_i'$ where $1 \leq i \leq (t-1)^2 + 1$ each of length two with vertex x_1 of $P_{(t-1)^2+1}$. This completes construction of G. We first show that $Dist(G) = t$. For $t = 2$, we have two twin stems attached with x_1, and hence $Dist(G) = 2$. For $t \geq 3$, we define a labeling $f : V(G) \rightarrow \{1, 2, 3, ..., t\}$ as follows: $f(x_i) = t$, for all i, where $1 \leq i \leq (t-1)^2 + 1$:

$$f(u_i) = \begin{cases} 1 & if \ 1 \le i \le t-1, \\ 2 & if \ t \le i \le 2(t-1), \\ 3 & if \ 2t-1 \le i \le 3(t-1), \\ \vdots & \vdots \\ t-1 & if \ (t-1)(t-2)+1 \le i \le (t-1)^2, \\ t & if \ i = (t-1)^2+1, \end{cases}$$

$$f(u_i') = \begin{cases} i \bmod(t-1) & if \ 1 \le i \bmod(t-1) \le t-2 \ and \ i \ne (t-1)^2+1, \\ t-1 & if \ i \bmod(t-1) = 0, \\ t & if \ i = (t-1)^2+1. \end{cases}$$

Using this labeling, one can see the unique automorphism preserving this labeling is the identity automorphism. Hence, f is a t-distinguishing. With permutations with a repetition of $t-1$ colors, when two of them are taken at a time, $2\binom{t-1}{2} + (t-1)$, $(t-1)^2+1$ twin stems can be labeled by at least t colors. Hence, t is the least integer such that G has a t-distinguishing labeling. Now, we denote the corresponding vertices of G_2 as v_i, v_i', y_i for all i, where $1 \le i \le (t-1)^2+1$ and construct a functigraph F_G by defining $g : A \to B$ as follows: $g(u_i) = g(u_i') = y_i$, for all i, where $1 \le i \le (t-1)^2+1$ and $g(x_i) = y_i$, for all i, where $1 \le i \le (t-1)^2+1$ as shown in Figure 4. Thus, F_G has only symmetries of $(t-1)^2+1$ twin stems attached with y_1. Hence, $Dist(F_G) = t$. $\square$

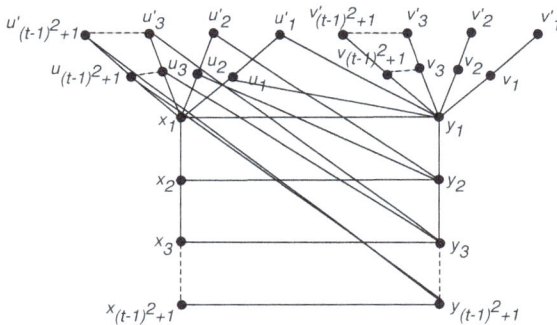

Figure 4. Graph with $Dist(G) = t = Dist(F_G)$.

Consider an integer $t \ge 4$. We construct a graph G similarly as in the proof of Lemma 4 by taking a path graph $P_{(t-3)^2+1} : x_1 x_2 ... x_{(t-3)^2+1}$ and attach $(t-3)^2+1$ twin stems $x_1 u_i u_i'$ where $1 \le i \le (t-3)^2+1$ with any one of its end vertex, say, x_1. Using the similar labeling and arguments as in the proof of Lemma 4, one can see that f is $t-2$ distinguishing and $t-2$ is the least integer such that G has $t-2$ distinguishing labeling. Define functigraph F_G, where $g : A \to B$ is defined by: $g(u_i) = g(u_i') = y_i$, for all i, where $1 \le i \le (t-3)^2+1$, $g(x_i) = v_i$, for all i, where $1 \le i \le (t-3)^2-1$, $g(x_i) = y_i$, for all i, where $(t-3)^2 \le i \le (t-3)^2+1$. From this construction, F_G has only symmetries in which two twin stems attached with y_1 can be mapped on each other under some automorphism of F_G, and hence $Dist(F_G) = 2$. Thus, we have the following result, which shows that $Dist(G) + Dist(F_G)$ can be arbitrarily large:

Lemma 5. *For any integer $t \ge 4$, there exists a connected graph G and a function g such that $Dist(G) + Dist(F_G) = t$.*

Consider $t \ge 3$. We construct a graph G similarly as in the proof of Lemma 4 by taking a path graph $P_{4(t-1)^2+1}: x_1 x_2 ... x_{4(t-1)^2+1}$ and attach $4(t-1)^2+1$ twin stems $x_1 u_i u_i'$, where $1 \le i \le 4(t-1)^2+1$ with

x_1. Using the similar labeling and arguments as in the proof of Lemma 4, one can see that f is $2t - 1$ distinguishing labeling and $2t - 1$ is the least integer such that G has $2t - 1$ distinguishing labeling. Let us now define g as $g(u_i) = g(u_i') = y_i$, for all i, where $1 \le i \le 4(t-1)^2 + 1$, $g(x_i) = v_i$, for all i, where $1 \le i \le 3t^2 - 4t$ and $g(x_i) = y_i$, for all i, where $3t^2 - 4t + 1 \le i \le 4(t-1)^2 + 1$. Thus, F_G has only symmetries of $(t-2)^2 + 1$ twin stems attached with y_1, and hence $Dist(F_G) = t - 1$. After making this type of construction, we have the following result which shows that $Dist(G) - Dist(F_G)$ can be arbitrarily large:

Lemma 6. *For any integer $t \ge 3$, there exists a connected graph G and a function g such that $Dist(G) - Dist(F_G) = t$.*

3. The Distinguishing Number of Functigraphs of Some Families of Graphs

In this section, we discuss a distinguishing number of functigraphs on complete graphs, edge deletion graphs of complete graph and joining of path graphs.

Let G be the complete graph of order $n \ge 3$. We use the following terminology for F_G in the proof of Theorem 1: Let $I = \{v_1, v_2, ..., v_s\}$ and $n_i = |\{u \in A : g(u) = v_i\}|$ for all i, where $1 \le i \le s$. In addition, let $l = \max\{n_i : 1 \le i \le s\}$ and $m = |\{n_i : n_i = 1, 1 \le i \le s\}|$. From the definitions of l and m, we note that $2 \le l \le n - s + 1$ and $0 \le m \le s - 1$.

In the next result, we find the distinguishing number of functigraphs of complete graphs, when g is bijective, in terms of function $\psi(m)$ as defined in the previous section.

Lemma 7. *Let G be the complete graph of order $n \ge 3$ and g be a bijective function; then, $Dist(F_G) = \psi(n)$.*

Proof. Let $A = \{u_1, u_2, ..., u_n\}$ and $I = \{g(u_1), g(u_2), ..., g(u_n)\} = B$. In addition, let $k = \psi(n)$. Let $f : V(F_G) \to \{1, 2, ..., k\}$ be a labeling in which $f(u_i)$ is defined as in Equation (2) and $f(g(u_i))$ as in Equation (3) in the proof of Lemma 3. Using this labeling, one can see that f is a k-distinguishing labeling for F_G. With permutations with repetitions of k colors, when two of them are taken at a time equal to $2\binom{k}{2} + k$, at least k colors are needed to label the vertices in F_G. Hence, k is the least integer for which F_G has k-distinguishing labeling. $\square$

Let G be a complete graph and let $g : A \to B$ be a function such that $2 \le m \le s$. Without loss of generality, assume $u_1, u_2, ..., u_m \in A$ are those vertices of A such that $g(u_i) \ne g(u_j)$, where $1 \le i \ne j \le m$ in B. In addition, $(u_i u_j)(g(u_i)g(u_j)) \in \Gamma(F_G)$ for all $i \ne j$, where $1 \le i, j \le m$. By using the similar labeling f as defined in Lemma 7, at least $\psi(m)$ colors are needed to break these automorphisms in F_G. Thus, we have the following proposition:

Proposition 3. *Let G be a complete graph of order $n \ge 3$ and g be a function such that $2 \le m \le s$; then, $Dist(F_G) \ge \psi(m)$.*

The following result gives the distinguishing number of functigraphs of complete graphs.

Theorem 1. *Let $G = K_n$ be the complete graph of order $n \ge 3$, and let $1 < s \le n - 1$; then,*

$$Dist(F_G) \in \{n - s, n - s + 1, \psi(m)\}.$$

Proof. We discuss the following cases for l:

1. If $l = n - s + 1 > 2$, then A contains $n - s + 1$ twin vertices and B contains $n - s$ twin vertices (except for $n = 3, 4$ where B contains no twin vertices). In addition, there are $m(= s - 1)$ vertices in A which have distinct images in B. By Proposition 3, these m vertices and their distinct images are labeled by at least $\psi(m)$ colors (only 1 color if $m = 1$). Since $n - s + 1$ is the largest

among $n - s + 1$, $n - s$ and $\psi(m)$, $n - s + 1$ is the least number such that F_G has $(n - s + 1)$—distinguishing labeling. Thus, $Dist(F_G) = n - s + 1$.

2. If $l = n - s + 1 = 2$, then $\psi(m) \geq \max\{n - s + 1, n - s\}$, and hence $Dist(F_G) = \psi(m)$.

3. If $l < n - s$, then B contains the largest set of $n - s$ twin vertices in F_G. In addition, there are $m(\leq s - 2)$ vertices in A, each of which have distinct images in B. Since $n - s \geq \psi(m)$, $Dist(F_G) = n - s$.

4. If $l = n - s > 2$, then both A and B contain the largest set of $n - s$ twin vertices in F_G. In addition, there are $m(= s - 2)$ vertices in A that have distinct images in B. Since $n - s \geq \psi(m)$, $Dist(F_G) = n - s$.

5. If $l = n - s = 2$, then we have the following two subcases:

   (a) If $1 < s \leq \lfloor \frac{n}{2} \rfloor + 1$, then both A and B contain the largest set of $n - s$ twin vertices in F_G. In addition, there are $m(= s - 2)$ vertices in A that have distinct images in B. Since $n - s \geq \psi(m)$ (if $\psi(m)$ exists), $Dist(F_G) = n - s$.

   (b) If $\lfloor \frac{n}{2} \rfloor + 1 < s \leq n - 1$, then $\psi(m) \geq \max\{n - s + 1, n - s\}$, and hence $Dist(F_G) = \psi(m)$.

□

We define a function $\phi : \mathbb{N} \to \mathbb{N} \setminus \{1\}$ as $\phi(i) = k$, where k is the least number such that $i \leq \binom{k}{2}$. For instance, $\phi(32) = 9$. Note that ϕ is well defined.

The following result gives the distinguishing number of functigraphs of a family of spanning subgraphs of complete graphs.

Theorem 2. *For a complete graph G of order $n \geq 5$ and G_i, where $1 \leq i \leq \lfloor \frac{n}{2} \rfloor$ is the graph deduced from G by deleting i edges with no common end vertices that join two saturated vertices of G for all i. If g is a constant function, then*

$$Dist(F_{G_i}) = \max\{n - 2i, \phi(i)\}.$$

Proof. On deleting i edges from G, we have $n - 2i$ saturated vertices and i twin-sets each of cardinality two (as shown in Figure 5 where G is the complete graph on 7 vertices, $i = 2$ and g is a constant function). We will now show that exactly $\phi(i)$ colors are required to label vertices of all i twin-sets. We observe that all vertices in twin sets of cardinality 2 are similar to each other in G. Since two vertices in a twin-set are labeled by a unique pair of colors out of $\binom{k}{2}$ pairs of k colors, at least k colors are required to label vertices of i twin-sets. Now, we discuss the following two cases for $\phi(i)$:

1. If $\phi(i) \leq n - 2i$, then the number of colors required to label $n - 2i$ saturated vertices is greater than or equal to the number of colors required to label the vertices of i twin-sets. Thus, we label $n - 2i$ saturated vertices with exactly $n - 2i$ colors and out of these $n - 2i$ colors, $\phi(i)$ colors will be used to label vertices of i twin-sets.

2. If $\phi(i) > n - 2i$, then the number of colors required to label $n - 2i$ saturated vertices is less than the number of colors required to label vertices of i twin-sets. Thus, we label vertices of i twin-sets with $\phi(i)$ colors and, out of these $\phi(i)$ colors, $n - 2i$ colors will be used to label saturated vertices in G_i.

Since g is constant, by using the same arguments as in the proof of Lemma 1, $Dist(F_{G_i}) = Dist(G_i)$. □

Suppose that $G = (V_1, E_1)$ and $G^* = (V_2, E_2)$ are two graphs with disjoint vertex sets V_1 and V_2 and disjoint edge sets E_1 and E_2. The *join* of G and G^* is the graph $G + G^*$, in which $V(G + G^*) = V_1 \cup V_2$ and $E(G + G^*) = E_1 \cup E_2 \cup \{uv : u \in V_1, v \in V_2\}$.

Proposition 4. *Let P_n be a path graph of order $n \geq 2$; then, for all $m, n \geq 2$ and $1 < s < m + n$, $1 \leq Dist(F_{P_m + P_n}) \leq 3$.*

Proof. Let $P_m : v_1, ..., v_m$ and $P_n : u_1, ..., u_n$. We discuss the following cases for m, n.

1. If $m = 2$ and $n = 2$, then $P_2 + P_2 = K_4$, and hence $1 \leq Dist(F_{K_4}) \leq 3$ by Theorem 1.
2. If $m = 2$ and $n = 3$, then $P_2 + P_3$ has three saturated vertices. Thus, $1 \leq Dist(F_{P_2+P_3}) \leq 4$ by Proposition 1. However, for all s where $2 \leq s \leq 4$ and all possible definitions of g in $F_{P_2+P_3}$, one can see $1 \leq Dist(F_{P_2+P_3}) \leq 3$.
3. If $m = 3$ and $n = 3$, then a labeling $f : V(P_3 + P_3) \to \{1, 2, 3\}$ defined as:

$$f(x) = \begin{cases} 1 & if & x = v_1, v_2, \\ 2 & if & x = v_3, u_3, \\ 3 & if & x = u_1, u_2, \end{cases}$$

is a distinguishing labeling for $P_3 + P_3$, and hence $Dist(P_3 + P_3) = 3$. Thus, $1 \leq Dist(F_{P_3+P_3}) \leq 4$ by Proposition 1. However, for all s where $2 \leq s \leq 5$ and all possible definitions of g in $F_{P_3+P_3}$, one can see $1 \leq Dist(F_{P_3+P_3}) \leq 3$.
4. If $m \geq 2$ and $n \geq 4$, then a labeling $f : V(P_m + P_n) \to \{1, 2\}$ defined as:

$$f(x) = \begin{cases} 1 & if & x = v_1, u_2, ..., u_n, \\ 2 & if & x = u_1, v_2, ..., v_m, \end{cases}$$

is a distinguishing labeling for $P_m + P_n$, and hence $Dist(P_m + P_n) = 2$. Thus, the result follows by Proposition 1.

$\square$

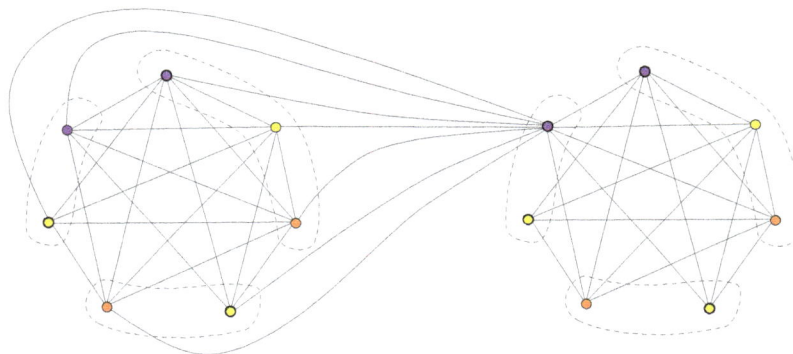

Figure 5. Graph $G = K_7$ and F_{G_2}, when g is a constant function. $Dist(F_{G_2}) = \phi(2) = n - 2i = 3$.

4. Conclusions

In this paper, we have studied the distinguishing number of functigraphs, which is an extension of the permutation graphs. We have given sharp lower and upper bounds of a distinguishing number of functigraphs. We have established the connection between the distinguishing number of graphs and its corresponding functigraph in the form of realizable results. We have computed the distinguishing number of functigraphs of a complete graph and the joining of path graphs. Furthermore, we have defined a function $\phi : \mathbb{N} \to \mathbb{N} \setminus \{1\}$ as $\phi(i) = k$, where k is the least number such that $i \leq \binom{k}{2}$. By using this function ϕ, we have found the distinguishing number of functigraphs of spanning subgraphs of complete graphs. In the future, it would be interesting to work on the distinguishing number of functigraphs of some well known families of graphs.

Symmetry **2018**, *10*, 332

Author Contributions: M.F., M.M., Z.U., U.A. and I.J. have contributed equally in the writing of this article.

Acknowledgments: The authors are very grateful to the anonymous reviewers for the evaluation and the constructive critiques for the improvement of this manuscript.

Conflicts of Interest: The authors declare no conflict of interest.

References

1. Rubin, F. Problem 729: The blind man's keys. *J. Recreat. Math.* **1980**, *12*, 128
2. Albertson, M.O.; Collins, K.L. Symmetry breaking in graphs. *Electron. J. Comb.* **1996**, *3*, R18.
3. Harary, F. Methods of destroying the symmetries of a graph. *Bull. Malasy. Math. Sci. Soc.* **2001**, *24*, 183–191.
4. Collins, K.L.; Trenk, A.N. The distinguishing chromatic number. *Electron. J. Comb.* **2006**, *13*, R16.
5. Kalinowski, R.; Sniak, M.P. Distinguishing graphs by edge colourings. *Eur. J. Comb.* **2015**, *45*, 124–131. [CrossRef]
6. Boutin, D. Identifying graph automorphisms using determining sets. *Electron. J. Comb.* **2006**, *13*, R78.
7. Albertson, M.O.; Boutin, D.L. Using determining sets to distinguish kneser graphs. *Electron. J. Comb.* **2007**, *14*, R20.
8. Albertson, M.O. Distinguishing Cartesian powers of graphs. *Electron. J. Comb.* **2005**, *12*, N17.
9. Bogstad, B.; Cowen, L.J. The distinguishing number of the hypercube. *Discret. Math.* **2004**, *283*, 29–35. [CrossRef]
10. Chan, M. The maximum distinguishing number of a group. *Electron. J. Comb.* **2006**, *13*, 70.
11. Cheng, C.T. On computing the distinguishing numbers of trees and forests. *Electron. J. Comb.* **2006**, *13*, 11.
12. Tymoczko, J. Distinguishing numbers for graphs and groups. *Electron. J. Comb.* **2004**, *11*, 63.
13. Schweitzer, P.; Schweitzer, P. Minimal Asymmetric Graphs. *arXiv* **2016**, arXiv:1605.01320v1.
14. Chartrand, G.; Harary, F. Planar permutation graphs. *Ann. Inst. Hneri Poincare* **1967**, *3*, 433–438.
15. Dorfler, W. On mapping graphs and permutation graphs. *Math. Slov.* **1978**, *28*, 277–288.
16. Chen, A.; Ferrero, D.; Gera, R.; Yi, E. Functigraphs: An extension of permutation graphs. *Math. Bohem.* **2011**, *136*, 27–37.
17. Eroh, L; Kang, C.X.; Yi, E. On metric dimension of functigraphs. *Discret. Math. Algorithms Appl.* **2013**, *5*, 1250060.
18. Eroh, L.; Gera, R.; Kang, C.X.; Larson, C.E.; Yi, E. Domination in functigraphs. *arXiv* **2011**, arXiv:1106.1147.
19. Kang, C.X.; Yi, E. On zero forcing number of functigraphs. *arXiv* **2012**, arXiv:1204.2238.
20. Fazil, M.; Javaid, I.; Murtaza, M. On fixing number of functigraphs. *arXiv* **2016**, arXiv:1611.03346.
21. Gosak, M.; Markovič, R.; Dolenšek, J.; Rupnik, M.S.; Marhl, M.; Stožer, A.; Perc, M. Network science of biological systems at different scales: A review. *Phys. Life Rev.* **2018**, *24*, 118–135. [CrossRef] [PubMed]
22. Jalili, M.; Perc, M. Information cascades in complex networks. *J. Complex Netw.* **2017**, *5*, 665–693. [CrossRef]

symmetry

MDPI

Article

Hyperbolicity on Graph Operators

J. A. Méndez-Bermúdez [1], **Rosalío Reyes** [2], **José M. Rodríguez** [2] and **José M. Sigarreta** [1,3,*]

1   Instituto de Física, Benemérita Universidad Autónoma de Puebla, Apartado Postal J-48, Puebla 72570, Mexico; jmendezb@ifuap.buap.mx
2   Departamento de Matemáticas, Universidad Carlos III de Madrid, Avenida de la Universidad 30, 28911 Leganés, Madrid, Spain; rreyes@math.uc3m.es (R.R.); jomaro@math.uc3m.es (J.M.R.)
3   Facultad de Matemáticas, Universidad Autónoma de Guerrero, Carlos E. Adame No.54 Col. Garita, Acapulco Gro. 39650, Mexico
*   Correspondence: josemariasigarretaalmira@hotmail.com; Tel.: +52-1-7441592272

Received: 24 July 2018; Accepted: 22 August 2018; Published: 24 August 2018

Abstract: A graph operator is a mapping $F : \Gamma \rightarrow \Gamma'$, where Γ and Γ' are families of graphs. The different kinds of graph operators are an important topic in Discrete Mathematics and its applications. The symmetry of this operations allows us to prove inequalities relating the hyperbolicity constants of a graph G and its graph operators: line graph, $\Lambda(G)$; subdivision graph, $S(G)$; total graph, $T(G)$; and the operators $R(G)$ and $Q(G)$. In particular, we get relationships such as $\delta(G) \leq \delta(R(G)) \leq \delta(G) + 1/2$, $\delta(\Lambda(G)) \leq \delta(Q(G)) \leq \delta(\Lambda(G)) + 1/2$, $\delta(S(G)) \leq 2\delta(R(G)) \leq \delta(S(G)) + 1$ and $\delta(R(G)) - 1/2 \leq \delta(\Lambda(G)) \leq 5\delta(R(G)) + 5/2$ for every graph which is not a tree. Moreover, we also derive some inequalities for the Gromov product and the Gromov product restricted to vertices.

Keywords: graph operators; gromov hyperbolicity; geodesics

---

1. Introduction

In [1], J. Krausz introduced the concept of graph operators. These operators have applications in studies of graph dynamics (see [2,3]) and topological indices (see [4–6]). Many large graphs can be obtained by applying graph operators on smaller ones, thus some of their properties are strongly related. Motivated by the above works, we study here the hyperbolicity constant of several graph operators.

Along this paper, we denote by $G = (V(G), E(G))$ a connected simple graph with edges of length 1 (unless edge lengths are explicitly given) and $V \neq \emptyset$. Given an edge $e = uv \in E(G)$ with endpoints u and v, we write $V(e) = \{u, v\}$. Next, we recall the definition of some of the main graph operators.

The *line graph*, $\Lambda(G)$, is the graph constructed from G with vertices the set of edges of G, and and two 19 vertices are adjacent if and only if their corresponding edges are incident in G.

The *subdivision graph*, $S(G)$, is the graph constructed from G substituting each of its edges by a path of length 2.

The graph $Q(G)$ is the graph constructed from $S(G)$ byadding edges between adjacent vertices in $\Lambda(G)$.

The graph $R(G)$ is constructed from $S(G)$ by adding edges between adjacent vertices in G.

The *total graph*, $T(G)$, is constructed from $S(G)$ by adding edges between adjacent vertices in G or $\Lambda(G)$.

We define:

$$E_E(G) := \{\{e_1, e_2\} : e_1, e_2 \in E(G), e_1 \neq e_2, |V(e_1) \cap V(e_2)| = 1\},$$

and

$$E_V(G) := \{\{e, u\} : e \in E(G), u \in V(e)\}.$$

So, we have the following:

$\Lambda(G) := (E(G), E_E(G))$.

$S(G) := (V(G) \cup E(G), E_V(G))$.

$T(G) := (V(G) \cup E(G), E(G) \cup E_E(G) \cup E_V(G))$.

$R(G) := (V(G) \cup E(G), E(G) \cup E_V(G))$.

$Q(G) := (V(G) \cup E(G), E_E(G) \cup E_V(G))$.

The Gromov hyperbolic spaces have multiple applications both theoretical and practical (see [7–10]). A space is geodesic if any two points in it can be joined by a curve whose length is the distance between them. In this paper we will consider a graph G as a geodesic metric space and any geodesic joining x and y will be denote by $[xy]$.

Let X be a geodesic metric space and $x, y, z \in X$. A *geodesic triangle* with vertices x, y, z, denoted by $T = \{x, y, z\}$, is the union of three geodesics $[xy]$, $[yz]$ and $[zx]$. We write also $T = \{[xy], [yz], [zx]\}$. If the δ-neighborhood of the union of any two sides of T contains the other side, we say that T is δ-*thin*. We define $\delta(T) := \inf\{\delta \geq 0 : T \text{ is } \delta\text{-thin}\}$. The space X is δ-*hyperbolic* if all geodesic triangles T in X are δ-thin. Let us denote the sharp hyperbolicity constant of X, by $\delta(X)$, i.e., $\delta(X) := \sup\{\delta(T) : T \text{ is a geodesic triangle in } X\}$. X is *Gromov hyperbolic* if X is δ-hyperbolic for some $\delta \geq 0$; then X is Gromov hyperbolic if and only if $\delta(X) < \infty$.

In this paper we prove inequalities relating the hyperbolicity constants of a graph G and its graph operators $\Lambda(G)$, $S(G)$, $T(G)$, $R(G)$ and $Q(G)$, using their symmetries.

2. Definitions and Background

There are several equivalent definitions for Gromov hyperbolicity (see, e.g., [11–13]), in particular, the definition that we use in this work has an important geometric meaning and serves as a basis for multiple applications (see [14–19]).

Given a graph G, the *Gromov product* of $q_1, q_2 \in G$ with base point $q_0 \in G$ is defined as

$$(q_1, q_2)_{q_0} := \frac{1}{2}\left(d(q_1, q_0) + d(q_2, q_0) - d(q_1, q_2)\right).$$

For every Gromov hyperbolic graph G, we have

$$(q_1, q_3)_{q_0} \geq \min\left\{(q_1, q_2)_{q_0}, (q_2, q_3)_{q_0}\right\} - \delta \tag{1}$$

for every $q_0, q_1, q_2, q_3 \in G$ and some constant $\delta \geq 0$ ([12,13]).

We denote by $\delta^*(G)$ the sharp constant for the inequality (1), i.e.,

$$\delta^*(G) := \sup\left\{\min\left\{(q_1, q_2)_{q_0}, (q_2, q_3)_{q_0}\right\} - (q_1, q_2)_{q_0} : q_0, q_1, q_2, q_3 \in G\right\}.$$

Indeed, our definition of Gromov hyperbolicity is equivalent to (1); furthermore, we have $\delta^*(G) \leq 4\delta(G)$ and $\delta(G) \leq 3\delta^*(G)$ ([12,13]). In [20] (Proposition II.20) we found the following improvement of the previous inequality: $\delta^*(G) \leq 2\delta(G)$.

We denote by $\delta_v^*(G)$ the constant of hyperbolicity of the Gromov product restricted to the vertices of G, i.e.,

$$\delta_v^*(G) := \sup\left\{\min\left\{(q_1, q_2)_{q_0}, (q_2, q_3)_{q_0}\right\} - (q_1, q_3)_{q_0} : q_0, q_1, q_2, q_3 \in V(G)\right\}.$$

3. Main Results

The following result is immediate from the definition of $S(G)$.

Proposition 1. *Let G be a graph. Then*

$$\delta(S(G)) = 2\delta(G), \qquad \delta^*(S(G)) = 2\delta^*(G).$$

We remark that the equality is not true for $\delta_v^*(G)$ (e.g., $S(C_5) = C_{10}$ but $2\delta_v^*(C_5) = 1 \neq 2 = \delta_v^*(S(G))$), but inequalities may apply. The next result appears in [21].

Theorem 1. *Let $B = (V_0 \cup V_1, E)$ be a bipartite graph. We have $\delta_B(V_i) \leq \delta_v^*(B) \leq \delta_B(V_i) + 2$, where*

$$\delta_B(V_i) = \sup\{\min\left\{(x,y)_w, (y,z)_w\right\} - (x,z)_w : x,y,z,w \in V_i\}$$

for every $i \in \{1,2\}$.

Corollary 1. *Let G be a graph. Then*

$$2\delta_v^*(G) \leq \delta_v^*(S(G)) \leq 2\delta_v^*(G) + 2.$$

Proof. Note that $S(G)$ can be considered as a bipartite graph, where $V(S(G)) = V(G) \cup V(\Lambda(G))$. Theorem 1 gives $\delta_{S(G)}(V(G)) \leq \delta_v^*(S(G)) \leq \delta_{S(G)}(V(G)) + 2$. Since $\delta_{S(G)}(V(G)) = 2\delta_v^*(G)$, the desired inequalities hold. □

Proposition 2. *Let G be a graph. Then*

$$\delta_v^*(G) \leq \delta^*(G) \leq \delta_v^*(G) + 3.$$

Proof. The inequality $\delta_v^*(G) \leq \delta^*(G)$ is direct. Let us prove the other inequality.

For every $q_0, q_1, q_2 \in G$ there are $q_0', q_1', q_2' \in V(G)$ such that $d(q_i, q_i') \leq 1/2$ for $i = 0, 1, 2$. Then

$$
\begin{aligned}
\left|(q_1,q_2)_{q_0} - (q_1',q_2')_{q_0'}\right| &= \frac{1}{2}\left|d(q_0,q_1) + d(q_0,q_2) - d(q_1,q_2) - d(q_0',q_1') - d(q_0',q_2') + d(q_1',q_2')\right| \\
&\leq \frac{1}{2}\left|d(q_0,q_1) - d(q_0',q_1')\right| + \frac{1}{2}\left|d(q_0,q_2) - d(q_0',q_2')\right| + \frac{1}{2}\left|d(q_1,q_2) - d(q_1',q_2')\right| \\
&\leq \frac{3}{2}.
\end{aligned}
$$

Given $q_0, q_1, q_2, q_3 \in G$, let $q_0', q_1', q_2', q_3' \in V(G)$, with $d(q_i, q_i') \leq 1/2$ for $i = 0, 1, 2, 3$. We have

$$
\begin{aligned}
(q_1,q_3)_{q_0} \geq (q_1',q_3')_{q_0'} - \frac{3}{2} &\geq \min\left\{(q_1',q_2')_{q_0'}, (q_2',q_3')_{q_0'}\right\} - \delta_v^*(G) - \frac{3}{2} \\
&\geq \min\left\{(q_1,q_2)_{q_0} - \frac{3}{2}, (q_2,q_3)_{q_0} - \frac{3}{2}\right\} - \delta_v^*(G) - \frac{3}{2} \\
&= \min\{(q_1,q_2)_{q_0}, (q_2,q_3)_{q_0}\} - \delta_v^*(G) - 3,
\end{aligned}
$$

and we conclude $\delta^*(G) \leq \delta_v^*(G) + 3$. □

Let H be a subgraph of G, H is *isometric* if $d_H(x,y) = d_G(x,y)$ for every $x,y \in H$. We will need the following well-known result.

Lemma 1. *Let H be an isometric subgraph of G. Then*

$$\delta(H) \leq \delta(G),$$

$$\delta^*(H) \leq \delta^*(G),$$

$$\delta_v^*(H) \leq \delta_v^*(G).$$

Since G is an isometric subgraph of $T(G)$ and $R(G)$, and $\Lambda(G)$ is an isometric subgraph of $T(G)$ and $Q(G)$, we have the following consequence of Lemma 1.

Corollary 2. *For any graph G, we have*

$$\delta(G) \leq \delta(T(G)), \qquad \delta^*(G) \leq \delta^*(T(G)), \qquad \delta_v^*(G) \leq \delta_v^*(T(G)),$$
$$\delta(G) \leq \delta(R(G)), \qquad \delta^*(G) \leq \delta^*(R(G)), \qquad \delta_v^*(G) \leq \delta_v^*(R(G)),$$
$$\delta(\Lambda(G)) \leq \delta(T(G)), \qquad \delta^*(\Lambda(G)) \leq \delta^*(T(G)), \qquad \delta_v^*(\Lambda(G)) \leq \delta_v^*(T(G)),$$
$$\delta(\Lambda(G)) \leq \delta(Q(G)), \qquad \delta^*(\Lambda(G)) \leq \delta^*(Q(G)), \qquad \delta_v^*(\Lambda(G)) \leq \delta_v^*(Q(G)).$$

The hyperbolicity of the line graph has been studied previously (see [21–23]). We have the following results.

Theorem 2. *[22] (Corollary 3.12) Let G be a graph. Then*

$$\delta(G) \leq \delta(\Lambda(G)) \leq 5\delta(G) + 5/2.$$

Furthermore, the first inequality is sharp: the equality is attained by every cycle graph.

Theorem 3. *[21] (Theorem 6) Let G be a graph. Then*

$$\delta_v^*(G) - 1 \leq \delta_v^*(\Lambda(G)) \leq \delta_v^*(G) + 1.$$

Theorem 4. *Let G be a graph. Then*

$$\delta^*(G) - 4 \leq \delta^*(\Lambda(G)) \leq \delta^*(G) + 4.$$

Proof. Proposition 2 and Theorem 3 give $\delta^*(G) \leq \delta_v^*(G) + 3 \leq \delta_v^*(\Lambda(G)) + 4 \leq \delta^*(\Lambda(G)) + 4$, and $\delta^*(\Lambda(G)) \leq \delta_v^*(\Lambda(G)) + 3 \leq \delta_v^*(G) + 4 \leq \delta^*(G) + 4$. $\square$

From Proposition 1, and Theorems 2 and 4 we have:

Corollary 3. *Let G be a graph. Then*

$$\delta(S(G)) \leq 2\delta(\Lambda(G)) \leq 5\delta(S(G)) + 5,$$
$$\delta^*(S(G)) - 8 \leq 2\delta^*(\Lambda(G)) \leq \delta^*(S(G)) + 8.$$

Corollary 2 and Theorems 2, 3 and 4 have the following consequence.

Corollary 4. *Let G be a graph. Then*

$$\delta(G) \leq \delta(Q(G)),$$
$$\delta_v^*(G) \leq \delta_v^*(Q(G)) + 1,$$
$$\delta^*(G) \leq \delta^*(Q(G)) + 4.$$

Theorem 4 improves the inequality $\delta^*(\Lambda(G)) \leq \delta^*(G) + 6$ in [23].

Given a graph G with multiple edges, we define the graph $B(G)$, obtained from G, substituting each multiple edge for one of its simple edges of shorter length (see [23]).

Remark 1. *By argument in the proof of [24](Theorem 8) we have: If in each multiple edge there is at most one edge with length greater than $j := \inf\{d(u, v) : u, v \text{ are joined by a multiple edge of } G\}$, then $\delta(G) \leq \max\left\{\delta(B(G)) + \frac{J-j}{2}, \frac{J+j}{4}\right\}$, where, $J := \sup\{L(e) : e \text{ is an edge contained in a multiple edge of } G\}$.*

Corollary 5. *Let G be a graph. Then*

$$\max\left\{\delta(G), \frac{3}{4}\right\} \le \delta(R(G)) \le \max\left\{\delta(G) + \frac{1}{2}, \frac{3}{4}\right\}.$$

Proof. Note that $R(G)$ can be obtained by adding an edge of length 2 to each pair of adjacent vertices in G, so the graph becomes a graph with multiple edges, with $j = 1$ and $J = 2$. Then [24] (Theorem 8) and Remark 1 give the result. $\square$

From [25] (Theorem 11), we have the following result.

Lemma 2. *Given the following graphs with edges of length 1, we have*

- *If P_n is a path graph, then $\delta(P_n) = 0$ for all $n \ge 1$.*
- *If C_n is a cycle graph, then $\delta(C_n) = n/4$ for all $n \ge 3$.*
- *If K_n is a complete graph, then $\delta(K_1) = \delta(K_2) = 0$, $\delta(K_3) = 3/4$ and $\delta(K_n) = 1$ for all $n \ge 4$.*

If G is not a tree, we define its *girth* $g(G)$ by

$$g(G) := \inf\{L(C) : C \text{ is a cycle in } G\}.$$

From [26] (Theorem 17), we have:

Theorem 5. *If G is not a tree, then*

$$\delta(G) \ge \frac{g(G)}{4}.$$

Corollary 6. *If G is not a tree, then*

$$\delta(G) \ge \frac{3}{4}.$$

Corollary 7. *If G is not a tree, then*

$$\delta(G) \le \delta(R(G)) \le \delta(G) + \frac{1}{2}.$$

Proof. Since G is not a tree, Corollary 6 gives $\delta(G) \ge 3/4$, and so

$$\max\left\{\delta(G), \frac{3}{4}\right\} = \delta(G), \qquad \max\left\{\delta(G) + \frac{1}{2}, \frac{3}{4}\right\} = \delta(G) + \frac{1}{2},$$

and Corollary 5 gives the inequalities. $\square$

Theorem 2 and Corollary 7 have the following consequence.

Corollary 8. *If G is not a tree, then*

$$\delta(R(G)) - \frac{1}{2} \le \delta(\Lambda(G)) \le 5\delta(R(G)) + \frac{5}{2}.$$

From Proposition 1 and Corollary 7 we have the following result.

Corollary 9. *If G is not a tree, then*

$$\delta(S(G)) \le 2\delta(R(G)) \le \delta(S(G)) + 1.$$

Theorem 6. *Let G be a graph. Then*

$$\delta^*(\Lambda(G)) \leq \delta^*(Q(G)) \leq \delta_v^*(\Lambda(G)) + 6 \leq \delta^*(\Lambda(G)) + 6,$$
$$\delta_v^*(\Lambda(G)) \leq \delta_v^*(Q(G)) \leq \delta_v^*(\Lambda(G)) + 6,$$
$$\delta^*(\Lambda(G)) \leq \delta^*(T(G)) \leq \delta_v^*(\Lambda(G)) + 9 \leq \delta^*(\Lambda(G)) + 9,$$
$$\delta_v^*(\Lambda(G)) \leq \delta_v^*(T(G)) \leq \delta_v^*(\Lambda(G)) + 6,$$
$$\delta^*(G) \leq \delta^*(R(G)) \leq \delta_v^*(G) + 6 \leq \delta^*(G) + 6,$$
$$\delta_v^*(G) \leq \delta_v^*(R(G)) \leq \delta_v^*(G) + 6,$$
$$\delta^*(G) \leq \delta^*(T(G)) \leq \delta_v^*(G) + 9 \leq \delta^*(G) + 9,$$
$$\delta_v^*(G) \leq \delta_v^*(T(G)) \leq \delta_v^*(G) + 6.$$

Proof. The lower bounds follow from Corollary 2. We consider the map $P : Q(G) \to \Lambda(G)$ such that $P(q) = q$ if $q \in \Lambda(G)$, $P(q) = v_q$ if $q \notin \Lambda(G)$, where $v_q \in V(\Lambda(G))$ and $d_{Q(G)}(q, v_q) \leq 1$. If $q_0, q_1, q_2, q_3 \in Q(G)$, then

$$\left| d_{Q(G)}(q_i, q_j) - d_{\Lambda(G)}(P(q_i), P(q_j)) \right| = \left| d_{Q(G)}(q_i, q_j) - d_{Q(G)}(P(q_i), P(q_j)) \right| \leq 2,$$

since $\Lambda(G)$ is an isometric subgraph of $Q(G)$ and

$$\left| (q_i, q_j)_{q_0} - (P(q_i), P(q_j))_{P(q_0)} \right|$$
$$= \frac{1}{2} \left| d_{Q(G)}(q_0, q_i) + d_{Q(G)}(q_0, q_j) - d_{Q(G)}(q_i, q_j) \right.$$
$$\left. - d_{\Lambda(G)}(P(q_0), P(q_i)) - d_{\Lambda(G)}(P(q_0), P(q_j)) + d_{\Lambda(G)}(P(q_i), P(q_j)) \right| \leq 3,$$

for $i, j \in \{1, 2, 3\}$. Thus,

$$(q_1, q_3)_{q_0} \geq (P(q_1), P(q_3))_{P(q_0)} - 3$$
$$\geq \min\{(P(q_1), P(q_2))_{P(q_0)}, (P(q_2), P(q_3))_{P(q_0)}\} - \delta_v^*(\Lambda(G)) - 3$$
$$\geq \min\{(q_1, q_2)_{q_0} - 3, (q_2, q_3)_{q_0} - 3\} - \delta_v^*(\Lambda(G)) - 3$$
$$= \min\{(q_1, q_2)_{q_0}, (q_2, q_3)_{q_0}\} - \delta_v^*(\Lambda(G)) - 6.$$

Therefore,

$$\delta^*(\Lambda(G)) + 6 \geq \delta_v^*(\Lambda(G)) + 6 \geq \delta^*(Q(G)) \geq \delta_v^*(Q(G)).$$

These inequalities allow us to obtain the result for upper bounds of $\delta^*(Q(G))$ and $\delta_v^*(Q(G))$. The other upper bounds can be obtained similarly. □

From Theorems 3 and 6 and Corollary 4 we have:

Corollary 10. *For all graph G, we have*

$$\delta_v^*(G) - 1 \leq \delta_v^*(Q(G)) \leq \delta_v^*(G) + 7,$$
$$\delta^*(G) - 4 \leq \delta^*(Q(G)) \leq \delta_v^*(G) + 7 \leq \delta^*(G) + 7.$$

From Corollaries 2, 4 and 10, Theorem 6 and the inequalities $\delta(G) \leq 3\delta^*(G)$ and $\delta^*(G) \leq 2\delta(G)$, we have:

Corollary 11. *Let G be a graph. Then*

$$\delta(\Lambda(G)) \le \delta(Q(G)) \le 6\delta(\Lambda(G)) + 18,$$
$$\delta(\Lambda(G)) \le \delta(T(G)) \le 6\delta(\Lambda(G)) + 27,$$
$$\delta(G) \le \delta(T(G)) \le 6\delta(G) + 27,$$
$$\delta(G) \le \delta(Q(G)) \le 6\delta(G) + 21.$$

Proof. Corollaries 2 and 4 give the lower bounds. On the other hand, Theorem 6 gives $\delta(Q(G)) \le 3\delta^*(Q(G)) \le 3\delta^*(\Lambda(G)) + 18 \le 6\delta(\Lambda(G)) + 18$, $\delta(T(G)) \le 3\delta^*(T(G)) \le 3(\delta^*(\Lambda(G)) + 9) \le 6\delta(\Lambda(G)) + 27$; we obtain the third upper bound in a similar way. Corollary 10 gives $3\delta^*(Q(G)) \le 3(\delta^*(G) + 7) \le 6\delta(G) + 21$, obtaining the last upper bound. $\square$

Let G be a graph, a family of subgraphs $\{G_s\}_s$ of G is a *T-decomposition* if $\cup_s G_s = G$ and $G_s \cap G_r$ is either a *cut-vertex* or the empty set for each $s \ne r$ (see [25]).

The following result was proved in [24] (Theorem 3).

Lemma 3. *Given a graph G and* $\{G_s\}_s$ *any T-decomposition of G, then*

$$\delta(G) = \sup_s \delta(G_s).$$

The following results improve the inequality $\delta(Q(G)) \le 6\delta(\Lambda(G)) + 18$ in Corollary 11.

Theorem 7. *Let G be a path graph, then*

$$0 = \delta(\Lambda(G)) \le \delta(Q(G)) \le 3/4.$$

Proof. Since G is a path graph, $\Lambda(G)$ is also a path graph, and so $0 = \delta(\Lambda(G)) \le \delta(Q(G))$.

Consider the T-decomposition $\{G_n\}$ of $Q(G)$. Since each connected component G_n is either a cycle C_3 or a path of length 1, we have $\delta(Q(G)) = \sup_n\{\delta(G_n)\} \le 3/4$, by Lemmas 2 and 3. $\square$

The union of the set of the midpoints of the edges of a graph G and the set of vertices, $V(G)$, will be denote by $N(G)$. Let $\mathbb{T}_1$ be the set of geodesic triangles T in G such that every vertex of T belong to $N(G)$ and $\delta_1(G) := \inf\{\lambda : \text{every triangle in } \mathbb{T}_1 \text{ is } \lambda\text{-thin}\}$.

Lemma 4. *[27] (Theorems 2.5 and 2.7) For every graph G, we have* $\delta_1(G) = \delta(G)$. *Furthermore, if G is hyperbolic, then there exists* $T \in \mathbb{T}_1$ *with* $\delta(T) = \delta(G)$.

The previous lemma allows to reduce the study of the hyperbolicity constant of a graph G to study only the geodetic triangles of G, whose vertices are vertices of G (i.e., belong to $V(G)$) or midpoints of the edges of G.

Theorem 8. *If G is not a path graph, then*

$$\delta(\Lambda(G)) \le \delta(Q(G)) \le \delta(\Lambda(G)) + 1/2.$$

Proof. By Corollary 2 we have the first inequality. We will prove the second one. If $\delta(Q(G)) = \infty$, then Theorem 6 gives $\delta(\Lambda(G)) = \infty$, and the second inequality holds. Assume now that $\delta(Q(G)) < \infty$ (and so, $\delta(\Lambda(G)) < \infty$ by Theorem 6). If G is not a path, then $\Lambda(G)$ is not a tree and Corollary 6 gives $\delta(\Lambda(G)) \ge 3/4$.

For each $v \in V(G)$, let us define $V_v := \{u \in V(Q(G)) : uv \in E(Q(G))\} = \{u \in V(\Lambda(G)) : uv \in E(Q(G))\}$. Denote by G_v and G_v^* the subgraphs of $Q(G)$ induced by the sets $V_v \cup \{v\}$ and V_v, respectively. Note that both G_v and G_v^* are complete graphs for every $v \in V(G)$, and if

G^* is a complete graph with r vertices, then G_v is a complete graph with $r+1$ vertices. Also, $Q(G) = \Lambda(G) \cup (\cup_{v \in V(G)} G_v)$.

By Lemma 4 there exists a geodesic triangle $T \in \mathbb{T}_1$ in $Q(G)$ with $\delta(T) = \delta(Q(G))$. Denote by $\gamma_1, \gamma_2, \gamma_3$ the sides of T. Without loss of generality we can assume that there exists $p \in \gamma_1$ with $d_{Q(G)}(p, \gamma_2 \cup \gamma_3) = \delta(T) = \delta(Q(G))$. Thus, T is a cycle and each vertex of T is either the midpoint of some edge of $Q(G)$ or a vertex of $Q(G)$.

If G_v contains to T for some $v \in V(G)$, then $\delta(Q(G)) = \delta(T) \leq \delta(G_v) \leq 1 < 3/4 + 1/2 \leq \delta(\Lambda(G)) + 1/2$ by Lemma 2, since G_v is an isometric subgraph of $Q(G)$.

If $\Lambda(G)$ contains to T, then $\delta(Q(G)) = \delta(T) \leq \delta(\Lambda(G))$ by Lemma 1, since $\Lambda(G)$ is isometric.

Suppose that T is not contained either in $\Lambda(G)$ nor G_v with $v \in V(G)$.

Note that if $T \cap (G_v \setminus G_v^*) \neq \emptyset$ for some $v \in V(G)$, then there exists at least one vertex of T in $G_v \setminus \Lambda(G)$. In order to form a triangle $T^* \subset \Lambda(G)$ from T, we define $\gamma_i^* := \gamma_i \cap \Lambda(G)$. Note that, for $i \in \{1, 2, 3\}$, γ_i^* is a geodesic, since $\Lambda(G)$ is a isometric subgraph of $Q(G)$.

We denote by $x_{i,j}$ the common vertex of γ_i and γ_j and by u_i and u_j the other vertices of γ_i and γ_j respectively.

We consider the following cases:

Case A. We assume that exactly one vertex of T belongs to $Q(G) \setminus \Lambda(G)$. Thus, there exists $v \in V(G)$ such that $T \cap (G_v \setminus G_v^*) \neq \emptyset$. By Lemma 4, we have two possibilities: the vertex of T is a vertex of G or a midpoint of an edge in $G_v \setminus G_v^*$.

We can suppose that $x_{i,j} \in T \setminus \Lambda(G)$. Let v be a vertex of $V(G)$ such that $x_{i,j} \in G_v \setminus \Lambda(G)$. Let x_i (respectively, x_j) be the closest point of γ_i^* (respectively, γ_j^*) to $x_{i,j}$. Thus, $x_i x_j \in E(\Lambda(G))$. Let v^* be the midpoint of the edge $x_i x_j$. Let T_1 be the connected component of $T \setminus \Lambda(G)$ joining x_i and x_j. Note that $L(T_1) = 2$. We analyze the two possibilities:

Case A1. Assume that $x_{i,j} \in V(Q(G))$. Let us define $\sigma_i := \gamma_i^* \cup [x_i v^*]$ and $\sigma_j := \gamma_j^* \cup [x_j v^*]$. We are going to prove that σ_i and σ_j are geodesics in $\Lambda(G)$. In fact, we prove now that if $\gamma_i^* = [z_j x_i]$, then $d_{Q(G)}(z_j, x_j) \leq d_{Q(G)}(z_j, x_i)$. Seeking for a contradiction assume that $d_{Q(G)}(z_j, x_j) > d_{Q(G)}(z_j, x_i)$. Thus,

$$d_{Q(G)}(z_j, x_i) + d_{Q(G)}(x_i, x_{i,j}) = d_{Q(G)}(z_j, x_i) + 1 \leq d_{Q(G)}(z_j, x_j) + d_{Q(G)}(x_j, x_{i,j})$$

therefore γ_j is not a geodesic obtaining the desired contradiction and we conclude $d_{Q(G)}(z_j, x_j) \leq d_{Q(G)}(z_j, x_i)$. Hence, σ_i is a geodesic in $\Lambda(G)$.

Case A2. There is an edge $e \in E(Q(G)) \setminus E(\Lambda(G))$ such that $x_{i,j}$ is the midpoint of e, thus without loss of generality we can assume that $e = x_i v$, and we define $\sigma_i := \gamma_i^*$ and $\sigma_j := \gamma_j^* \cup x_j x_i$. Thus, σ_i is a geodesic in $\Lambda(G)$.

Note that $\gamma_j^* \cup x_j v \cup [v x_{i,j}]$ and $\sigma_j \cup [x_i x_{i,j}] = \gamma_j^* \cup x_j x_i \cup [x_i x_{i,j}]$ have the same endpoints and length; therefore, σ_j is also a geodesic in $\Lambda(G)$.

Case B. Assume that there are two vertices of T in some connected component of $T \setminus \Lambda(G)$. Thus, there exists $v \in V(G)$ such that $T \cap (G_v \setminus G_v^*) \neq \emptyset$. By Lemma 4, we have two possibilities again: both vertices of T are midpoints of edges or one vertex of T is a vertex of G and the other is a midpoint of an edge.

We can assume that $u_i, u_j \in G_v \setminus G_v^*$ for some v. We denote by x_i' (respectively, x_j') the closest point in γ_i^* (respectively, γ_j^*) to u_i (respectively, u_j); then $x_i' x_j' \in E(\Lambda(G))$. Let v' be the midpoint of the edge $x_i' x_j'$. Let T_2 be the connected component of $T \setminus \Lambda(G)$ joining x_i' and x_j'. Note that $L(T_2) = 2$.

We analyze the two possibilities again:

Case B1. The vertices u_i, u_j of T are the midpoints of $x_i' v$ and $x_j' v$. Thus, $\sigma_i := \gamma_i^*$, $\sigma_j := \gamma_j^*$ and $\sigma_k := x_i' x_j'$ are geodesics in $\Lambda(G)$.

Case B2. Otherwise, we can assume without loss of generality that $u_j = v$ and u_i is the midpoint of $x_i v$. We have $d_{Q(G)}(u_i, x_j) = d_{Q(G)}(u_i, x_i) + 1$ and so, $\sigma_i := \gamma_i^*$ and $\sigma_j := \gamma_j^* \cup x_i' x_i'$ are geodesics in $\Lambda(G)$. In this case we define $\sigma_k := \{x_i'\}$.

Note that the most general possible case is the following: there are at most three vertices $v_1, v_2, v_3 \in V(G)$ such that $T \cap (G_{v_i} \setminus G_{v_i^*}) \neq \emptyset$, for $i = 1, 2, 3$. Repeating the previous process at most three times we obtain a geodesic triangle T^* in $\Lambda(G)$ with sides γ'_1, γ'_2 and γ'_3 containing γ^*_1, γ^*_2 and γ^*_3, respectively.

If $p \in \Lambda(G)$, then one can check that $\delta(Q(G)) = d_{Q(G)}(p, \gamma_2 \cup \gamma_3) \leq d_{Q(G)}(p, \gamma'_2 \cup \gamma'_3) + 1/2 \leq \delta(\Lambda(G)) + 1/2$. If $p \notin \Lambda(G)$, then $\delta(Q(G)) = d_{Q(G)}(p, \gamma_2 \cup \gamma_3) \leq 5/4$; since $\delta(\Lambda(G)) \geq 3/4$, we have $\delta(\Lambda(G)) + 1/2 \geq 5/4 \geq \delta(Q(G))$. This finishes the proof. □

Proposition 1, Theorems 2 and 8, and Corollary 3 have the following consequence.

Corollary 12. *Let G be a graph. If G is not a path graph, then*

$$\delta(S(G)) \leq 2\delta(Q(G)) \leq 5\delta(S(G)) + 6.$$

4. Conclusions

In this paper, we obtained several inequalities and closed formulas relating the hyperbolicity constants of a graph G and its graph operators $\Lambda(G)$, $S(G)$, $T(G)$, $R(G)$ and $Q(G)$, by the use of their symmetries. As a first step, as the basis of our research, we found relations among the Gromov hyperbolicity constant (satisfying the Rips condition), the Gromov product and the Gromov product restricted to vertices. In the same direction, we derived inequalities between Gromov products and graph operators; as examples we mention: $\delta^*_v(G) \leq \delta^*(G) \leq \delta^*_v(G) + 3$, $\delta^*_v(G) \leq \delta^*_v(Q(G)) + 1$ and $\delta^*(G) \leq \delta^*(R(G)) \leq \delta^*_v(G) + 6 \leq \delta^*(G) + 6$.

Then, we studied relations between the Gromov hyperbolicity constant of a graph and the application of given operators to that graph. In this context, we obtained inequalities such as: $\delta(G) \leq \delta(R(G)) \leq \delta(G) + 1/2$, $\delta(\Lambda(G)) \leq \delta(Q(G)) \leq \delta(\Lambda(G)) + 1/2$, $\delta(S(G)) \leq 2\delta(R(G)) \leq \delta(S(G)) + 1$ and $\delta(R(G)) - 1/2 \leq \delta(\Lambda(G)) \leq 5\delta(R(G)) + 5/2$, where G not a tree.

We believe that our work may motivate the investigation of related open problems such as: (i) the computation of the hyperbolicity constant on geometric graphs; (ii) the analysis of hyperbolicity on the graph operators reported here (i.e., $\Lambda(G)$, $S(G)$, $T(G)$, $R(G)$ and $Q(G)$) when applied to geometric graphs; (iii) the study of the hyperbolicity constants of additional graph operators; and (iv) the identification of the properties of graph operations that break or preserve hyperbolicity.

Author Contributions: The authors contributed equally to this work.

Funding: Supported in part by two grants from Ministerio de Economía y Competitividad, Agencia Estatal de Investigación (AEI) and Fondo Europeo de Desarrollo Regional (FEDER) (MTM2016-78227-C2-1-P and MTM2015-69323-REDT), Spain.

Acknowledgments: The authors would like to thank the editor and the anonymous referees whose comments and suggestions greatly improved the presentation of this paper.

References

1. Krausz, J. Démonstration nouvelle d'un théorème de Whitney sur les réseaux. *Mat. Fiz. Lapok* **1943**, *50*, 75–85.
2. Harary, F.; Norman, R.Z. Some properties of line digraphs. *Rend. Circ. Mat. Palermo* **1960**, *9*, 161–168. [CrossRef]
3. Prisner, E. *Graph Dynamics*; Chapman and Hall/CRC: Boca Raton, FL, USA, 1995; Volume 338.
4. Bindusree, A.R.; Naci Cangul, I.; Lokesh, V.; Sinan Cevik, A. Zagreb polynomials of three graph operators. *Filomat* **2016**, *30*, 1979–1986. [CrossRef]
5. Ranjini, P.S.; Lokesha, V. Smarandache-Zagreb index on three graph operators. *Int. J. Math. Comb.* **2010**, *3*, 1–10.
6. Yan, W.; Yang, B.-Y.; Yeh, Y.-N. The behavior of Wiener indices and polynomials of graphs under five graph decorations. *Appl. Math. Lett.* **2007**, *20*, 290–295. [CrossRef]

7. Gromov, M. Hyperbolic groups. In *Essays in Group Theory*; Gersten, S.M., Ed.; Mathematical Sciences Research Institute Publications; Springer: Berlin, Germany, 1987; Volume 8, pp. 75–263.

8. Oshika, K. *Discrete Groups*; AMS Bookstore: Providence, RI, USA, 2002.

9. Jonckheere, E.A. Contrôle du traffic sur les réseaux à géométrie hyperbolique–Vers une théorie géométrique de la sécurité l'acheminement de l'information. *J. Eur. Syst. Autom.* **2002**, *8*, 45–60.

10. Jonckheere, E.A.; Lohsoonthorn, P. Geometry of network security. In Proceedings of the 2004 American Control Conference, Boston, MA, USA, 30 June–2 July 2004; pp. 111–151.

11. Bowditch, B.H. Notes on Gromov's hyperbolicity criterion for path-metric spaces. In *Group Theory from a Geometrical Viewpoint*; Ghys, E., Haefliger, A., Verjovsky, A., Eds.; World Scientific: River Edge, NJ, USA, 1991; pp. 64–167.

12. Alonso, J.; Brady, T.; Cooper, D.; Delzant, T.; Ferlini, V.; Lustig, M.; Mihalik, M.; Shapiro, M.; Short, H. Notes on word hyperbolic groups. In *Group Theory from a Geometrical Viewpoint*; Ghys, E., Haefliger, A., Verjovsky, A., Eds.; World Scientific: Singapore, 1992; pp. 3–63.

13. Ghys, E.; de la Harpe, P. *Sur les Groupes Hyperboliques d'après Mikhael Gromov*; Progress in Mathematics 83; Birkhäuser Boston Inc.: Boston, MA, USA, 1990.

14. Chepoi, V.; Dragan, F.F.; Vaxès, Y. Core congestion is inherent in hyperbolic networks. In Proceedings of the Twenty-Eighth Annual ACM-SIAM Symposium on Discrete Algorithms, Barcelona, Spain, 16–19 January 2017; pp. 2264–2279.

15. Grippo, E.; Jonckheere, E.A. Effective resistance criterion for negative curvature: Application to congestion control. In Proceedings of the 2016 IEEE Conference on Control Applications (CCA), Buenos Aires, Argentina, 19–22 September 2016.

16. Li, S.; Tucci, G.H. Traffic Congestion in Expanders, (p, δ)-Hyperbolic Spaces and Product of Trees. *Int. Math.* **2015**, *11*, 134–142.

17. Shang, Y. Lack of Gromov-hyperbolicity in colored random networks. *Pan-Am. Math. J.* **2011**, *21*, 27–36.

18. Shang, Y. Lack of Gromov-hyperbolicity in small-world networks. *Cent. Eur. J. Math.* **2012**, *10*, 1152–1158. [CrossRef]

19. Shang, Y. Non-hyperbolicity of random graphs with given expected degrees. *Stoch. Models* **2013**, *29*, 451–462. [CrossRef]

20. Soto, M. Quelques Propriétés Topologiques des Graphes et Applications a Internet et aux Réseaux. Ph.D. Thesis, Université Paris Diderot, Paris, France, 2011.

21. Coudert, D.; Ducoffe, G. On the hyperbolicity of bipartite graphs and intersection graphs. *Discret. Appl. Math.* **2016**, *214*, 187–195. [CrossRef]

22. Carballosa, W.; Rodríguez, J.M.; Sigarreta, J.M. New inequalities on the hyperbolicity constant of line graphs. *ARS Comb.* **2014**, *129*, 367–386.

23. Carballosa, W.; Rodríguez, J.M.; Sigarreta, J.M.; Villeta, M. On the hyperbolicity constant of line graphs. *Electron. J. Comb.* **2011**, *18*, 210.

24. Bermudo, S.; Rodríguez, J.M.; Sigarreta, J.M.; Vilaire, J.-M. Gromov hyperbolic graphs. *Discret. Math.* **2013**, *313*, 1575–1585. [CrossRef]

25. Rodríguez, J.M.; Sigarreta, J.M.; Vilaire, J.-M.; Villeta, M. On the hyperbolicity constant in graphs. *Discret. Math.* **2011**, *311*, 211–219. [CrossRef]

26. Michel, J.; Rodríguez, J.M.; Sigarreta, J.M.; Villeta, M. Hyperbolicity and parameters of graphs. *ARS Comb.* **2011**, *100*, 43–63.

27. Bermudo, S.; Rodríguez, J.M.; Sigarreta, J.M. Computing the hyperbolicity constant. *Comput. Math. Appl.* **2011**, *62*, 4592–4595. [CrossRef]

symmetry

MDPI

Article

General $(\alpha, 2)$-Path Sum-Connectivirty Indices of One Important Class of Polycyclic Aromatic Hydrocarbons

Haiying Wang

School of Science, China University of Geosciences (Beijing), Beijing 100083, China; whycht@cugb.edu.cn

Received: 14 August 2018; Accepted: 6 September 2018; Published: 21 September 2018

Abstract: The general (α, t)-path sum-connectivity index of a molecular graph originates from many practical problems, such as the three-dimensional quantitative structure–activity relationships (3D QSAR) and molecular chirality. For arbitrary nonzero real number α and arbitrary positive integer t, it is defined as ${}^t\chi_\alpha(G) = \sum_{P^t = v_{i_1} v_{i_2} \cdots v_{i_{t+1}} \subseteq G} [d_G(v_{i_1}) d_G(v_{i_2}) \cdots d_G(v_{i_{t+1}})]^\alpha$, where we take the sum over all possible paths of length t of G and two paths $v_{i_1} v_{i_2} \cdots v_{i_{t+1}}$ and $v_{i_{t+1}} \cdots v_{i_2} v_{i_1}$ are considered to be one path. In this work, one important class of polycyclic aromatic hydrocarbons and their structures are firstly considered, which play a role in organic materials and medical sciences. We try to compute the exact general $(\alpha, 2)$-path sum-connectivity indices of these hydrocarbon systems. Furthermore, we exactly derive the monotonicity and the extremal values of these polycyclic aromatic hydrocarbons for any real number α. These valuable results could produce strong guiding significance to these applied sciences.

Keywords: topological indices; general (α, t)-path sum-connectivity index; polycyclic aromatic hydrocarbons

1. Introduction

1.1. Application Background

In many fields (e.g., physics, chemistry, and electrical networks), the boiling point, the melting point, the chemical bonds, and the bond energy are all important quantifiable parameters. To understand the physico-chemical properties of chemical compounds or network structures, we abstractly define different concepts, collectively named the topological descriptors or the topological indices after mathematical modelings. We called them different names, such as Randić index and Zagreb index [1–3]. Different index represents its corresponding chemical structures in graph-theoretical terms via arbitrary molecular graph.

In the past decades, these two-dimensional topological indices have been used as a powerful approach to discover many new drugs, such as anticonvulsants, anineoplastics, antimalarials, and antiallergics and Silico generation [4–8]. Therefore, the practice has proven that the topological indices and the quantitative structure-activity relationships (QSAR) have moved from an attractive possibility to representing a foundation stone in the process of drug discovery and other research areas [9–12].

Most importantly, with the further study of chemical indices and drug design and discovery, three-dimensional molecular features (topographic indices) and molecular chirality are also presented. It is increasingly urgent to study the three-dimensional quantitative structure-activity relationships such as molecular chirality. However, so far there have been few results, except for one related definition that is mentioned generally in [8].

1.2. Definitions and Notations

In the whole paper, we always let $G = (V_G, E_G)$ be a simple molecular graph, in which V_G and E_G are the vertex set and edge set of G, respectively. We denote $|V_G|$ and $|E_G|$ as the numbers of vertices and edges of G, respectively. In physico-chemical graph theory, the atoms and the bonds represent the vertices and edges, respectively. Two vertices are called adjacent if there is an edge between them in G. For any vertex $u \in V_G$, the number of its adjacent vertices is called its degree in G and denote $d_G(u)$. The set of all neighbors of u is denoted by $N_G(u)$, and a vertex of G is called a pendant if its degree is 1. Similarly, the minimum and maximum degree of G are denoted by δ_G and Δ_G, respectively. All other notations and terminologies are referred to [13].

In 1975, Randić index was introduced by the chemist M. Randić during his study of alkanes [1]. As a molecular structure-descriptor and a graphical description of molecular structure, Randić index is most commonly used in the quantitative structure-property and structure-activity studies [6,14]. Randić index is defined as the sum over all edges $uv \in E_G$ of a molecular graph of the terms $[d_G(u)d_G(v)]^{-\frac{1}{2}}$. That is,

$$R(G) = \sum_{uv \in E_G} [d_G(u)d_G(v)]^{-\frac{1}{2}}.$$

The first Zagreb index was introduced more than forty years ago by Gutman and Trinajestić [15,16], and is defined as

$$M_1(G) = \sum_{x \in V_G} d_G(x) = \sum_{uv \in E_G} [d_G(u) + d_G(v)].$$

Later [17], some researchers began to define another new index of a graph G as

$$\chi(G) = \sum_{uv \in E_G} [d_G(u) + d_G(v)]^{-\frac{1}{2}},$$

which is named the sum-connectivity index and denoted by $\chi(G)$.

In 2008, Zhou and Trinajestic [17] proposed the sum-connectivity index, which is a closely related variant of Randić connectivity index of G. Now we define the general sum-connectivity index $\chi_\alpha(G)$ as

$$\chi_\alpha(G) = \sum_{uv \in E_G} [d_G(u) + d_G(v)]^\alpha.$$

With the intention of extending the applicability of the general sum-connectivity index, we begin to consider the general (α, t)-path sum-connectivity index of G as where we take the sum over all possible paths of length t of G:

$$^t\chi_\alpha(G) = \sum_{P^t = v_{i_1}v_{i_2}\cdots v_{i_{t+1}} \subseteq G} [d_G(v_{i_1}) + d_G(v_{i_2}) + \cdots + d_G(v_{i_{t+1}})]^\alpha,$$

with any nonzero real number α and any positive integer t, and two paths $v_{i_1}v_{i_2}\cdots v_{i_{t+1}}$ and $v_{i_{t+1}}\cdots v_{i_2}v_{i_1}$ are considered to be one path.

According to the above definition, the *general (α, t)-path sum-connectivity index* of an arbitrary graph is one real constant and an important invariant under graph automorphism. It is closely related to the structures of a molecular graph. For any molecular material, only by mastering its structure can we calculate the exact value of its *general (α, t)-path sum-connectivity index*.

In this work, one important class of polycyclic aromatic hydrocarbons and their structures are considered which play a role in organic materials and medical sciences. Then, we try to compute the exact general $(\alpha, 2)$-path sum-connectivity indices of these hydrocarbon systems. Furthermore, we exactly derive its monotonicity and extremal values for these polycyclic aromatic hydrocarbons for any real number α. These valuable results could produce strong guiding significance to these applied sciences.

For convenience, it is necessary to simplify some basic concepts and notations in polycyclic aromatic hydrocarbons. A vertex with degree i is called an i-vertex. An edge between a j-vertex and a k-vertex is called a (j,k)-edge. Besides, the numbers of i-vertices and (j,k)-edges are denoted as n_i and m_{jk}, respectively.

Let $v_{i_0}v_{i_1}\cdots v_{i_t}$ be a path P^t of length t in polycyclic aromatic hydrocarbons, denoted $P^t = v_{i_0}v_{i_1}\cdots v_{i_t}$. $(d_G(v_{i_0}), d_G(v_{i_1}), \cdots, d_G(v_{i_t}))$ is called its degree sequence. Obviously, there are in total two types (i.e., $(1,3,3)$ and $(3,3,3)$) of degree sequences of different 2-paths in these polycyclic aromatic hydrocarbons in Figure 1. Let m_{133} and m_{333} denote the numbers of all 2-paths of the degree sequence types $(1,3,3)$ and $(3,3,3)$ in polycyclic aromatic hydrocarbons, respectively.

2. Polycyclic Aromatic Hydrocarbons

Polycyclic aromatic hydrocarbons are important and ubiquitous combustion materials. They belong to one class of hydrocarbon molecules. Polycyclic aromatic hydrocarbons have been considered as an important class of carcinogens. They also play a role in the graphitisation of medical science and organic materials [18,19].

In the field of chemical materials, polycyclic aromatic hydrocarbons have become molecular analogues of graphite for interstellar species and building blocks of functional materials for device applications [20–22]. Thus, detailed descriptions of all these molecular properties are necessary for the available synthetic routes to polycyclic aromatic hydrocarbons and their specific applications.

In essence, polycyclic aromatic hydrocarbons can be considered as small pieces of graphene sheets, in which the free valences of the dangling bonds are saturated by hydrocarbons. Vice versa, a graphene sheet can be interpreted as an infinite polycyclic aromatic hydrocarbon molecule [22]. Many scientists have reported many successful applications of polycyclic aromatic hydrocarbons in graphite surface modeling. As we know, *benzenoid systems* are a very famous family of hydrocarbon molecules and belong to the *circumcoronene homologous series* of *benzenoid*, and polycyclic aromatic hydrocarbons have very similar properties to them.

One important class of polycyclic aromatic hydrocarbons shown in Figure 1 belong to linear and regular circular polycyclic aromatic hydrocarbons [22]. However, the class of *symmetrical poly-aromatic hydrocarbons* is important in sciencesw. For an arbitrary positive integer n, let PAH_n be the general expression of this class of polycyclic aromatic hydrocarbons shown in Figure 1.

Obviously, the first three members of this hydrocarbon family are given in Figure 2, where PAH_1 is called *benzene*, PAH_2 *coronene*, and PAH_3 *circumcoronene* . Obviously, *benzene* has 6 carbon atoms and 6 hydrogen atoms, *coronene* has 24 carbon atoms and 12 hydrogen atoms, and *circumcoronene* has 54 carbon atoms and 18 hydrogen atoms.

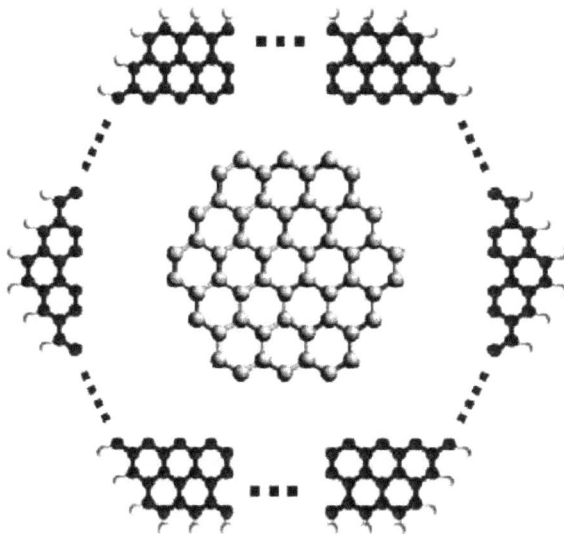

Figure 1. General representation of a polycyclic aromatic hydrocarbon.

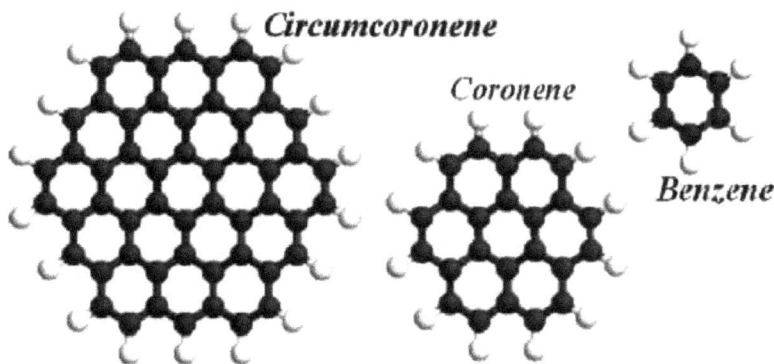

Figure 2. The first three graphs of polycyclic aromatic hydrocarbons.

From Figure 1 above, we know that the class *polycyclic aromatic hydrocarbon* PAH_n contains $6n^2$ carbon atoms and $6n$ are hydrogen atoms. Thus, this molecular graph has $6n^2 + 6n$ vertices or atoms such that $6n^2$ of them are carbon atoms and $6n$ are hydrogen atoms. Each hydrogen atom is 1-vertex and each carbon atom is 3-vertex in PAH_n. Therefore, this hydrocarbon molecule PAH_n satisfies that $|V_{PAH_n}| = 6n^2 + 6n$. In this hydrocarbon molecule, we have

$$|E_{PAH_n}| = \frac{3 \times 6n^2 + 1 \times 6n}{2} = 9n^2 + 3n,$$

in which $|E_{PAH_n}|$ means its number of edges (actually chemical bonds).

According to Figure 1, each hydrogen atoms has just one edge/bond between only one carbon atom in the class of polycyclic aromatic hydrocarbon system. Any other carbon atoms just have three bonds with carbon atoms or hydrogen atoms. From the structure of Figure 1, it is clear that we can divide the edge set of the class of polycyclic aromatic hydrocarbons into two partitions: the $(1,3)$-edge subset and the $(3,3)$-edge subset. Thus,

$$m_{13} = n_1 = 6n$$

and

$$m_{33} = |E_{PAH_n}| - n_1 = 9n^2 - 3n.$$

3. Main Results on the General $(\alpha, 2)$-Path Sum-Connectivity Indices of PAH_n

In this section, let PAH_n be the general representation of the class of polycyclic aromatic hydrocarbon molecules in Figure 1 for any positive integer n. Then, there are $6n$ hydrogen atoms and $6n^2$ carbon atoms in PAH_n. We compute *the general $(\alpha, 2)$-path sum-connectivity index* of a family of polycyclic aromatic hydrocarbons as follows. The indices should directly reflect the material's natural properties.

Theorem 1. *For an arbitrary real number α, the general $(\alpha, 2)$-path sum-connectivity index of PAH_n is equal to*

$$^2\chi_\alpha(PAH_n) = 6 \cdot n \cdot [3^{2\alpha+1} \cdot n - 2(9^\alpha - 7^\alpha)]. \tag{1}$$

Proof of Theorem 1. According to the structures of PAH_n, consider any $(1,3)$-edge e. Then, there are in total two different 2-paths, and each path contains this edge e. Consider any $(3,3)$-edge e'. There are in total four different 2-paths, and each path contains this edge e'. Since we do not distinguish between the paths $v_{i_1} v_{i_2} \cdots v_{i_{l+1}}$ and $v_{i_{l+1}} \cdots v_{i_2} v_{i_1}$, each 2-path of PAH_n will compute twice. Then, the total number of different 2-paths, denoted $N(P^2)$, is

$$N(P^2) = \frac{N_H \cdot 2 + m_{33} \cdot 4}{2} = \frac{6n \cdot 2 + (9n^2 - 3n) \cdot 4}{2} = 18n^2.$$

□

If the degree sequence of a 2-path is the type $(1,3,3)$, then this path begins or ends with one hydrogen atom. Obviously, each hydrogen atom can produce two different P^2, and there are $6n$ hydrogen atoms in PAH_n. Then,

$$m_{133} = 2 \cdot m_{13} = 2 \cdot 6n = 12n.$$

Since there are in total two types $(1,3,3)$ and $(3,3,3)$ of degree sequences of 2-paths in PAH_n, we have

$$m_{333} + m_{133} = N(P^2),$$

which induces that

$$m_{333} = N(P^2) - m_{133} = 18n^2 - 12n.$$

By usage of the definitions of *the general $(\alpha, 2)$-path sum-connectivity index*, we can compute it of the polycyclic aromatic hydrocarbon in Figure 1 as follows:

$$^2\chi_\alpha(PAH_n) = \sum_{P^2 = v_{i_1} v_{i_2} v_{i_3} \in G} [d_G(v_{i_1}) + d_G(v_{i_1}) + d_G(v_{i_1})]^\alpha$$

$$= m_{133} \cdot (1 + 3 + 3)^\alpha + m_{333} \cdot (3 + 3 + 3)^\alpha$$

$$= (12n) \cdot 7^\alpha + (18n^2 - 12n) \cdot 9^\alpha$$

$$= 6 \cdot n \cdot [3^{2\alpha+1} \cdot n - 2(9^\alpha - 7^\alpha)].$$

4. The Monotonicity and the Extremal Values of $^2\chi_\alpha(PAH_n)$

Let PAH_n be the general representation of the class of polycyclic aromatic hydrocarbon molecules shown in Figure 1 for any positive integer n. In this section, we approach the monotonicity and the extremal values of $^2\chi_\alpha(PAH_n)$ for any real number α.

By Equation (1), we can see that $^2\chi_\alpha(PAH_n)$ is a strictly increasing function on n. That is, the larger n is, the larger $^2\chi_\alpha(PAH_n)$ is.

Let

$$^2\chi_\alpha(PAH_n) = 6 \cdot n \cdot [3^{2\alpha+1} \cdot n - 2(9^\alpha - 7^\alpha)] = 0. \tag{2}$$

Then, Equation (2) has two real zeroes $n_1 = 0$ and $n_2 = \frac{2}{3}[1 - (\frac{7}{9})^\alpha]$.

It is clear that

$$n_2 = \frac{2}{3}[1 - (\frac{7}{9})^\alpha] < \frac{2}{3}$$

for any real number α. Thus, $^2\chi_\alpha(PAH_n)$ is a strictly increasing function on the positive number n and for any real number α.

Thus, we can conclude the theorem as follows.

Theorem 2. *Let PAH_n be the general representation of the class of polycyclic aromatic hydrocarbon molecules shown in Figure 1. Then*

1. *For any real number α, we have $^2\chi_\alpha(PAH_n)$ is strictly increasing with respect to all positive integers n.*
2. *The smallest general $(\alpha, 2)$-path sum-connectivity index of Polycyclic aromatic hydrocarbons is*

$$^2\chi_\alpha(PAH_n)_{min} = ^2\chi_\alpha(PAH_1) = 6[9^\alpha + 2 \cdot 7^\alpha] \tag{3}$$

when and only when $n = 1$. Of course, PAH_1 is benzene (see Figure 1).

5. Conclusions

The general sum-connectivity index $\chi_\alpha(PAH_n)$ and its minimum value of the class of polycyclic aromatic hydrocarbons can be obtained by substituting the specific value $t = 1$ in the results above.

6. Further Research

In this article, we only consider one important class of *symmetrical poly-aromatic hydrocarbons* PAH_n (see Figure 1), which belong to linear and regular circular polycyclic aromatic hydrocarbons [22]. However, there are broader and more useful polycyclic aromatic hydrocarbons in the world. There are many linearly fused circular *PAHs* with different structures, such as *naphthalene, anthracene, tetracene,* and *pentacene*. On the other hand, there are great nonlinear and irregular or non-symmetrical aromatic hydrocarbons, such as *pyrene, benzopyrene,* derivatives of *azulene* and *pentahelicene*. In the future, we intend to conduct scientific research on the relationship between the complicated aromatic hydrocarbons and their *general (α, t)-path sum-connectivity indices*. This research will be very meaningful, interesting and worthwhile.

Funding: This research was funded by National Science Foundation of China [No. 11701530] and Fundamental Research Funds for the Central Universities [No. 2652015193] and [No. 2652017146].

Acknowledgments: The author is indebted to Assistant Editor Jocelyn He and the anonymous referees for their valuable comments and remarks that improved the presentation of the paper.

Conflicts of Interest: The author declares no conflict of interest.

References

1. Randić, M. On characterization of molecular branching. *J. Am. Chem. Soc.* **1975**, *97*, 6609–6615. [CrossRef]
2. Lu, M.; Zhang, L.; Tian, F. On the Randić index of cacti. *MATCH Commun. Math. Comput. Chem.* **2006**, *56*, 551–556.
3. Wang, S.; Wei, B. Sharp bounds of multiplicative Zagreb indices of *k*-Trees. *Discret. Appl. Math.* **2015**, *180*, 168–175. [CrossRef]
4. Kier, L.B.; Hall, L.H. *Medicinal Chemistry: Molecular Connectivity in Chemistry and Drug Research*; Academic Press: New York, NY, USA, 1976.
5. Kier, L.B.; Hall, L.H.; Murray, W.J.; Randić, M. Molecular connectivity V: The connectivity concept applied to density. *J. Pharm. Sci.* **1976**, *65*, 1226–1227. [CrossRef] [PubMed]
6. Kier, L.B.; Hall, L.H. *Molecular Connectivity in Structure-Activity Analysis*; Wiley: New York, NY, USA, 1986.
7. Zanni, R.; Galvez-Liompart, M.; García-Domenech, R.; Galvez, J. Latest advances in molecular topology applications for drug discovery. *Expert Opin. Drug Discov.* **2015**, *9*, 945–957. [CrossRef] [PubMed]
8. Estrada, E.; Patlewicz, G.; Uriarte, E. From molecular graphs to drugs: A review on the use of topological indices in drug design and discovery. *Indian J. Chem.* **2003**, *42*, 1315–1329.
9. Gutman, I. Advances in the theory of Benzenoid hydrocarbons II. In *Topics in Current Chemistry*; Springer: Berlin, Germany, 1992.
10. Gutman, I. Extremal hexagonal chains. *J. Math. Chem.* **1993**, *12*, 197–210. [CrossRef]
11. Gutman, I.; Cyvin, S.J. *Introduction to the Theory of Benzenoid Hydrocarbons*; Springer: Berlin, Germany, 1989.
12. Gutman, I.; Cyvin, S.J. Advances in the theory of Benzenoid hydrocarbons. In *Topics in Current Chemistry*; Springer: Berlin, Germany, 1990.
13. Bondy, J.A.; Murty, U.S.R. *Graph Theory with Applications*; MacMillan: New York, NY, USA, 1976.
14. Devillers, J.; Balaban, A.T. *Topological Indices and Related Descriptors in QSAR and QSPR*; Gordon and Breach Science Publishers: New York, NY, USA, 1999.
15. Gutman, I.; Trinajestić, N. Graph theory and molecular orbitals. III. Total φ-electron energy of alternant hydrocarbons. *Chem. Phys. Lett.* **1972**, *17*, 535–538. [CrossRef]
16. Gutman, I.; Das, K.C. The first Zagreb index 30 years after. *MATCH Commun. Math. Comput. Chem.* **2004**, *50*, 83–92.
17. Zhou, B.; Trinajestić, N. On a novel connectivity index. *J. Math. Chem.* **2009**, *46*, 1252–1270. [CrossRef]
18. Gupta, S.P. QSAR studies on local anesthetics. *Chem. Rev.* **1991**, *91*, 1109–1119. [CrossRef]
19. Wiersum, U.E.; Jenneskens, L.W. *Jenneskens in Gas Phase Reactions in Organic Synthesis*; Gordon and Breach Science Publishers: Amsterdam, The Netherlands, 1997; pp. 143–194.
20. Berresheim, A.J.; Muller, M.; Mollen, K. Polyphenylene nanostructures. *Chem. Rev.* **1999**, *99*, 1747–1785. [CrossRef] [PubMed]
21. Bauschlicher, C.W.; Bakes, E.L.O. Infrared spectra of polycyclic aromatic hydrocarbons. *Chem. Phys.* **2000**, *262*, 285–291. [CrossRef]
22. Anthony, J.E. Functionalized acenes and heteroacenes for organic electronics. *Chem. Rev.* **2006**, *106*, 5028–5048. [CrossRef] [PubMed]

MDPI

Article

Secure Resolving Sets in a Graph

Hemalathaa Subramanian [1,*] and Subramanian Arasappan [2]

[1] Research Scholar, Register No. 10445, The M.D.T. Hindu College, Tirunelveli 627 010 Affiliated to Manonmaniam Sundaranar University, Abishekapatti, Tirunelveli 627 012, Tamilnadu, India
[2] Principal, Department of Mathematics, The M.D.T. Hindu College, Tirunelveli 627 010, India; asmani1963@gmail.com
*   Correspondence: hemarmath@gmail.com

Received: 5 September 2018; Accepted: 19 September 2018; Published: 27 September 2018

Abstract: Let $G = (V, E)$ be a simple, finite, and connected graph. A subset $S = \{u_1, u_2, \ldots, u_k\}$ of $V(G)$ is called a resolving set (locating set) if for any $x \in V(G)$, the code of x with respect to S that is denoted by $C_S(x)$, which is defined as $C_S(x) = (d(u_1, x), d(u_2, x), .., d(u_k, x))$, is different for different x. The minimum cardinality of a resolving set is called the dimension of G and is denoted by $dim(G)$. A security concept was introduced in domination. A subset D of $V(G)$ is called a dominating set of G if for any v in $V - D$, there exists u in D such that u and v are adjacent. A dominating set D is secure if for any u in $V - D$, there exists v in D such that $(D - \{v\}) \cup \{u\}$ is a dominating set. A resolving set R is secure if for any $s \in V - R$, there exists $r \in R$ such that $(R - \{r\}) \cup \{s\}$ is a resolving set. The secure resolving domination number is defined, and its value is found for several classes of graphs. The characterization of graphs with specific secure resolving domination number is also done.

Keywords: resolving set; domination; secure resolving set and secure resolving domination

1. Introduction

Let $G = (V, E)$ be a simple, finite, and connected graph. Let $S = \{u_1, u_2, \ldots, u_k\}$ on which the ordering $(u_1, u_2, \ldots, u_k)$ is imposed. For any $w \in V(G)$, the ordered k-tuples $r(w \mid S) = (d(u_1, w), d(u_2, w), \ldots, d(u_k, w))$ is known as the metric description of w with respect to S. The set S is called a resolving set of G if $r(u \mid S) = r(w \mid S)$ implies $u = w$ for all $u, w \in V(G)$. A resolving set of G of minimum cardinality is called a minimum resolving set or a basis, and the cardinality of a minimum resolving set is called the dimension of G, which is denoted by $dim(G)$ [1].

The idea of locating sets in a connected graph is already available in the literature [2,3]. Slater initiated the concept of locating sets (resolving sets) and a reference set (metric dimension) nearly four decades ago. Later, Harary and Melter found the above-mentioned theory [4] independently. They adopted the term metric dimension for locating number. Several papers have been published on resolving sets, resolving dominating sets, independent resolving sets, etc.

Security is a concept that is associated with several types of sets in a graph. For example, a dominating set D of G is secure set if for any $v \in V - D$ there exist $u \in D$ such that $(D - \{u\}) \cup \{v\}$ is a dominating set [5,6]. Secure independent sets, secure equitable sets etc., have been defined and discussed. In this paper, secure resolving sets and secure resolving dominating sets are introduced and studied.

In this paper, G refers to a simple, finite, and connected graph. The abbreviations used in this paper are as follows:

- SR set: Secure resolving set
- SRD set: Secure resolving dominating set

The role of symmetry in the following study:

Regarding the symmetry role, a complete graph has vertex transitivity, which is a symmetry. $K_{m,n}$ has vertices that have degree symmetry in the partite sets. One more important aspect of symmetry is present in the concept of the SR set, as well as in SRD sets. In SR sets, every vertex has the opportunity of being a member of a resolving set. Thus, a symmetry is achieved in the presence of vertices. A similar thing happens in domination. In practical application, in any Executive Council, equal opportunity is to be given to all of the members of the General Council for inclusion in the Executive Council. Thus, the spirit of symmetry is present in the form of equality. That is, there is a symmetry in the treatment of vertices.

2. Secure Resolving Dimension

Definition 1. *A subset T of G is a SR set of G if T is resolving and for any $x \in V - T$, there exists $y \in T$ such that $(T - \{y\}) \cup \{x\}$ is a resolving set of G. The minimum cardinality of a SR set of G is known as the secure resolving dimension of G, and is marked by sdim(G).*

Remark 1. *The existence of a SR set is guaranteed.For, in any graph, the vertex set V(G) is a secure set as well as a resolving set.*

Remark 2. *dim(G) ≤ sdim (G).*

3. Secure Resolving Dimension for Some Well-Known Graphs

1. $sdim (K_n) = n - 1 = dim (K_n)$
2. $sdim (K_{1,n}) = n > dim (K_{1,n})$
3. $sdim (K_{m,n}) = m + n - 2 = dim(K_{m,n}) (m, n \geq 2)$
4. $sdim (P_n) = 2 > dim (P_n) = 1 (n \geq 3)$
5. $sdim(C_n) = 2 = dim(C_n)$

6. $sdim (K_m (a_1, a_2, \dots , a_m)) = \begin{cases} dim(K_m(a_1, a_2, \dots, a_m)) + 1 \\ \quad if\ a_i \geq 2\ for\ atleast\ one\ i \\ dim\ (K_m(a_1, a_2, \dots, a_m)) \\ \quad if\ a_i = 1\ for\ all\ i. \end{cases}$

where $(K_m(a_1, a_2, \dots , a_m))$ is the multi-star graph formed by joining $a_i \geq 1 (1 \leq i \leq m)$ pendant vertices to each vertex x_i of a complete graph K_m with $V(K_m) = \{x_1, x_2, \dots , x_m\}$.

Illustration 1. *Consider C_5.*

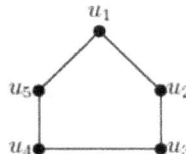

let $H = \{u_1, u_3\}$. Then, H is resolving, and for any $u \in V - H$, there exists $v \in H$ such that $(H - \{v\}) \cup \{u\}$ is a resolving set of C_5. It can be easily seen that $sdim(G) = 2$.

4. Secure Resolving Dimension for Special Classes of Graphs

Observation 1. *Let order of $G \geq 3$. Suppose sdim (G) = 1. Then, dim (G) = (since sdim(G) ≥ dim(G)). Therefore, $G = P_n$. However, $sdim(P_n) = 2$, which is a contradiction. Therefore, $sdim(G) \geq 2$.*

Observation 2. *sdim(G) = 1 if and only if $G = P_1$ or P_2.*

Theorem 1. $sdim(G) = 2$, G is a tree if and only if $G = P_n$ $(n \geq 3)$.

Proof. If $G = P_n$ $(n \geq 3)$, then $sdim(G) = 2$.

Conversely, suppose $n \geq 4$. Suppose there are two pendant vertices v_1, v_2 adjacent with w of G. Take a vertex t, which is adjacent to w $(t \neq v_1, v_2)$. $\{v_1, w\}$ is not a resolving set, since t and v_2 will not have distinct codes with respect to $\{v_1, w\}$. Assume that $\{v_1, v_2\}$ is a resolving set of G. Then, it is not secure, since $\{v_1, w\}$ and $\{v_2, w\}$ are not resolving sets. Suppose $\{v_1, t\}$ is resolving. Then, it is not secure (since $\{v_1, w\}$, $\{w, t\}$ are not resolving sets of G). Let $\{a, b\}$ be a resolving set of G. $a, b \notin \{v_1, v_2, w\}$. Then, $\{a, b\}$ is not secure, since neither $\{w, b\}$ nor $\{w, a\}$ is a resolving set of P_n (since $d(v_1, a) = d(w, a) + 1 = d(v_2, a))$. Therefore, no vertex of G supports two or more pendant vertices. Suppose that w is a vertex of G that supports one pendant vertex and there exists at least two neighbors of w having degrees greater than or equal to two. Then, we will not get any resolving set with cardinality two containing w. Therefore, any vertex of G with a pendant neighbor has at most one neighbor of degree greater than or equal to two. Therefore, G is a path. Suppose that $n = 3$. Since G is acyclic and connected, $G = P_3$. □

Theorem 2. $sdim(C_n) = dim(C_n) = 2$.

Proof. Let $V(C_n) = \{m_1, m_2, \ldots, m_n\}$.

Case (i): $n = 2k + 1$.

Let $M = \{m_1, m_2\}$. Then, M is a resolving set of C_n. It can be verified that $\{m_1, m_i\}$ is a resolving set where $3 \leq i \leq n$.

Case (ii): $n = 2k$.

Then, m_1 and m_k are diametrically opposite vertices. Let $M = \{m_1, m_2\}$. Clearly, $\{m_1, m_2\}$ is a resolving set of C_n. It can be substantiated that $\{m_1, m_i\}$ is resolving when $3 \leq i \leq n, i \neq k$. Also, $\{m_k, m_2\}$ is a resolving set of C_n. Therefore, $sdim\,(G) = 2 = dim(G)$.

Remark 3. $sdim(G) \leq 1 + dim(G)$.

Proof. $sdim(G) \geq dim(G)$. Suppose *that* $dim(G) < sdim(G)$. Let $T = \{u_1, u_2, \ldots ., u_k\}$ be a basis of G. Let $W = \{u_1, u_2, \ldots, u_k, v\}$. Then, W is a SR set of G. Therefore, $sdim(G) \leq k + 1$. However, $sdim(G) > dim(G) = k$. Therefore, $sdim(G) \leq 1 + dim(G)$. Hence the remark. □

Theorem 3. $sdim(G) = n - 1$ if and only if $G = K_n$ or $K_{1, n-1}$.

Proof. Let $G = K_n$ or $K_{1,n-1}$. Then, $sdim(G) = n - 1$. Suppose that $sdim(G) = n - 1$. Then, $dim(G)$ is $n - 1$ or $n - 2$. If $dim(G) = n - 1$, then $G = K_n$. Suppose that $dim(G)$ is $n - 2$. Then, $G = K_{a, b}$ $(a, b \geq 1)$, $Ka + \overline{Kb}$ $(b \geq 2, a \geq 1)$, $K_a + (K_1 \cup K_b)$ $(b, a \geq 1)$. Suppose that $G = K_{a,b}$ $(a, b \geq 2)$, $sdim(G) = dim(G) = a + b - 2$ [1]. Suppose that $G = Ka + \overline{Kb}$ $(b \geq 2, a \geq 1)$, then $dim(G) = a + b - 2$. If $a = 1$, then $Ka + \overline{Kb} = K_{1,b}$. In this case, $sdim(G) = b$ and $dim(G) = b - 1$. If $a \geq 2$, then $sdim(G) = a + b - 2 = dim(G)$.

Let $G = K_a + (K_1 \cup K_b)$, $(a, b \geq 1)$. When $a = 1$ and $b = 1$, $G = P_3$ and $sdim(P_3) = 2$ and $dim(P_3) = 1$. Clearly, G is a star. When $a = 1$ and $b \geq 2$, $sdim(G) = a + b - 1 = dim(G)$. Suppose that $a > 1$ and $b = 1$. Then, $sdim(G) = a = dim(G)$. Suppose that $a, b > 1$. Then, $sdim(G) = a + b - 1 = dim(G)$. Except when G is a star, $sdim(G) = dim(G) = n - 2$. Therefore, $G = K_{1, n-1}$. □

Theorem 3. *Let T be a connected graph. Let* $G = TK_2$. *Then,* $sdim(T) \leq sdim(TK_2) \leq sdim(T) + 1$.

Proof. Refer to Theorem 7 [1]. Let $G = T K_2, T_1$, and T_2 be the transcripts of T in G. Let X be a basis of T, and let $X_1 = \{x_1, x_2, \ldots, x_k\}$ and $X_2 = \{y_1, y_2, \ldots, y_k\}$ be the basis of T_1 and T_2 respectively, corresponding to X. Let $S = X_1 \cup \{y_1\}$. Then, S is a SR set of G. Therefore, $sdim(G) \leq sdim(T) + 1$. Let V_1,

V_2 be the vertex sets of T_1 and T_2 respectively. Then, $V(G) = V_1 \cup V_2$. Let X be a secure basis of G. Let $X_1 = X \cap V_1$, $X_2 = X \cap V_2$. Let $S_1 \subseteq V(T_1)$ be the union of X_1 and the set X'_2 consisting of those vertices of V_1 corresponding to X_2. Then, S_1 is a SR set of T_1. Therefore, $sdim(T) = sdim(T_1) \leq |S_1| = |X_1 \cup X'_2| \leq |X_1| + |X'_2| = |X| = sdim(G)$. Hence, the theorem. $\square$

Corollary 1. *Let $\epsilon > 0$. Then, $\frac{sdim(G)}{sdim(T)} < \epsilon$ where T is a connected induced subgraph of G.*

Proof. Let $T = K_{1,2^n+1}$. $sdim(T) = 2^n+1$. Let $G = T K_2$. Then, we get a graph G containing T as an induced subgraph [1]. Further, $sdim(G) \leq 2n$. Therefore, $\frac{sdim(G)}{sdim(T)} \leq \frac{2n}{2^{n+1}} \to 0$ as $n \to 0$. Hence, the corollary. $\square$

5. Secure Resolving Domination Number

Definition 2. *Let U be a subset of G. $U = \{u_1, u_2, \ldots, u_k\}$ of $V(G)$ is said to be a SRD set of G if U is a dominating set of G, U is resolving, and U is secure. The minimum cardinality of a SRD set of G is known as a secure resolving domination number of G, and is represented by $\gamma_{sr}(G)$.*

Remark 4. *V is a SRD set of G.*

6. Secure Resolving Domination Number for Some Well-known Graphs

1. $\gamma_{sr}(K_n) = n - 1, n \geq 2$.
2. $\gamma_{sr}(K_{1,n-1}) = n - 1, n \geq 2$.
3. $\gamma_{sr}(P_n) = \begin{cases} 2 & \text{if } n = 3, 4 \\ \left\lceil \frac{n}{3} \right\rceil + 1 & \text{if } n \geq 5. \end{cases}$
4. $\gamma_{sr}(C_n) = \begin{cases} 2 & \text{if } n = 3, 4 \\ \left\lceil \frac{n}{3} \right\rceil + 1 & \text{if } n \geq 5. \end{cases}$
5. $\gamma_{sr}(K_{a_1,a_2,\ldots,a_m}) = (a_1 + a_2 + \ldots + a_m) - m$.
6. $\gamma_{sr}(K_m(a_1, a_2, \ldots, a_m)) = \begin{cases} m + a_{k+1} + \ldots + a_m - k \\ \text{if } a_1 = \ldots = a_k = 1 \ a_i \geq 2, k+1 \leq i \leq m \\ m \text{ if } a_i = 1 \text{ for all } i \end{cases}$

where $(K_m(a_1, a_2, \ldots, a_m))$ is the multi-star graph formed by joining $a_i \geq 1$ $(1 \leq i \leq m)$ pendant vertices to each vertex x_i of a complete graph K_m with $V(K_m) = \{x_1, x_2, \ldots, x_m\}$.

Illustration 2. *Consider the following graph $K_3(1, 1, 1)$.*

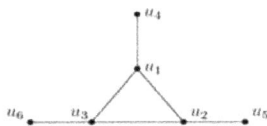

let $N = \{u_1, u_2, u_3\}$. Then N is a secure, dominating, and resolving set of $K_3(1, 1, 1)$. It can be easily seen that. $\gamma_{sr}(K_3(1, 1, 1)) = 3$.

Proposition 1. *Let γ_s be the minimum cardinality of a secure dominating set of G. Then, $\max\{\gamma_s(G), dim(G), \gamma_r(G)\} \leq \gamma_{sr}(G) \leq \gamma_s(G) + dim(G)$.*

Proof. Let L be a minimum secure dominating set of G and W be a basis of G. Then, $L \cup W$ is a SRD set of G. Hence, $\gamma_{sr}(G) \leq \gamma_s(G) + dim(G)$. The first inequality is obvious. $\square$

Remark 5. *$P \cup \{u\}$ is a SRD set of G, P is a minimum resolving dominating set of G.*

Illustration 3. *Consider the given graph G.*

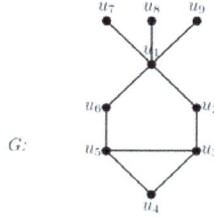

Here, $\gamma(G) = 2$ and $dim(G) = 3$ (since $\{u_1, u_4\}$ is a minimum dominating set, $\{u_5, u_7, u_8\}$ is a minimum resolving set of G). $\gamma_r(G) = 4$ (since $\{u_1, u_5, u_7, u_8\}$ is a minimum resolving dominating set of G. $\{u_1, u_4, u_5, u_7, u_8\}$ is a SRD set of G. Let S be a minimum SRD set of G. Consequently, $\gamma_{sr}(G) \leq 5$. Since S is resolving, S must contain two of the pendant vertices. If S contains u_2, then u_6 and the remaining pendant vertices are not resolved. If S contains u_6, then u_2 and the remaining pendant vertices are not resolved. If S contains u_4, then u_5 and u_3 are not resolved. Therefore, either S contains u_5 and two of the pendant vertices or u_3 and two of the pendant vertices. If S contains u_5 and two of the pendant vertices, then the remaining pendant vertex is not resolved. Therefore, the resolving dominating set contains u_1. Therefore, the possibilities of the resolving dominating sets are $\{u_1, u_5, u_7, u_8\}$, $\{u_1, u_5, u_8, u_9\}$, $\{u_1, u_5, u_7, u_9\}$, $\{u_1, u_3, u_7, u_8\}$, $\{u_1, u_3, u_8, u_9\}$, and $\{u_1, u_3, u_7, u_9\}$. None of these is secure, since u_2 and u_6 cannot be replaced in all of these sets. Therefore, $\gamma_{sr}(G) \geq 5$. Hence, $\gamma_{sr}(G) = 5$. Thus, $\gamma(G) < dim(G) < \gamma_r(G) < \gamma_{sr}(G)$. Also, $\gamma_{sr}(G) = \gamma(G) + dim(G)$. $\gamma_s(G) = 3$, since G has no secure dominating set with two vertices, and $\{u_1, u_4, u_3\}$ is a secure dominating set. Therefore, $\gamma_{sr}(G) = 5 < \gamma_s(G) + dim(G) = 6$ and $max\{\gamma_s(G), dim(G), \gamma_r(G)\} = 4 < \gamma_{sr}(G) = 5$.

Remark 6. When $G = K_n$, $\gamma(G) = 1$, $\gamma_s(G) = 1$, $dim(G) = n - 1$, $\gamma_r(G) = n - 1$, $\gamma_{sr}(G) = n - 1$. Therefore, $max\{\gamma_s(G), dim(G), \gamma_r(G)\} = \gamma_{sr}(G)$.

Observation 3. $\gamma_{sr}(G) \geq g(m, d)$, where $g(m, d) = min\left\{ t : t + \sum_{i=1}^{t} \binom{t}{i} (d - 1)^{t-i} \geq m \right\}$, d is a diameter of G, the order of G is $m \geq 2$, and d and m are positive integers with $d < m$. This follows from proposition 2.1 [6] and that $\gamma_{sr}(G) \geq \gamma_r(G)$.

Observation 4. For every positive integer k, there are only finitely many connected graphs with secure resolving domination number k.

Proof. Consider a graph G with order $m \geq 2$ and $\gamma_{sr}(G) = k$. From corollary 2.2 [6] $m \leq k + \sum_{i=1}^{k} \binom{k}{i} (d - 1)^{l-i}$. $\gamma(G) \leq \gamma_{sr}(G) = k$. Therefore, the diameter of G is not more than $3k - 1$. Therefore, $m \leq k + \sum_{i=1}^{k} \binom{k}{i} (3k - 2)^{k-i}$. Therefore, there are only finitely many connected graphs with $\gamma_{sr}(G) = k$. $\square$

Remark 7. Suppose that $\gamma_{sr}(G) = 2$. Then, the number of connected graphs with $\gamma_{sr}(G) = 2$ has an order of at most 11.

Proof. By the above observation, $n \leq 2 + \sum_{i=1}^{2} \binom{2}{i} (6 - 2)^{2-i} = 2 + \binom{2}{1} 4 + \binom{2}{2} 4 = 2 + 8 + 1 = 11$.

In fact, the above bound for n can be improved. $\square$

Observation 5. For any G with $\gamma_{sr}(G) = 2$, the order of G is not more than 4.

Proof. Let $\gamma_{sr}(G) = 2$. Let $X = \{p, q\}$ be a γ_{sr}—set of G. If $d(p, q) \geq 4$, then p and q cannot dominate the point at a distance 2 from p in the shortest path joining p and q. Therefore, $d(p, q) \leq 3$. $\square$

Case (i): Distance between p and q is 1.

As every single vertex in $V(G) - X$ is adjacent with either both p and q or one of them, the distances of the vertices in $V(G) - X$ from p and q are $(1, 2)$, $(2, 1)$, and $(1, 1)$. Then, G is as follows:

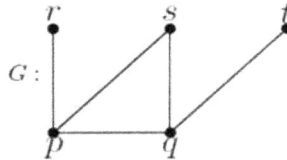

Here, s cannot enter X by removing a vertex of X, since such a resulting set is not a dominating set. Therefore, $G = P_4$. If both pendants r and t are removed, then the resulting set is K_3, for which $\gamma_{sr}(G) = 2$. That is, $G = K_3$.

If r, s, and t are present and r is adjacent with s, then the graph is:

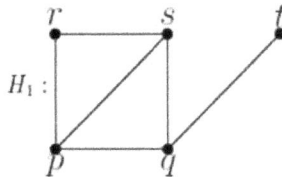

Here, r cannot enter X, since resolving fails.

If r, s, and t are existing, r and t are adjacent with s, then the graph is:

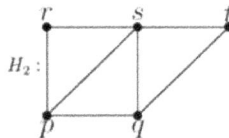

Here, s cannot enter X, since resolution fails.

If r and t are adjacent, then the graph H_3 is as follows:

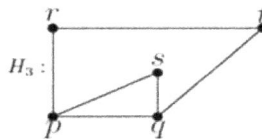

In H_3, s cannot enter X, since domination fails. If r, s, and t are mutually adjacent, then the graph H_4 is as follows:

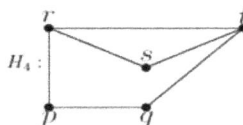

In the above graph, s cannot enter X, since resolution fails. The remaining cases are: (i) s is not present, and r and t are non-adjacent. In this case, $G = P_4$. (ii) r and t are available, s is not present and r and t are adjacent. We get $G = C_4$. (iii) r and s are alone present and r and s adjacent. We get C_4 with a diagonal. (iv) r and s are alone and present, and they are not adjacent. We get K_3 with a pendant vertex. Thus, in this case, $G = P_3, P_4, C_4, C_4$ with a diagonal and K_3 with a pendant vertex.

Case (ii): $d(p,q) = 2$.

Since every vertex in $V(G) - X$ is adjacent with at least one of p and q, the distances of the vertices 3 in $V(G) - X$ from p and q are $(1, 3), (3, 1), (1, 2), (2, 1),$ and $(1, 1)$. Therefore, the graph is as follows:

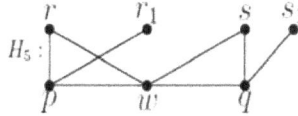

For security in H_5, r_1 cannot enter $\{p, q\}$ by removing p or q, since domination fails. Therefore, only one of r and r_1 can be present. Similarly, one of s and s_1 can be present. Therefore, the graphs are as follows:

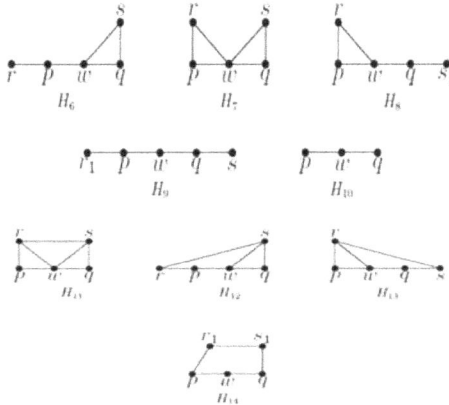

In graph H_9, w cannot enter $X = \{p, q\}$. In H_6, if w enters X by removing p or q, then the resulting set is not resolving, although it is dominating. In graph H_7, if w enters X, then for domination, q should be replaced by w. However, the resulting set is not resolving. Same is the graph H_8. In graphs $H_{11}, H_{12},$ and H_{13}, w cannot enter X, since resolution fails. In graph H_{14}, w cannot enter X, since domination fails.

Case (iii): $d(p,q) = 3$.

Since vertices in $V(G) - X$ are adjacent with one of p and q, the distances of the vertices in $V(G) - X$ from p and q are $(1, 1), (1, 2), (2, 1), (1, 3), (3, 1), (1, 4),$ and $(4, 1)$.

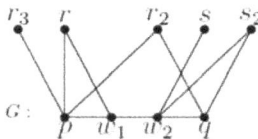

Only one of $r, r_2,$ and r_3 can be present, since $\{p, q\}$ is an SRD set. Similarly, only one of s and s_2 can be present. If any number of edges among the vertices $r_3, r, x_2, s,$ and s_2 are inserted, then w_1 cannot enter X by replacing p or q, since domination fails.

Subcase (i): r_3 is present.

In this case, w_1 cannot enter X by replacing p, q, since domination fails.

Subcase (ii): s_2 is exist.

In this case, w_2 cannot enter X by replacing p, q. (since domination fails).

Subcase (iii): One of r, r_2, and s is present.

Then, the graphs are as follows:

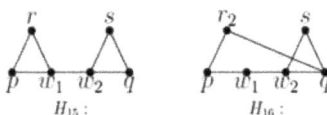

In H_{15}, either w_1 or w_2 cannot enter X by replacing p, q, since resolution fails. In H_{16}, w_1 cannot enter X, since domination fails.

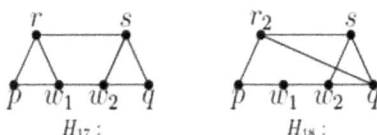

In H_{17}, w_1 cannot enter X, since resolution fails. In H_{18}, w_1 cannot enter X, since domination fails.

Subcase (iv): Only one of r, r_2 is present, and none of s, s_2 is present.

Then, the graphs are as follows:

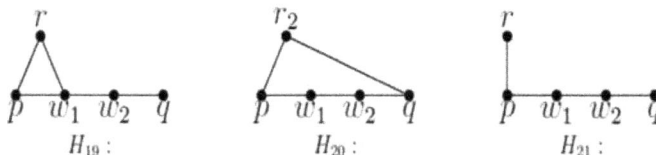

In H_{19}, w_1 cannot enter X, since resolution fails. In H_{20}, w_2 cannot enter X, since domination fails. In H_{21}, w_1 cannot enter X, since domination fails. Similarly, if only one of s and s_2 is present, and none of r, r_1, and r_2 is present, then w_2 cannot enter X. Therefore, $G = P_4$.

Subcase (v): *None of x, x_2, x_3, y, and y_2 is present. Then, $G = P_4$.* □

Corollary 2. $\gamma_{sr}(G) = 2$ *if and only if $G = P_4$, P_3, C_3, C_4, and K_3 with a pendant vertex and $K_4 - \{e\}$.*

Proposition 2. *Let $l \geq 1$, $m \geq 2$, and $n = l + m$ be three integers. Then, there exists G with $\gamma(G) = l$, $dim(G) = q$ and $\gamma_{sr}(G) = n$.*

Proof. We follow the proof given in proposition 3.1 [6]. Construct a graph G from the path P_{3l-1}: $v_1, v_2, \ldots, v_{3l-1}$ of order $3l - 1$. Join m—pairs of vertices x_j, y_j, $1 \leq j \leq m$ and join x_j and y_j for each j. Consider, F_j—a copy of the path P_2: x_j y_j. Join the vertex of F_j, $1 \leq j \leq m$ to the vertex v_{3t-1}. For $l = m = 2$, the graph is as follows:

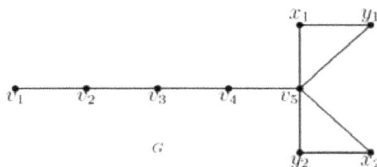

G

Let $V = \{v_1, v_2, \ldots, v_{3l-1}\}$, $T = \{x_1, x_2, \ldots, x_m\}$, and $W = \{y_1, y_2, \ldots, y_m\}$. Then, $\gamma(G) = l$ and $dim(G) = m$ (since $\{v_2, v_5, \ldots, v_{3l-1}\}$ is dominating, and T is a basis of G. Each resolving set of G has at least one vertex from each set, $\{x_j, y_j\}, 1 \le j \le m$. All of the vertices x_j, y_j, and v_{3l-1} are dominated by them. We need at least $\frac{3l-2}{3} = l$ vertices to dominate $V - \{v_{3l-1}\}$. As a result, $\gamma_r(G) \ge l + m$. However, $K = \{v_2, v_5, \ldots, v_{3l-1}\} \cup X$ is a resolving dominating set for G. Hence, $\gamma_r(G) \le |K| = l + m$. Therefore, $\gamma_r(G) = l + m$. Clearly, K is a SRD set of G. Therefore, $\gamma_{sr}(G) \le l + m$. However, $\gamma_{sr}(G) \ge \gamma_r(G) = l + m$. Therefore, $\gamma_{sr}(G) = l + m = n$. $\square$

Theorem 4. *Let G be a graph of order $n \ge 2$. $\gamma_{sr}(G) = n - 1$ if and only if $G = K_n$ or $K_{1, n-1}$.*

Proof. $\gamma_{sr}(G) = n - 1$. Consequently, no $(n - 2)$ subset of $V(G)$ is a SRD set of G. Suppose that there exists an $(n - 2)$ resolving subset S of $V(G)$ that is not a secure dominating set of G. Let $V(G) - S = \{u, v\}$. Suppose that S is not a dominating set of G. Since G is connected, exactly one of u and v is not dominated by S, say u. Clearly, u is a pendant of v. $\square$

Claim: v is adjacent with every vertex of S.
Suppose that v is not adjacent with a vertex w of S. Let $T' = (S - \{w\}) \cup \{v\} = V(G) - \{u, w\}$. Since G is connected and w is not adjacent with u and v, w is adjacent with some vertex of S. T' is a dominating set of G. Therefore, there exists an $(n - 2)$ subset that is a resolving and dominating set of G.
$S_1 = (T' \cup \{u\}) - \{v\}$ is a dominating set of G. Clearly, S is a resolving set, since $d(u, v) = 1$, $d(u, w) \ge 2$. Therefore, S_1 is a secure resolving domination set of G. Therefore, $\gamma_{sr}(G) \le n - 2$, which is a contradiction. w is adjacent with some vertex x in S. Therefore, $S_2 = (S \cup \{w\}) - \{x\}$ is a dominating set of G. $d(u, v) \ge 2$ and $d(x, w) = 1$. Therefore, S_2 is a secure resolving domination set of G. Therefore, $\gamma_{sr}(G) \le n - 2$, which is a contradiction. Suppose that S is a dominating set of G, but not a secure dominating set of G. Suppose that u cannot enter S by replacing a vertex of S. Then, any neighbor of u is either an isolate of S or has private neighbor v. Suppose that every neighbor x of u is an isolate of S. In this case, if u is not adjacent with v, then G is disconnected, which is a contradiction. If u is adjacent with v, then $(S - N[u]) \cup \{v\}$ is connected. Then, $(S - \{x\}) \cup \{u\}$ is a dominating set of G.
Suppose that $(S - N(u)) = \phi$. Then, either G is a star or G is of the form:

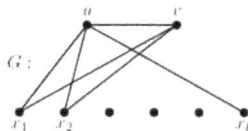

where v is adjacent with some or all of $x_1, x_2, \ldots, x_k$. If G is a star, then $\gamma_{sr}(G) = n - 1$. If G is not a star, then the above graph has $\gamma_{sr}(G) \le n - 2$, which is a contradiction.
Suppose that $(S - N(u)) \ne \phi$. Then, v is adjacent with at least one vertex, say z of $(S - N(u))$. $d(v, z) = 1$, $d(x_i, z) \ne 1$. Therefore, $(S - \{x_i\}) \cup \{u\}$ is resolving. Therefore, there exists an $(n - 2)$ SRD set of G, which is a contradiction.
Suppose that there exists a neighbor x of u which has private neighbor v. Let x be an isolate of S. Then, G is of the form H_1 or of the form H_2, where u and v are made adjacent in H_1. However, H_1 and H_2 have an $(n - 2)$ secure dominating set of G, which is a contradiction.

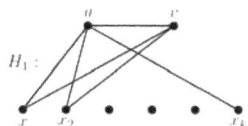

If x is not an isolate of S, then either G is complete, or G has $(n-2)$ SRD set of G, which is a contradiction. Similarly, v can enter S by replacing a vertex of S. Therefore, any $(n-2)$ resolving subset of $V(G)$ is a secure dominating set of G, provided that G is not a star or G is not K_n.

Therefore, the theorem follows.

7. Discussion and Conclusions

A study of SR sets and SRD sets is initiated in this paper. Further work may be done on (i) conditions for the minimality of SR sets (SRD sets), (ii) uniform SR set (SRD set) (that is to find the least positive integer t such that every subset of $V(G)$ of cardinality t is an SR set (SRD set)), (iii) a study of secure metric resolving sets (metric resolving dominating sets) in a graph, and (iv) secure independent resolving sets (secure independent resolving dominating sets).

Author Contributions: H.S. first author is responsible for the preparation of an article, review, editing and visualization under the guidance of S.A.

Funding: This research is carried out without any external financial support.

Conflicts of Interest: The authors confirm no conflict of interest.

References

1. Chartrand, G.; Eroh, L.; Johnson, M.A.; Oellermann, O.R. Resolvability in graphs and the metric dimension of a graph. *Discret. Appl. Math.* **2000**, *105*, 99–113. [CrossRef]
2. Slater, P.J. Leaves of trees. In Proceedings of the 6th Southeast Conference on Combinatorics, Graph Theory and Computing, Boca Raton, FL, USA, 17–20 February 1975; pp. 549–559.
3. Slater, P.J. Dominating and reference sets in graphs. *J. Math. Phys.* **1988**, *22*, 445–455.
4. Brigham, R.C.; Chartrand, G.; Dutton, R.D.; Zhang, P. Resolving Domination in Graphs. *Math. Bohem.* **2003**, *128*, 25–36.
5. Harary, F.; Melter, R.A. On the metric dimension of graph. *Ars Comb.* **1976**, *24*, 191–195.
6. Cockayne, E.J.; Favaron, O.; Mynhardt, C.M. Secure domination, weak roman domination and forbidden subgraph. *Bull. Inst. Comb. Appl.* **2003**, *39*, 87–100.

symmetry

MDPI

Article

Harmonic Index and Harmonic Polynomial on Graph Operations

Juan C. Hernández-Gómez [1] , **J. A. Méndez-Bermúdez** [2,*] , **José M. Rodríguez** [3]
and José M. Sigarreta [1,2]

[1] Facultad de Matemáticas, Universidad Autónoma de Guerrero, Carlos E. Adame No.54 Col. Garita, Acalpulco Gro. 39650, Mexico; jcarloshg@gmail.com (J.C.H.-G.); josemariasigarretaalmira@hotmail.com (J.M.S.)
[2] Instituto de Física, Benemérita Universidad Autónoma de Puebla, Apartado Postal J-48, Puebla 72570, Mexico
[3] Departamento de Matemáticas, Universidad Carlos III de Madrid, Avenida de la Universidad 30, Leganés, 28911 Madrid, Spain; jomaro@math.uc3m.es
* Correspondence: jmendezb@ifuap.buap.mx; Tel.: +52-222-229-5610

Received: 3 August 2018; Accepted: 28 September 2018; Published: 1 October 2018

Abstract: Some years ago, the harmonic polynomial was introduced to study the harmonic topological index. Here, using this polynomial, we obtain several properties of the harmonic index of many classical symmetric operations of graphs: Cartesian product, corona product, join, Cartesian sum and lexicographic product. Some upper and lower bounds for the harmonic indices of these operations of graphs, in terms of related indices, are derived from known bounds on the integral of a product on nonnegative convex functions. Besides, we provide an algorithm that computes the harmonic polynomial with complexity $O(n^2)$.

Keywords: harmonic index; harmonic polynomial; inverse degree index; products of graphs; algorithm

1. Introduction

A single number representing a chemical structure, by means of the corresponding molecular graph, is known as topological descriptor. Topological descriptors play a prominent role in mathematical chemistry, particularly in studies of quantitative structure–property and quantitative structure–activity relationships. Moreover, a topological descriptor is called a topological index if it has a mutual relationship with a molecular property. Thus, since topological indices encode some characteristics of a molecule in a single number, they can be used to study physicochemical properties of chemical compounds.

After the seminal work of Wiener [1], many topological indices have been defined and analysed. Among all topological indices, probably the most studied is the Randić connectivity index (R) [2]. Several hundred papers and, at least, two books report studies of R (see, for example, [3–7] and references therein). Moreover, with the aim of improving the predictive power of R, many additional topological descriptors (similar to R) have been proposed. In fact, the first and second Zagreb indices, M_1 and M_2, respectively, can be considered as the main successors of R. They are defined as

$$M_1(G) = \sum_{uv \in E(G)} (d_u + d_v) = \sum_{u \in V(G)} d_u^2, \qquad M_2(G) = \sum_{uv \in E(G)} d_u d_v,$$

where uv is the edge of G between vertices u and v, and d_u is the degree of vertex u. Both M_1 and M_2 have recently attracted much interest (see, e.g., [8–11]) (in particular, they are included in algorithms used to compute topological indices).

Another remarkable topological descriptor is the *harmonic* index, defined in [12] as

$$H(G) = \sum_{uv \in E(G)} \frac{2}{d_u + d_v}.$$

This index has attracted a great interest in the lasts years (see, e.g., [13–18]). In particular, in [16] appear relations for the harmonic index of some operations of graphs.

In [19], the *harmonic polynomial* of a graph G is defined as

$$H(G, x) = \sum_{uv \in E(G)} x^{d_u + d_v - 1},$$

and the harmonic polynomials of some graphs are computed. For more information on the study of polynomials associated with topological indices and their practical applications, see, e.g., [20–23].

This polynomial owes its name to the fact that $2 \int_0^1 H(G, x)\, dx = H(G)$.

The characterization of any graph by a polynomial is one of the open important problems in graph theory. In recent years, there have been many works on graph polynomials (see, e.g., [21,24] and the references therein). The research in this area has been largely driven by the advantages offered by the use of computers: it is simpler to represent a graph by a polynomial (a vector with dimension $O(n)$) than by the adjacency matrix (an $n \times n$ matrix). Some parameters of a graph allow to define polynomials related to a graph. Although several polynomials are interesting since they compress information about the graphs structure; unfortunately, the well-known polynomials do not solve the problem of the characterization of any graph, since there are often non-isomorphic graphs with the same polynomial.

Polynomials have proved to be useful in the study of several topological indices. There are many papers studying topological indices on graph operations (see, e.g., [25–27]).

Along this work, $G = (V, E) = (V(G), E(G))$ indicates a finite, undirected and simple (i.e., without multiple edges and loops) graph with $E \neq \varnothing$. The main aim of this paper is to obtain several computational properties of the harmonic polynomial. In Section 2, we obtain closed formulas to compute the harmonic polynomial of many classical symmetric operations of graphs: Cartesian product, corona product, join, Cartesian sum and lexicographic product. These formulas are interesting by themselves and, furthermore, allow to obtain new inequalities for the harmonic index of these operations of graphs. Besides, we provide in the last section an algorithm that computes this polynomial with complexity $O(n^2)$.

We would like to stress that the symmetry property present in the operations on graphs studied here (Cartesian product, corona product, join, Cartesian sum and lexicographic product) was an essential tool in the study of the topological indexes, because it allowed us to obtain closed formulas for the harmonic polynomial and to deduce the optimal bounds for that index.

2. Definitions and Background

The following result appears in Proposition 1 of [19].

Proposition 1. *If G is a k-regular graph with m edges, then $H(G, x) = mx^{2k-1}$.*

Propositions 2, 4, 5, 7 in [19] have the following consequences on the graphs: K_n (the complete graph with n vertices), C_n (the cycle with $n \geq 3$ vertices), Q_n (the n-dimensional hypercube), K_{n_1,n_2} (the complete bipartite graph with $n_1 + n_2$ vertices), P_n (the path graph with n vertices), and W_n (the wheel graph with $n \geq 4$ vertices).

Proposition 2. *We have*

$$H(K_n, x) = \frac{1}{2} n(n-1)x^{2n-3}, \qquad H(C_n, x) = nx^3,$$
$$H(Q_n, x) = n2^{n-1}x^{2n-1}, \qquad H(K_{n_1,n_2}, x) = n_1 n_2 x^{n_1+n_2-1},$$
$$H(P_n, x) = 2x^2 + (n-3)x^3, \qquad H(W_n, x) = (n-1)(x^{n+1} + x^5).$$

In Propositions 2.3 and 2.6 in [28] appear the following result.

Proposition 3. *If G is a graph with m edges, then:*

- $H^{(k)}(G, x) \geq 0$ *for every* $k \geq 0$ *and* $x \in [0, \infty)$;
- $H(G, x) > 0$ *on* $(0, \infty)$ *and* $H(G, x)$ *is strictly increasing on* $[0, \infty)$;
- $H(G, x)$ *is strictly convex on* $[0, \infty)$ *if and only if G is not isomorphic to a union of path graphs* P_2; *and*
- $0 = H(G, 0) \leq H(G, x) \leq H(G, 1) = m$ *for every* $x \in [0, 1]$.

Considering the Zagreb indices, Fath-Tabar [29] defined the first Zagreb polynomial as

$$M_1(G, x) := \sum_{uv \in E(G)} x^{d_u + d_v}.$$

The harmonic and the first Zagreb indices are related by several inequalities (see [30], Theorem 2.5 [31] and [32], p. 234). Moreover, the harmonic and the first Zagreb polynomials are related by the equality $M_1(G, x) = x H(G, x)$,

In [33], Shuxian defined the following polynomial related to the first Zagreb index as

$$M_1^*(G, x) := \sum_{u \in V(G)} d_u x^{d_u}.$$

Given a graph G, let us denote by $S(G)$ its *subdivision graph*. $S(G)$ is constructed from G by inserting an additional vertex into each of its edges. Concerning $S(G)$, in Theorem 2.1 of [25], the following result appears.

Theorem 1. *For the subdivision graph S(G) of G, the first Zagreb polynomial is*

$$M_1(S(G), x) = x^2 M_1^*(G, x).$$

Since the harmonic and the first Zagreb polynomials are related by the equality $M_1(G, x) = x H(G, x)$, we have the following result for the harmonic polynomial of the subdivision graph.

Proposition 4. *Given a graph G, the harmonic polynomial of its subdivision graph S(G) is*

$$H(S(G), x) = x M_1^*(G, x).$$

Similarly, we can obtain the harmonic polynomial for the other operations on graphs appearing in [25].

Next, we obtain the harmonic polynomial for other classical operations: Cartesian product, corona product, join, Cartesian sum and lexicographic product. It is important to stress that, since large graphs are composed by smaller ones by the use of products of graphs (and, as a consequence, their properties are strongly related), the study of products of graphs is a relevant and timely research subject.

Let us recall the definitions of these classical products in graph theory.

The *Cartesian product* $G_1 \times G_2$ of the graphs G_1 and G_2 has the vertex set $V(G_1 \times G_2) = V(G_1) \times V(G_2)$ and $(u_i, v_j)(u_k, v_l)$ is an edge of $G_1 \times G_2$ if $u_i = u_k$ and $v_j v_l \in E(G_2)$, or $u_i u_k \in E(G_1)$ and $v_j = v_l$.

Given two graphs G_1 and G_2, we define the *corona product* $G_1 \circ G_2$ as the graph obtained by adding to G_1, $|V(G_1)|$ copies of G_2 and joining each vertex of the i-th copy with the vertex $v_i \in V(G_1)$.

The *join* $G_1 + G_2$ is defined as the graph obtained by taking one copy of G_1 and one copy of G_2, and joining by an edge each vertex of G_1 with each vertex of G_2.

The *Cartesian sum* $G_1 \oplus G_2$ of the graphs G_1 and G_2 has the vertex set $V(G_1 \oplus G_2) = V(G_1) \times V(G_2)$ and $(u_i, v_j)(u_k, v_l)$ is an edge of $G_1 \oplus G_2$ if $u_i u_k \in E(G_1)$ or $v_j v_l \in E(G_2)$.

The *lexicographic product* $G_1 \odot G_2$ of the graphs G_1 and G_2 has $V(G_1) \times V(G_2)$ as vertex set, so that two distinct vertices $(u_i, v_j), (u_k, v_l)$ of $V(G_1 \odot G_2)$ are adjacent if either $u_i u_k \in E(G_1)$, or $u_i = u_k$ and $v_j v_l \in E(G_2)$.

Let us introduce another topological index that will be very useful in this work.

The *inverse degree* $ID(G)$ of a graph G is defined by

$$ID(G) := \sum_{u \in V(G)} \frac{1}{d_u} = \sum_{uv \in E(G)} \left(\frac{1}{d_u^2} + \frac{1}{d_v^2} \right).$$

It is relevant to mention that the surmises inferred through the computer program Graffiti [12] attracted the attention of researchers. Thus, since then, several studies (see, e.g., [34–38]) focusing on relationships between $ID(G)$ and other graph invariants (such as diameter, edge-connectivity, matching number and Wiener index) have appeared in the literature.

Let us define the *inverse degree polynomial* of a graph G as

$$ID(G, x) = \sum_{u \in V(G)} x^{d_u - 1}.$$

Thus, we have $\int_0^1 ID(G, x)\, dx = ID(G)$. Note that $x(x ID(G, x))' = M_1^*(G, x)$.

The following result summarizes some interesting properties of the inverse degree polynomial. Recall that a vertex of a graph is said to be *pendant* if it has degree 1.

Proposition 5. *If G is a graph with n vertices and k pendant vertices, then:*

- $ID^{(j)}(G, x) \geq 0$ *for every $j \geq 0$ and $x \in [0, \infty)$;*
- $ID(G, x) > 0$ *on $(0, \infty)$;*
- $ID(G, x)$ *is strictly increasing on $[0, \infty)$ if and only if G is not isomorphic to a union of path graphs P_2;*
- $ID(G, x)$ *is strictly convex on $[0, \infty)$ if and only if G is not isomorphic to a union of path graphs; and*
- $k = ID(G, 0) \leq ID(G, x) \leq ID(G, 1) = n$ *for every $x \in [0, 1]$.*

Proof. Since every coefficient of the polynomial $ID(G, x)$ is non-negative, the first statement holds.

Since every coefficient of the polynomial $ID(G, x)$ is non-negative and $ID(G, x)$ is not identically zero, we have $ID(G, x) > 0$ on $(0, \infty)$.

Since every coefficient of the polynomial $ID(G, x)$ is non-negative, we have $ID'(G, x) > 0$ on $(0, \infty)$ if and only if there exists a vertex $u \in V(G)$ with $d_u \geq 2$, and this holds if and only if G is not isomorphic to a union of path graphs P_2.

Similarly, $ID(G, x)$ is strictly convex on $[0, \infty)$ if and only if there exists a vertex $u \in V(G)$ with $d_u \geq 3$, and this holds if and only if G is not isomorphic to a union of path graphs.

Finally, if $x \in [0, 1]$, then

$$k = ID(G,0) \leq \sum_{u \in V(G)} x^{d_u - 1} \leq \sum_{u \in V(G)} 1 = ID(G,1) = n.$$

$\square$

Proposition 4 has the following consequence, which illustrates how these polynomials associated to topological indices provide information about the topological indices themselves.

Corollary 1. *Given a graph G with maximum degree Δ, the harmonic index of the subdivision graph $S(G)$ satisfies*

$$H(S(G)) \leq 2\Delta\, ID(G).$$

Proof. Proposition 4 gives

$$H(S(G)) = 2\int_0^1 H(S(G),x)\,dx = 2\int_0^1 x\, M_1^*(G,x)\,dx = 2\int_0^1 x \sum_{u \in V(G)} d_u x^{d_u}\,dx$$

$$\leq 2\Delta \int_0^1 \sum_{u \in V(G)} x^{d_u - 1}\,dx = 2\Delta \int_0^1 ID(G,x)\,dx = 2\Delta\, ID(G).$$

$\square$

3. Computation of the Harmonic Index of Graph Operations

Let us start with the formula of the harmonic polynomial of the Cartesian product.

Theorem 2. *Given two graphs G_1 and G_2, the harmonic polynomial of the Cartesian product $G_1 \times G_2$ is*

$$H(G_1 \times G_2, x) = x^2 H(G_1, x)\, ID(G_2, x^2) + x^2 H(G_2, x)\, ID(G_1, x^2).$$

Proof. Denote by n_1 and n_2 the cardinality of the vertices of G_1 and G_2, respectively.

Note that if $(u_i, v_j) \in V(G_1 \times G_2)$, then $d_{(u_i, v_j)} = d_{u_i} + d_{v_j}$.

If $(u_i, v_k)(u_j, v_k) \in E(G_1 \times G_2)$, then the corresponding monomial of the harmonic polynomial is

$$x^{d_{u_i} + d_{v_k} + d_{u_j} + d_{v_k} - 1} = x^{2d_{v_k}} x^{d_{u_i} + d_{u_j} - 1}.$$

Hence,

$$\sum_{k=1}^{n_2} \sum_{u_i u_j \in E(G_1)} x^{2d_{v_k}} x^{d_{u_i} + d_{u_j} - 1} = x^2 \sum_{k=1}^{n_2} (x^2)^{d_{v_k} - 1} \sum_{u_i u_j \in E(G_1)} x^{d_{u_i} + d_{u_j} - 1} = x^2 ID(G_2, x^2)\, H(G_1, x).$$

The same argument gives that the sum of the monomials corresponding to $(u_k, v_i)(u_k, v_j) \in E(G_1 \times G_2)$ is $x^2 H(G_2, x)\, ID(G_1, x^2)$, and the equality holds. $\square$

Next, we present two useful improvements (for convex functions) of the well-known Chebyshev's inequalities.

Lemma 1 ([39]). *Let $f_1, \dots, f_k$ be non-negative convex functions defined on the interval $[0,1]$. Then,*

$$\int_0^1 \prod_{i=1}^k f_i(x)\,dx \geq \frac{2^k}{k+1} \prod_{i=1}^k \int_0^1 f_i(x)\,dx.$$

Lemma 2 (Corollary 5.2 [40]). *Let $f_1, \ldots, f_k$ be non-negative convex functions defined on the interval* $[0, 1]$. *Then*

$$\int_0^1 \prod_{i=1}^k f_i(x)\, dx \leq \frac{2}{k+1} \left(\prod_{i=1}^k \int_0^1 f_i(x)\, dx \right)^{1/k} \left(\prod_{i=1}^k (f_i(0) + f_i(1)) \right)^{1-1/k}.$$

Theorem 3. *Given two graphs G_1 and G_2 with n_1 and n_2 vertices, and m_1 and m_2 edges, respectively, the harmonic index of the Cartesian product $G_1 \times G_2$ satisfies*

$$H(G_1 \times G_2) \geq \frac{1}{2} H(G_1)\, ID(G_2) + \frac{1}{2} H(G_2)\, ID(G_1),$$

$$H(G_1 \times G_2) \leq \min \left\{ \frac{2}{3} \left(m_1 n_2 H(G_1)\, ID(G_2) \right)^{1/2}, \frac{1}{2} \left(m_1^2 n_2^2 H(G_1)\, ID(G_2) \right)^{1/3} \right\}$$

$$+ \min \left\{ \frac{2}{3} \left(m_2 n_1 H(G_2)\, ID(G_1) \right)^{1/2}, \frac{1}{2} \left(m_2^2 n_1^2 H(G_2)\, ID(G_1) \right)^{1/3} \right\}.$$

Proof. Propositions 3 and 5 give that $H(G_1, x), ID(G_2, x^2), H(G_2, x), ID(G_1, x^2)$ are non-negative convex functions. Thus, Lemma 1 gives

$$\int_0^1 2x^2 H(G_1, x)\, ID(G_2, x^2)\, dx \geq \frac{2^3}{3+1} \int_0^1 x\, dx \int_0^1 H(G_1, x)\, dx \int_0^1 2x\, ID(G_2, x^2)\, dx$$

$$= 2\frac{1}{2} \int_0^1 H(G_1, x)\, dx \int_0^1 ID(G_2, x)\, dx = \frac{1}{2} H(G_1)\, ID(G_2).$$

Similarly,

$$\int_0^1 2x^2 H(G_2, x)\, ID(G_1, x^2)\, dx \geq \frac{1}{2} H(G_2)\, ID(G_1).$$

These inequalities, Theorem 2 and $H(G_1 \times G_2) = 2\int_0^1 H(G_1 \times G_2, x)\, dx$ give the lower bound. Lemma 2 and Propositions 3 and 5 give

$$\int_0^1 2x^2 H(G_1, x)\, ID(G_2, x^2)\, dx \leq \int_0^1 2x\, H(G_1, x)\, ID(G_2, x^2)\, dx$$

$$\leq \frac{2}{3} \left(\int_0^1 H(G_1, x)\, dx \int_0^1 2x\, ID(G_2, x^2)\, dx \right)^{1/2} \left(2H(G_1, 1)\, ID(G_2, 1) \right)^{1/2}$$

$$= \frac{2}{3} \left(m_1 n_2 H(G_1)\, ID(G_2) \right)^{1/2}.$$

In addition, Lemma 2 and Propositions 3 and 5 give

$$\int_0^1 2x^2 H(G_1, x)\, ID(G_2, x^2)\, dx \leq \frac{1}{2} \left(\int_0^1 x\, dx \int_0^1 H(G_1, x)\, dx \int_0^1 2x\, ID(G_2, x^2)\, dx \right)^{1/3} \left(2H(G_1, 1)\, ID(G_2, 1) \right)^{2/3}$$

$$= \frac{1}{2} \left(m_1^2 n_2^2 H(G_1)\, ID(G_2) \right)^{1/3}.$$

These inequalities give

$$\int_0^1 2x^2 H(G_1, x)\, ID(G_2, x^2)\, dx \leq \min \left\{ \frac{2}{3} \left(m_1 n_2 H(G_1)\, ID(G_2) \right)^{1/2}, \frac{1}{2} \left(m_1^2 n_2^2 H(G_1)\, ID(G_2) \right)^{1/3} \right\}.$$

Similarly,

$$\int_0^1 2x^2 H(G_2, x)\, ID(G_1, x^2)\, dx \leq \min \left\{ \frac{2}{3} \left(m_2 n_1 H(G_2)\, ID(G_1) \right)^{1/2}, \frac{1}{2} \left(m_2^2 n_1^2 H(G_2)\, ID(G_1) \right)^{1/3} \right\}.$$

These inequalities, Theorem 2 and $H(G_1 \times G_2) = 2 \int_0^1 H(G_1 \times G_2, x) \, dx$ give the upper bound. $\qquad\square$

Theorem 4. *Given two graphs G_1 and G_2, with n_1 and n_2 vertices, respectively, the harmonic polynomial of the corona product $G_1 \circ G_2$ is*

$$H(G_1 \circ G_2, x) = x^{2n_2} H(G_1, x) + n_1 x^2 H(G_2, x) + x^{n_2+2} ID(G_1, x) \, ID(G_2, x).$$

Proof. The degree of $u \in V(G_1)$, considered as a vertex of $G_1 \circ G_2$, is $d_u + n_2$. The degree of any copy v' of $v \in V(G_2)$, considered as a vertex of $G_1 \circ G_2$, is $d_v + 1$.

If $u_i u_j \in E(G_1)$, then the corresponding monomial of the harmonic polynomial of $G_1 \circ G_2$ is

$$x^{d_{u_i}+n_2+d_{u_j}+n_2-1} = x^{2n_2} x^{d_{u_i}+d_{u_j}-1}.$$

Hence,

$$\sum_{u_i u_j \in E(G_1)} x^{2n_2} x^{d_{u_i}+d_{u_j}-1} = x^{2n_2} \sum_{u_i u_j \in E(G_1)} x^{d_{u_i}+d_{u_j}-1} = x^{2n_2} H(G_1, x).$$

If $v_i v_j \in E(G_2)$, then each corresponding monomial of the harmonic polynomial of $G_1 \circ G_2$ is

$$x^{d_{v_i}+1+d_{v_j}+1-1} = x^2 x^{d_{v_i}+d_{v_j}-1}.$$

Therefore,

$$\sum_{v_i v_j \in E(G_2)} x^2 x^{d_{v_i}+d_{v_j}-1} = x^2 \sum_{v_i v_j \in E(G_2)} x^{d_{v_i}+d_{v_j}-1} = x^2 H(G_2, x).$$

If we add the corresponding polynomials of the n_1 copies of G_2, then we obtain $n_1 x^2 H(G_2, x)$.

If $u_i v_j' \in E(G_1 \circ G_2)$ with $u_i \in V(G_1)$ and $v_j \in V(G_2)$, then the corresponding monomial of the harmonic polynomial is

$$x^{d_{u_i}+n_2+d_{v_j}+1-1} = x^{n_2+2} x^{d_{u_i}-1} x^{d_{v_j}-1}.$$

Hence,

$$\sum_{i=1}^{n_1} \sum_{j=1}^{n_2} x^{n_2+2} x^{d_{u_i}-1} x^{d_{v_j}-1} = x^{n_2+2} \sum_{i=1}^{n_1} x^{d_{u_i}-1} \sum_{j=1}^{n_2} x^{d_{v_j}-1} = x^{n_2+2} ID(G_1, x) \, ID(G_2, x).$$

Thus, the equality holds. $\qquad\square$

Theorem 5. *Given two graphs G_1 and G_2 with n_1 and n_2 vertices, m_1 and m_2 edges, and k_1 and k_2 pendant vertices, respectively, the harmonic index of the corona product $G_1 \circ G_2$ satisfies*

$$H(G_1 \circ G_2) \geq \frac{4}{3(2n_2+1)} H(G_1) + \frac{4n_1}{9} H(G_1) + \frac{4}{n_2+3} ID(G_1) \, ID(G_2),$$

$$H(G_1 \circ G_2) \leq \frac{2}{3} \left(\frac{2m_1}{2n_2+1} H(G_1) \right)^{1/2} + \frac{2n_1}{3} \left(\frac{2m_2}{3} H(G_2) \right)^{1/2}$$

$$+ \left(\frac{1}{n_2+3} ID(G_1) \, ID(G_2)(n_1+k_1)^2(n_2+k_2)^2 \right)^{1/3}.$$

Proof. Lemma 1 gives

$$\int_0^1 2x^{2n_2} H(G_1, x) \, dx \geq \frac{4}{3} \int_0^1 x^{2n_2} dx \int_0^1 2H(G_1, x) \, dx = \frac{4}{3(2n_2+1)} H(G_1),$$

$$\int_0^1 2n_1 x^2 H(G_2, x) \, dx \geq \frac{4n_1}{3} \int_0^1 x^2 dx \int_0^1 2H(G_1, x) \, dx = \frac{4n_1}{9} H(G_1),$$

$$\int_0^1 2x^{n_2+2} ID(G_1,x)\, ID(G_2,x)\, dx \geq \frac{8}{4} \int_0^1 2x^{n_2+2} dx \int_0^1 ID(G_1,x)\, dx \int_0^1 ID(G_2,x)\, dx$$

$$= \frac{4}{n_2+3}\, ID(G_1)\, ID(G_2).$$

These inequalities, Theorem 4 and $H(G_1 \circ G_2) = 2\int_0^1 H(G_1 \circ G_2, x)\, dx$ give the lower bound. Lemma 2 and Proposition 3 give

$$\int_0^1 2x^{2n_2} H(G_1,x)\, dx \leq \frac{2}{3}\left(\int_0^1 x^{2n_2} dx \int_0^1 2H(G_1,x)\, dx\right)^{1/2} (2H(G_1,1))^{1/2}$$

$$= \frac{2}{3}\left(\frac{2m_1}{2n_2+1} H(G_1)\right)^{1/2}.$$

In addition, Lemma 2 and Proposition 3 give

$$\int_0^1 2n_1 x^2 H(G_2,x)\, dx \leq \frac{2n_1}{3}\left(\int_0^1 x^2 dx \int_0^1 2H(G_2,x)\, dx\right)^{1/2} (2H(G_2,1))^{1/2}$$

$$= \frac{2n_1}{3}\left(\frac{2m_2}{3} H(G_2)\right)^{1/2}.$$

Lemma 2 and Proposition 5 give

$$\int_0^1 2x^{n_2+2} ID(G_1,x)\, ID(G_2,x)\, dx \leq \frac{2}{4} 2 \left(\int_0^1 x^{n_2+2} dx \int_0^1 ID(G_1,x)\, dx \int_0^1 ID(G_2,x)\, dx\right)^{1/3}$$

$$\cdot \left((ID(G_1,1)+ID(G_1,0))(ID(G_2,1)+ID(G_2,0))\right)^{2/3}$$

$$= \left(\frac{1}{n_2+3} ID(G_1)\, ID(G_2)(n_1+k_1)^2(n_2+k_2)^2\right)^{1/3}.$$

These inequalities, Theorem 4 and $H(G_1 \circ G_2) = 2\int_0^1 H(G_1 \circ G_2, x)\, dx$ give the upper bound. □

Theorem 6. *Given two graphs G_1 and G_2, with n_1 and n_2 vertices, respectively, the harmonic polynomial of the join $G_1 + G_2$ is*

$$H(G_1+G_2,x) = x^{2n_2} H(G_1,x) + x^{2n_1} H(G_2,x) + x^{n_1+n_2+1} ID(G_1,x)\, ID(G_2,x).$$

Proof. The degree of $u \in V(G_1)$, considered as a vertex of $G_1 + G_2$, is $d_u + n_2$. The degree of $v \in V(G_2)$, considered as a vertex of $G_1 + G_2$, is $d_v + n_1$.

If $u_i u_j \in E(G_1)$, then the corresponding monomial of the harmonic polynomial of $G_1 + G_2$ is

$$x^{d_{u_i}+n_2+d_{u_j}+n_2-1} = x^{2n_2} x^{d_{u_i}+d_{u_j}-1}.$$

Hence,

$$\sum_{u_i u_j \in E(G_1)} x^{2n_2} x^{d_{u_i}+d_{u_j}-1} = x^{2n_2} \sum_{u_i u_j \in E(G_1)} x^{d_{u_i}+d_{u_j}-1} = x^{2n_2} H(G_1,x).$$

If $v_i v_j \in E(G_2)$, then the corresponding monomial of the harmonic polynomial of $G_1 + G_2$ is

$$x^{d_{v_i}+n_1+d_{v_j}+n_1-1} = x^{2n_1} x^{d_{v_i}+d_{v_j}-1}.$$

Therefore,

$$\sum_{v_i v_j \in E(G_2)} x^{2n_1} x^{d_{v_i}+d_{v_j}-1} = x^{2n_1} \sum_{v_i v_j \in E(G_2)} x^{d_{v_i}+d_{v_j}-1} = x^{2n_1} H(G_2,x).$$

If $u_i v_j \in E(G_1 + G_2)$ with $u_i \in V(G_1)$ and $v_j \in V(G_2)$, then the corresponding monomial of the harmonic polynomial is

$$x^{d_{u_i} + n_2 + d_{v_j} + n_1 - 1} = x^{n_1 + n_2 + 1} x^{d_{u_i} - 1} x^{d_{v_j} - 1}.$$

Hence,

$$\sum_{i=1}^{n_1} \sum_{j=1}^{n_2} x^{n_1 + n_2 + 1} x^{d_{u_i} - 1} x^{d_{v_j} - 1} = x^{n_1 + n_2 + 1} \sum_{i=1}^{n_1} x^{d_{u_i} - 1} \sum_{j=1}^{n_2} x^{d_{v_j} - 1} = x^{n_1 + n_2 + 1} ID(G_1, x) \, ID(G_2, x),$$

Thus, the equality holds. $\quad \square$

Theorem 7. *Given two graphs G_1 and G_2 with n_1 and n_2 vertices, m_1 and m_2 edges, and k_1 and k_2 pendant vertices, respectively, the harmonic index of the join $G_1 + G_2$ satisfies*

$$H(G_1 + G_2) \geq \frac{4}{3(2n_2 + 1)} H(G_1) + \frac{4}{3(2n_1 + 1)} H(G_2) + \frac{4}{n_1 + n_2 + 2} ID(G_1) \, ID(G_2),$$

$$H(G_1 + G_2) \leq \frac{2}{3} \left(\frac{2m_1}{2n_2 + 1} H(G_1) \right)^{1/2} + \frac{2}{3} \left(\frac{2m_2}{2n_1 + 1} H(G_2) \right)^{1/2}$$

$$+ \left(\frac{1}{n_1 + n_2 + 2} ID(G_1) \, ID(G_2)(n_1 + k_1)^2 (n_2 + k_2)^2 \right)^{1/3}.$$

Proof. We have seen in the proof of Theorem 5 that

$$\frac{4}{3(2n_2 + 1)} H(G_1) \leq \int_0^1 2x^{2n_2} H(G_1, x) \, dx \leq \frac{2}{3} \left(\frac{2m_1}{2n_2 + 1} H(G_1) \right)^{1/2}.$$

Similarly, we obtain

$$\frac{4}{3(2n_1 + 1)} H(G_2) \leq \int_0^1 2x^{2n_1} H(G_2, x) \, dx \leq \frac{2}{3} \left(\frac{2m_2}{2n_1 + 1} H(G_2) \right)^{1/2}.$$

Lemma 1 gives

$$\int_0^1 2x^{n_1 + n_2 + 1} ID(G_1, x) \, ID(G_2, x) \, dx \geq \frac{8}{4} \int_0^1 2x^{n_1 + n_2 + 1} dx \int_0^1 ID(G_1, x) \, dx \int_0^1 ID(G_2, x) \, dx$$

$$= \frac{4}{n_1 + n_2 + 2} ID(G_1) \, ID(G_2).$$

Lemma 2 and Proposition 5 give

$$\int_0^1 2x^{n_1 + n_2 + 1} ID(G_1, x) \, ID(G_2, x) \, dx \leq \frac{2}{4} 2 \left(\int_0^1 x^{n_1 + n_2 + 1} dx \int_0^1 ID(G_1, x) \, dx \int_0^1 ID(G_2, x) \, dx \right)^{1/3}$$

$$\cdot \left((ID(G_1, 1) + ID(G_1, 0))(ID(G_2, 1) + ID(G_2, 0)) \right)^{2/3}$$

$$= \left(\frac{1}{n_1 + n_2 + 2} ID(G_1) \, ID(G_2)(n_1 + k_1)^2 (n_2 + k_2)^2 \right)^{1/3}.$$

These inequalities, Theorem 6 and $H(G_1 + G_2) = 2 \int_0^1 H(G_1 + G_2, x) \, dx$ give the bounds. $\quad \square$

Theorem 8. *Given two graphs G_1 and G_2, with n_1 and n_2 vertices, respectively, the harmonic polynomial of the Cartesian sum $G_1 \oplus G_2$ is*

$$H(G_1 \oplus G_2, x) = x^{2n_1 + n_2 - 1} H(G_1, x^{n_2}) \, ID^2(G_2, x^{n_1}) + x^{n_1 + 2n_2 - 1} H(G_2, x^{n_1}) \, ID^2(G_1, x^{n_2})$$

$$- x^{n_1 + n_2 - 1} H(G_1, x^{n_2}) \, H(G_2, x^{n_1}).$$

Proof. Note that if $(u_i, v_j) \in V(G_1 \oplus G_2)$, then $d_{(u_i,v_j)} = n_2 d_{u_i} + n_1 d_{v_j}$.

If $(u_i, v_j)(u_k, v_l) \in E(G_1 \oplus G_2)$, then the corresponding monomial of the harmonic polynomial is

$$x^{n_2 d_{u_i} + n_1 d_{v_j} + n_2 d_{u_k} + n_1 d_{v_l} - 1} = x^{2n_1 + n_2 - 1} (x^{n_2})^{d_{u_i} + d_{u_k} - 1} (x^{n_1})^{d_{v_j} - 1} (x^{n_1})^{d_{v_l} - 1}$$
$$= x^{n_1 + n_2 - 1} (x^{n_2})^{d_{u_i} + d_{u_k} - 1} (x^{n_1})^{d_{v_j} + d_{v_l} - 1}.$$

Hence, the sum of the corresponding monomials with $u_i u_k \in E(G_1)$ is

$$\sum_{j,l=1}^{n_2} \sum_{u_i u_k \in E(G_1)} x^{2n_1 + n_2 - 1} (x^{n_2})^{d_{u_i} + d_{u_k} - 1} (x^{n_1})^{d_{v_j} - 1} (x^{n_1})^{d_{v_l} - 1}$$
$$= x^{2n_1 + n_2 - 1} \sum_{j=1}^{n_2} (x^{n_1})^{d_{v_j} - 1} \sum_{l=1}^{n_2} (x^{n_1})^{d_{v_l} - 1} \sum_{u_i u_k \in E(G_1)} (x^{n_2})^{d_{u_i} + d_{u_k} - 1}$$
$$= x^{2n_1 + n_2 - 1} H(G_1, x^{n_2}) ID^2(G_2, x^{n_1}).$$

Similarly, the sum of the corresponding monomials with $v_j v_l \in E(G_2)$ is

$$x^{n_1 + 2n_2 - 1} H(G_2, x^{n_1}) ID^2(G_1, x^{n_2}).$$

If we add these two terms, then we take into account twice the corresponding monomials with $u_i u_k \in E(G_1)$ and $v_j v_l \in E(G_2)$:

$$\sum_{u_i u_k \in E(G_1)} \sum_{v_j v_l \in E(G_2)} x^{n_1 + n_2 - 1} (x^{n_2})^{d_{u_i} + d_{u_k} - 1} (x^{n_1})^{d_{v_j} + d_{v_l} - 1}$$
$$= x^{n_1 + n_2 - 1} \sum_{u_i u_k \in E(G_1)} (x^{n_2})^{d_{u_i} + d_{u_k} - 1} \sum_{v_j v_l \in E(G_2)} (x^{n_1})^{d_{v_j} + d_{v_l} - 1}$$
$$= x^{n_1 + n_2 - 1} H(G_1, x^{n_2}) H(G_2, x^{n_1}).$$

Hence, the equality holds. $\square$

Theorem 9. *Given two graphs G_1 and G_2 with n_1 and n_2 vertices, and m_1 and m_2 edges, respectively, the harmonic index of the Cartesian sum $G_1 \oplus G_2$ satisfies*

$$H(G_1 \oplus G_2) \geq \frac{16}{15 n_1^2 n_2} H(G_1) ID^2(G_2) + \frac{16}{15 n_1 n_2^2} H(G_2) ID^2(G_1)$$
$$- \frac{2}{3} \left(\frac{m_1 m_2}{n_1 n_2} H(G_1) H(G_2) \right)^{1/2},$$

$$H(G_1 \oplus G_2) \leq \frac{n_2}{2} \left(\frac{4 m_1^2}{n_1^2} H(G_1) ID^2(G_2) \right)^{1/3} + \frac{n_1}{2} \left(\frac{4 m_2^2}{n_2^2} H(G_2) ID^2(G_1) \right)^{1/3}$$
$$- \frac{1}{2 n_1 n_2} H(G_1) H(G_2).$$

Proof. Lemma 1 gives

$$\int_0^1 2 x^{2n_1 + n_2 - 1} H(G_1, x^{n_2}) ID^2(G_2, x^{n_1}) \, dx \geq \frac{16}{5} \int_0^1 x^2 dx \int_0^1 2 x^{n_2 - 1} H(G_1, x^{n_2}) \, dx$$
$$\cdot \int_0^1 x^{n_1 - 1} ID(G_2, x^{n_1}) \, dx \int_0^1 x^{n_1 - 1} ID(G_2, x^{n_1}) \, dx$$
$$= \frac{16}{15 n_1^2 n_2} H(G_1) ID^2(G_2),$$

$$\int_0^1 2x^{n_1+n_2-1} H(G_1, x^{n_2}) H(G_2, x^{n_1}) \, dx \geq \frac{8}{4} \int_0^1 x \, dx \int_0^1 2x^{n_2-1} H(G_1, x^{n_2}) \, dx \int_0^1 x^{n_1-1} H(G_2, x^{n_1}) \, dx$$

$$= \frac{1}{2n_1 n_2} H(G_1) H(G_2).$$

The same argument gives

$$\int_0^1 2x^{n_1+2n_2-1} H(G_2, x^{n_1}) ID^2(G_1, x^{n_2}) \, dx \geq \frac{16}{15 n_1 n_2^2} H(G_2) ID^2(G_1).$$

Lemma 2 and Propositions 3 and 5 give

$$\int_0^1 2x^{2n_1+n_2-1} H(G_1, x^{n_2}) ID^2(G_2, x^{n_1}) \, dx \leq \int_0^1 2x^{n_2-1} H(G_1, x^{n_2}) x^{2n_1-2} ID^2(G_2, x^{n_1}) \, dx$$

$$\leq \frac{2}{4} \left(\int_0^1 2x^{n_2-1} H(G_1, x^{n_2}) \, dx \int_0^1 x^{n_1-1} ID(G_2, x^{n_1}) \, dx \int_0^1 x^{n_1-1} ID(G_2, x^{n_1}) \, dx \right)^{1/3}$$

$$\cdot \left(2H(G_1,1) ID(G_2,1) ID(G_2,1) \right)^{2/3} = \frac{1}{2} \left(\frac{1}{n_2 n_1^2} H(G_1) ID^2(G_2) \right)^{1/3} \left(2m_1 n_2^2 \right)^{2/3}$$

$$= \frac{n_2}{2} \left(\frac{4m_1^2}{n_1^2} H(G_1) ID^2(G_2) \right)^{1/3}.$$

The same argument gives

$$\int_0^1 2x^{n_1+2n_2-1} H(G_2, x^{n_1}) ID^2(G_1, x^{n_2}) \, dx \leq \frac{n_1}{2} \left(\frac{4m_2^2}{n_2^2} H(G_2) ID^2(G_1) \right)^{1/3}.$$

In addition, Lemma 2 and Proposition 3 give

$$\int_0^1 2x^{n_1+n_2-1} H(G_1, x^{n_2}) H(G_2, x^{n_1}) \, dx \leq \frac{1}{2} \int_0^1 2x^{n_2-1} H(G_1, x^{n_2}) 2x^{n_1-1} H(G_2, x^{n_1}) \, dx$$

$$\leq \frac{1}{2} \frac{2}{3} \left(\int_0^1 2x^{n_2-1} H(G_1, x^{n_2}) \, dx \int_0^1 2x^{n_1-1} H(G_2, x^{n_1}) \, dx \right)^{1/2} \left(2H(G_1,1) 2H(G_2,1) \right)^{1/2}$$

$$= \frac{2}{3} \left(\frac{m_1 m_2}{n_1 n_2} H(G_1) H(G_2) \right)^{1/2}.$$

These inequalities, Theorem 8 and $H(G_1 \oplus G_2) = 2 \int_0^1 H(G_1 \oplus G_2, x) \, dx$ give the desired bounds. $\square$

Theorem 10. *Given two graphs G_1 and G_2, with n_1 and n_2 vertices, respectively, the harmonic polynomial of the lexicographic product $G_1 \odot G_2$ is*

$$H(G_1 \odot G_2, x) = x^{2n_2} ID(G_1, x^{2n_2}) H(G_2, x) + x^{n_2+1} H(G_1, x^{n_2}) ID^2(G_2, x).$$

Proof. Note that if $(u_i, v_j) \in V(G_1 \odot G_2)$, then $d_{(u_i, v_j)} = n_2 d_{u_i} + d_{v_j}$.

If $(u_i, v_j)(u_i, v_k) \in E(G_1 \odot G_2)$, then the corresponding monomial of the harmonic polynomial is

$$x^{n_2 d_{u_i} + d_{v_j} + n_2 d_{u_i} + d_{v_k} - 1} = x^{2n_2} \left(x^{2n_2} \right)^{d_{u_i} - 1} x^{d_{v_j} + d_{v_k} - 1}.$$

Hence,

$$\sum_{i=1}^{n_1} \sum_{v_j v_k \in E(G_2)} x^{2n_2} \left(x^{2n_2} \right)^{d_{u_i} - 1} x^{d_{v_j} + d_{v_k} - 1} = x^{2n_2} \sum_{i=1}^{n_1} \left(x^{2n_2} \right)^{d_{u_i} - 1} \sum_{v_j v_k \in E(G_2)} x^{d_{v_j} + d_{v_k} - 1}$$

$$= x^{2n_2} ID(G_1, x^{2n_2}) H(G_2, x).$$

If $(u_i, v_j)(u_k, v_l) \in E(G_1 \odot G_2)$ with $u_i u_k \in E(G_1)$, then the corresponding monomial of the harmonic polynomial is

$$x^{n_2 d_{u_i} + d_{v_j} + n_2 d_{u_k} + d_{v_l} - 1} = x^{n_2 + 1} (x^{n_2})^{d_{u_i} + d_{u_k} - 1} x^{d_{v_j} - 1} x^{d_{v_l} - 1}.$$

Hence, the sum of their corresponding monomials is

$$\sum_{u_i u_k \in E(G_1)} \sum_{j,l=1}^{n_2} x^{n_2 + 1} (x^{n_2})^{d_{u_i} + d_{u_k} - 1} x^{d_{v_j} - 1} x^{d_{v_l} - 1}$$

$$= x^{n_2 + 1} \sum_{u_i u_k \in E(G_1)} (x^{n_2})^{d_{u_i} + d_{u_k} - 1} \sum_{j=1}^{n_2} x^{d_{v_j} - 1} \sum_{l=1}^{n_2} x^{d_{v_l} - 1}$$

$$= x^{n_2 + 1} H(G_1, x^{n_2}) \, ID^2(G_2, x).$$

We obtain the desired equality by adding these two terms. $\square$

Theorem 11. *Given two graphs G_1 and G_2 with n_1 and n_2 vertices, m_1 and m_2 edges, and k_1 and k_2 pendant vertices, respectively, the harmonic index of the lexicographic product $G_1 \odot G_2$ satisfies*

$$H(G_1 \odot G_2) \geq \frac{1}{2n_2} ID(G_1) \, H(G_2) + \frac{16}{15n_2} H(G_1) \, ID^2(G_2)$$

$$H(G_1 \odot G_2) \leq \frac{2}{3} \left(\frac{n_1 m_2}{n_2} ID(G_1) \, H(G_2) \right)^{1/2} + \frac{1}{2} \left(\frac{4m_1^2}{n_2} H(G_1) \, ID^2(G_2)(n_2 + k_2)^4 \right)^{1/3}.$$

Proof. Lemma 1 gives

$$\int_0^1 2x^{2n_2} ID(G_1, x^{2n_2}) \, H(G_2, x) \, dx \geq \frac{8}{4} \int_0^1 x \, dx \int_0^1 x^{2n_2 - 1} ID(G_1, x^{2n_2}) \, dx \int_0^1 2H(G_2, x) \, dx$$

$$= \frac{1}{2n_2} ID(G_1) \, H(G_2),$$

$$\int_0^1 2x^{n_2 + 1} H(G_1, x^{n_2}) \, ID^2(G_2, x) \, dx \geq \frac{16}{5} \int_0^1 x^2 \, dx \int_0^1 2x^{n_2 - 1} H(G_1, x^{n_2}) \, dx \int_0^1 ID(G_2, x) \, dx \int_0^1 ID(G_2, x) \, dx$$

$$= \frac{16}{15n_2} H(G_1) \, ID^2(G_2).$$

Lemma 2 and Propositions 3 and 5 give

$$\int_0^1 2x^{2n_2} ID(G_1, x^{2n_2}) \, H(G_2, x) \, dx \leq \int_0^1 x^{2n_2 - 1} ID(G_1, x^{2n_2}) \, 2H(G_2, x) \, dx$$

$$\leq \frac{2}{3} \left(\int_0^1 x^{2n_2 - 1} ID(G_1, x^{2n_2}) \, dx \int_0^1 2H(G_2, x) \, dx \right)^{1/2} (ID(G_1, 1) \, 2H(G_2, 1))^{1/2}$$

$$= \frac{2}{3} \left(\frac{n_1 m_2}{n_2} ID(G_1) \, H(G_2) \right)^{1/2},$$

$$\int_0^1 2x^{n_2 + 1} H(G_1, x^{n_2}) \, ID^2(G_2, x) \, dx \leq \int_0^1 2x^{n_2 - 1} H(G_1, x^{n_2}) \, ID^2(G_2, x) \, dx$$

$$\leq \frac{2}{4} \left(\int_0^1 2x^{n_2 - 1} H(G_1, x^{n_2}) \, dx \int_0^1 ID(G_2, x) \, dx \int_0^1 ID(G_2, x) \, dx \right)^{1/3}$$

$$\cdot (2H(G_1, 1) \, (ID(G_2, 1) + ID(G_2, 0))^2)^{2/3}$$

$$= \frac{1}{2} \left(\frac{4m_1^2}{n_2} H(G_1) \, ID^2(G_2)(n_2 + k_2)^4 \right)^{1/3}.$$

These inequalities, Theorem 10 and $H(G_1 \odot G_2) = 2 \int_0^1 H(G_1 \odot G_2, x) \, dx$ give the bounds. $\square$

4. Algorithm for the Computation of the Harmonic Polynomial

The procedure shown in Algorithm 1 allows to compute the harmonic polynomial of a graph G with n vertices. This algorithm for computing the harmonic polynomial of a graph shows a complexity $O(n^2)$.

Algorithm 1 procedure Harmonic-Polynomial

Require: AM(G)—Adjacency matrix of G.
1: n = order(AM(G))
2: HPolynomial = $[0] * (2 * (n - 1))$
3: let D be a list with the degree of each vertex
4: **for** all i with $i \in \{1, 2, ...n - 1\}$ **do**
5:    **for** all j with $j \in \{i + 1, i + 2, ...n\}$ **do**
6:       **if** AM[i][j] == 1 **then**
7:          $v = D[i]$
8:          $u = D[j]$
9:          HPolynomial[$v + u - 1$] = HPolynomial [$v + u - 1$] + 1
10:       **end if**
11:    **end for**
12: **end for**
13: **return** HPolynomial

Author Contributions: The authors contributed equally to this work.

Funding: This work was supported in part by two grants from Ministerio de Economía y Competititvidad, Agencia Estatal de Investigación (AEI) and Fondo Europeo de Desarrollo Regional (FEDER) (MTM2016-78227-C2-1-P and MTM2017-90584-REDT), Spain.

Conflicts of Interest: The authors declare no conflict of interest. The founding sponsors had no role in the design of the study; in the collection, analyses, or interpretation of data; in the writing of the manuscript, and in the decision to publish the results.

References

1. Wiener, H. Structural determination of paraffin boiling points. *J. Am. Chem. Soc.* **1947**, *69*, 17–20. [CrossRef] [PubMed]
2. Randić, M. On characterization of molecular branching. *J. Am. Chem. Soc.* **1975**, *97*, 6609–6615. [CrossRef]
3. Gutman, I.; Furtula, B. (Eds.) *Recent Results in the Theory of Randić Index*; University Kragujevac: Kragujevac, Serbia, 2008.
4. Li, X.; Gutman, I. *Mathematical Aspects of Randić Type Molecular Structure Descriptors*; University Kragujevac: Kragujevac, Serbia, 2006.
5. Li, X.; Shi, Y. A survey on the Randić index. *Match Commun. Math. Comput. Chem.* **2008**, *59*, 127–156.
6. Rodríguez-Velázquez, J.A.; Sigarreta, J.M. On the Randić index and condicional parameters of a graph. *Match Commun. Math. Comput. Chem.* **2005**, *54*, 403–416.
7. Rodríguez-Velázquez, J.A.; Tomás-Andreu, J. On the Randić index of polymeric networks modelled by generalized Sierpinski graphs. *Match Commun. Math. Comput. Chem.* **2015**, *74*, 145–160.
8. Borovicanin, B.; Furtula, B. On extremal Zagreb indices of trees with given domination number. *Appl. Math. Comput.* **2016**, *279*, 208–218. [CrossRef]
9. Das, K.C. On comparing Zagreb indices of graphs. *Match Commun. Math. Comput. Chem.* **2010**, *63*, 433–440.
10. Das, K.C.; Gutman, I.; Furtula, B. Survey on geometric-arithmetic indices of graphs. *Match Commun. Math. Comput. Chem.* **2011**, *65*, 595–644.
11. Liu, M. A simple approach to order the first Zagreb indices of connected graphs. *Match Commun. Math. Comput. Chem.* **2010**, *63*, 425–432.
12. Fajtlowicz, S. On conjectures of Graffiti-II. *Congr. Numer.* **1987**, *60*, 187–197.
13. Deng, H.; Balachandran, S.; Ayyaswamy, S.K.; Venkatakrishnan, Y.B. On the harmonic index and the chromatic number of a graph. *Discret. Appl. Math.* **2013**, *161*, 2740–2744. [CrossRef]

14. Favaron, O.; Mahéo, M.; Saclé, J.F. Some eigenvalue properties in graphs (conjectures of Graffiti-II). *Discr. Math.* **1993**, *111*, 197–220. [CrossRef]
15. Rodríguez, J.M.; Sigarreta, J.M. New results on the harmonic index and Its generalizations. *Match Commun. Math. Comput. Chem.* **2017**, *78*, 387–404.
16. Shwetha Shetty, B.; Lokesha, V.; Ranjini, P.S. On the harmonic index of graph operations. *Trans. Combin.* **2015**, *4*, 5–14.
17. Wua, R.; Tanga, Z.; Deng, H. A lower bound for the harmonic index of a graph with minimum degree at least two. *Filomat* **2013**, *27*, 51–55. [CrossRef]
18. Zhong, L.; Xu, K. Inequalities between vertex-degree-based topological indices. *Match Commun. Math. Comput. Chem.* **2014**, *71*, 627–642.
19. Iranmanesh, M.A.; Saheli, M. On the harmonic index and harmonic polynomial of Caterpillars with diameter four. *Iran. J. Math. Chem.* **2014**, *5*, 35–43.
20. Nazir, R.; Sardar, S.; Zafar, S.; Zahid, Z. Edge version of harmonic index and harmonic polynomial of some classes of graphs. *J. Appl. Math. Inform.* **2016**, *34*, 479–486. [CrossRef]
21. Reza Farahani, M. Zagreb index, zagreb polynomial of circumcoronene series of benzenoid. *Adv. Mater. Corros.* **2013**, *2*, 16–19.
22. Reza Farahani, M. On the Schultz polynomial and Hosoya polynomial of circumcoronene series of benzenoid. *J. Appl. Math. Inform.* **2013**, *31*, 595–608. [CrossRef]
23. Reza Farahani, M.; Gao, W.; Rajesh Kanna, M.R.; Pradeep Kumar, R.; Liu, J.-B. General Randic, sum-connectivity, hyper-Zagreb and harmonic indices, and harmonic polynomial of molecular graphs. *Adv. Phys. Chem.* **2016**, *2016*, 2315949. [CrossRef]
24. Carballosa, W.; Hernández-Gómez, J.C.; Rosario, O.; Torres, Y. Computing the strong alliance polynomial of a graph. *Inv. Oper.* **2016**, *37*, 115–123.
25. Bindusree, A.R.; Naci Cangul, I.; Lokesha, V.; Sinan Cevik, A. Zagreb polynomials of three graph operators. *Filomat* **2016**, *30*, 1979–1986. [CrossRef]
26. Khalifeh, M.H.; Yousefi-Azari, H.; Ashrafi, A.R. The first and second Zagreb indices of some graph operations. *Discret. Appl. Math.* **2009**, *157*, 804–811. [CrossRef]
27. Loghman, A. PI polynomials of product graphs. *Appl. Math. Lett.* **2009**, *22*, 975–979. [CrossRef]
28. Nápoles, J.E.; Rodríguez, J.M.; Sigarreta, J.M.; Zohrevand, M. On the properties of the harmonic polynomial. Submitted.
29. Fath-Tabar, G.H. Zagreb polynomial and Pi Indices of some nano structures. *Digest J. Nanomat. Biostruct.* **2009**, *4*, 189–191.
30. Ilić, A. Note on the harmonic index of a graph. *Appl. Math. Lett.* **2012**, *25*, 561–566.
31. Xu, X. Relationships between harmonic index and other topological indices. *Appl. Math. Sci.* **2012**, *6*, 2013–2018.
32. Gutman, I.; Furtula, B.; Das, K.C.; Milovanovic, E.; Milovanovic, I. (Eds.) *Bounds in Chemical Graph Theory Basics (Volume 1). Mathematical Chemistry Monograph No. 19*; University Kragujevac: Kragujevac, Serbia, 2017.
33. Shuxian, L. Zagreb polynomials of thorn graphs. *Kragujevac J. Sci.* **2011**, *33*, 33–38.
34. Dankelmann, P.; Hellwig, A.; Volkmann, L. Inverse degree and edge-connectivity. *Discret. Math.* **2008**, *309*, 2943–2947. [CrossRef]
35. Zhang, Z.; Zhang, J.; Lu, X. The relation of matching with inverse degree of a graph. *Discret. Math.* **2005**, *301*, 243–246. [CrossRef]
36. Erdös, P.; Pach, J.; Spencer, J. On the mean distance between points of a graph. *Congr. Numer.* **1988**, *64*, 121–124.
37. Entringer, R. Bounds for the average distance-inverse degree product in trees. In *Combinatorics, Graph Theory, and Algorithms*; New Issues Press: Kalamazoo, MI, USA, 1996; Volume I, II, pp. 335–352.
38. Rodríguez, J.M.; Sánchez, J.L.; Sigarreta, J.M. Inequalities on the inverse degree index. Submitted.

39. Anderson, B.J. An inequality for convex functions. *Nordisk Mat. Tidsk.* **1958**, *6*, 25–26.
40. Csiszár, V.; Móri, T.F. Sharp integral inequalities for products of convex functions. *J. Inequal. Pure Appl. Math.* **2007**, *8*, 94.

symmetry

MDPI

Article

On Degree-Based Topological Indices of Symmetric Chemical Structures

Jia-Bao Liu [1], **Haidar Ali** [2] , **Muhammad Kashif Shafiq** [2,*] **and Usman Munir** [2]

1   School of Mathematics and Physics, Anhui Jianzhu University, Hefei 230601, China; liujiabaoad@163.com
2   Department of Mathematics, Government College University, Faisalabad 38000, Pakistan;
     haidarali@gcuf.edu.pk (H.A.); mianusman24120@gmail.com (U.M.)
*   Correspondence: kashif4v@gmail.com; Tel.: +92-321-4866779

Received: 18 October 2018; Accepted: 6 November 2018; Published: 9 November 2018

Abstract: A Topological index also known as connectivity index is a type of a molecular descriptor that is calculated based on the molecular graph of a chemical compound. Topological indices are numerical parameters of a graph which characterize its topology and are usually graph invariant. In QSAR/QSPR study, physico-chemical properties and topological indices such as Randić, atom-bond connectivity (ABC) and geometric-arithmetic (GA) index are used to predict the bioactivity of chemical compounds. Graph theory has found a considerable use in this area of research. In this paper, we study HDCN1(m,n) and HDCN2(m,n) of dimension m, n and derive analytical closed results of general Randić index $R_\alpha(\mathcal{G})$ for different values of α. We also compute the general first Zagreb, ABC, GA, ABC_4 and GA_5 indices for these Hex derived cage networks for the first time and give closed formulas of these degree-based indices.

Keywords: general randić index; atom-bond connectivity (ABC) index; geometric-arithmetic (GA) index; Hex-Derived Cage networks; $HDCN1(m, n)$, $HDCN2(m, n)$

1. Introduction

A graph is formed by vertices and edges connecting the vertices. A network is a connected simple graph having no multiple edges and loops. A topological index is a function Top : $\sum \rightarrow \mathbb{R}$ where $\mathbb{R}$ is the set of real numbers and $\sum$ is the finite simple graph with property that $Top(G_1) = Top(G_2)$ if G_1 and G_2 are isomorphic. A topological index is a numerical value associated with chemical constitution for correlation of chemical structure with various physical properties, chemical reactivity or biological activity. Many tools, such as topological indices has provided by graph theory to the chemists. *Cheminformatics* is new subject which is a combination of chemistry, mathematics and information science. It studies Quantitative structure-activity (QSAR) and structure-property (QSPR) relationships that are used to predict the biological activities and properties of chemical compounds. In the QSAR /QSPR study, physico-chemical properties and topological indices such as Wiener index, Szeged index, Randić index, Zagreb indices and ABC index are used to predict bioactivity of the chemical compounds. "In terms of graph theory, the structural formula of a chemical compound represents the molecular graph, in which vertices are represents to atoms and edges as chemical bonds". A molecular descriptor is a numeric number, which represents the properties of a chemical graph. Basically, a molecular descriptor and topological descriptor are different from each other. A molecular descriptor represents the underlying chemical graph but a topological descriptor are the representation of physico-chemical properties of underlying chemical graph in addition to show the whole structure. Topological indices have many applications in the field of nanobiotechnology and QSAR/QSPR study. Topological indices were firstly introduced by Wiener [1], he named the resulting index as path number while he was working on boiling point of Paraffin. Later on, it renamed as Wiener index [2]. Consider "n" Hex-Derived networks $(HDN1(1,1))$,$(HDN1(2,2))$ and so on to $(HDN1(m,n))$. Connect every boundary vertices of $(HDN1(1,1))$ to its mirror image vertices in $(HDN1(2,2))$ by an edge and so on to $(HDN1(m,n))$. As a result, we found a graph, which is called Hex-Derived Cage networks with "n" layers. In this article, the notations which we used take from the books [3,4].

In this article, Graph $(\mathcal{G})$ is considered to be a graph with vertex set $V(\mathcal{G})$ and edge set $E(\mathcal{G})$, the $d(a)$ is the degree of vertex $a \in V(\mathcal{G})$ and $S(a) = \sum\limits_{b \in N_\mathcal{G}(a)} d(b)$ where $N_\mathcal{G}(a) = \{b \in V(\mathcal{G}) \mid ab \in E(\mathcal{G})\}$.

Let $\mathcal{G}$ be a graph. Then the Wiener index is written as

$$W(\mathcal{G}) = \frac{1}{2} \sum_{(a,b)} d(a,b) \tag{1}$$

The Randić index [5] is the oldest degree-based topological index invented by Milan Randić, denoted as $R_{-\frac{1}{2}}(\mathcal{G})$ and defined as

$$R_{-\frac{1}{2}}(\mathcal{G}) = \sum_{ab \in E(\mathcal{G})} \frac{1}{\sqrt{d(a)d(b)}} \tag{2}$$

$R_\alpha(\mathcal{G})$ is a general Randić index and it is defined as

$$R_\alpha(\mathcal{G}) = \sum_{ab \in E(\mathcal{G})} (d(a)d(b))^\alpha \quad \text{for } \alpha \in \mathbb{R} \tag{3}$$

A topological index which has a great importance was introduced by Ivan Gutman and *Trinajstić* is Zagreb index and defined as

$$M_1(\mathcal{G}) = \sum_{ab \in E(\mathcal{G})} (d(a) + d(b)) \tag{4}$$

Estrada et al. in [6] invented a very famous degree-based topological index ABC and defined as

$$ABC(\mathcal{G}) = \sum_{ab \in E(\mathcal{G})} \sqrt{\frac{d(a) + d(b) - 2}{d(a)d(b)}} \tag{5}$$

GA index is also a very famous connectivity topological descriptor, which invented by Vukičević et al. [7] and denoted as

$$GA(\mathcal{G}) = \sum_{ab \in E(\mathcal{G})} \frac{2\sqrt{d(a)d(b)}}{(d(a) + d(b))} \tag{6}$$

ABC_4 and GA_5 indices find only if we find the edge partition of interconnection networks each edge in the graphs depend on sum of the degrees of end vertices. ABC_4 index invented by Ghorbani et al. [8] and written as

$$ABC_4(\mathcal{G}) = \sum_{ab \in E(\mathcal{G})} \sqrt{\frac{S(a) + S(b) - 2}{S(a)S(b)}} \tag{7}$$

The latest version of index is GA_5 invented by Graovac et al. [9] and defined as

$$GA_5(\mathcal{G}) = \sum_{ab \in E(\mathcal{G})} \frac{2\sqrt{S(a)S(b)}}{(S(a) + S(b))} \tag{8}$$

For any graph G for $\alpha = 1$, the general Randić index is second Zagreb index.

2. Main Results

Hex-Derived Cage networks $HDCN1(m, n)$ (show in Figure 1) and $HDCN2(m, n)$ (show in Figure 2) give closed formulas of that indices, we study the general Randić, first Zagreb, ABC, GA, ABC_4 and GA_5 indices of certain graphs in [10]. These days there is a broad research activity on ABC and GA indices and their variants, for additionally investigation of topological indices of different families see, [1,11–23].

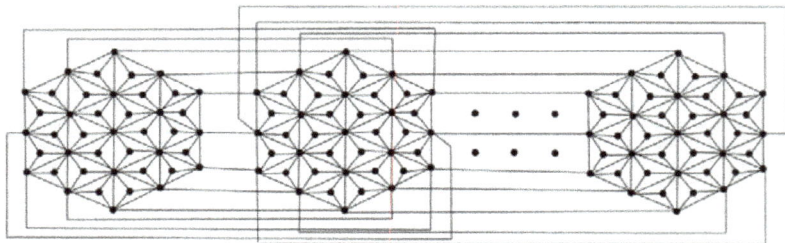

Figure 1. Hex-Derived Network ($HDCN1(3, n)$).

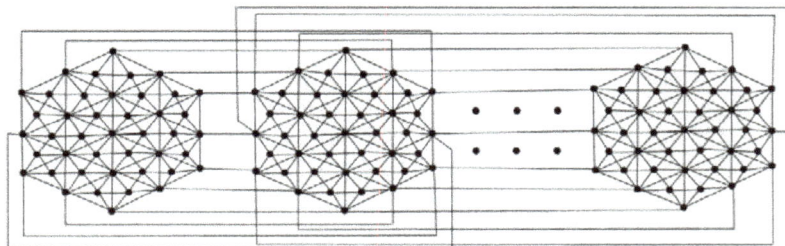

Figure 2. Hex-Derived Network ($HDCN2(3, n)$).

2.1. Results for Hex-Derived Cage Networks

We compute specific degree-based topological indices of Hex-Derived Cage networks. In this paper, we calculate Randić index $R_\alpha(\mathcal{G})$ with $\alpha = 1, -1, \frac{1}{2}, -\frac{1}{2}$, M_1, ABC, GA, ABC_4 and GA_5 for Hex-Derived Cage networks $HDCN1(m, n)$ and $HDCN2(m, n)$.

Theorem 1. *Let* $\mathcal{G}_1 \cong HDCN1(m, n)$ *be the Hex-Derived Cage network, then its general Randić index is equal to*

$$R_\alpha(\mathcal{G}_1) = \begin{cases} 18(108n^3 - 219n^2 + 25n + 91), & \alpha = 1; \\ 6(36n^3 + 3(7\sqrt{3} - 34)n^2 + (4\sqrt{21} + 6\sqrt{7} + \\ 28\sqrt{6} - 84\sqrt{3} + 12\sqrt{2} + 35)n + 2\sqrt{42} - \\ 8\sqrt{21} - 12\sqrt{7} - 56\sqrt{6} + 100\sqrt{3} + 39), & \alpha = \frac{1}{2}; \\ \frac{11907n^3 - 17003n^2 + 12343n - 3051}{21168}, & \alpha = -1; \\ \frac{15n^3}{4} + (\frac{8}{\sqrt{3}} - \frac{125}{12})n^2 + (\frac{4}{\sqrt{7}} + \\ 4\sqrt{6} - \frac{32}{\sqrt{3}} + \sqrt{2} + 5\sqrt{\frac{3}{7}} + \frac{109}{21})n - \\ \frac{8}{\sqrt{7}} - 8\sqrt{6} + \frac{38}{\sqrt{3}} + 3\sqrt{2} + 2\sqrt{\frac{6}{7}} - \\ 10\sqrt{\frac{3}{7}} + \frac{13}{14}, & \alpha = -\frac{1}{2}. \end{cases}$$

Proof. Let $\mathcal{G}_1$ be the Hex-Derived Cage network $(HDCN1(m,n))$ where $m = n \geq 5$. The edge set of $HDCN1(m,n)$ are divided into seventeen partitions based on the degree of end vertices shows in Table 1. Thus from Equation (3), is follows that

$$R_\alpha(\mathcal{G}_1) = \sum_{ab \in E(\mathcal{G})} (d(a)d(b))^\alpha$$

For $\alpha = 1$

$$R_1(\mathcal{G}_1) = \sum_{j=1}^{17} \sum_{ab \in E_j(\mathcal{G})} deg(u) \cdot deg(v)$$

By using the edge partition given in Table 1, we have

$R_1(\mathcal{G}_1) = 18|E_1(\mathcal{G}_1)| + 21|E_2(\mathcal{G}_1)| + 24|E_3(\mathcal{G}_1)| + 27|E_4(\mathcal{G}_1)| + 36|E_5(\mathcal{G}_1)| + 42|E_6(\mathcal{G}_1)| + 48|E_7(\mathcal{G}_1)| + 72|E_8(\mathcal{G}_1)| + 49|E_9(\mathcal{G}_1)| + 63|E_{10}(\mathcal{G}_1)| + 84|E_{11}(\mathcal{G}_1)| + 64|E_{12}(\mathcal{G}_1)| + 72|E_{13}(\mathcal{G}_1)| + 96|E_{14}(\mathcal{G}_1)| + 81|E_{15}(\mathcal{G}_1)| + 108|E_{16}(\mathcal{G}_1)| + 144|E_{17}(\mathcal{G}_1)|$

After simplification, we have

$$R_1(\mathcal{G}_1) = 18(108n^3 - 219n^2 + 25n + 91)$$

For $\alpha = \frac{1}{2}$

$$R_{\frac{1}{2}}(\mathcal{G}_1) = \sum_{j=1}^{17} \sum_{ab \in E_j(\mathcal{G})} \sqrt{d(a) \cdot d(b)}$$

Using the edge partition from Table 1, we have

$R_{\frac{1}{2}}(\mathcal{G}_1) = 3\sqrt{2}|E_1(\mathcal{G}_1)| + \sqrt{21}|E_2(\mathcal{G}_1)| + 2\sqrt{6}|E_3(\mathcal{G}_1)| + 3\sqrt{3}|E_4(\mathcal{G}_1)| + 6|E_5(\mathcal{G}_1)| + \sqrt{42}|E_6(\mathcal{G}_1)| + 4\sqrt{3}|E_7(\mathcal{G}_1)| + 6\sqrt{2}|E_8(\mathcal{G}_1)| + 7|E_9(\mathcal{G}_1)| + 3\sqrt{7}|E_{10}(\mathcal{G}_1)| + 2\sqrt{21}|E_{11}(\mathcal{G}_1)| + 8|E_{12}(\mathcal{G}_1)| + 6\sqrt{2}|E_{13}(\mathcal{G}_1)| + 4\sqrt{6}|E_{14}(\mathcal{G}_1)| + 9|E_{15}(\mathcal{G}_1)| + 6\sqrt{3}|E_{16}(\mathcal{G}_1)| + 12|E_{17}(\mathcal{G}_1)|$

After simplification, we have

$R_{\frac{1}{2}}(\mathcal{G}_1) = 6(36n^3 + 3(7\sqrt{3} - 34)n^2 + (4\sqrt{21} + 6\sqrt{7} + 28\sqrt{6} - 84\sqrt{3} + 12\sqrt{2} + 35)n + 2\sqrt{42} - 8\sqrt{21} - 12\sqrt{7} - 56\sqrt{6} + 100\sqrt{3} + 39)$

For $\alpha = -1$

$$R_{-1}(\mathcal{G}_1) = \sum_{j=1}^{17} \sum_{ab \in E_j(\mathcal{G})} \frac{1}{d(a) \cdot d(b)}$$

$R_{-1}(\mathcal{G}_1) = \frac{1}{18}|E_1(\mathcal{G}_1)| + \frac{1}{21}|E_2(\mathcal{G}_1)| + \frac{1}{24}|E_3(\mathcal{G}_1)| + \frac{1}{27}|E_4(\mathcal{G}_1)| + \frac{1}{36}|E_5(\mathcal{G}_1)| + \frac{1}{42}|E_6(\mathcal{G}_1)| + \frac{1}{48}|E_7(\mathcal{G}_1)| + \frac{1}{72}|E_8(\mathcal{G}_1)| + \frac{1}{49}|E_9(\mathcal{G}_1)| + \frac{1}{63}|E_{10}(\mathcal{G}_1)| + \frac{1}{84}|E_{11}(\mathcal{G}_1)| + \frac{1}{64}|E_{12}(\mathcal{G}_1)| + \frac{1}{72}|E_{13}(\mathcal{G}_1)| + \frac{1}{96}|E_{14}(\mathcal{G}_1)| + \frac{1}{81}|E_{15}(\mathcal{G}_1)| + \frac{1}{108}|E_{16}(\mathcal{G}_1)| + \frac{1}{144}|E_{17}(\mathcal{G}_1)|$

After simplification, we have

$$R_{-1}(\mathcal{G}_1) = \frac{11907n^3 - 17003n^2 + 12343n - 3051}{21168}$$

For $\alpha = -\frac{1}{2}$

$$R_{-\frac{1}{2}}(\mathcal{G}_1) = \sum_{j=1}^{17} \sum_{ab \in E_j(\mathcal{G})} \frac{1}{\sqrt{d(a) \cdot d(b)}}$$

$R_{-\frac{1}{2}}(\mathcal{G}_1) = \frac{\sqrt{2}}{6}|E_1(\mathcal{G}_1)| + \frac{\sqrt{21}}{21}|E_2(\mathcal{G}_1)| + \frac{\sqrt{6}}{12}|E_3(\mathcal{G}_1)| + \frac{\sqrt{3}}{9}|E_4(\mathcal{G}_1)| + \frac{1}{6}|E_5(\mathcal{G}_1)| + \frac{\sqrt{42}}{42}|E_6(\mathcal{G}_1)| +$
$\frac{\sqrt{3}}{12}|E_7(\mathcal{G}_1)| + \frac{\sqrt{2}}{12}|E_8(\mathcal{G}_1)| + \frac{1}{7}|E_9(\mathcal{G}_1)| + \frac{\sqrt{7}}{21}|E_{10}(\mathcal{G}_1)| + \frac{\sqrt{21}}{42}|E_{11}(\mathcal{G}_1)| + \frac{1}{8}|E_{12}(\mathcal{G}_1)| + \frac{\sqrt{2}}{12}|E_{13}(\mathcal{G}_1)| +$
$\frac{\sqrt{6}}{24}|E_{14}(\mathcal{G}_1)| + \frac{1}{9}|E_{15}(\mathcal{G}_1)| + \frac{\sqrt{3}}{18}|E_{16}(\mathcal{G}_1)| + \frac{1}{12}|E_{17}(\mathcal{G}_1)|$

After simplification, we have

$R_{-\frac{1}{2}}(\mathcal{G}_1) = \frac{15n^3}{4} + (\frac{8}{\sqrt{3}} - \frac{125}{12})n^2 + (\frac{4}{\sqrt{7}} + 4\sqrt{6} - \frac{32}{\sqrt{3}} + \sqrt{2} + 5\sqrt{\frac{3}{7}} + \frac{109}{21})n - \frac{8}{\sqrt{7}} - 8\sqrt{6} + \frac{38}{\sqrt{3}} + 3\sqrt{2} + 2\sqrt{\frac{6}{7}} - 10\sqrt{\frac{3}{7}} + \frac{13}{14}$ □

In the below theorem, we calculate the Zagreb index of $\mathcal{G}_1(m,n)$.

Theorem 2. *The first Zagreb index of hex-derived cage network HDCN1(m, n) is equal to*

$$M_1(\mathcal{G}_1) = 18(27n^3 - 51n^2 + 10n + 14)$$

Proof. With the help of Table 1, we calculate the Zagreb index as

$$M_1(\mathcal{G}_1) = \sum_{ab \in E(\mathcal{G})} (d(a) + d(b)) = \sum_{j=1}^{17} \sum_{ab \in E_j(\mathcal{G})} (d(a) + d(b))$$

$M_1(\mathcal{G}_1) = 9|E_1(\mathcal{G}_1)| + 10|E_2(\mathcal{G}_1)| + 11|E_3(\mathcal{G}_1)| + 12|E_4(\mathcal{G}_1)| + 15|E_5(\mathcal{G}_1)| + 13|E_6(\mathcal{G}_1)| + 14|E_7(\mathcal{G}_1)| + 18|E_8(\mathcal{G}_1)| + 14|E_9(\mathcal{G}_1)| + 16|E_{10}(\mathcal{G}_1)| + 19|E_{11}(\mathcal{G}_1)| + 16|E_{12}(\mathcal{G}_1)| + 17|E_{13}(\mathcal{G}_1)| + 20|E_{14}(\mathcal{G}_1)| + 18|E_{15}(\mathcal{G}_1)| + 21|E_{16}(\mathcal{G}_1)| + 24|E_{17}(\mathcal{G}_1)|$

After some calculations, we get

$$M_1(\mathcal{G}_1) = 18(27n^3 - 51n^2 + 10n + 14)$$

□

Table 1. Edge partition of Hex-Derived Cage network (*HDCN1*) based on degrees of end vertices of each edge.

(d_u, d_v) where $ab \in E(\mathcal{G}_1)$	Number of Edges	(d_u, d_v) where $ab \in E(\mathcal{G}_1)$	Number of Edges
$E_1 = (3, 6)$	24	$E_{10} = (7, 7)$	$6n - 18$
$E_2 = (3, 7)$	$2(6n - 12)$	$E_{11} = (7, 9)$	$2(6n - 12)$
$E_3 = (3, 8)$	$6(6n - 12)$	$E_{12} = (7, 12)$	$6n - 12$
$E_4 = (3, 9)$	$18n^2 - 72n + 72$	$E_{13} = (8, 8)$	$2(6n - 18)$
$E_5 = (3, 12)$	$18n^3 - 54n^2 + 42n$	$E_{14} = (8, 9)$	$2(6n - 12)$
$E_6 = (6, 7)$	12	$E_{15} = (8, 12)$	$4(6n - 12)$
$E_7 = (6, 8)$	24	$E_{16} = (9, 9)$	$12n^2 - 60n + 72$
$E_8 = (6, 12)$	12	$E_{17} = (9, 12)$	$12n^2 - 48n + 48$
$E_9 = (12, 12)$	$9n^3 - 33n^2 + 30n$		

In the next theorem, we calculate the *ABC*, *GA*, ABC_4 and GA_5 indices of Hex-Derived Cage network *HDCN1*(m, n).

Theorem 3. *Let HDCN1(m,n) be Hex-Derived Cage network, then we have*

- $ABC(\mathcal{G}_1) = \frac{3}{4}(4\sqrt{13} + \sqrt{22})n^3 + \frac{1}{12}(8\sqrt{57} + 24\sqrt{30} - 33\sqrt{22} - 108\sqrt{13} + 64)n^2 + (-\frac{80}{3} + 8\sqrt{\frac{6}{7}} + 4\sqrt{2} + \frac{54\sqrt{3}}{7} + 3\sqrt{\frac{7}{2}} + 5\sqrt{\frac{11}{2}} + 9\sqrt{6} - 8\sqrt{\frac{19}{3}} + \sqrt{\frac{51}{7}} + 7\sqrt{13} - 7\sqrt{30})n + 44 - 16\sqrt{\frac{6}{7}} - 4\sqrt{2} - \frac{120\sqrt{3}}{7} - \sqrt{\frac{7}{2}} - 18\sqrt{6} + 8\sqrt{\frac{19}{3}} - 2\sqrt{\frac{51}{7}} + 2\sqrt{\frac{66}{7}} + 6\sqrt{30}.$

- $GA(\mathcal{G}_1)) = \frac{117n^3}{5} + \frac{3}{35}(185\sqrt{3} - 749)n^2 + (\frac{108}{5} + \frac{144\sqrt{2}}{17} - \frac{444\sqrt{3}}{7} + \frac{1248\sqrt{6}}{55} + \frac{9\sqrt{7}}{2} + \frac{348\sqrt{21}}{95})n + \frac{24\sqrt{42}}{13} - \frac{696\sqrt{21}}{95} - 9\sqrt{7} - \frac{2496\sqrt{6}}{55} + \frac{540\sqrt{3}}{7} + \frac{120\sqrt{2}}{17} + 18.$

- $ABC_4(\mathcal{G}_1) = \frac{6}{5}\sqrt{\frac{38}{7}}(n-4)^2 + 2\sqrt{\frac{69}{79}}(n-2) + \frac{1}{3}\sqrt{\frac{62}{5}}n(3n^2 - 15n + 19) + \frac{1}{15}\sqrt{\frac{89}{2}}n(3n^2 - 17n + 24) + \sqrt{\frac{177}{14}}(n^2 - 5n + 6) + 2\sqrt{\frac{115}{77}}(n^2 - 5n + 6) + 2\sqrt{\frac{86}{105}}(n^2 - 5n + 6) + 2\sqrt{\frac{2}{5}}(n^2 - 5n + 6) + \frac{1}{7}\sqrt{\frac{83}{2}}(n^2 - 6n + 8) + \frac{8}{23}\sqrt{34}(n^2 - 9n + 20) + 2\sqrt{\frac{113}{79}}(n-2) + \sqrt{\frac{334}{395}}(n-2) + 4\sqrt{\frac{30}{79}}(n-2) + 6\sqrt{\frac{15}{79}}(n-2) + 6\sqrt{\frac{58}{41}}(n-3) + 4\sqrt{\frac{34}{41}}(n-3) + 12\sqrt{\frac{7}{41}}(n-3) + 18\sqrt{\frac{6}{287}}(n-3) + \frac{20(n-4)^2}{\sqrt{253}} + 2\sqrt{\frac{194}{115}}(n-4)^2 + 2\sqrt{\frac{151}{161}}(n-4)^2 + \frac{144(n-4)}{\sqrt{5293}} + \frac{48(n-4)}{\sqrt{85}} + \frac{3}{5}\sqrt{\frac{254}{79}}(n-4) + \frac{54}{41}\sqrt{2}(n-4) + 2\sqrt{\frac{114}{67}}(n-4) + 2\sqrt{\frac{447}{469}}(n-4) + 3\sqrt{\frac{22}{29}}(n-4) + 4\sqrt{\frac{6}{23}}(n-4) + 12\sqrt{\frac{6}{29}}(n-4) + 6\sqrt{\frac{93}{469}}(n-4) + 12\sqrt{\frac{138}{1189}}(n-4) + 30\sqrt{\frac{2}{119}}(n-4) + 18\sqrt{\frac{5}{391}}(n-4) + 28\sqrt{\frac{6}{737}}(n-4) + \frac{3}{17}\sqrt{134}(n-5) + \frac{6}{29}\sqrt{114}(n-5) + \frac{24}{67}\sqrt{33}(n-5) + \frac{21}{25}\sqrt{2}(n-5) + \frac{206\sqrt{194}}{385} + 4\frac{\sqrt{\frac{678}{11}}}{7} + \frac{6\sqrt{41}}{7} + \frac{24}{\sqrt{29}} + 2\sqrt{\frac{110}{19}} + 3\sqrt{\frac{43}{14}} + 3\sqrt{\frac{53}{19}} + 4\sqrt{\frac{138}{77}} + 4\sqrt{\frac{94}{55}} + 4\sqrt{\frac{786}{737}} + 4\sqrt{\frac{78}{79}} + 4\sqrt{\frac{74}{77}} + 2\sqrt{\frac{82}{95}} + 6\sqrt{\frac{190}{287}} + \frac{24\sqrt{\frac{22}{41}}}{7} + 6\sqrt{\frac{65}{133}} + \frac{60\sqrt{\frac{3}{7}}}{7} + 12\sqrt{\frac{10}{41}} + 6\sqrt{\frac{115}{779}} + 12\sqrt{\frac{10}{77}} + 12\sqrt{\frac{34}{287}} + 12\sqrt{\frac{2}{17}} + 12\sqrt{\frac{78}{779}} + 24\sqrt{\frac{5}{91}} + 60\sqrt{\frac{2}{247}} + 36\sqrt{\frac{2}{553}}.$

- $GA_5(\mathcal{G}_1) = \frac{40}{13}\sqrt{14}(n-4)^2 + \frac{48}{163}\sqrt{1659}(n-2) + \frac{12}{7}\sqrt{10}n(3n^2 - 15n + 19) + \frac{24}{29}\sqrt{210}(n^2 - 5n + 6) + \frac{16}{13}\sqrt{77}(n^2 - 5n + 6) + \frac{12}{19}\sqrt{70}(n^2 - 5n + 6) + \frac{18}{5}\sqrt{21}(n^2 - 5n + 6) + 9n^3 - 33n^2 - 66n + \frac{3}{14}\sqrt{2607}(n-2) + \frac{36}{169}\sqrt{790}(n-2) + \frac{48}{115}\sqrt{553}(n-2) + \frac{144}{115}\sqrt{79}(n-2) + \frac{24}{29}\sqrt{574}(n-3) + \frac{72}{43}\sqrt{205}(n-3) + \frac{216}{59}\sqrt{82}(n-3) + \frac{64}{19}\sqrt{41}(n-3) + \frac{6}{11}\sqrt{253}(n-4)^2 + \frac{8}{11}\sqrt{230}(n-4)^2 + \frac{16}{17}\sqrt{161}(n-4)^2 + \frac{12}{73}\sqrt{5293}(n-4) + \frac{3}{17}\sqrt{4623}(n-4) + \frac{6}{25}\sqrt{2211}(n-4) + \frac{24}{97}\sqrt{2010}(n-4) + \frac{48}{151}\sqrt{1407}(n-4) + \frac{24}{35}\sqrt{1189}(n-4) + \frac{48}{137}\sqrt{1173}(n-4) + \frac{8}{21}\sqrt{986}(n-4) + \frac{48}{101}\sqrt{561}(n-4) + \frac{48}{49}\sqrt{510}(n-4) + \frac{48}{95}\sqrt{469}(n-4) + \frac{48}{43}\sqrt{406}(n-4) + \frac{24}{19}\sqrt{357}(n-4) + \frac{20}{43}\sqrt{158}(n-4) + \frac{40}{39}\sqrt{134}(n-4) + \frac{32}{15}\sqrt{29}(n-4) + 12(n-4) + 36(n-5) + \frac{48\sqrt{5214}}{145} + \frac{48\sqrt{4422}}{133} + \frac{48\sqrt{1558}}{79} + \frac{48\sqrt{1122}}{67} + \frac{48\sqrt{1066}}{67} + \frac{16\sqrt{779}}{39} + \frac{96\sqrt{574}}{97} + \frac{16\sqrt{494}}{17} + \frac{48\sqrt{462}}{47} + \frac{32\sqrt{287}}{23} + \frac{16\sqrt{266}}{11} + \frac{96\sqrt{231}}{61} + \frac{32\sqrt{203}}{19} + \frac{72\sqrt{190}}{83} + \frac{48\sqrt{154}}{25} + \frac{96\sqrt{91}}{41} + \frac{21\sqrt{79}}{16} + \frac{336\sqrt{66}}{115} + 3\sqrt{55} + \frac{28\sqrt{41}}{15} + \frac{32\sqrt{38}}{9} + \frac{36\sqrt{19}}{7} + \frac{192\sqrt{7}}{11} + \frac{1336\sqrt{2}}{33} + 258.$

Proof. From Table 1 we calculate the $ABC(\mathcal{G}_1)$ as

$$ABC(\mathcal{G}_1) = \sum_{ab \in E(\mathcal{G})} \sqrt{\frac{d(a) + d(b) - 2}{d(a) \cdot d(b)}} = \sum_{j=1}^{17} \sum_{ab \in E_j(\mathcal{G})} \sqrt{\frac{d(a) + d(b) - 2}{d(a) \cdot d(b)}}$$

$ABC(\mathcal{G}_1) = \frac{\sqrt{14}}{6}|E_1(\mathcal{G}_1)| + \frac{2\sqrt{42}}{21}|E_2(\mathcal{G}_1)| + \frac{\sqrt{6}}{4}|E_3(\mathcal{G}_1)| + \frac{\sqrt{30}}{9}|E_4(\mathcal{G}_1)| + \frac{\sqrt{13}}{6}|E_5(\mathcal{G}_1)| + \frac{\sqrt{462}}{42}|E_6(\mathcal{G}_1)| + \frac{1}{2}|E_7(\mathcal{G}_1)| + \frac{\sqrt{2}}{3}|E_8(\mathcal{G}_1)| + \frac{2\sqrt{3}}{7}|E_9(\mathcal{G}_1)| + \frac{\sqrt{2}}{3}|E_{10}(\mathcal{G}_1)| + \frac{\sqrt{357}}{42}|E_{11}(\mathcal{G}_1)| + \frac{\sqrt{14}}{8}|E_{12}(\mathcal{G}_1)| + \frac{\sqrt{30}}{12}|E_{13}(\mathcal{G}_1)| + \frac{\sqrt{3}}{4}|E_{14}(\mathcal{G}_1)| + \frac{4}{9}|E_{15}(\mathcal{G}_1)| + \frac{\sqrt{57}}{18}|E_{16}(\mathcal{G}_1)| + \frac{\sqrt{22}}{12}|E_{17}(\mathcal{G}_1)|.$

After simplification, we have

$ABC(\mathcal{G}_1) = \frac{3}{4}(4\sqrt{13} + \sqrt{22})n^3 + \frac{1}{12}(8\sqrt{57} + 24\sqrt{30} - 33\sqrt{22} - 108\sqrt{13} + 64)n^2 + (-\frac{80}{3} + 8\sqrt{\frac{6}{7}} + 4\sqrt{2} + \frac{54\sqrt{3}}{7} + 3\sqrt{\frac{7}{2}} + 5\sqrt{\frac{11}{2}} + 9\sqrt{6} - 8\sqrt{\frac{19}{3}} + \sqrt{\frac{51}{7}} + 7\sqrt{13} - 7\sqrt{30})n + 44 - 16\sqrt{\frac{6}{7}} - 4\sqrt{2} - \frac{120\sqrt{3}}{7} - \sqrt{\frac{7}{2}} - 18\sqrt{6} + 8\sqrt{\frac{19}{3}} - 2\sqrt{\frac{51}{7}} + 2\sqrt{\frac{66}{7}} + 6\sqrt{30}.$

Now we calculate GA from Equation (6) as

$$GA(\mathcal{G}_1) = \sum_{ab \in E(\mathcal{G})} \frac{2\sqrt{d(a)d(b)}}{(d(a) + d(b))} = \sum_{j=1}^{17} \sum_{ab \in E_j(\mathcal{G})} \frac{2\sqrt{d(a)d(b)}}{(d(a) + d(b))}$$

From Table 1 calculate $GA(\mathcal{G}_1)$ as

$$GA(\mathcal{G}_1) = \frac{2\sqrt{2}}{3}|E_1(\mathcal{G}_1)| + \frac{\sqrt{21}}{5}|E_2(\mathcal{G}_1)| + \frac{4\sqrt{6}}{11}|E_3(\mathcal{G}_1)| + \frac{\sqrt{3}}{2}|E_4(\mathcal{G}_1)| + \frac{4}{5}|E_5(\mathcal{G}_1)| + \frac{2\sqrt{42}}{13}|E_6(\mathcal{G}_1)| + \frac{4\sqrt{3}}{7}|E_7(\mathcal{G}_1)| + \frac{2\sqrt{2}}{7}|E_8(\mathcal{G}_1)| + 1|E_9(\mathcal{G}_1)| + \frac{3\sqrt{7}}{8}|E_{10}(\mathcal{G}_1)| + \frac{4\sqrt{21}}{19}|E_{11}(\mathcal{G}_1)| + 1|E_{12}(\mathcal{G}_1)| + \frac{12\sqrt{2}}{17}|E_{13}(\mathcal{G}_1)| + \frac{2\sqrt{6}}{5}|E_{14}(\mathcal{G}_1)| + 1|E_{15}(\mathcal{G}_1)| + \frac{4\sqrt{3}}{7}|E_{16}(\mathcal{G}_1)| + 1|E_{17}(\mathcal{G}_1)|.$$

After simplification, we have

$$GA(\mathcal{G}_1) = \frac{117n^3}{5} + \frac{3}{35}(185\sqrt{3} - 749)n^2 + \left(\frac{108}{5} + \frac{144\sqrt{2}}{17} - \frac{444\sqrt{3}}{7} + \frac{1248\sqrt{6}}{55} + \frac{9\sqrt{7}}{2} + \frac{348\sqrt{21}}{95}\right)n + \frac{24\sqrt{42}}{13} - \frac{696\sqrt{21}}{95} - 9\sqrt{7} - \frac{2496\sqrt{6}}{55} + \frac{540\sqrt{3}}{7} + \frac{120\sqrt{2}}{17} + 18.$$

If we consider an edge partition based on degree sum of neighbors of end vertices; then the edge set $E(HDCN1(m,n))$ are divided into sixtynine edge partition $E_j(HDCN1(m,n))$, $18 \le j \le 86$ shows in Table 2.

From Equation (7), we have

$$ABC_4(\mathcal{G}_1) = \sum_{ab \in E(\mathcal{G})} \sqrt{\frac{S(a) + S(b) - 2}{S(a)S(b)}} = \sum_{j=18}^{86} \sum_{ab \in E_j(\mathcal{G})} \sqrt{\frac{S(a) + S(b) - 2}{S(a)S(b)}}.$$

From Table 2 we use edge partition, we get

$$ABC_4(\mathcal{G}_1) = \frac{2\sqrt{1066}}{67}|E_{18}(\mathcal{G}_1)| + \frac{\sqrt{1456}}{41}|E_{19}(\mathcal{G}_1)| + \frac{\sqrt{1976}}{51}|E_{20}(\mathcal{G}_1)| + \frac{2\sqrt{1372}}{77}|E_{21}(\mathcal{G}_1)| + \frac{\sqrt{1400}}{39}|E_{22}(\mathcal{G}_1)| + \frac{\sqrt{1568}}{42}|E_{23}(\mathcal{G}_1)| + \frac{\sqrt{1624}}{43}|E_{24}(\mathcal{G}_1)| + \frac{\sqrt{1848}}{47}|E_{25}(\mathcal{G}_1)| + \frac{2\sqrt{1876}}{95}|E_{26}(\mathcal{G}_1)| + \frac{2\sqrt{2212}}{107}|E_{27}(\mathcal{G}_1)| + \frac{\sqrt{2296}}{55}|E_{28}(\mathcal{G}_1)| + \frac{\sqrt{1980}}{48}|E_{29}(\mathcal{G}_1)| + \frac{2\sqrt{2010}}{97}|E_{30}(\mathcal{G}_1)| + \frac{\sqrt{2040}}{49}|E_{31}(\mathcal{G}_1)| + \frac{2\sqrt{2070}}{99}|E_{32}(\mathcal{G}_1)| + \frac{\sqrt{2520}}{57}|E_{33}(\mathcal{G}_1)| + \frac{\sqrt{1792}}{44}|E_{34}(\mathcal{G}_1)| + \frac{\sqrt{1856}}{45}|E_{35}(\mathcal{G}_1)| + \frac{\sqrt{2432}}{54}|E_{36}(\mathcal{G}_1)| + \frac{\sqrt{2624}}{57}|E_{37}(\mathcal{G}_1)| + \frac{2\sqrt{2178}}{99}|E_{38}(\mathcal{G}_1)| + \frac{\sqrt{2211}}{50}|E_{39}(\mathcal{G}_1)| + \frac{2\sqrt{2244}}{101}|E_{40}(\mathcal{G}_1)| + \frac{\sqrt{2277}}{51}|E_{41}(\mathcal{G}_1)| + \frac{\sqrt{2607}}{3}|E_{56}(\mathcal{G}_1)| + \frac{2\sqrt{2772}}{117}|E_{43}(\mathcal{G}_1)| + \frac{\sqrt{2736}}{56}|E_{44}(\mathcal{G}_1)| + \frac{2\sqrt{2844}}{115}|E_{45}(\mathcal{G}_1)| + \frac{\sqrt{2952}}{59}|E_{46}(\mathcal{G}_1)| + \frac{\sqrt{3024}}{60}|E_{47}(\mathcal{G}_1)| + \frac{\sqrt{3240}}{63}|E_{48}(\mathcal{G}_1)| + \frac{\sqrt{2009}}{45}|E_{49}(\mathcal{G}_1)| + \frac{2\sqrt{2296}}{97}|E_{50}(\mathcal{G}_1)| + \frac{2\sqrt{3116}}{117}|E_{51}(\mathcal{G}_1)| + \frac{2\sqrt{2450}}{99}|E_{52}(\mathcal{G}_1)| + \frac{2\sqrt{3234}}{115}|E_{53}(\mathcal{G}_1)| + \frac{\sqrt{3871}}{64}|E_{54}(\mathcal{G}_1)| + 1|E_{55}(\mathcal{G}_1)| + \frac{2\sqrt{3350}}{117}|E_{56}(\mathcal{G}_1)| + \frac{2\sqrt{3950}}{129}|E_{57}(\mathcal{G}_1)| + \frac{\sqrt{3248}}{57}|E_{58}(\mathcal{G}_1)| + \frac{\sqrt{3696}}{56}|E_{59}(\mathcal{G}_1)| + \frac{\sqrt{4256}}{66}|E_{60}(\mathcal{G}_1)| + \frac{\sqrt{4292}}{69}|E_{61}(\mathcal{G}_1)| + \frac{\sqrt{3364}}{58}|E_{62}(\mathcal{G}_1)| + \frac{\sqrt{3944}}{63}|E_{63}(\mathcal{G}_1)| + \frac{\sqrt{4756}}{70}|E_{64}(\mathcal{G}_1)| + \frac{2\sqrt{4422}}{133}|E_{65}(\mathcal{G}_1)| + \frac{\sqrt{4488}}{67}|E_{66}(\mathcal{G}_1)| + \frac{2\sqrt{5214}}{145}|E_{67}(\mathcal{G}_1)| + \frac{\sqrt{5544}}{75}|E_{68}(\mathcal{G}_1)| + \frac{\sqrt{4489}}{67}|E_{69}(\mathcal{G}_1)| + \frac{\sqrt{4623}}{68}|E_{70}(\mathcal{G}_1)| + \frac{\sqrt{5293}}{73}|E_{71}(\mathcal{G}_1)| + \frac{2\sqrt{5628}}{151}|E_{72}(\mathcal{G}_1)| + \frac{\sqrt{4624}}{68}|E_{73}(\mathcal{G}_1)| + \frac{2\sqrt{4692}}{137}|E_{74}(\mathcal{G}_1)| + \frac{\sqrt{5712}}{76}|E_{75}(\mathcal{G}_1)| + \frac{\sqrt{4761}}{69}|E_{76}(\mathcal{G}_1)| + \frac{2\sqrt{5796}}{153}|E_{77}(\mathcal{G}_1)| + \frac{\sqrt{6232}}{79}|E_{78}(\mathcal{G}_1)| + \frac{\sqrt{6840}}{83}|E_{79}|(\mathcal{G}_1) + \frac{2\sqrt{6636}}{163}|E_{80}(\mathcal{G}_1)| + \frac{2\sqrt{7110}}{169}|E_{81}(\mathcal{G}_1)| + \frac{\sqrt{4724}}{82}|E_{82}(\mathcal{G}_1)| + \frac{\sqrt{7380}}{86}|E_{83}(\mathcal{G}_1)| + \frac{\sqrt{7056}}{84}|E_{84}(\mathcal{G}_1)| + \frac{\sqrt{7560}}{87}|E_{85}(\mathcal{G}_1)| + 1|E_{86}(\mathcal{G}_1)|.$$

After simplification, we get

$$ABC_4(\mathcal{G}_1) = \frac{6}{5}\sqrt{\frac{38}{7}}(n-4)^2 + 2\sqrt{\frac{69}{79}}(n-2) + \frac{1}{3}\sqrt{\frac{62}{5}}n(3n^2 - 15n + 19) + \frac{1}{15}\sqrt{\frac{89}{2}}n(3n^2 - 17n + 24) + \sqrt{\frac{177}{14}}(n^2 - 5n + 6) + 2\sqrt{\frac{115}{77}}(n^2 - 5n + 6) + 2\sqrt{\frac{86}{105}}(n^2 - 5n + 6) + 2\sqrt{\frac{2}{5}}(n^2 - 5n + 6) + \frac{1}{7}\sqrt{\frac{83}{2}}(n^2 - 6n + 8) + \frac{8}{23}\sqrt{34}(n^2 - 9n + 20) + 2\sqrt{\frac{113}{79}}(n-2) + \sqrt{\frac{334}{395}}(n-2) + 4\sqrt{\frac{30}{79}}(n-2) + 6\sqrt{\frac{15}{79}}(n-2) + 6\sqrt{\frac{58}{41}}(n-3) + 4\sqrt{\frac{34}{41}}(n-3) + 12\sqrt{\frac{7}{41}}(n-3) + 18\sqrt{\frac{6}{287}}(n-3) + \frac{20(n-4)^2}{\sqrt{253}} + 2\sqrt{\frac{194}{115}}(n-4)^2 + 2\sqrt{\frac{151}{161}}(n-4)^2 + \frac{144(n-4)}{\sqrt{5293}} + \frac{48(n-4)}{\sqrt{85}} + \frac{3}{5}\sqrt{\frac{254}{79}}(n-4) + \frac{54}{41}\sqrt{2}(n-4) + 2\sqrt{\frac{114}{67}}(n-4) + 2\sqrt{\frac{447}{469}}(n-4) + 3\sqrt{\frac{22}{29}}(n-4) + 4\sqrt{\frac{6}{23}}(n-4) + 12\sqrt{\frac{6}{29}}(n-4) + 6\sqrt{\frac{93}{469}}(n-4) + 12\sqrt{\frac{138}{1189}}(n-4) + 30\sqrt{\frac{2}{119}}(n-4) + 18\sqrt{\frac{5}{391}}(n-4) + 28\sqrt{\frac{6}{737}}(n-4) + \frac{3}{17}\sqrt{134}(n-5) + \frac{6}{29}\sqrt{114}(n-5) + \frac{24}{67}\sqrt{33}(n-5) + \frac{21}{25}\sqrt{2}(n-5) + \frac{206\sqrt{194}}{385} + 4\frac{\sqrt{\frac{678}{11}}}{7} + \frac{6\sqrt{41}}{7} + \frac{24}{\sqrt{29}} + 2\sqrt{\frac{110}{19}} + 3\sqrt{\frac{43}{14}} + 3\sqrt{\frac{53}{19}} + 4\sqrt{\frac{138}{77}} + 4\sqrt{\frac{94}{55}} + 4\sqrt{\frac{786}{737}} + 4\sqrt{\frac{78}{79}} + 4\sqrt{\frac{74}{77}} + 2\sqrt{\frac{82}{95}} + 6\sqrt{\frac{190}{287}} + \frac{24\sqrt{\frac{22}{41}}}{7} + 6\sqrt{\frac{65}{133}} + \frac{60\sqrt{\frac{3}{7}}}{7} + 12\sqrt{\frac{10}{41}} + 6\sqrt{\frac{115}{779}} + 12\sqrt{\frac{10}{77}} + 12\sqrt{\frac{34}{287}} + 12\sqrt{\frac{2}{17}} + 12\sqrt{\frac{78}{779}} + 24\sqrt{\frac{5}{91}} + 60\sqrt{\frac{2}{247}} + 36\sqrt{\frac{2}{553}}$$

Now we find $GA_5(\mathcal{G}_1)$ as

$$GA_5(\mathcal{G}_1) = \sum_{ab \in E(\mathcal{G})} \frac{2\sqrt{S(a)S(b)}}{(S(a)+S(b))} = \sum_{j=18}^{86} \sum_{ab \in E_j(\mathcal{G})} \frac{2\sqrt{S(a)S(b)}}{(S(a)+S(b))}.$$

Using the edge partition from Table 2, we get

$$GA_5(\mathcal{G}_1) = \sqrt{\tfrac{65}{1066}}|E_{18}(\mathcal{G}_1)| + \sqrt{\tfrac{5}{91}}|E_{19}(\mathcal{G}_1)| + \tfrac{5}{\sqrt{494}}|E_{20}(\mathcal{G}_1)| + \sqrt{\tfrac{75}{1372}}|E_{21}(\mathcal{G}_1)| + \sqrt{\tfrac{19}{350}}|E_{22}(\mathcal{G}_1)| +$$

$$\sqrt{\tfrac{41}{784}}|E_{23}(\mathcal{G}_1)| + \sqrt{\tfrac{3}{58}}|E_{24}(\mathcal{G}_1)| + \sqrt{\tfrac{23}{462}}|E_{25}(\mathcal{G}_1)| + \sqrt{\tfrac{93}{1876}}|E_{26}(\mathcal{G}_1)| + \sqrt{\tfrac{105}{2212}}|E_{27}(\mathcal{G}_1)| +$$

$$\sqrt{\tfrac{27}{574}}|E_{28}(\mathcal{G}_1)| + \sqrt{\tfrac{47}{990}}|E_{29}(\mathcal{G}_1)| + \sqrt{\tfrac{19}{402}}|E_{30}(\mathcal{G}_1)| + \sqrt{\tfrac{4}{85}}|E_{31}(\mathcal{G}_1)| + \sqrt{\tfrac{97}{2070}}|E_{32}(\mathcal{G}_1)| +$$

$$\sqrt{\tfrac{14}{315}}|E_{33}(\mathcal{G}_1)| + \sqrt{\tfrac{43}{896}}|E_{34}(\mathcal{G}_1)| + \sqrt{\tfrac{11}{232}}|E_{35}(\mathcal{G}_1)| + \sqrt{\tfrac{53}{1216}}|E_{36}(\mathcal{G}_1)| + \sqrt{\tfrac{7}{164}}|E_{37}(\mathcal{G}_1)| +$$

$$\sqrt{\tfrac{97}{2178}}|E_{38}(\mathcal{G}_1)| + \sqrt{\tfrac{98}{2211}}|E_{39}(\mathcal{G}_1)| + \sqrt{\tfrac{9}{204}}|E_{40}(\mathcal{G}_1)| + \sqrt{\tfrac{10}{2277}}|E_{41}(\mathcal{G}_1)| + \sqrt{\tfrac{110}{2607}}|E_{56}(\mathcal{G}_1)| +$$

$$\sqrt{\tfrac{115}{2772}}|E_{43}(\mathcal{G}_1)| + \sqrt{\tfrac{55}{1368}}|E_{44}(\mathcal{G}_1)| + \sqrt{\tfrac{113}{2844}}|E_{45}(\mathcal{G}_1)| + \sqrt{\tfrac{29}{738}}|E_{46}(\mathcal{G}_1)| + \sqrt{\tfrac{59}{1512}}|E_{47}(\mathcal{G}_1)| +$$

$$\sqrt{\tfrac{31}{810}}|E_{48}(\mathcal{G}_1)| + \sqrt{\tfrac{88}{2009}}|E_{49}(\mathcal{G}_1)| + \sqrt{\tfrac{95}{2296}}|E_{50}(\mathcal{G}_1)| + \sqrt{\tfrac{115}{3116}}|E_{51}(\mathcal{G}_1)| + \sqrt{\tfrac{97}{2450}}|E_{52}(\mathcal{G}_1)| +$$

$$\sqrt{\tfrac{113}{3234}}|E_{53}(\mathcal{G}_1)| + \sqrt{\tfrac{126}{3871}}|E_{54}(\mathcal{G}_1)| + \tfrac{\sqrt{98}}{50}|E_{55}(\mathcal{G}_1)| + \sqrt{\tfrac{23}{670}}|E_{56}(\mathcal{G}_1)| + \sqrt{\tfrac{127}{3950}}|E_{57}(\mathcal{G}_1)| +$$

$$\sqrt{\tfrac{7}{203}}|E_{58}(\mathcal{G}_1)| + \sqrt{\tfrac{5}{154}}|E_{59}(\mathcal{G}_1)| + \sqrt{\tfrac{65}{2128}}|E_{60}(\mathcal{G}_1)| + \sqrt{\tfrac{17}{574}}|E_{61}(\mathcal{G}_1)| + \sqrt{\tfrac{57}{1682}}|E_{62}(\mathcal{G}_1)| +$$

$$\sqrt{\tfrac{31}{952}}|E_{63}(\mathcal{G}_1)| + \sqrt{\tfrac{69}{2378}}|E_{64}(\mathcal{G}_1)| + \sqrt{\tfrac{131}{4422}}|E_{65}(\mathcal{G}_1)| + \sqrt{\tfrac{33}{1122}}|E_{66}(\mathcal{G}_1)| + \sqrt{\tfrac{143}{5214}}|E_{67}(\mathcal{G}_1)| +$$

$$\sqrt{\tfrac{37}{1344}}|E_{68}(\mathcal{G}_1)| + \sqrt{\tfrac{132}{4489}}|E_{69}(\mathcal{G}_1)| + \sqrt{\tfrac{134}{4623}}|E_{70}(\mathcal{G}_1)| + \sqrt{\tfrac{144}{5293}}|E_{71}(\mathcal{G}_1)| + \sqrt{\tfrac{149}{5628}}|E_{72}(\mathcal{G}_1)| +$$

$$\sqrt{\tfrac{67}{2312}}|E_{73}(\mathcal{G}_1)| + \sqrt{\tfrac{135}{4692}}|E_{74}(\mathcal{G}_1)| + \sqrt{\tfrac{75}{4856}}|E_{75}(\mathcal{G}_1)| + \sqrt{\tfrac{136}{4761}}|E_{76}(\mathcal{G}_1)| + \sqrt{\tfrac{151}{5796}}|E_{77}(\mathcal{G}_1)| +$$

$$\sqrt{\tfrac{159}{6232}}|E_{78}(\mathcal{G}_1)| + \sqrt{\tfrac{41}{1710}}|E_{79}|(\mathcal{G}_1) + \sqrt{\tfrac{161}{6384}}|E_{80}(\mathcal{G}_1)| + \sqrt{\tfrac{167}{7110}}|E_{81}(\mathcal{G}_1)| + \sqrt{\tfrac{81}{6724}}|E_{82}(\mathcal{G}_1)| +$$

$$\sqrt{\tfrac{17}{738}}|E_{83}(\mathcal{G}_1)| + \sqrt{\tfrac{83}{3528}}|E_{84}(\mathcal{G}_1)| + \sqrt{\tfrac{86}{3785}}|E_{85}(\mathcal{G}_1)| + \sqrt{\tfrac{89}{450}}|E_{86}(\mathcal{G}_1)|.$$

After simplification, we get

$$GA_5(\mathcal{G}_1) = \tfrac{40}{13}\sqrt{14}(n-4)^2 + \tfrac{48}{163}\sqrt{1659}(n-2) + \tfrac{12}{7}\sqrt{10}n(3n^2-15n+19) + \tfrac{24}{29}\sqrt{210}(n^2-5n+6) +$$
$$\tfrac{16}{13}\sqrt{77}(n^2-5n+6) + \tfrac{12}{19}\sqrt{70}(n^2-5n+6) + \tfrac{18}{5}\sqrt{21}(n^2-5n+6) + 9n^3-33n^2-66n + \tfrac{3}{14}\sqrt{2607}(n-2) + \tfrac{36}{169}\sqrt{790}(n-2) + \tfrac{48}{107}\sqrt{553}(n-2) + \tfrac{144}{115}\sqrt{79}(n-2) + \tfrac{24}{55}\sqrt{574}(n-3) + \tfrac{72}{43}\sqrt{205}(n-3) +$$
$$\tfrac{216}{59}\sqrt{82}(n-3) + \tfrac{64}{19}\sqrt{41}(n-3) + \tfrac{6}{17}\sqrt{253}(n-4)^2 + \tfrac{8}{11}\sqrt{230}(n-4)^2 + \tfrac{16}{17}\sqrt{161}(n-4)^2 + \tfrac{12}{73}\sqrt{5293}(n-4) + \tfrac{3}{17}\sqrt{4623}(n-4) + \tfrac{6}{25}\sqrt{2211}(n-4) + \tfrac{24}{97}\sqrt{2010}(n-4) + \tfrac{48}{151}\sqrt{1407}(n-4) + \tfrac{24}{35}\sqrt{1189}(n-4) +$$
$$\tfrac{48}{137}\sqrt{1173}(n-4) + \tfrac{8}{21}\sqrt{986}(n-4) + \tfrac{48}{101}\sqrt{561}(n-4) + \tfrac{48}{49}\sqrt{510}(n-4) + \tfrac{48}{95}\sqrt{469}(n-4) + \tfrac{48}{43}\sqrt{406}(n-4) + \tfrac{24}{19}\sqrt{357}(n-4) + \tfrac{20}{43}\sqrt{158}(n-4) + \tfrac{40}{39}\sqrt{134}(n-4) + \tfrac{32}{15}\sqrt{29}(n-4) + 12(n-4) + 36(n-5) +$$
$$\tfrac{48\sqrt{5214}}{145} + \tfrac{48\sqrt{4422}}{133} + \tfrac{48\sqrt{1558}}{79} + \tfrac{48\sqrt{1122}}{67} + \tfrac{48\sqrt{1066}}{67} + \tfrac{16\sqrt{779}}{39} + \tfrac{96\sqrt{574}}{97} + \tfrac{16\sqrt{494}}{17} + \tfrac{48\sqrt{462}}{47} + \tfrac{32\sqrt{287}}{23} +$$
$$\tfrac{16\sqrt{266}}{11} + \tfrac{96\sqrt{231}}{61} + \tfrac{32\sqrt{203}}{19} + \tfrac{72\sqrt{190}}{83} + \tfrac{48\sqrt{154}}{25} + \tfrac{96\sqrt{91}}{41} + \tfrac{21\sqrt{79}}{16} + \tfrac{336\sqrt{66}}{115} + 3\sqrt{55} + \tfrac{28\sqrt{41}}{15} + \tfrac{32\sqrt{38}}{9} +$$
$$\tfrac{36\sqrt{19}}{7} + \tfrac{192\sqrt{7}}{11} + \tfrac{1336\sqrt{2}}{33} + 258 \quad \square$$

Table 2. Edge partition of Hex-Derived Cage network (*HDCN1*) based on degrees of end vertices of each edge.

(S_u, S_v) where $ab \in E(\mathcal{G}_1)$	Number of Edges	(S_u, S_v) where $ab \in E(\mathcal{G}_1)$	Number of Edges
$E_{18} = (26, 41)$	24	$E_{53} = (49, 66)$	24
$E_{19} = (26, 56)$	24	$E_{54} = (49, 79)$	12
$E_{20} = (26, 76)$	24	$E_{55} = (50, 50)$	$6n - 30$
$E_{21} = (28, 49)$	24	$E_{56} = (50, 67)$	$2(6n - 24)$
$E_{22} = (28, 50)$	$2(6n - 24)$	$E_{57} = (50, 79)$	$6n - 24$
$E_{23} = (28, 56)$	24	$E_{58} = (56, 58)$	24
$E_{24} = (28, 58)$	$4(6n - 24)$	$E_{59} = (56, 66)$	24
$E_{25} = (28, 66)$	24	$E_{60} = (56, 76)$	24
$E_{26} = (28, 67)$	$2(6n - 24)$	$E_{61} = (56, 82)$	24
$E_{27} = (28, 79)$	$2(6n - 12)$	$E_{62} = (58, 58)$	$2(6n - 30)$
$E_{28} = (28, 82)$	$2(6n - 18)$	$E_{63} = (58, 68)$	$2(6n - 24)$
$E_{29} = (30, 66)$	24	$E_{64} = (58, 82)$	$4(6n - 24)$
$E_{30} = (30, 67)$	$2(6n - 24)$	$E_{65} = (66, 67)$	24
$E_{31} = (30, 68)$	$4(6n - 24)$	$E_{66} = (66, 68)$	24
$E_{32} = (30, 69)$	$12n^2 - 96n + 192$	$E_{67} = (66, 79)$	24
$E_{33} = (30, 84)$	$6n^2 - 30n + 36$	$E_{68} = (66, 84)$	24
$E_{34} = (32, 56)$	24	$E_{69} = (67, 67)$	$2(6n - 30)$
$E_{35} = (32, 58)$	$2(6n - 24)$	$E_{70} = (67, 69)$	$2(6n - 24)$
$E_{36} = (32, 76)$	24	$E_{71} = (67, 79)$	$2(6n - 24)$
$E_{37} = (32, 82)$	$4(6n - 18)$	$E_{72} = (67, 84)$	$2(6n - 24)$
$E_{38} = (33, 66)$	24	$E_{73} = (68, 68)$	$2(6n - 30)$
$E_{39} = (33, 67)$	$2(6n - 24)$	$E_{74} = (68, 69)$	$2(6n - 24)$
$E_{40} = (33, 68)$	$2(6n - 24)$	$E_{75} = (68, 84)$	$4(6n - 24)$
$E_{41} = (33, 69)$	$6n^2 - 48n + 96$	$E_{76} = (69, 69)$	$12n^2 - 108n + 240$
$E_{42} = (33, 79)$	$2(6n - 12)$	$E_{77} = (69, 84)$	$12n^2 - 96n + 192$
$E_{43} = (33, 84)$	$12n^2 - 60n + 72$	$E_{78} = (76, 82)$	24
$E_{44} = (36, 76)$	24	$E_{79} = (76, 90)$	12
$E_{45} = (36, 79)$	$2(6n - 12)$	$E_{80} = (79, 84)$	$2(6n - 12)$
$E_{46} = (36, 82)$	$6(6n - 18)$	$E_{81} = (79, 90)$	$6n - 12$
$E_{47} = (36, 84)$	$18n^2 - 90n + 108$	$E_{82} = (82, 82)$	$2(6n - 24)$
$E_{48} = (36, 90)$	$18n^3 - 90n^2 + 114n$	$E_{83} = (82, 90)$	$4(6n - 18)$
$E_{49} = (41, 49)$	12	$E_{84} = (84, 84)$	$6n^2 - 36n + 48$
$E_{50} = (41, 56)$	24	$E_{85} = (84, 90)$	$12n^2 - 60n + 72$
$E_{51} = (41, 76)$	12	$E_{86} = (90, 90)$	$9n^3 - 51n^2 + 72n$
$E_{52} = (49, 50)$	12		

2.2. Results for Hex-Derived Cage Network (HDCN2(m,n))

In this portion, we find some degree-based topological indices for Hex-Derived Cage network $(HDCN2(m, n))$. We calculate the general Randić index $R_\alpha(\mathcal{G})$ with $\alpha = \{1, -1, \frac{1}{2}, -\frac{1}{2}\}$, ABC, GA, ABC_4 and GA_5 in the the below theorems for $(HDCN2(m, n))$.

Theorem 4. *Let* $G_2 \cong HDCN2(m,n)$ *be the Hex-Derived Cage network, then its general Randić index is equal to*

$$
R_\alpha(\mathcal{G}_2) = \begin{cases}
6(486n^3 - 1068n^2 + 312n + 293), & \alpha = 1; \\
6(9(2\sqrt{2}+3)n^3 + (2\sqrt{30}+2\sqrt{15}+3\sqrt{6}+6\sqrt{5}+ \\
\quad 12\sqrt{3}-60\sqrt{2}-81)n^2 + 2(\sqrt{35}-2\sqrt{30}+\sqrt{21}- \\
\quad \sqrt{15}+4\sqrt{10}+3\sqrt{7}+2\sqrt{6}-12\sqrt{5}-20\sqrt{3} \\
\quad +30\sqrt{2}+14)n + 2\sqrt{42}-4\sqrt{35}+4\sqrt{30}-4\sqrt{21}- \\
\quad 16\sqrt{10}-12\sqrt{7}-20\sqrt{6}+24\sqrt{5}+48\sqrt{3}- \\
\quad 12\sqrt{2}+39), & \alpha = \frac{1}{2}; \\
\frac{297675n^3 - 445655n^2 + 282283n - 40155}{529200}, & \alpha = -1; \\
\frac{3}{4}(2\sqrt{2}+3)n^3 + (\frac{4}{\sqrt{5}}+\frac{2}{\sqrt{3}}-5\sqrt{2}+2\sqrt{\frac{6}{5}}+\sqrt{\frac{2}{3}}+ \\
\quad \sqrt{\frac{3}{5}}-\frac{83}{12})n^2 + (\frac{461}{105})+6\sqrt{\frac{2}{5}}+\sqrt{\frac{3}{7}}-\sqrt{\frac{3}{5}}-\sqrt{\frac{2}{3}}- \\
\quad 4\sqrt{\frac{6}{5}}+5\sqrt{2}-\frac{5}{\sqrt{3}}-\frac{16}{\sqrt{5}}+\frac{4}{\sqrt{7}}+\frac{12}{\sqrt{35}})n - \frac{24}{\sqrt{35}}- \\
\quad \frac{8}{\sqrt{7}}+\frac{16}{\sqrt{5}}+\frac{8}{\sqrt{3}}-\sqrt{2}+4\sqrt{\frac{6}{5}}+2\sqrt{\frac{6}{7}}-2\sqrt{\frac{2}{3}}- \\
\quad 2\sqrt{\frac{3}{7}}-12\sqrt{\frac{2}{5}}+\frac{13}{14}, & \alpha = -\frac{1}{2}.
\end{cases}
$$

Proof. Let $\mathcal{G}_2$ be the Hex-Derived Cage network $(HDCN2(m,n))$ where $m = n \geq 5$. The edge set of $HDCN2(m,n)$ is divided into twenty partitions based on the degree of end vertices. Table 3 shows these edge partition of $HDCN2(m,n)$.

$$
R_\alpha(\mathcal{G}_2) = \sum_{ab \in E(\mathcal{G})} (d(a)d(b))^\alpha
$$

For $\alpha = 1$

$$
R_1(\mathcal{G}_2) = \sum_{j=1}^{20} \sum_{ab \in E_j(\mathcal{G})} deg(u) \cdot deg(v)
$$

Using the edge partition from Table 3, we get

$R_1(\mathcal{G}_2) = 25|E_1(\mathcal{G}_2)| + 30|E_2(\mathcal{G}_2)| + 35|E_3(\mathcal{G}_2)| + 40|E_4(\mathcal{G}_2)| + 45|E_5(\mathcal{G}_2)| + 60|E_6(\mathcal{G}_2)| + 36|E_7(\mathcal{G}_2)| + 42|E_8(\mathcal{G}_2)| + 48|E_9(\mathcal{G}_2)| + 54|E_{10}(\mathcal{G}_2)| + 72|E_{11}(\mathcal{G}_2)| + 49|E_{12}(\mathcal{G}_2)| + 63|E_{13}(\mathcal{G}_2)| + 84|E_{14}(\mathcal{G}_2)| + 64|E_{15}(\mathcal{G}_2)| + 72|E_{16}(\mathcal{G}_2)| + 96|E_{17}(\mathcal{G}_2)| + 81|E_{18}(\mathcal{G}_2)| + 108|E_{19}(\mathcal{G}_2)| + 144|E_{20}(\mathcal{G}_2)|$

After simplification, we get

$$
R_1(\mathcal{G}_2) = 6(486n^3 - 1068n^2 + 312n + 293)
$$

For $\alpha = \frac{1}{2}$

$$
R_{\frac{1}{2}}(\mathcal{G}_2) = \sum_{j=1}^{20} \sum_{ab \in E_j(\mathcal{G})} \sqrt{d(a) \cdot d(b)}
$$

Using edge partition from Table 3, we get

$R_{\frac{1}{2}}(\mathcal{G}_2) = 5|E_1(\mathcal{G}_2)| + \sqrt{30}|E_2(\mathcal{G}_2)| + \sqrt{35}|E_3(\mathcal{G}_2)| + 2\sqrt{10}|E_4(\mathcal{G}_2)| + 3\sqrt{5}|E_5(\mathcal{G}_2)| + 2\sqrt{15}|E_6(\mathcal{G}_2)| + 6|E_7(\mathcal{G}_2)| + \sqrt{42}|E_8(\mathcal{G}_2)| + 4\sqrt{3}|E_9(\mathcal{G}_2)| + 3\sqrt{6}|E_{10}(\mathcal{G}_2)| + 6\sqrt{2}|E_{11}(\mathcal{G}_2)| + 7|E_{12}(\mathcal{G}_2)| + 3\sqrt{7}|E_{13}(\mathcal{G}_2)| + 2\sqrt{21}|E_{14}(\mathcal{G}_2)| + 8|E_{15}(\mathcal{G}_2)| + 6\sqrt{2}|E_{16}(\mathcal{G}_2)| + 4\sqrt{6}|E_{17}(\mathcal{G}_2)| + 9|E_{18}(\mathcal{G}_2)| + 6\sqrt{3}|E_{19}(\mathcal{G}_2)| + 12|E_{20}(\mathcal{G}_2)|$

After simplification, we get

$R_{\frac{1}{2}}(\mathcal{G}_2) = 6(9(2\sqrt{2}+3)n^3 + (2\sqrt{30}+2\sqrt{15}+3\sqrt{6}+6\sqrt{5}+12\sqrt{3}-60\sqrt{2}-81)n^2 + 2(\sqrt{35}-2\sqrt{30}+\sqrt{21}-\sqrt{15}+4\sqrt{10}+3\sqrt{7}+2\sqrt{6}-12\sqrt{5}-20\sqrt{3}+30\sqrt{2}+14)n + 2\sqrt{42}-4\sqrt{35}+4\sqrt{30}-4\sqrt{21}-16\sqrt{10}-12\sqrt{7}-20\sqrt{6}+24\sqrt{5}+48\sqrt{3}-12\sqrt{2}+39)$

For $\alpha = -1$

$$R_{-1}(\mathcal{G}_2) = \sum_{j=1}^{20} \sum_{ab \in E_j(\mathcal{G})} \frac{1}{d(a) \cdot d(b)}$$

$R_{-1}(\mathcal{G}_2) = \frac{1}{25}|E_1(\mathcal{G}_2)| + \frac{1}{30}|E_2(\mathcal{G}_2)| + \frac{1}{35}|E_3(\mathcal{G}_2)| + \frac{1}{40}|E_4(\mathcal{G}_2)| + \frac{1}{45}|E_5(\mathcal{G}_2)| + \frac{1}{60}|E_6(\mathcal{G}_2)| + \frac{1}{36}|E_7(\mathcal{G}_2)| + \frac{1}{42}|E_8(\mathcal{G}_2)| + \frac{1}{48}|E_9(\mathcal{G}_2)| + \frac{1}{54}|E_{10}(\mathcal{G}_2)| + \frac{1}{72}|E_{11}(\mathcal{G}_2)| + \frac{1}{49}|E_{12}(\mathcal{G}_2)| + \frac{1}{63}|E_{13}(\mathcal{G}_2)| + \frac{1}{84}|E_{14}(\mathcal{G}_2)| + \frac{1}{64}|E_{15}(\mathcal{G}_2)| + \frac{1}{72}|E_{16}(\mathcal{G}_2)| + \frac{1}{96}|E_{17}(\mathcal{G}_2)| + \frac{1}{81}|E_{18}(\mathcal{G}_2)| + \frac{1}{108}|E_{19}(\mathcal{G}_2)| + \frac{1}{144}|E_{20}(\mathcal{G}_2)|$

After simplification, we get

$$R_{-1}(\mathcal{G}_2) = \frac{297675n^3 - 445655n^2 + 282283n - 40155}{529200}$$

For $\alpha = -\frac{1}{2}$

$$R_{-\frac{1}{2}}(\mathcal{G}_2) = \sum_{j=1}^{20} \sum_{ab \in E_j(\mathcal{G})} \frac{1}{\sqrt{d(a) \cdot d(b)}}$$

$R_{-\frac{1}{2}}(\mathcal{G}_2) = \frac{1}{5}|E_1(\mathcal{G}_2)| + \frac{1}{\sqrt{30}}|E_2(\mathcal{G}_2)| + \frac{1}{\sqrt{35}}|E_3(\mathcal{G}_2)| + \frac{1}{2\sqrt{10}}|E_4(\mathcal{G}_2)| + \frac{1}{3\sqrt{5}}|E_5(\mathcal{G}_2)| + \frac{1}{2\sqrt{15}}|E_6(\mathcal{G}_2)| + \frac{1}{6}|E_7(\mathcal{G}_2)| + \frac{1}{\sqrt{42}}|E_8(\mathcal{G}_2)| + \frac{1}{4\sqrt{3}}|E_9(\mathcal{G}_2)| + \frac{1}{3\sqrt{6}}|E_{10}(\mathcal{G}_2)| + \frac{1}{6\sqrt{2}}|E_{11}(\mathcal{G}_2)| + \frac{1}{7}|E_{12}(\mathcal{G}_2)| + \frac{1}{3\sqrt{7}}|E_{13}(\mathcal{G}_2)| + \frac{1}{2\sqrt{21}}|E_{14}(\mathcal{G}_2)| + \frac{1}{8}|E_{15}(\mathcal{G}_2)| + \frac{1}{6\sqrt{2}}|E_{16}(\mathcal{G}_2)| + \frac{1}{\sqrt{17}}|E_{17}(\mathcal{G}_2)| + \frac{1}{9}|E_{18}(\mathcal{G}_2)| + \frac{1}{6\sqrt{3}}|E_{19}(\mathcal{G}_2)| + \frac{1}{12}|E_{20}(\mathcal{G}_2)|$

After simplification, we get

$R_{-\frac{1}{2}}(\mathcal{G}_2) = \frac{3}{4}(2\sqrt{2}+3)n^3 + (\frac{4}{\sqrt{5}} + \frac{2}{\sqrt{3}} - 5\sqrt{2} + 2\sqrt{\frac{6}{5}} + \sqrt{\frac{2}{3}} + \sqrt{\frac{3}{5}} - \frac{83}{12})n^2 + (\frac{461}{105}) + 6\sqrt{\frac{2}{5}} + \sqrt{\frac{3}{5}} - \sqrt{\frac{3}{5}} - \sqrt{\frac{2}{3}} - 4\sqrt{\frac{6}{5}} + 5\sqrt{2} - \frac{5}{\sqrt{3}} - \frac{16}{\sqrt{5}} + \frac{4}{\sqrt{7}} + \frac{12}{\sqrt{35}})n - \frac{24}{\sqrt{35}} - \frac{8}{\sqrt{7}} + \frac{16}{\sqrt{5}} + \frac{8}{\sqrt{3}} - \sqrt{2} + 4\sqrt{\frac{6}{5}} + 2\sqrt{\frac{6}{7}} - 2\sqrt{\frac{2}{3}} - 2\sqrt{\frac{3}{7}} - 12\sqrt{\frac{2}{5}} + \frac{13}{14}$ □

In this theorem, we find the first Zagreb index for hex-derived cage network $\mathcal{G}_2$.

Theorem 5. *For Hex-Derived Cage Network* $(\mathcal{G}_2)$, *the first Zagreb index is equal to*

$$M_1(\mathcal{G}_2) = 12(54n^3 - 109n^2 + 34n + 21)$$

Proof. Let $\mathcal{G}_2$ be the Hex-Derived Cage Network $(\mathcal{G}_2)$. Using the edge partition from Table 3, we have

$$M_1(\mathcal{G}_2) = \sum_{ab \in E(\mathcal{G})} (d(a) + d(b)) = \sum_{j=1}^{20} \sum_{ab \in E_j(\mathcal{G})} (d(a) + d(b))$$

$M_1(\mathcal{G}_2) = 10|E_1(\mathcal{G}_2)| + 11|E_2(\mathcal{G}_2)| + 12|E_3(\mathcal{G}_2)| + 13|E_4(\mathcal{G}_2)| + 14|E_5(\mathcal{G}_2)| + 17|E_6(\mathcal{G}_2)| + 12|E_7(\mathcal{G}_2)| + 13|E_8(\mathcal{G}_2)| + 14|E_9(\mathcal{G}_2)| + 15|E_{10}(\mathcal{G}_2)| + 18|E_{11}(\mathcal{G}_2)| + 14|E_{12}(\mathcal{G}_2)| + 16|E_{13}(\mathcal{G}_2)| + 19|E_{14}(\mathcal{G}_2)| + 16|E_{15}(\mathcal{G}_2)| + 17|E_{16}(\mathcal{G}_2)| + 20|E_{17}(\mathcal{G}_2)| + 18|E_{18}(\mathcal{G}_2)| + 21|E_{19}(\mathcal{G}_2)| + 24|E_{20}(\mathcal{G}_2)|.$

After simplification, we get

$$M_1(\mathcal{G}_2) = 12(54n^3 - 109n^2 + 34n + 21)$$

□

In below theorem, we calculate the ABC, GA, ABC_4 and GA_5 indices of Hex-Derived Cage Network $\mathcal{G}_2$.

Theorem 6. *Let* $\mathcal{G}_2$ *be the Hex-Derived Cage Network for every positive integer* $m = n \geq 5$; *then we have*

- $ABC(\mathcal{G}_2) = 2\sqrt{2}(3n^3 - 10n^2 + 8n + 2) + \frac{1}{2}\sqrt{\frac{11}{2}}n(3n^2 - 11n + 10) + \sqrt{\frac{5}{2}}n(3n^2 - 11n + 10) + 6\sqrt{\frac{6}{5}}(n^2 - 2n + 2) + \frac{16}{3}(n^2 - 5n + 6) + 3n(n-1) + \frac{12\sqrt{2n}}{5} + 6n + \sqrt{\frac{26}{3}}(n-2)^2 + 2\sqrt{\frac{19}{3}}(n-2)^2 + 8\sqrt{\frac{3}{5}}(n-2)^2 + \sqrt{30}(n-2) + \sqrt{\frac{51}{7}}(n-2) + 6\sqrt{\frac{22}{5}}(n-2) + 6\sqrt{3}(n-2) + 4\sqrt{2}(n-2) + 12\sqrt{\frac{2}{7}}(n-2) + 3\sqrt{\frac{7}{2}}(n-3) + \frac{12}{7}\sqrt{3}(n-3) + 2\sqrt{\frac{66}{7}}$

- $GA(\mathcal{G}_2) = 4\sqrt{2}(3n^3 - 10n^2 + 8n + 2) + \frac{24}{11}\sqrt{30}(n^2 - 2n + 2) + 18n^3 - 54n^2 + \frac{24}{17}\sqrt{15}(n-1)n + \frac{48\sqrt{3}n}{7} + 12n + \frac{12}{5}\sqrt{6}(n-2)^2 + \frac{36}{7}\sqrt{5}(n-2)^2 + \frac{48}{7\sqrt{3}}(n-2)^2 + 2\sqrt{35}(n-2) + \frac{24}{19}\sqrt{21}(n-2) + \frac{96}{13}\sqrt{10}(n-2) + \frac{9}{2}\sqrt{7}(n-2) + \frac{48}{13}\sqrt{6}(n-2) + \frac{144}{17}\sqrt{2}(n-2) + 12(n-3) + \frac{24\sqrt{42}}{13} + 54.$

- $ABC_4(\mathcal{G}_2) = \frac{1}{18}\sqrt{\frac{107}{2}}n(3n^2 - 17n + 24) + \frac{1}{9}\sqrt{\frac{53}{2}}(3n^3 - 13n^2 + 16n - 10) + \frac{4}{9}\sqrt{5}n(3n^2 - 15n + 19) + \frac{2}{7}\sqrt{\frac{534}{7}}(n^2 - 5n + 6) + 3\sqrt{\frac{102}{101}}(n^2 - 5n + 6) + 2\sqrt{\frac{69}{101}}(n^2 - 5n + 6) + \frac{24}{7}\sqrt{\frac{37}{101}}(n^2 - 5n + 6) + 3\sqrt{\frac{94}{707}}(n^2 - 5n + 6) + \frac{60}{101}\sqrt{2}(n^2 - 6n + 8) + \frac{15}{19}\sqrt{6}(n^2 - 9n + 20) + 7\sqrt{\frac{2}{95}}(3n - 8) + \frac{1}{7}\sqrt{\frac{202}{13}}(n-2)^2 + \frac{4}{7}\sqrt{\frac{258}{13}}(n-2) + \frac{4}{13}\sqrt{19}(n-2) + \frac{5}{3}\sqrt{2}(n-2) + \frac{12}{7}\sqrt{\frac{142}{95}}(n-2) + \sqrt{\frac{67}{95}}(n-2) + 12\sqrt{\frac{194}{9595}}(n-2) + 24\sqrt{\frac{11}{1235}}(n-2) + \sqrt{\frac{129}{5}}(n-3) + 2\sqrt{\frac{302}{33}}(n-3) + \frac{4}{3}\sqrt{\frac{205}{33}}(n-3) + \sqrt{\frac{274}{55}}(n-3) + 2\sqrt{\frac{145}{33}}(n-3) + \frac{3}{7}\sqrt{\frac{123}{19}}(n-4)^2 + 2\sqrt{\frac{174}{133}}(n-4)^2 + 30\sqrt{\frac{7}{1919}}(n-4)^2 + \frac{4}{35}\sqrt{366}(n-4) + \frac{2}{5}\sqrt{\frac{447}{19}}(n-4) + \frac{6}{5}\sqrt{\frac{206}{13}}(n-4) + \frac{2}{3}\sqrt{14}(n-4) + 4\sqrt{\frac{46}{35}}(n-4) + 2\sqrt{\frac{42}{37}}(n-4) + \frac{12}{5}\sqrt{\frac{46}{65}}(n-4) + \frac{24}{5}\sqrt{\frac{58}{101}}(n-4) + 3\sqrt{\frac{37}{65}}(n-4) + 6\sqrt{\frac{2}{13}}(n-4) + 6\sqrt{\frac{38}{259}}(n-4) + 6\sqrt{\frac{2}{19}}(n-4) + 6\sqrt{\frac{334}{3515}}(n-4) + 6\sqrt{\frac{346}{3737}}(n-4) + \frac{66}{7}\sqrt{\frac{2}{37}}(n-4) + 72\sqrt{\frac{2}{715}}(n-4) + \frac{56}{33}(n-4) + \frac{6}{37}\sqrt{146}(n-5) + \frac{8}{25}\sqrt{37}(n-5) + \frac{96}{65}\sqrt{2}(n-5) + \frac{32}{\sqrt{37}} + 2\sqrt{\frac{249}{37}} + \frac{8\sqrt{6}}{5} + 2\sqrt{\frac{202}{35}} + 2\sqrt{\frac{109}{21}} + \frac{8\sqrt{\frac{106}{35}}}{3} + \sqrt{\frac{66}{23}} + \frac{183\sqrt{2}}{37} + 4\sqrt{\frac{678}{511}} + 2\sqrt{\frac{70}{53}} + \frac{48\sqrt{\frac{30}{73}}}{7} + 8\sqrt{\frac{14}{37}} + 8\sqrt{\frac{330}{949}} + 8\sqrt{\frac{134}{511}} + 8\sqrt{\frac{6}{23}} + 12\sqrt{\frac{127}{851}} + 6\sqrt{\frac{3}{23}} + \frac{32\sqrt{\frac{10}{77}}}{3} + 12\sqrt{\frac{290}{2701}} + 12\sqrt{\frac{17}{161}} + 12\sqrt{\frac{21}{253}} + 24\sqrt{\frac{30}{689}} + 24\sqrt{\frac{2}{65}} + 12\sqrt{\frac{146}{5035}} + 24\sqrt{\frac{166}{6935}} + 16\sqrt{\frac{6}{265}} + 48\sqrt{\frac{31}{3869}} + 48\sqrt{\frac{43}{7373}}.$

- $GA_5(\mathcal{G}_2) = 3n(3n^2 - 17n + 24) + 4\sqrt{2}n(3n^2 - 15n + 19) + \frac{12}{143}\sqrt{4242}(n^2 - 5n + 6) + \frac{28}{25}\sqrt{101}(n^2 - 5n + 6) + \frac{24}{13}\sqrt{42}(n^2 - 5n + 6) + \frac{2424}{209}(n^2 - 5n + 6) + \frac{36}{149}\sqrt{570}(3n - 8) + 9n^3 - 21n^2 - 90n + \frac{252}{103}\sqrt{6}(n-2)^2 + \frac{6}{49}\sqrt{9595}(n-2) + \frac{12}{67}\sqrt{3705}(n-2) + \frac{72}{203}\sqrt{285}(n-2) + \frac{7}{6}\sqrt{95}(n-2) + \frac{21}{11}\sqrt{39}(n-2) + \frac{144}{17}\sqrt{2}(n-2) + \frac{144}{139}\sqrt{110}(n-3) + \frac{72}{17}\sqrt{66}(n-3) + \frac{9120\sqrt{33}(n-3)}{1127} + \frac{48}{11}\sqrt{30}(n-3) + \frac{16}{59}\sqrt{1919}(n-4)^2 + \frac{24}{59}\sqrt{798}(n-4)^2 + \frac{168}{125}\sqrt{19}(n-4)^2 + \frac{24}{175}\sqrt{7474}(n-4) + \frac{24}{169}\sqrt{7030}(n-4) + \frac{24}{113}\sqrt{2886}(n-4) + \frac{8}{25}\sqrt{1406}(n-4) + \frac{12}{29}\sqrt{777}(n-4) + \frac{36}{41}\sqrt{715}(n-4) + \frac{15}{11}\sqrt{303}(n-4) + \frac{1350}{791}\sqrt{195}(n-4) + \frac{9}{8}\sqrt{111}(n-4) + \frac{56}{41}\sqrt{74}(n-4) + \frac{240}{151}\sqrt{57}(n-4) + \frac{1496}{217}\sqrt{26}(n-4) + \frac{80}{13}\sqrt{14}(n-4) + \frac{210}{31}\sqrt{3}(n-4) + 12(n-4) + 36(n-5) + \frac{8\sqrt{7373}}{29} + \frac{2\sqrt{6935}}{7} + \frac{16\sqrt{5402}}{49} + \frac{6\sqrt{5035}}{37} + \frac{8\sqrt{3869}}{21} + \frac{48\sqrt{3066}}{115} + \frac{3\sqrt{2847}}{7} + \frac{12\sqrt{2067}}{23} + \frac{32\sqrt{851}}{43} + \frac{18\sqrt{511}}{17} + \frac{9\sqrt{455}}{25} + \frac{72\sqrt{318}}{107} + \frac{36\sqrt{265}}{49} + \frac{36\sqrt{259}}{25} + \frac{288\sqrt{253}}{191} + \frac{60\sqrt{219}}{37} + \frac{72\sqrt{185}}{41} + \frac{288\sqrt{161}}{155} + \frac{144\sqrt{138}}{73} + \frac{144\sqrt{115}}{137} + \frac{192\sqrt{111}}{85} + \frac{8\sqrt{77}}{3} + \frac{168\sqrt{73}}{61} + \frac{288\sqrt{70}}{103} + \frac{732\sqrt{69}}{175} + 4\sqrt{35} + \frac{192\sqrt{21}}{37} + 258.$

Proof. Using the edge partition from Table 3, we find ABC as

$$ABC(\mathcal{G}_2) = \sum_{ab \in E(\mathcal{G})} \sqrt{\frac{d(a) + d(b) - 2}{d(a) \cdot d(b)}} = \sum_{j=1}^{20} \sum_{ab \in E_j(\mathcal{G})} \sqrt{\frac{d(a) + d(b) - 2}{d(a) \cdot d(b)}}$$

$ABC(\mathcal{G}_2) = \frac{2\sqrt{2}}{5}|E_1(\mathcal{G}_2)| + \sqrt{\frac{3}{10}}|E_2(\mathcal{G}_2)| + \sqrt{\frac{2}{7}}|E_3(\mathcal{G}_2)| + \frac{\sqrt{11}}{2\sqrt{10}}|E_4(\mathcal{G}_2)| + \frac{2\sqrt{3}}{3\sqrt{5}}|E_5(\mathcal{G}_2)| + \frac{1}{2}|E_6(\mathcal{G}_2)| + \sqrt{\frac{5}{18}}|E_7(\mathcal{G}_2)| + \sqrt{\frac{11}{42}}|E_8(\mathcal{G}_2)| + \frac{1}{2}|E_9(\mathcal{G}_2)| + \sqrt{\frac{13}{54}}|E_{10}(\mathcal{G}_2)| + \frac{\sqrt{2}}{3}|E_{11}(\mathcal{G}_2)| + \frac{2\sqrt{3}}{7}|E_{12}(\mathcal{G}_2)| +$

$\sqrt{\frac{14}{63}}|E_{13}(\mathcal{G}_2)| + \sqrt{\frac{17}{84}}|E_{14}(\mathcal{G}_2)| + \sqrt{\frac{7}{32}}|E_{15}(\mathcal{G}_2)| + \sqrt{\frac{5}{24}}|E_{16}(\mathcal{G}_2)| + \sqrt{\frac{3}{16}}|E_{17}(\mathcal{G}_2)| + \sqrt{\frac{16}{81}}|E_{18}(\mathcal{G}_2)| + \sqrt{\frac{19}{108}}|E_{19}(\mathcal{G}_2)| + \sqrt{\frac{11}{72}}|E_{20}(\mathcal{G}_2)|.$

After simplification, we get

$ABC(\mathcal{G}_2) = 2\sqrt{2}(3n^3 - 10n^2 + 8n + 2) + \frac{1}{2}\sqrt{\frac{11}{2}}n(3n^2 - 11n + 10) + \sqrt{\frac{5}{2}}n(3n^2 - 11n + 10) + 6\sqrt{\frac{6}{5}}(n^2 - 2n + 2) + \frac{16}{3}(n^2 - 5n + 6) + 3n(n-1) + \frac{12\sqrt{2}n}{5} + 6n + \sqrt{\frac{26}{3}}(n-2)^2 + 2\sqrt{\frac{19}{3}}(n-2)^2 + 8\sqrt{\frac{3}{5}}(n-2)^2 + \sqrt{30}(n-2) + \sqrt{\frac{51}{7}}(n-2) + 6\sqrt{\frac{22}{5}}(n-2) + 6\sqrt{3}(n-2) + 4\sqrt{2}(n-2) + 12\sqrt{\frac{2}{7}}(n-2) + 3\sqrt{\frac{7}{2}}(n-3) + \frac{12}{7}\sqrt{3}(n-3) + 2\sqrt{\frac{66}{7}}$

Using the edge partition from Table 3, we find GA as

$GA(\mathcal{G}_2) = 1|E_1(\mathcal{G}_2)| + \frac{2\sqrt{30}}{11}|E_2(\mathcal{G}_2)| + \frac{\sqrt{35}}{6}|E_3(\mathcal{G}_2)| + \frac{4\sqrt{10}}{13}|E_4(\mathcal{G}_2)| + \frac{3\sqrt{5}}{7}|E_5(\mathcal{G}_2)| + \frac{4\sqrt{15}}{17}|E_6(\mathcal{G}_2)| + 1|E_7(\mathcal{G}_2)| + \frac{2\sqrt{42}}{13}|E_8(\mathcal{G}_2)| + \frac{2\sqrt{12}}{7}|E_9(\mathcal{G}_2)| + \frac{2\sqrt{54}}{15}|E_{10}(\mathcal{G}_2)| + \frac{2\sqrt{2}}{3}|E_{11}(\mathcal{G}_2)| + 1|E_{12}(\mathcal{G}_2)| + \frac{3\sqrt{7}}{8}|E_{13}(\mathcal{G}_2)| + \frac{4\sqrt{21}}{19}|E_{14}(\mathcal{G}_2)| + 1|E_{15}(\mathcal{G}_2)| + \frac{12\sqrt{2}}{17}|E_{16}(\mathcal{G}_2)| + \frac{\sqrt{23}}{5}|E_{17}(\mathcal{G}_2)| + 1|E_{18}(\mathcal{G}_2)| + \frac{2\sqrt{3}}{7}|E_{19}(\mathcal{G}_2)| + 1|E_{20}(\mathcal{G}_2)|.$

After simplification, we get

$GA(\mathcal{G}_2) = 4\sqrt{2}(3n^3 - 10n^2 + 8n + 2) + \frac{24}{11}\sqrt{30}(n^2 - 2n + 2) + 18n^3 - 54n^2 + \frac{24}{17}\sqrt{15}(n-1)n + \frac{48\sqrt{3}n}{7} + 12n + \frac{12}{5}\sqrt{6}(n-2)^2 + \frac{36}{7}\sqrt{5}(n-2)^2 + \frac{48}{7\sqrt{3}}(n-2)^2 + 2\sqrt{35}(n-2) + \frac{24}{19}\sqrt{21}(n-2) + \frac{96}{13}\sqrt{10}(n-2) + \frac{9}{2}\sqrt{7}(n-2) + \frac{48}{5}\sqrt{6}(n-2) + \frac{144}{17}\sqrt{2}(n-2) + 12(n-3) + \frac{24\sqrt{42}}{13} + 54.$

If we suppose an edge partition based on degree sum of neighbors of end vertices, then the edge set $E(\mathcal{G}_2)$ can be divided into seventy six edge partition $E_j(\mathcal{G}_2)$, $21 \le j \le 96$. Table 4 shows these edge partitions.

From Equation (7), we get

$$ABC_4(\mathcal{G}_2) = \sum_{ab \in E(\mathcal{G})} \sqrt{\frac{S(a) + S(b) - 2}{S(a)S(b)}} = \sum_{j=21}^{96} \sum_{ab \in E_j(\mathcal{G})} \sqrt{\frac{S(a) + S(b) - 2}{S(a)S(b)}}.$$

Using the edge partition from Table 3, we get

$ABC_4(\mathcal{G}_2) = \sqrt{\frac{72}{1369}}|E_{21}(\mathcal{G}_2)| + \frac{4}{\sqrt{333}}|E_{22}(\mathcal{G}_2)| + \sqrt{\frac{83}{1776}}|E_{23}(\mathcal{G}_2)| + \sqrt{\frac{98}{2321}}|E_{24}(\mathcal{G}_2)| + \sqrt{\frac{127}{3404}}|E_{25}(\mathcal{G}_2)| +$

$\sqrt{\frac{74}{1443}}|E_{26}(\mathcal{G}_2)| + \sqrt{\frac{86}{1911}}|E_{27}(\mathcal{G}_2)| + \sqrt{\frac{90}{2067}}|E_{28}(\mathcal{G}_2)| + \sqrt{\frac{42}{2106}}|E_{29}(\mathcal{G}_2)| + \sqrt{\frac{110}{2847}}|E_{30}(\mathcal{G}_2)| +$

$\sqrt{\frac{111}{2886}}|E_{31}(\mathcal{G}_2)| + \sqrt{\frac{44}{1235}}|E_{32}(\mathcal{G}_2)| + \sqrt{\frac{43}{940}}|E_{33}(\mathcal{G}_2)| + \sqrt{\frac{101}{2520}}|E_{34}(\mathcal{G}_2)| + \sqrt{\frac{103}{2600}}|E_{35}(\mathcal{G}_2)| +$

$\sqrt{\frac{137}{3960}}|E_{36}(\mathcal{G}_2)| + \sqrt{\frac{89}{2058}}|E_{37}(\mathcal{G}_2)| + \sqrt{\frac{113}{3066}}|E_{38}(\mathcal{G}_2)| + \sqrt{\frac{17}{1554}}|E_{39}(\mathcal{G}_2)| + \sqrt{\frac{115}{3150}}|E_{40}(\mathcal{G}_2)| +$

$\sqrt{\frac{29}{798}}|E_{41}(\mathcal{G}_2)| + \sqrt{\frac{141}{4242}}|E_{42}(\mathcal{G}_2)| + \sqrt{\frac{96}{2385}}|E_{43}(\mathcal{G}_2)| + \sqrt{\frac{106}{2835}}|E_{44}(\mathcal{G}_2)| + \sqrt{\frac{27}{558}}|E_{45}(\mathcal{G}_2)| +$

$\frac{5}{\sqrt{648}}|E_{46}(\mathcal{G}_2)| + \sqrt{\frac{109}{3024}}|E_{47}(\mathcal{G}_2)| + \sqrt{\frac{111}{3120}}|E_{48}(\mathcal{G}_2)| + \sqrt{\frac{69}{228}}|E_{49}(\mathcal{G}_2)| + \sqrt{\frac{145}{4752}}|E_{50}(\mathcal{G}_2)| +$

$\sqrt{\frac{101}{2646}}|E_{51}(\mathcal{G}_2)| + \sqrt{\frac{122}{3577}}|E_{52}(\mathcal{G}_2)| + \frac{11}{\sqrt{3626}}|E_{53}(\mathcal{G}_2)| + \sqrt{\frac{122}{3675}}|E_{54}(\mathcal{G}_2)| + \sqrt{\frac{123}{3724}}|E_{55}(\mathcal{G}_2)| +$

$\sqrt{\frac{142}{4655}}|E_{56}(\mathcal{G}_2)| + \sqrt{\frac{148}{4949}}|E_{57}(\mathcal{G}_2)| + \sqrt{\frac{105}{2862}}|E_{58}(\mathcal{G}_2)| + \sqrt{\frac{124}{3869}}|E_{59}(\mathcal{G}_2)| + \sqrt{\frac{146}{5035}}|E_{60}(\mathcal{G}_2)| +$

$\sqrt{\frac{53}{1458}}|E_{61}(\mathcal{G}_2)| + \sqrt{\frac{63}{1998}}|E_{62}(\mathcal{G}_2)| + \sqrt{\frac{105}{621}}|E_{63}(\mathcal{G}_2)| + \sqrt{\frac{147}{5130}}|E_{64}(\mathcal{G}_2)| + \sqrt{\frac{151}{5346}}|E_{65}(\mathcal{G}_2)| +$

$\sqrt{\frac{153}{5454}}|E_{66}(\mathcal{G}_2)| + \sqrt{\frac{20}{729}}|E_{67}(\mathcal{G}_2)| + \sqrt{\frac{126}{4095}}|E_{68}(\mathcal{G}_2)| + \sqrt{\frac{134}{4599}}|E_{69}(\mathcal{G}_2)| + \sqrt{\frac{153}{5796}}|E_{70}(\mathcal{G}_2)| +$

$\sqrt{\frac{160}{6237}}|E_{71}(\mathcal{G}_2)| + \sqrt{\frac{128}{4225}}|E_{72}(\mathcal{G}_2)| + \sqrt{\frac{46}{1625}}|E_{73}(\mathcal{G}_2)| + \sqrt{\frac{54}{2145}}|E_{74}(\mathcal{G}_2)| + \sqrt{\frac{145}{5402}}|E_{75}(\mathcal{G}_2)| +$

$\sqrt{\frac{116}{5475}}|E_{76}(\mathcal{G}_2)| + \sqrt{\frac{166}{6935}}|E_{77}(\mathcal{G}_2)| + \sqrt{\frac{172}{7373}}|E_{78}(\mathcal{G}_2)| + \sqrt{\frac{73}{2738}}|E_{79}(\mathcal{G}_2)| + \sqrt{\frac{37}{1406}}|E_{80}(\mathcal{G}_2)| +$

$\sqrt{\frac{167}{7030}}|E_{81}(\mathcal{G}_2)| + \sqrt{\frac{173}{7474}}|E_{82}(\mathcal{G}_2)| + \sqrt{\frac{148}{5625}}|E_{83}(\mathcal{G}_2)| + \sqrt{\frac{149}{5700}}|E_{84}(\mathcal{G}_2)| + \sqrt{\frac{58}{2525}}|E_{85}(\mathcal{G}_2)| +$

$\sqrt{\frac{75}{2888}}|E_{86}(\mathcal{G}_2)| + \sqrt{\frac{175}{7676}}|E_{87}(\mathcal{G}_2)| + \sqrt{\frac{189}{9108}}|E_{88}(\mathcal{G}_2)| + \sqrt{\frac{99}{4948}}|E_{89}(\mathcal{G}_2)| + \sqrt{\frac{194}{9595}}|E_{90}(\mathcal{G}_2)| +$

$\sqrt{\frac{201}{10260}}|E_{91}(\mathcal{G}_2)| + \frac{14}{\sqrt{9801}}|E_{92}(\mathcal{G}_2)| + \sqrt{\frac{205}{10692}}|E_{93}(\mathcal{G}_2)| + \sqrt{\frac{200}{10201}}|E_{94}(\mathcal{G}_2)| + \sqrt{\frac{207}{10908}}|E_{95}(\mathcal{G}_2)| +$
$\sqrt{\frac{3}{162}}|E_{96}(\mathcal{G}_2)|.$

After simplification, we have

$ABC_4(\mathcal{G}_2) = \frac{1}{18}\sqrt{\frac{107}{2}}n(3n^2 - 17n + 24) + \frac{1}{9}\sqrt{\frac{53}{2}}(3n^3 - 13n^2 + 16n - 10) + \frac{4}{9}\sqrt{5}n(3n^2 - 15n +$
$19) + \frac{2}{7}\sqrt{\frac{534}{7}}(n^2 - 5n + 6) + 3\sqrt{\frac{102}{101}}(n^2 - 5n + 6) + 2\sqrt{\frac{69}{101}}(n^2 - 5n + 6) + \frac{24}{7}\sqrt{\frac{37}{101}}(n^2 - 5n + 6) +$
$3\sqrt{\frac{94}{707}}(n^2 - 5n + 6) + \frac{60}{101}\sqrt{2}(n^2 - 6n + 8) + \frac{15}{19}\sqrt{6}(n^2 - 9n + 20) + 7\sqrt{\frac{2}{95}}(3n - 8) + \frac{1}{7}\sqrt{\frac{202}{3}}(n - 2)^2 +$
$\frac{4}{7}\sqrt{\frac{258}{13}}(n - 2) + \frac{4}{13}\sqrt{19}(n - 2) + \frac{5}{3}\sqrt{2}(n - 2) + \frac{12}{7}\sqrt{\frac{142}{95}}(n - 2) + \sqrt{\frac{67}{95}}(n - 2) + 12\sqrt{\frac{194}{9595}}(n - 2) +$
$24\sqrt{\frac{11}{1235}}(n - 2) + \sqrt{\frac{129}{5}}(n - 3) + 2\sqrt{\frac{302}{33}}(n - 3) + \frac{4}{3}\sqrt{\frac{205}{33}}(n - 3) + \sqrt{\frac{274}{55}}(n - 3) + 2\sqrt{\frac{145}{33}}(n - 3) +$
$\frac{3}{7}\sqrt{\frac{123}{19}}(n - 4)^2 + 2\sqrt{\frac{174}{133}}(n - 4)^2 + 30\sqrt{\frac{7}{1919}}(n - 4)^2 + \frac{4}{35}\sqrt{366}(n - 4) + \frac{2}{5}\sqrt{\frac{447}{19}}(n - 4) + \frac{6}{5}\sqrt{\frac{206}{13}}(n -$
$4) + \frac{2}{3}\sqrt{14}(n - 4) + 4\sqrt{\frac{46}{35}}(n - 4) + 2\sqrt{\frac{42}{37}}(n - 4) + \frac{12}{5}\sqrt{\frac{46}{65}}(n - 4) + \frac{24}{5}\sqrt{\frac{58}{101}}(n - 4) + 3\sqrt{\frac{37}{65}}(n -$
$4) + 6\sqrt{\frac{2}{13}}(n - 4) + 6\sqrt{\frac{38}{259}}(n - 4) + 6\sqrt{\frac{2}{19}}(n - 4) + 6\sqrt{\frac{334}{3515}}(n - 4) + 6\sqrt{\frac{346}{3737}}(n - 4) + \frac{66}{7}\sqrt{\frac{2}{37}}(n -$
$4) + 72\sqrt{\frac{2}{715}}(n - 4) + \frac{56}{33}(n - 4) + \frac{6}{37}\sqrt{146}(n - 5) + \frac{8}{25}\sqrt{37}(n - 5) + \frac{96}{65}\sqrt{2}(n - 5) + \frac{32}{\sqrt{37}} + 2\sqrt{\frac{249}{37}} +$
$\frac{8\sqrt{6}}{5} + 2\sqrt{\frac{202}{35}} + 2\sqrt{\frac{109}{21}} + \frac{8\sqrt{\frac{106}{35}}}{3} + \sqrt{\frac{66}{23}} + \frac{183\sqrt{2}}{37} + 4\sqrt{\frac{678}{511}} + 2\sqrt{\frac{70}{53}} + \frac{48\sqrt{\frac{30}{73}}}{7} + 8\sqrt{\frac{14}{37}} + 8\sqrt{\frac{330}{949}} + 8\sqrt{\frac{134}{511}} +$
$8\sqrt{\frac{6}{23}} + 12\sqrt{\frac{127}{851}} + 6\sqrt{\frac{3}{23}} + \frac{32\sqrt{\frac{10}{77}}}{3} + 12\sqrt{\frac{290}{2701}} + 12\sqrt{\frac{17}{161}} + 12\sqrt{\frac{21}{253}} + 24\sqrt{\frac{30}{689}} + 24\sqrt{\frac{2}{65}} + 12\sqrt{\frac{146}{5035}} +$
$24\sqrt{\frac{166}{6935}} + 16\sqrt{\frac{6}{265}} + 48\sqrt{\frac{31}{3869}} + 48\sqrt{\frac{43}{7373}}.$

From Equation (8), we get

$$GA_5(\mathcal{G}_2) = \sum_{ab \in E(\mathcal{G})} \frac{2\sqrt{S(a)S(b)}}{(S(a) + S(b))} = \sum_{j=21}^{96} \sum_{ab \in E_j(\mathcal{G})} \frac{2\sqrt{S(a)S(b)}}{(S(a)S(b))}.$$

Table 3. Edge partition of Hex-Derived Cage network ($HDCN2$) based on degrees of end vertices of each edge.

(d_u, d_v) where $ab \in E(\mathcal{G}_2)$	Number of Edges	(d_u, d_v) where $ab \in E(\mathcal{G}_2)$	Number of Edges
$E_1 = (5,5)$	$6n$	$E_{11} = (6,12)$	$18n^3 - 60n^2 + 48n + 12$
$E_2 = (5,6)$	$12n^2 - 24n + 24$	$E_{12} = (7,7)$	$6n - 18$
$E_3 = (5,7)$	$2(6n - 12)$	$E_{13} = (7,9)$	$2(6n - 12)$
$E_4 = (5,8)$	$4(6n - 12)$	$E_{14} = (7,12)$	$6n - 12$
$E_5 = (5,9)$	$12n^2 - 48n + 48$	$E_{15} = (8,8)$	$2(6n - 18)$
$E_6 = (5,12)$	$6n^2 - 6n$	$E_{16} = (8,9)$	$2(6n - 12)$
$E_7 = (6,6)$	$9n^3 - 33n^2 + 30n$	$E_{17} = (8,12)$	$4(6n - 12)$
$E_8 = (6,7)$	12	$E_{18} = (9,9)$	$12n^2 - 60n + 72$
$E_9 = (6,8)$	$12n$	$E_{19} = (9,12)$	$12n^2 - 48n + 48$
$E_{10} = (6,9)$	$6n^2 - 24n + 24$	$E_{20} = (12,12)$	$9n^3 - 33n^2 + 30n$

Using the edge partition from Table 4, we get

$GA_5(\mathcal{G}_2) = 1|E_{21}(\mathcal{G}_2)| + \frac{\sqrt{1665}}{41}|E_{22}(\mathcal{G}_2)| + \frac{2\sqrt{1776}}{85}|E_{23}(\mathcal{G}_2)| + \frac{\sqrt{2331}}{50}|E_{24}(\mathcal{G}_2)| + \frac{2\sqrt{3404}}{129}|E_{25}(\mathcal{G}_2)| +$
$\frac{\sqrt{1443}}{38}|E_{26}(\mathcal{G}_2)| + \frac{\sqrt{1911}}{44}|E_{27}(\mathcal{G}_2)| + \frac{\sqrt{2067}}{46}|E_{28}(\mathcal{G}_2)| + \frac{2\sqrt{2106}}{93}|E_{29}(\mathcal{G}_2)| + \frac{\sqrt{2847}}{56}|E_{30}(\mathcal{G}_2)| +$
$\frac{2\sqrt{2886}}{113}|E_{31}(\mathcal{G}_2)| + \frac{\sqrt{3705}}{67}|E_{32}(\mathcal{G}_2)| + \frac{\sqrt{1920}}{44}|E_{33}(\mathcal{G}_2)| + \frac{2\sqrt{2520}}{103}|E_{34}(\mathcal{G}_2)| + \frac{2\sqrt{2600}}{105}|E_{35}(\mathcal{G}_2)| +$
$\frac{2\sqrt{3960}}{139}|E_{36}(\mathcal{G}_2)| + \frac{2\sqrt{2058}}{91}|E_{37}(\mathcal{G}_2)| + \frac{2\sqrt{3066}}{115}|E_{38}(\mathcal{G}_2)| + \frac{\sqrt{3108}}{58}|E_{39}(\mathcal{G}_2)| + \frac{2\sqrt{3150}}{117}|E_{40}(\mathcal{G}_2)| +$
$\frac{\sqrt{3192}}{59}|E_{41}(\mathcal{G}_2)| + \frac{2\sqrt{4242}}{143}|E_{42}(\mathcal{G}_2)| + \frac{2\sqrt{2385}}{49}|E_{43}(\mathcal{G}_2)| + \frac{\sqrt{2835}}{54}|E_{44}(\mathcal{G}_2)| + \frac{2\sqrt{4140}}{137}|E_{45}(\mathcal{G}_2)| +$
$\frac{\sqrt{2592}}{51}|E_{46}(\mathcal{G}_2)| + \frac{2\sqrt{3024}}{111}|E_{47}(\mathcal{G}_2)| + \frac{2\sqrt{3120}}{113}|E_{48}(\mathcal{G}_2)| + \frac{\sqrt{4416}}{70}|E_{49}(\mathcal{G}_2)| + \frac{2\sqrt{4752}}{147}|E_{50}(\mathcal{G}_2)| +$
$\frac{2\sqrt{2646}}{103}|E_{51}(\mathcal{G}_2)| + \frac{\sqrt{3577}}{61}|E_{52}(\mathcal{G}_2)| + \frac{2\sqrt{3626}}{123}|E_{53}(\mathcal{G}_2)| + \frac{\sqrt{3675}}{62}|E_{54}(\mathcal{G}_2)| + \frac{2\sqrt{3724}}{125}|E_{55}(\mathcal{G}_2)| +$
$\frac{\sqrt{4655}}{72}|E_{56}(\mathcal{G}_2)| + \frac{\sqrt{4949}}{75}|E_{57}(\mathcal{G}_2)| + \frac{2\sqrt{2862}}{107}|E_{58}(\mathcal{G}_2)| + \frac{\sqrt{3869}}{63}|E_{59}(\mathcal{G}_2)| + \frac{\sqrt{5035}}{74}|E_{60}(\mathcal{G}_2)| + 1|E_{61}(\mathcal{G}_2)| +$

$\frac{\sqrt{3996}}{64}|E_{62}(\mathcal{G}_2)| + \frac{\sqrt{4968}}{73}|E_{63}(\mathcal{G}_2)| + \frac{2\sqrt{5130}}{149}|E_{64}(\mathcal{G}_2)| + \frac{2\sqrt{5346}}{153}|E_{65}(\mathcal{G}_2)| + \frac{2\sqrt{5454}}{155}|E_{66}(\mathcal{G}_2)| +$
$\frac{\sqrt{5832}}{81}|E_{67}(\mathcal{G}_2)| + \frac{\sqrt{4095}}{64}|E_{68}(\mathcal{G}_2)| + \frac{\sqrt{4599}}{68}|E_{69}(\mathcal{G}_2)| + \frac{2\sqrt{5796}}{155}|E_{70}(\mathcal{G}_2)| + \frac{\sqrt{6237}}{81}|E_{71}(\mathcal{G}_2)| +$
$1|E_{72}(\mathcal{G}_2)| + \frac{\sqrt{4875}}{70}|E_{73}(\mathcal{G}_2)| + \frac{\sqrt{6435}}{82}|E_{74}(\mathcal{G}_2)| + \frac{2\sqrt{5402}}{147}|E_{75}(\mathcal{G}_2)| + \frac{\sqrt{5475}}{74}|E_{76}(\mathcal{G}_2)| + \frac{\sqrt{6935}}{84}|E_{77}(\mathcal{G}_2)| +$
$\frac{\sqrt{7373}}{87}|E_{78}(\mathcal{G}_2)| + 1|E_{79}(\mathcal{G}_2)| + \frac{\sqrt{5624}}{75}|E_{80}(\mathcal{G}_2)| + \frac{2\sqrt{7030}}{169}|E_{81}(\mathcal{G}_2)| + \frac{2\sqrt{7474}}{175}|E_{82}(\mathcal{G}_2)| + 1|E_{83}(\mathcal{G}_2)| +$
$\frac{2\sqrt{5700}}{151}|E_{84}(\mathcal{G}_2)| + \frac{\sqrt{7575}}{88}|E_{85}(\mathcal{G}_2)| + 1|E_{86}(\mathcal{G}_2)| + \frac{2\sqrt{7676}}{177}|E_{87}(\mathcal{G}_2)| + \frac{2\sqrt{9108}}{191}|E_{88}(\mathcal{G}_2)| + \frac{\sqrt{9936}}{100}|E_{89}(\mathcal{G}_2)| +$
$\frac{\sqrt{9595}}{98}|E_{90}(\mathcal{G}_2)| + \frac{2\sqrt{10260}}{203}|E_{91}(\mathcal{G}_2)| + 1|E_{92}(\mathcal{G}_2)| + \frac{2\sqrt{10692}}{207}|E_{93}(\mathcal{G}_2)| + 1|E_{94}(\mathcal{G}_2)| + \frac{2\sqrt{10908}}{209}|E_{95}(\mathcal{G}_2)| +$
$1|E_{96}(\mathcal{G}_2)|.$

After simplification, we get

$GA_5(\mathcal{G}_2) = 3n(3n^2 - 17n + 24) + 4\sqrt{2}n(3n^2 - 15n + 19) + \frac{12}{143}\sqrt{4242}(n^2 - 5n + 6) + \frac{28}{25}\sqrt{101}(n^2 - 5n + 6) + \frac{24}{13}\sqrt{42}(n^2 - 5n + 6) + \frac{2424}{209}(n^2 - 5n + 6) + \frac{36}{149}\sqrt{570}(3n - 8) + 9n^3 - 21n^2 - 90n + \frac{252}{103}\sqrt{6}(n - 2)^2 + \frac{6}{49}\sqrt{9595}(n - 2) + \frac{12}{67}\sqrt{3705}(n - 2) + \frac{72}{203}\sqrt{285}(n - 2) + \frac{7}{6}\sqrt{95}(n - 2) + \frac{21}{11}\sqrt{39}(n - 2) + \frac{144}{17}\sqrt{2}(n - 2) + \frac{144}{139}\sqrt{110}(n - 3) + \frac{72}{17}\sqrt{66}(n - 3) + \frac{9120\sqrt{33}(n-3)}{1127} + \frac{48}{11}\sqrt{30}(n - 3) + \frac{16}{59}\sqrt{1919}(n - 4)^2 + \frac{24}{59}\sqrt{798}(n - 4)^2 + \frac{168}{125}\sqrt{19}(n - 4)^2 + \frac{24}{175}\sqrt{7474}(n - 4) + \frac{24}{169}\sqrt{7030}(n - 4) + \frac{24}{113}\sqrt{2886}(n - 4) + \frac{8}{25}\sqrt{1406}(n - 4) + \frac{12}{29}\sqrt{777}(n - 4) + \frac{36}{41}\sqrt{715}(n - 4) + \frac{15}{11}\sqrt{303}(n - 4) + \frac{1350}{791}\sqrt{195}(n - 4) + \frac{9}{8}\sqrt{111}(n - 4) + \frac{56}{41}\sqrt{74}(n - 4) + \frac{240}{151}\sqrt{57}(n - 4) + \frac{1496}{217}\sqrt{26}(n - 4) + \frac{80}{13}\sqrt{14}(n - 4) + \frac{210}{31}\sqrt{3}(n - 4) + 12(n - 4) + 36(n - 5) + \frac{8\sqrt{7373}}{29} + \frac{2\sqrt{6935}}{49} + \frac{16\sqrt{5402}}{37} + \frac{6\sqrt{5035}}{21} + \frac{8\sqrt{3869}}{7} + \frac{48\sqrt{3066}}{115} + \frac{3\sqrt{2847}}{7} + \frac{12\sqrt{2067}}{23} + \frac{32\sqrt{851}}{43} + \frac{18\sqrt{511}}{17} + \frac{9\sqrt{455}}{8} + \frac{72\sqrt{318}}{107} + \frac{36\sqrt{265}}{49} + \frac{36\sqrt{259}}{25} + \frac{288\sqrt{253}}{191} + \frac{60\sqrt{219}}{37} + \frac{72\sqrt{185}}{41} + \frac{288\sqrt{161}}{155} + \frac{144\sqrt{138}}{73} + \frac{144\sqrt{115}}{137} + \frac{192\sqrt{111}}{85} + \frac{8\sqrt{77}}{3} + \frac{168\sqrt{73}}{61} + \frac{288\sqrt{70}}{103} + \frac{732\sqrt{69}}{175} + 4\sqrt{35} + \frac{192\sqrt{21}}{37} + 258. \quad \square$

The Comparison graphs for ABC, GA, ABC_4 and GA_5 in case of a Hex Derived Cage networks $HDCN1(m, n)$ and $HDCN2(m, n)$ of dimension m and n are shown in Figures 3 and 4 respectively.

Table 4. Edge partition of Hex-Derived Cage network ($HDCN2$) based on sum of degrees of end vertices of each edge.

(S_u, S_v) where $ab \in E(\mathcal{G}_2)$	Number of Edges	(S_u, S_v) where $ab \in E(\mathcal{G}_2)$	Number of Edges
$E_{21} = (37, 37)$	12	$E_{59} = (53, 73)$	24
$E_{22} = (37, 45)$	24	$E_{60} = (53, 95)$	12
$E_{23} = (37, 48)$	24	$E_{61} = (54, 54)$	$9n^3 - 39n^2 + 48n - 30$
$E_{24} = (37, 63)$	24	$E_{62} = (54, 74)$	$2(6n - 24)$
$E_{25} = (37, 92)$	24	$E_{63} = (54, 92)$	24
$E_{26} = (37, 39)$	$6n - 12$	$E_{64} = (54, 95)$	$3(6n - 16)$
$E_{27} = (39, 49)$	$2(6n - 12)$	$E_{65} = (54, 99)$	$6(6n - 18)$
$E_{28} = (39, 53)$	24	$E_{66} = (54, 101)$	$18n^2 - 90n + 108$
$E_{29} = (39, 54)$	$2(6n - 24)$	$E_{67} = (54, 108)$	$18n^3 - 90n^2 + 114n$
$E_{30} = (39, 73)$	24	$E_{68} = (63, 65)$	24
$E_{31} = (39, 74)$	$2(6n - 24)$	$E_{69} = (63, 73)$	24
$E_{32} = (39, 95)$	$2(6n - 12)$	$E_{70} = (63, 92)$	24
$E_{33} = (40, 48)$	$4(6n - 18)$	$E_{71} = (63, 99)$	24
$E_{34} = (40, 63)$	24	$E_{72} = (65, 65)$	$2(6n - 30)$
$E_{35} = (40, 65)$	$4(6n - 24)$	$E_{73} = (65, 75)$	$2(6n - 24)$
$E_{36} = (40, 99)$	$2(6n - 18)$	$E_{74} = (65, 99)$	$4(6n - 24)$
$E_{37} = (42, 49)$	$12n^2 - 60n + 72$	$E_{75} = (73, 74)$	24
$E_{38} = (42, 73)$	24	$E_{76} = (73, 75)$	24
$E_{39} = (42, 74)$	$2(6n - 24)$	$E_{77} = (73, 95)$	24
$E_{40} = (42, 75)$	$4(6n - 24)$	$E_{78} = (73, 101)$	24
$E_{41} = (42, 76)$	$12n^2 - 96n + 192$	$E_{79} = (74, 74)$	$2(6n - 30)$
$E_{42} = (42, 101)$	$6n^2 - 30n + 36$	$E_{80} = (74, 76)$	$2(6n - 24)$
$E_{43} = (45, 53)$	12	$E_{81} = (74, 95)$	$2(6n - 24)$
$E_{44} = (45, 63)$	24	$E_{82} = (74, 101)$	$2(6n - 24)$
$E_{45} = (45, 92)$	12	$E_{83} = (75, 75)$	$2(6n - 30)$

Table 4. *Cont.*

(S_u, S_v) where $ab \in E(\mathcal{G}_2)$	Number of Edges	(S_u, S_v) where $ab \in E(\mathcal{G}_2)$	Number of Edges
$E_{46} = (48, 54)$	$2(6n - 12)$	$E_{84} = (75, 76)$	$2(6n - 24)$
$E_{47} = (48, 63)$	24	$E_{85} = (75, 101)$	$4(6n - 24)$
$E_{48} = (48, 65)$	$2(6n - 24)$	$E_{86} = (76, 76)$	$12n^2 - 108n + 240$
$E_{49} = (48, 92)$	24	$E_{87} = (76, 101)$	$12n^2 - 96n + 192$
$E_{50} = (48, 99)$	$4(6n - 18)$	$E_{88} = (92, 99)$	24
$E_{51} = (49, 54)$	$6n^2 - 24n + 24$	$E_{89} = (92, 108)$	12
$E_{52} = (49, 73)$	24	$E_{90} = (95, 101)$	$2(6n - 12)$
$E_{53} = (49, 74)$	$2(6n - 24)$	$E_{91} = (95, 108)$	$6n - 12$
$E_{54} = (49, 75)$	$2(6n - 24)$	$E_{92} = (99, 99)$	$2(6n - 24)$
$E_{55} = (49, 76)$	$6n^2 - 48n + 96$	$E_{93} = (99, 108)$	$4(6n - 18)$
$E_{56} = (49, 95)$	$2(6n - 12)$	$E_{94} = (101, 101)$	$6n^2 - 36n + 48$
$E_{57} = (49, 101)$	$12n^2 - 60n + 72$	$E_{95} = (101, 108)$	$12n^2 - 60n + 72$
$E_{58} = (53, 54)$	12	$E_{96} = (108, 108)$	$9n^3 - 51n^2 + 72n$

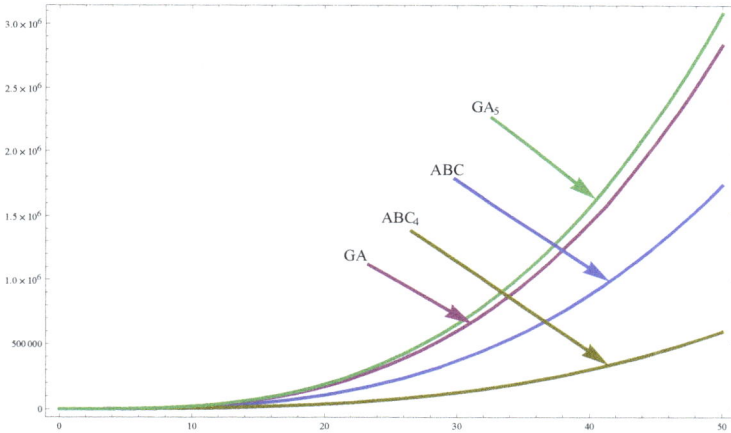

Figure 3. Comparison of ABC, GA, ABC_4 and GA_5 for $HDCN1(m, n)$.

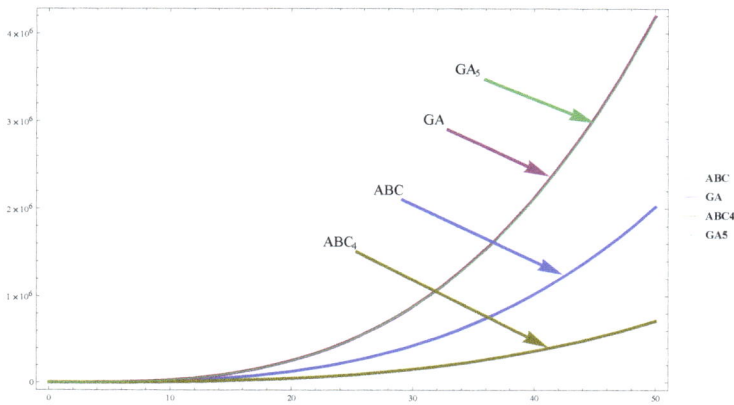

Figure 4. Comparison of ABC, GA, ABC_4 and GA_5 for $HDCN2(m, n)$.

3. Conclusions

In this paper, certain degree-based topological indices, namely the general Randić index, atomic-bond connectivity index (ABC), geometric-arithmetic index (GA) and first Zagreb index were studied for the first time and analytical closed formulas for $HDCN1(m,n)$ and $HDCN2(m,n)$ cage networks were determined which will help the people working in network science to understand and explore the underlying topologies of these networks.

For the future, we are interested in designing some new architectures/networks and then study their topological indices which will be quite helpful to understand their underlying topologies.

Author Contributions: Data curation, U.M.; Funding acquisition, J.-B.L.; Methodology, J.-B.L.; Software, U.M.; Supervision, M.K.S.; Writing—original draft, H.A.

Funding: The work was partially supported by China Postdoctoral Science Foundation under grant No. 2017M621579 and Postdoctoral Science Foundation of Jiangsu Province under grant No. 1701081B, Project of Anhui Jianzhu University under Grant no. 2016QD116 and 2017dc03, Anhui Province Key Laboratory of Intelligent Building & Building Energy Saving.

Acknowledgments: The authors would like to thank all the respected reviewers for their suggestions and useful comments, which resulted in an improved version of this paper.

Conflicts of Interest: The authors declare no conflict of interest.

References

1. Wiener, H. Structural determination of paraffin boiling points. *J. Am. Chem. Soc.* **1947**, *69*, 17–20. [CrossRef] [PubMed]
2. Deza, M.; Fowler, P.W.; Rassat, A.; Rogers, K.M. Fullerenes as tiling of surfaces. *J. Chem. Inf. Comput. Sci.* **2000**, *40*, 550–558. [CrossRef] [PubMed]
3. Diudea, M.V.; Gutman, I.; Lorentz, J. *Molecular Topology*; Huntington: Columbus, OH, USA, 2001.
4. Gutman, I.; Polansky, O.E. *Mathematical Concepts in Organic Chemistry*; Springer: New York, NY, USA, 1986.
5. Randić, M. On Characterization of molecular branching. *J. Am. Chem. Soc.* **1975**, *97*, 6609–6615. [CrossRef]
6. Estrada, E.; Torres, L.; Rodríguez, L.; Gutman, I. An atom-bond connectivity index: Modelling the enthalpy of formation of alkanes. *Indian J. Chem.* **1998**, *37A*, 849–855.
7. Vukičević, D.; Furtula, B. Topological index based on the ratios of geometrical and arithmetical means of end-vertex degrees of edges. *J. Math. Chem.* **2009**, *46*, 1369–1376. [CrossRef]
8. Ghorbani, M.; Hosseinzadeh, M.A. Computing ABC_4 index of nanostar dendrimers. *Optoelectron. Adv. Mater. Rapid Commun.* **2010**, *4*, 1419–1422.
9. Graovac, A.; Ghorbani, M.; Hosseinzadeh, M.A. Computing fifth geometric-arithmetic index for nanostar dendrimers. *J. Math. Nanosci.* **2011**, *1*, 33–42.
10. Hayat, S.; Imran, M. Computation of topological indices of certain graphs. *Appl. Math. Comput.* **2014**, *240*, 213–228.
11. Hussain, Z.; Munir, M.; Rafique, S.; Min Kang, S. Topological Characterizations and Index-Analysis of New Degree-Based Descriptors of Honeycomb Networks. *Symmetry* **2018**, *10*, 478. [CrossRef]
12. Amic, D.; Beslo, D.; Lucic, B.; Nikolic, S.; Trinajstić, N. The vertex-connectivity index revisited. *J. Chem. Inf. Comput. Sci.* **1998**, *38*, 819–822. [CrossRef]
13. Bača, M.; Horváthová, J.; Mokrišová, M.; Suhányiovă, A. On topological indices of fullerenes. *Appl. Math. Comput.* **2015**, *251*, 154–161. [CrossRef]
14. Baig, A.Q.; Imran, M.; Ali, H. Computing Omega, Sadhana and PI polynomials of benzoid carbon nanotubes, *Optoelectron. Adv. Mater. Rapid Commun.* **2015**, *9*, 248–255.
15. Baig, A.Q.; Imran, M.; Ali, H. On Topological Indices of Poly Oxide, Poly Silicate, DOX and DSL Networks. *Can. J. Chem.* **2015**, *93*, 730–739. [CrossRef]
16. Caporossi, G.; Gutman, I.; Hansen, P.; Pavlovíc, L. Graphs with maximum connectivity index. *Comput. Biol. Chem.* **2003**, *27*, 85–90. [CrossRef]
17. Imran, M.; Baig, A.Q.; Ali, H. On topological properties of dominating David derived graphs. *Can. J. Chem.* **2016**, *94*, 137–148. [CrossRef]

18. Imran, M.; Baig, A.Q.; Ali, H. On molecular topological properties of hex-derived graphs. *J. Chemom.* **2016**, *30*, 121–129. [CrossRef]
19. Imran, M.; Baig, A.Q.; Ali, H.; Rehman, S.U. On topological properties of poly honeycomb graphs. *Period. Math. Hung.* **2016**, *73*, 100–119. [CrossRef]
20. Iranmanesh, A.; Zeraatkar, M. Computing GA index for some nanotubes. *Optoelectron. Adv. Mater. Rapid Commun.* **2010**, *4*, 1852–1855.
21. Lin, W.; Chen, J.; Chen, Q.; Gao, T.; Lin, X.; Cai, B. Fast computer search for trees with minimal ABC index based on tree degree sequences. *MATCH Commun. Math. Comput. Chem.* **2014**, *72*, 699–708.
22. Manuel, P.D.; Abd-El-Barr, M.I.; Rajasingh, I.; Rajan, B. An efficient representation of Benes networks and its applications. *J. Discret. Algorithms* **2008**, *6*, 11–19. [CrossRef]
23. Palacios, J.L. A resistive upper bound for the ABC index. *MATCH Commun. Math. Comput. Chem.* **2014**, *72*, 709–713.

symmetry

MDPI

Article

Binary Locating-Dominating Sets in Rotationally-Symmetric Convex Polytopes

Hassan Raza [1], Sakander Hayat [2],*[ID] and Xiang-Feng Pan [1]

[1] School of Mathematical Sciences, Anhui University, Hefei 230601, China; hassan_raza783@yahoo.com (H.R.); xfpan@ahu.edu.cn (X.-F.P.)
[2] Faculty of Engineering Sciences, GIK Institute of Engineering Sciences and Technology, Topi, Swabi 23460, Pakistan
* Correspondence: sakander1566@gmail.com; Tel.: +92-342-4431402

Received: 16 November 2018; Accepted: 4 December 2018; Published: 6 December 2018

Abstract: A convex polytope or simply polytope is the convex hull of a finite set of points in Euclidean space $\mathbb{R}^d$. Graphs of convex polytopes emerge from geometric structures of convex polytopes by preserving the adjacency-incidence relation between vertices. In this paper, we study the problem of binary locating-dominating number for the graphs of convex polytopes which are symmetric rotationally. We provide an integer linear programming (ILP) formulation for the binary locating-dominating problem of graphs. We have determined the exact values of the binary locating-dominating number for two infinite families of convex polytopes. The exact values of the binary locating-dominating number are obtained for two rotationally-symmetric convex polytopes families. Moreover, certain upper bounds are determined for other three infinite families of convex polytopes. By using the ILP formulation, we show tightness in the obtained upper bounds.

Keywords: dominating set; binary locating-domination number; rotationally-symmetric convex polytopes; ILP models

MSC: 05C69; 05C90

1. Introduction

Graphs considered in this paper are all simple, finite and undirected.

We consider a graph $G = (V, E)$ having no isolated vertices. For any vertex $x \in V$, the set $N_G(x) = \{y \in V | (x, y) \in E\}$ is called the *open neighborhood* of x. Moreover, $N_G[x] = N_G(x) \cup \{x\}$ is called the *closed neighborhood* of x. Cardinality of the open neighborhood of a vertex is called its *degree/valency*. Whenever it is cleared from the context, we omit G from the notations $V(G)$, $E(G)$, $N_G(v)$, $N_G[v]$ and $d_G(v)$. A subset $D \subseteq V$ is said to be a *dominating set* of G, if for any $x \in V \setminus D$, we have $N[x] \cap S \neq \emptyset$. The minimum cardinality of a dominating set in G is called its *domination number* denote by $\gamma(G)$. The book by Haynes et al. [1] covers all the literature on domination related parameters of graphs until 1980.

An alternative approach to study a dominating set is a binary assignment of 1 (resp. 0) to a vertex if it belongs (resp. does not belong) to D. In this terminology, D is called dominating set if the sum of weights of closed neighborhoods of any vertex in G is at least one. In other words, any vertex $x \in V$ satisfies $|D \cap N[x]| \geq 1$. For a dominating set S, if additionally every pair of distinct vertices $x, y \in V \setminus S$ satisfies $N(x) \cap S \neq N(y) \cap S$, then S is called a *binary locating-dominating set*. In a similar fashion, the minimum cardinality of a binary-locating set is called the *binary locating-dominating number* of G usually denoted by $\gamma_{l-d}(G)$. It is important to notice that the concept of locating-dominating number in the literature is similar to the binary locating-dominating number. Locating-domination related parameters have been studied relatively more than the other varieties of dominations.

Haynes et al. [2] have studied the problems of locating-dominating number and total dominating numbers for trees. Charon et al. [3] studied the minimum cardinalities of r-locating-dominating and r-identifying codes for cycles and chains. Moreover, they characterized the extremal values for these parameters. For more details on this study, we refer the reader to [4,5]. The concepts of fault-tolerant locating-dominating and open neighborhood locating-dominating sets in trees have been studied by Seo et al. [6,7] and Salter [8]. For more on locating-dominating sets and related parameters, we suggest the reader to [5,9–11].

Note that computational complexity of the binary locating-dominating and the identifying code problems is NP-hard—see, for example, [12,13]. For a positive integer k and a graph G, Charon et al. [12] showed that the problem of finding an r-locating-dominating code and r-identifying code is NP-complete, where r is a positive integer. We refer the interested readers to [14] by Lobstein where a comprehensive list of references on identifying codes and binary locating-dominating sets is provided.

The following result by Slater [11] gives us a tight lower bound for the binary locating-dominating number for regular graphs.

Theorem 1. [11] *Let G be a k-regular graph on n vertices. Then,*

$$\gamma_{l-d}(G) \geq \left\lceil \frac{2n}{k+3} \right\rceil.$$

A graph of a convex polytope is formed from its vertices and edges having the same incidence relation. Graphs of convex polytopes were first considered by Bača in [15,16]. He studied graceful and anti-graceful labeling problems for these geometrically important graphs. Imran et al. [17–19] studied the problem of minimum metric dimension for different infinite families of convex polytopes. Malik et al. [20] also computed the metric dimension of two infinite families of convex polytopes. Kratica et al. [21] considered minimal double resolving sets and the strong metric dimension problem for some families of convex polytopes. Samlan et al. [22] considered three optimization problems, known as the local metric, the fault-tolerant metric and the strong metric dimension problem, for two infinite families of convex polytopes. Simić et al. [23] studied the problem of binary locating-dominating number of some convex polytopes. The ILP model presented in the next section was essentially given by Simić et al. [23]. Other graph-theoretic parameters having potential applications in chemistry are studied in [24–27].

2. An Integer Linear Programming Model

In this section, we present an integer linear programming (ILP) model of minimum binary-locating domination problem. This model will be used to show tightness in upper bounds for different families of graphs which are studied in the next sections.

Bange et al. [28] provided an ILP formulation of minimum identifying code problem. For an identifying set S, the decision variables v_i are defined as:

$$v_i = \begin{cases} 1, & i \in S; \\ 0, & i \notin S. \end{cases}$$

Then, the ILP formulation by Bange et al. [28] for minimum identifying code problem is as follows:

$$\min \sum_{i \in V} v_i, \tag{1}$$

subject to the following constraints

$$\sum_{j \in N[i]} v_i \geq 1, \qquad i \in V, \tag{2}$$

$$\sum_{j \in N[i]\nabla N[k]} v_i \geq 1, \qquad i,k \in V, \ i \neq k, \tag{3}$$

$$v_i \in \{0,1\}, \qquad i \in V. \tag{4}$$

In the above formulation, the minimal cardinality for the identifying code set is ensured by the objective function (1). Dominating set S is defined by constraints (2), constraints (3) represent identifying feature, whereas constraints (4) provide the binary nature of decision variables v_i.

Next, we modify this formulation for the binary-locating domination problem. We achieve this goal by changing constrains (3) into the following constraints:

$$v_i + v_k + \sum_{j \in N[i]\nabla N[k]} v_i \geq 1, \qquad i,k \in V, \ i \neq k. \tag{5}$$

Note that constraints (3) and (5) are the same when vertices i and k are not adjacent, e.g., $N[i]\nabla N[k] = \{i,j\} \cup (N(i)\nabla N(k))$. We can only see the change between constraints (3) and (5), when i and k are adjacent, i.e., $i \in N(k)$. Then, by constraints (5), at least one of vertices i, k or some $j \in N(i)\nabla N(k)$ must be in S. When i and k are not neighbors, then $N[i]\nabla N[k] = \{i,j\} \cup (N(i)\nabla N(k))$, so constraints (3) and (5) are equal.

Sweigart et al. [29] showed that, for any two vertices u and v if $d(u,v) \geq 3$, then both u and v have no common neighbors. This implies that we do not need to check the set $N(u) \cap S \neq N(v) \cap S$ for equivalence, since it permits us to reduce the number of constraints that the locating requirements generate. Therefore, this becomes computationally important for large graphs. By employing this idea, we improve constraints (5) as follows:

$$v_i + v_k + \sum_{j \in N(i)\nabla N(k)} v_i \geq 1, \qquad i,k \in V, \ i \neq k, \ d(i,k) \leq 2. \tag{6}$$

Note that, by using the proposed formulation comprising a reduced number of constraints, we can find exact optimal values for problems with small dimensions. Furthermore, in order to obtain suboptimal solutions for large dimensions, ILP formulation can be optimized by efficient metaheuristic approaches (see, for example, [30]).

3. The Exact Values

In this section, we find the exact values of the binary locating-dominating number of two infinite families of convex polytopes.

3.1. The Graph of Convex Polytope H_n

3.1.1. Construction

In 1999, Bača [31] studied the labeling problem of a family of convex polytopes denoted by $\mathbb{B}_n$ ($n \geq 3$). Figure 1 depicts the graph of convex polytope $\mathbb{B}_n$. Imran and Siddiqui [32] studied a variation of $\mathbb{B}_n$ by generalizing it to the family of two parametric convex polytope denoted by $\mathbb{Q}_n^m$, see [32], Figure 1. Note that the $\mathbb{B}_n$ is a special case of $\mathbb{Q}_n^m$ with $m = 2$.

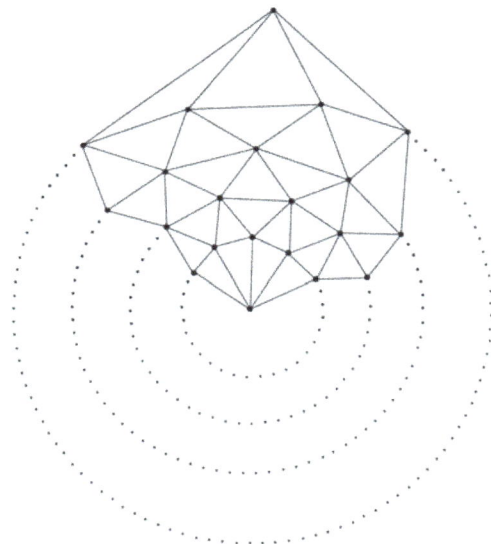

Figure 1. The graph of convex polytope $\mathbb{B}_n$.

For a given planar graph G, the dual of G denoted by $du(G)$ is obtained by adding a vertex in each internal face of G and then joining any two of them if their corresponding faces share an edge. Miller et al. [33] considered another variation of $\mathbb{B}_n$ by defining its dual. They denoted this new family of polytopes with R_n. Figure 2 shows the graph of R_n.

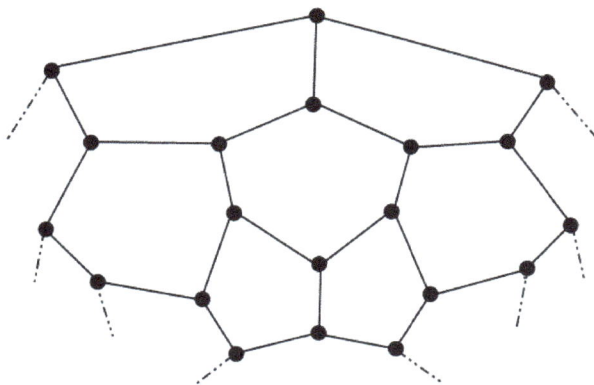

Figure 2. The graph of convex polytope R_n.

Note that the family R_n can also be obtained by adding a layer of hexagons between two pentagonal layers in the graph of D_n. The graph of D_n can be viewed in Figure 3. Miller et al. [33] studied the vertex-magic total labeling of R_n. Imran et al. [34] studied the minimum metric dimension problem for the family of R_n.

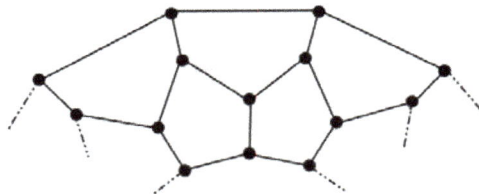

Figure 3. The graph of convex polytope D_n.

In this paper, we propose two further variations of D_n and study their binary locating-dominating number. In a similar fashion to Miller et al. [33], we add an extra layer of hexagons between the lower hexagonal layer and the outer pentagonal layer. We denote this new family of convex polytope with H_n. Figure 4 depicts the graph of convex polytope H_n. The weights' assignment to the vertices in Figure 4 helps to trace the binary locating-dominating sets in this family of convex polytopes.

The graph of convex polytope H_n comprises $2n$ pentagonal faces, $2n$ hexagonal faces and a pair of n-gonal faces.

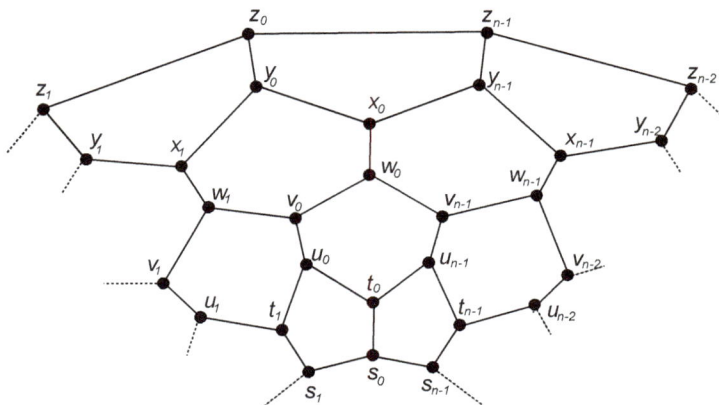

Figure 4. The graph of convex polytope H_n.

Mathematically, the graph of convex polytope H_n consists of the vertex set

$$V(H_n) = \{s_j, t_j, u_j, v_j, w_j, x_j, y_j, z_j \mid j = 0, \ldots, n-1\} \tag{7}$$

and the edge set

$$E(H_n) = \{s_j s_{j+1}, s_j t_j, t_j u_j, u_j t_{j+1}, u_j v_j, v_j w_j, v_j w_{j+1}, w_j x_j, x_j y_j, x_{j+1} y_j, y_j z_j, z_j z_{j+1} \mid j = 0, \ldots, n-1\}. \tag{8}$$

Note that arithmetic in the subscripts is performed modulo n.

Next, we validate the vertex and edge sets of the convex polytope H_n. In order to do that, we fix $n = 6$ and draw the graph H_6. According to expressions (7) and (8), we obtain the following vertex and edge sets for H_6:

$$V(H_6) = \{s_0, \ldots, s_5, t_0, \ldots, t_5, u_0, \ldots, u_5, v_0, \ldots, v_5, w_0, \ldots, w_5, x_0, \ldots, x_5, y_0, \ldots, y_5, z_0, \ldots, z_5\},$$

$$
\begin{aligned}
E(H_6) \;=\; & \{s_0s_1, s_1s_2, s_2s_3, s_3s_4, s_4s_5, s_5s_0, s_0t_0, s_1t_1, s_2t_2, s_3t_3, s_4t_4, s_5t_5, t_0u_0, t_1u_1, t_2u_2, t_3u_3, t_4u_4, t_5u_5, \\
& u_0t_1, u_1t_2, u_2t_3, u_3t_4, u_4t_5, u_5t_0, u_0v_0, u_1v_1, u_2v_2, u_3v_3, u_4v_4, u_5v_5, v_0w_0, v_1w_1, v_2w_2, v_3w_3, \\
& v_4w_4, v_5w_5, v_0w_1, v_1w_2, v_2w_3, v_3w_4, v_4w_5, v_5w_0, w_0x_0, w_1x_1, w_2x_2, w_3x_3, w_4x_4, w_5x_5, x_0y_0, \\
& x_1y_1, x_2y_2, x_3y_3, x_4y_4, x_5y_5, y_0x_1, y_1x_2, y_2x_3, y_3x_4, y_4x_5, y_5x_0, y_0z_0, y_1z_1, y_2z_2, y_3z_3, \\
& y_4z_4, y_5z_5, z_0z_1, z_1z_2, z_2z_3, z_3z_4, z_4z_5, z_5z_0 \}.
\end{aligned}
$$

By using these vertex and edge sets, we construct the graph of the convex polytope H_6. Figure 5 shows the graph of H_6. This validates the vertex and edge sets presented in Equations (7) and (8).

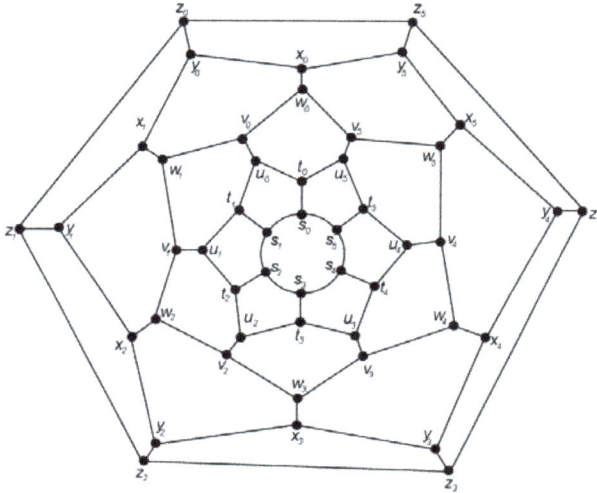

Figure 5. The graph of convex polytope H_6.

The following problems are open for this newly proposed family of convex polytopes.

Problem 1. *Let G be the family of convex polytopes H_n, where $n \geq 3$ is an integer. Then,*

(1)   Study vertex-face magic, edge-face magic, vertex-face anti-magic, edge-face anti-magic and vertex/edge total labeling of G. See the references [15,16,31,33] for similar research on other family of convex polytopes.
(2)   Study the minimum metric dimension problem for G. This problem is studied in [17–19,32,34] for other families of regular and non-regular convex polytopes.
(3)   Study fault-tolerant resolvability of G. A similar study for other classes of convex polytopes is conducted by Raza et al. [35] and Salman et al. [22].

3.1.2. Rotational Symmetry of the Convex Polytopes

The convex polytopes considered in this paper possess two kind of rotational symmetries: one is geometrical symmetry and the other is structural symmetry. By geometrical symmetry, we mean the symmetry possessed by the underlying geometrical convex polytopes. By structural symmetry, we mean the symmetry of the graphs of the underlying convex polytopes. We discuss both of these symmetries in details.

Erickson and Kim [36] studied various geometrical properties of certain convex polytopes. One of the perspectives of his study is different symmetries possessed by certain classes of convex polytopes. In particular, they showed the following result:

Theorem 2. *For any integer positive integer n, there is a neighborly family of n congruent convex 3-polytopes, each with a plane of bilateral symmetry, a line of 180° rotational symmetry, and a point of central symmetry.*

Let $\mathcal{H}_n$ denote the infinite point set $\{h_n(t) \mid t \in \mathbb{Z}\}$. The rotational symmetry is based on the fact that: a 180^0 rotation about the y-axis maps $h_n(t)$ to $h_n(-t)$ and thus preserves the point set $\mathcal{H}_n$. This implies that the Voronoi region of the underlying polytope is rotationally symmetric about the y-axis. Erickson and Kim [36] used the symmetry group of the convex polytope to show Theorem 2. In this scenario, the underlying geometrical shapes of convex polytopes considered in this paper possess rotational symmetry studied by Erickson and Kim [36].

Now, we discuss the structural symmetry possessed by the graphs of the convex polytopes considered in this paper. By structure-wise rotational symmetry, we mean that a fixed unit of a convex polytope can be rotated along a circle, by following the structural similarity, to obtain the complete graph of the convex polytope. Let us fixed a convex polytope, say H_n studied in the next subsubsection. In Figure 6, a unit of the graph of convex polytope H_n is presented. By rotating this unit along the dotted circle with center O, we can obtain the whole graph H_n. The part with bold edges shows the unit of this convex polytope, which is rotated along the dotted circle. The complete graph is obtained by completing one revolution of the unit (bold part) along the dotted circle having center O.

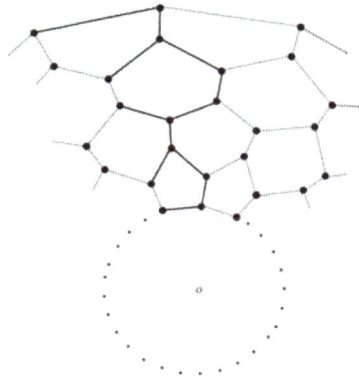

Figure 6. Unit of convex polytope H_n.

Note that this graph-theoretic structural similarity is common among all the families of convex polytopes considered in the subsequent subsections.

3.1.3. Binary Locating-Dominating Number of H_n

In this subsubsection, we present the main result for the family of convex polytope H_n. We find the exact value of the binary locating-dominating number for this family of convex polytope.

The following theorem presents the exact value of the binary locating-dominating number of H_n.

Theorem 3. *The binary locating-dominating number of H_n is given by the following expression:*

$$\gamma_{l-d}(H_n) = \left\lceil \frac{8n}{3} \right\rceil.$$

Proof. Note that H_n is a family of regular graphs of degree 3 on $8n$ vertices. By Theorem 1, we find the following lower bound on the binary locating-dominating number of H_n:

$$\gamma_{l-d}(H_n) \geq \left\lceil \frac{2(8n)}{6} \right\rceil = \left\lceil \frac{8n}{3} \right\rceil. \tag{9}$$

Let S be a subset of the vertex set of H_n, such that

$$
S = \begin{cases}
\{s_{3j+1}, t_{3j}, u_{3j+1}, v_{3j+2}, w_{3j+1}, x_{3j+2}, y_{3j}, z_{3j+2} \mid j = 0, ..., m-1\}, & n = 3m; \\
\{s_{3j+2}, t_{3j}, u_{3j+1}, v_{3j}, w_{3j+2}, x_{3j+1}, y_{3j+2}, z_{3j}\} \cup \\
\{t_{3m}, v_{3m}, y_{3m} \mid j = 0, ..., m-1\}, & n = 3m+1; \\
\{s_{3j}, t_{3j+1}, u_{3j+2}, v_{3j}, w_{3j+2}, x_{3j+1}, y_{3j+2}, z_{3j+1}\} \cup \\
\{s_{3m}, t_{3m+1}, v_{3m}, w_{3m+1}, y_{3m+1}, z_{3m} \mid j = 0, ..., m-1\}, & n = 3m+2.
\end{cases}
$$

Next, we show that S is a binary locating-dominating set of H_n. In order to prove that, we need to discuss the following three possible cases:

Case 1:   When $n = 3m$.

In order to show S to be a binary locating-dominating set, we need to show that the neighborhoods of all vertices in $V\backslash S$ are non-empty and distinct. Table 1 shows these neighborhoods and their intersections. Although some formulas for some intersections can be somewhat similar, but they are distinct.

Case 2:   When $n = 3m + 1$.

As in the previous case, the neighborhoods of all vertices in $V\backslash S$ are non-empty and distinct shown in Table 1.

Case 3:   When $n = 3m + 2$.

Similar to the previous two cases, Table 1 shows that the neighborhoods of all vertices in $V\backslash S$ are non-empty and distinct.

It is easily seen that $|S| = \lceil \frac{8n}{3} \rceil$. This shows that

$$
\gamma_{l-d}(H_n) \leq \left\lceil \frac{8n}{3} \right\rceil. \tag{10}
$$

By combining Inequalities (9) and (10), we obtain the result. $\quad\square$

Table 1. Binary locating-dominating vertices in H_n.

n	$v \in V\backslash S$	$S \cap N[v]$	$v \in V\backslash S$	$S \cap N[v]$
$3m$	s_{3j}	$\{s_{3j+1}, t_{3j}\}$	s_{3j+2}	$\{s_{3j+1}\}$
	t_{3j+1}	$\{s_{3j+1}, u_{3j+1}\}$	t_{3j+2}	$\{u_{3j+1}\}$
	u_{3j}	$\{t_{3j}\}$	u_{3j+2}	$\{t_{3(j+1)}, v_{3j+2}\}$
	v_{3j}	$\{w_{3j+1}\}$	v_{3j+1}	$\{w_{3j+1}, u_{3j+1}\}$
	w_{3j}	$\{v_{3j-1}\}$	w_{3j+2}	$\{v_{3j+2}, x_{3j+2}\}$
	x_{3j}	$\{y_{3j}\}$	x_{3j+1}	$\{y_{3j}, w_{3j+1}\}$
	y_{3j+1}	$\{x_{3j+2}\}$	y_{3j+2}	$\{x_{3j+2}, z_{3j+2}\}$
	z_{3j}	$\{y_{3j}, z_{3j-1}\}$	z_{3j+1}	$\{z_{3j+2}\}$
$3m+1$	s_{3j+1}	$\{s_{3j+2}\}$	$s_{3(j+1)}$	$\{s_{3j+2}, t_{3(j+1)}\}$
	t_{3j+1}	$\{u_{3j+1}\}$	t_{3j+2}	$\{u_{3j+1}, s_{3j+2}\}$
	u_{3j}	$\{t_{3j}, v_{3j}\}$	u_{3j+2}	$\{t_{3(j+1)}\}$
	v_{3j+1}	$\{u_{3j+1}, w_{3j+2}\}$	v_{3j+2}	$\{w_{3j+2}\}$
	w_{3j+1}	$\{v_{3j}, x_{3j+1}\}$	$w_{3(j+1)}$	$\{v_{3(j+1)}\}$
	x_{3j+2}	$\{w_{3j+2}, y_{3j+2}\}$	$x_{3(j+1)}$	$\{y_{3j+2}\}$
	y_{3j}	$\{x_{3j+1}, z_{3j}\}$	y_{3j+1}	$\{x_{3j+1}\}$
	z_{3j+2}	$\{y_{3j+2}, z_{3j+3}\}$	z_{3j+1}	$\{z_{3j}\}$
	s_0	$\{t_0\}$	u_{3m}	$\{t_0, t_{3m}, v_{3m}\}$
	w_0	$\{v_0, v_{3m}\}$	x_0	$\{y_{3m}\}$
	z_{3m}	$\{y_{3m}, z_0\}$		

Table 1. *Cont.*

n	$v \in V \backslash S$	$S \cap N[v]$	$v \in V \backslash S$	$S \cap N[v]$
$3m+2$	s_{3j+1}	$\{s_{3j}, t_{3j+1}\}$	s_{3j+2}	$\{s_{3(j+1)}\}$
	t_{3j+2}	$\{u_{3j+2}\}$	$t_{3(j+1)}$	$\{s_{3(j+1)}, u_{3j+2}\}$
	u_{3j}	$\{t_{3j+1}, v_{3j}\}$	u_{3j+1}	$\{t_{3j+1}\}$
	v_{3j+1}	$\{w_{3j+2}\}$	v_{3j+2}	$\{u_{3j+2}, w_{3j+2}\}$
	w_{3j}	$\{v_{3j}\}$	w_{3j+1}	$\{v_{3j}, x_{3j+1}\}$
	x_{3j+2}	$\{w_{3j+2}, y_{3j+2}\}$	$x_{3(j+1)}$	$\{y_{3j+2}\}$
	y_{3j+1}	$\{x_{3j+1}, z_{3j+1}\}$	y_{3j}	$\{x_{3j+1}\}$
	z_{3j}	$\{z_{3j+1}\}$	z_{3j+2}	$\{y_{3j+2}, z_{3j+1}\}$
	s_{3m+1}	$\{s_{3m}, s_0, t_{3m+1}\}$	t_0	$\{s_0\}$
	u_{3m+1}	$\{t_{3m+1}\}$	u_{3m}	$\{t_{3m+1}, v_{3m}\}$
	v_{3m+1}	$\{w_{3m+1}\}$	w_{3m}	$\{v_{3m}\}$
	x_0	$\{y_{3m+1}\}$	x_{3m+1}	$\{w_{3m+1}, y_{3m+1}\}$
	y_{3m}	$\{z_{3m}\}$	z_{3m+1}	$\{y_{3m+1}, z_{3m}\}$

3.2. The Graph of Convex Polytope H'_n

3.2.1. Construction

By following the same construction as for H_n, we define another variation of convex polytopes R_n and D_n. We add an extra layer of hexagons between the outer pentagonal layer and the next hexagonal layer of H_n. In other words, H'_n can be obtained by adding three hexagonal layers in R_n between outer pentagonal and inner hexagonal layers and four hexagonal layers in D_n between the two pentagonal layers.

The graph of convex polytope H_n comprises $2n$ pentagonal faces, $4n$ hexagonal faces and a pair of n-gonal faces. Figure 7 shows the graph of this family of convex polytopes. Mathematically, it has the vertex set

$$V(H'_n) = \{o_j, p_j, q_j, r_j, s_j, t_j, u_j, v_j, w_j, x_j, y_j, z_j \mid j = 0, \dots, n-1\}, \tag{11}$$

and the edge set

$$E(H'_n) = \{o_j o_{j+1}, o_j p_j, q_j p_j, q_j p_{j+1}, q_j r_j, r_j s_j, r_j s_{j+1}, s_j t_j, t_j u_j, t_{j+1} u_j, u_j v_j, \tag{12}$$
$$v_j w_j, v_j w_{j+1}, w_j x_j, x_j y_j, x_{j+1} y_j, y_j z_j, z_j z_{j+1} \mid j = 0, \dots, n-1\}.$$

Note that arithmetic in the subscripts is performed modulo n.

Next, we validate the vertex and edge cardinalities of the graph of convex polytope H'_n. We do that by fixing a value of $n = 6$, and we construct the graph of H'_6 from (11) and (12). We obtain the following vertex and edge set cardinalities for H'_6:

$$V(H'_6) = \{o_0, \dots, o_5 p_0, \dots, p_5 q_0, \dots, q_5 r_0, \dots, r_5 s_0, \dots, s_5, t_0, \dots, t_5, u_0, \dots, u_5, v_0, \dots, v_5, w_0, \dots, w_5,$$
$$x_0, \dots, x_5, y_0, \dots, y_5, z_0, \dots, z_5\},$$

$$E(H'_6) = \{o_0 o_1, o_1 o_2, o_2 o_3, o_3 o_4, o_4 o_5, o_5 o_0, o_0 p_0, o_1 p_1, o_2 p_2, o_3 p_3, o_4 p_4, o_5 p_5, p_0 q_0, p_1 q_1, p_2 q_2, p_3 q_3,$$
$$p_4 q_4, p_5 q_5, q_0 p_1, q_1 p_2, q_2 p_3, q_3 p_4, q_4 p_5, q_5 p_0, q_0 r_0, q_1 r_1, q_2 r_2, q_3 r_3, q_4 r_4, q_5 r_5, s_0 r_0, s_1 r_1, s_2 r_2,$$
$$s_3 r_3, s_4 r_4, s_5 r_5, r_0 s_1, r_1 s_2, r_2 s_3, r_3 s_4, r_4 s_5, r_4 s_0, s_0 t_0, s_1 t_1, s_2 t_2, s_3 t_3, s_4 t_4, s_5 t_5, t_0 u_0, t_1 u_1, t_2 u_2,$$
$$t_3 u_3, t_4 u_4, t_5 u_5, u_0 t_1, u_1 t_2, u_2 t_3, u_3 t_4, u_4 t_5, u_5 t_0, u_0 v_0, u_1 v_1, u_2 v_2, u_3 v_3, u_4 v_4, u_5 v_5, v_0 w_0,$$
$$v_1 w_1, v_2 w_2, v_3 w_3, v_4 w_4, v_5 w_5, v_0 w_1, v_1 w_2, v_2 w_3, v_3 w_4, v_4 w_5, v_5 w_0, w_0 x_0, w_1 x_1, w_2 x_2, w_3 x_3,$$
$$w_4 x_4, w_5 x_5, x_0 y_0, x_1 y_1, x_2 y_2, x_3 y_3, x_4 y_4, x_5 y_5, y_0 x_1, y_1 x_2, y_2 x_3, y_3 x_4, y_4 x_5, y_5 x_0, y_0 z_0,$$
$$y_1 z_1, y_2 z_2, y_3 z_3, y_4 z_4, y_5 z_5, z_0 z_1, z_1 z_2, z_2 z_3, z_3 z_4, z_4 z_5, z_5 z_0\}.$$

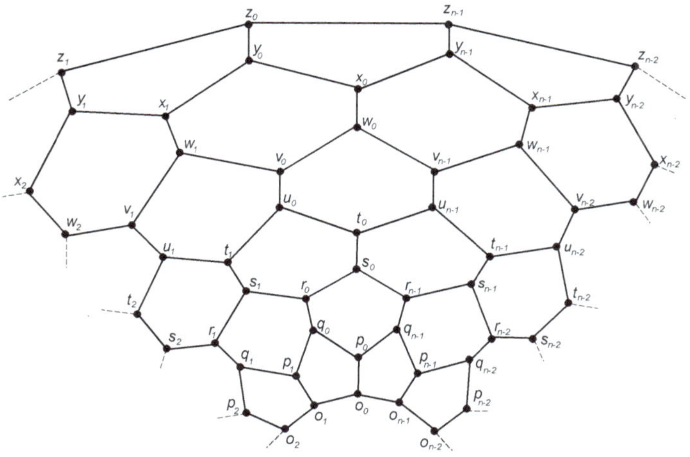

Figure 7. The graph of convex polytope $H'n$.

By using these vertex and edge sets, we construct the graph of the convex polytope H'_6. Figure 8 shows the graph of H'_6. This validates the vertex and edge sets presented in (11) and (12).

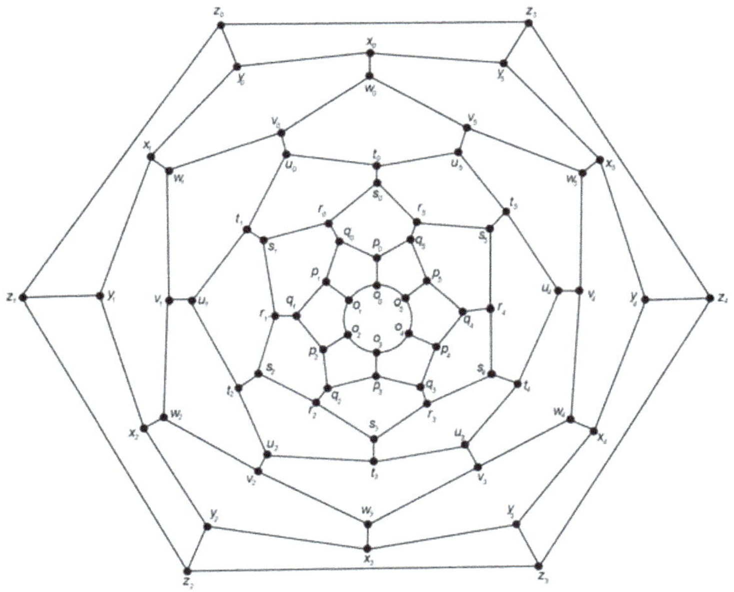

Figure 8. The graph of convex polytope H'_6.

3.2.2. Binary Locating-Dominating Number of H'_n

This subsubsection presents the main result for H'_n. We find the exact value of the the binary locating-dominating number of H'_n. In the following theorem, it is shown that the binary locating-dominating number of the family H'_n is exactly $4n$.

Theorem 4. *The binary locating-dominating number of H'_n is exactly $4n$, i.e.,*

$$\gamma_{l-d}(H'_n) = 4n.$$

Proof. As the graph H'_n is regular with degree 3. By Theorem 1, we obtain

$$\gamma_{l-d} \geq \left\lceil \frac{2(12n)}{6} \right\rceil = 4n.$$

Let $S \subset V(H'_n)$ such that $S = \{p_j, s_j, v_j, y_j \mid j = 0, \ldots, n-1\}$. Next, we show that S is a binary locating-dominating number of H'_n. It can be seen that

$$S \cap N[o_j] = [p_j], \ S \cap N[q_j] = [p_{j-1}, p_j], \ S \cap N[r_j] = [s_j, s_{j+1}], \ S \cap N[t_j] = [s_j], \ S \cap N[u_j] = [v_j],$$
$$S \cap N[w_j] = [v_{j-1}, v_j], \ S \cap N[x_j] = [y_{j-1}, y_j] \text{ and } S \cap N[z_j] = [y_j].$$

Note that all these intersections have at least one element and they are distinct as well. This shows that S is a binary locating dominating set of (H'_n) and therefore $\gamma_{l-d}(H'_n) \leq 4n$. By combining it with the fact $\gamma_{l-d}(H'_n) \geq 4n$, we obtain that $\gamma_{l-d}(H'_n) = 4n$. □

4. Tight Upper Bounds

In this section, we find tight upper bounds on the binary locating-dominating number of three infinite families of convex polytopes.

4.1. The Graph of Convex Polytope S_n

The graph of convex polytope S_n consists of $2n$ trigonal faces, $2n$ 4-gonal faces and a pair of n-sided faces (see Figure 9). Mathematically, it has the vertex set

$$V(S_n) = \{w_j, x_j, y_j, z_j \mid j = 0, \ldots, n-1\},$$

and the edge set

$$E(S_n) = \{w_j w_{j+1}, x_j x_{j+1}, y_j y_{j+1}, z_j z_{j+1} \mid j = 0, \ldots, n-1\} \cup \{w_{j+1} x_j, w_j x_j, x_j y_j, y_j z_j \mid j = 0, \ldots, n-1\}.$$

Imran et al. [19] showed that the metric dimension of S_n is 3. The graph of convex polytope S_n can also be obtained from the graph of convex polytope Q_n, defined in [16], by adding the edges $w_{j+1} x_j, y_j y_{j+1}$ and then deleting the edges x_{j+1}, y_j i.e., $V(S_n) = V(Q_n)$ and $E(S_n) = (E(Q_n) \cup \{w_{j+1} x_j, y_j y_{j+1} \mid j = 0, \ldots, n-1\}) \setminus \{x_{j+1} y_j \mid j = 0, \ldots, n-1\}$.

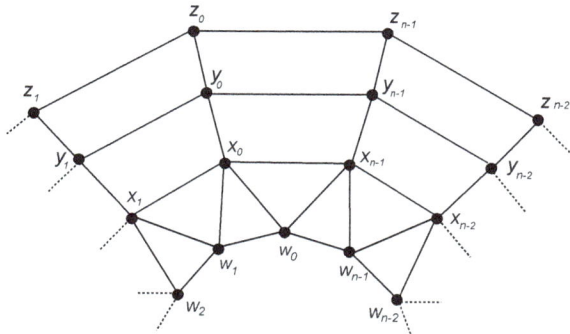

Figure 9. The graph of convex polytope S_n.

The following theorem gives a tight upper bound on the binary locating-dominating number of S_n.

Theorem 5. *Let G be the graph of convex polytope S_n. Then,*

$$\gamma_{l-d}(G) \leq \left\lceil \frac{7n}{5} \right\rceil,$$

and this upper bound is tight.

Proof. Let $S \subset V$ be a proper subset of the vertex set of S_n such that

$$S = \begin{cases} \{x_{5j}, x_{5j+1}, x_{5j+2}, x_{5j+3}, x_{5j+4}, z_{5j+1}, z_{5j+3} \mid j = 0, \ldots, m-1\}, & n = 5m; \\ \{x_{5j}, x_{5j+1}, x_{5j+2}, x_{5j+3}, x_{5j+4}, z_{5j+1}, z_{5j+3} \cup \\ \{x_{5m}, z_{5m}\} \mid j = 0, \ldots, m-1\}, & n = 5m+1; \\ \{x_{5j}, x_{5j+1}, x_{5j+2}, x_{5j+3}, x_{5j+4}, z_{5j+1}, z_{5j+3} \cup \\ \{x_{5m}, x_{5m+1}, z_{5m+1}\} \mid j = 0, \ldots, m-1\}, & n = 5m+2; \\ \{x_{5j}, x_{5j+1}, x_{5j+2}, x_{5j+3}, x_{5j+4}, z_{5j+1}, z_{5j+3} \cup \\ \{x_{5m}, x_{5m+1}, x_{5m+2}, z_{5m}, z_{5m+2}\}, & n = 5m+3; \\ \{x_{5m}, x_{5m+1}, x_{5m+2}, x_{5m+3}, z_{5m+1}, z_{5m+3}\} \cup \\ \{x_{5j}, x_{5j+1}, x_{5j+2}, x_{5j+3}, x_{5j+4}, z_{5j+1}, z_{5j+3} \mid j = 0, \ldots, m-1\}, & n = 5m+4. \end{cases}$$

Next, we show that S is a locating-dominating set of G. To do that, we discuss the following five possible cases:

Case 1: When $n = 5m$.

Table 2 depicts all vertices in $V \setminus S$ and the intersections of their closed neighborhoods with S. From the second column, we can see that all these intersections are nonempty and distinct. Thus, for any two vertices $u, v \in V \setminus S$, we have $S \cap N[v] \neq S \cap N[u] \neq \emptyset$. This shows that S is a binary locating-dominating set of S_n.

Case 2: When $n = 5m + 1$.

Similar to the argument in Case 1, we see from Table 2 that all the intersections are nonempty and distinct. This shows that S is a binary locating-dominating set for S_n, if $n = 5m + 1$.

Case 3: When $n = 5m + 2$.

Similar to the argument in Case 1 and Case 2, we see from Table 2 that all the intersections are nonempty and distinct. This shows that S is a binary locating-dominating set for S_n, if $n = 5m + 2$.

Thus, from the above discussion, we can say that Case 4 and Case 5 are analogous to above mentioned cases.

Note that $|S| = \lceil \frac{7n}{5} \rceil$; therefore, we have $\gamma_{l-d}(G) \le \lceil \frac{7n}{5} \rceil$.

In order to show tightness in the upper bound from Theorem 5, we use the CPLEX solver for the ILP formulation with constraints (1), (2), (4) and (6). As a result, we obtain the following optimal solutions: $\gamma_{l-d}(S_6) = 9$, $\gamma_{l-d}(S_7) = 10$, $\gamma_{l-d}(S_8) = 12$, $\gamma_{l-d}(S_9) = 13$, ..., $\gamma_{l-d}(S_{21}) = 30$, ..., $\gamma_{l-d}(S_{29}) = 41$. This shows the upper bound in Theorem 5 is tight. □

Table 2. Binary locating-dominating vertices in S_n.

n	$v \in V \backslash S$	$S \cap N[v]$	$v \in V \backslash S$	$S \cap N[v]$
$5m$	w_{5j}	$\{x_{5j}, x_{5(j-1)+4}\}$	w_{5j+1}	$\{x_{5j}, x_{5j+1}\}$
	w_{5j+2}	$\{x_{5j+1}, x_{5j+2}\}$	w_{5j+3}	$\{x_{5j+2}, x_{5j+3}\}$
	w_{5j+4}	$\{x_{5j+3}, x_{5j+4}\}$	y_{5j}	$\{x_{5j}\}$
	y_{5j+1}	$\{x_{5j+1}, z_{5j+1}\}$	y_{5j+2}	$\{x_{5j+2}\}$
	y_{5j+3}	$\{x_{5j+3}, z_{5j+3}\}$	y_{5j+4}	$\{x_{5j+4}\}$
	z_{5j}	$\{z_{5j+1}\}$	z_{5j+2}	$\{z_{5j+1}, z_{5j+3}\}$
	z_{5j+4}	$\{z_{5j+3}\}$		
$5m+1$	w_{5j+1}	$\{x_{5j}, x_{5j+1}\}$	w_{5j+2}	$\{x_{5j+1}, x_{5j+2}\}$
	w_{5j+3}	$\{x_{5j+2}, x_{5j+3}\}$	w_{5j+4}	$\{x_{5j+3}, x_{5j+4}\}$
	$w_{5(j+1)}$	$\{x_{5j+4}, x_{5(j+1)}\}$	y_{5j}	$\{x_{5j}\}$
	y_{5j+1}	$\{x_{5j+1}, z_{5j+1}\}$	y_{5j+2}	$\{x_{5j+2}\}$
	y_{5j+3}	$\{x_{5j+3}, z_{5j+3}\}$	y_{5j+4}	$\{x_{5j+4}\}$
	z_{5j}	$\{z_{5j+1}\}$	z_{5j+2}	$\{z_{5j+1}, z_{5j+3}\}$
	z_{5j+4}	$\{z_{5j+3}\}$	w_0	$\{x_0, x_{5m}\}$
	y_{5m}	$\{x_{5m}, z_{5m}\}$		
$5m+2$	w_{5j+1}	$\{x_{5j}, x_{5j+1}\}$	w_{5j+2}	$\{x_{5j+1}, x_{5j+2}\}$
	w_{5j+3}	$\{x_{5j+2}, x_{5j+3}\}$	w_{5j+4}	$\{x_{5j+3}, x_{5j+4}\}$
	$w_{5(j+1)}$	$\{x_{5j+4}, x_{5(j+1)}\}$	y_{5j}	$\{x_{5j}\}$
	y_{5j+1}	$\{x_{5j+1}, z_{5j+1}\}$	y_{5j+2}	$\{x_{5j+2}\}$
	y_{5j+3}	$\{x_{5j+3}, z_{5j+3}\}$	y_{5j+4}	$\{x_{5j+4}\}$
	z_{5j}	$\{z_{5j+1}\}$	z_{5j+2}	$\{z_{5j+1}, z_{5j+3}\}$
	z_{5j+4}	$\{z_{5j+3}\}$	w_0	$\{x_0, x_{5m+1}\}$
	w_{5m+1}	$\{x_{5m}, x_{5m+1}\}$	y_{5m}	$\{x_{5m}\}$
	y_{5m+1}	$\{x_{5m+1}, z_{5m+1}\}$	z_{5m}	$\{z_{5m+1}\}$
$5m+3$	w_{5j+1}	$\{x_{5j}, x_{5j+1}\}$	w_{5j+2}	$\{x_{5j+1}, x_{5j+2}\}$
	w_{5j+3}	$\{x_{5j+2}, x_{5j+3}\}$	w_{5j+4}	$\{x_{5j+3}, x_{5j+4}\}$
	$w_{5(j+1)}$	$\{x_{5j+4}, x_{5(j+1)}\}$	y_{5j}	$\{x_{5j}\}$
	y_{5j+1}	$\{x_{5j+1}, z_{5j+1}\}$	y_{5j+2}	$\{x_{5j+2}\}$
	y_{5j+3}	$\{x_{5j+3}, z_{5j+3}\}$	y_{5j+4}	$\{x_{5j+4}\}$
	z_{5j}	$\{z_{5j+1}\}$	z_{5j+2}	$\{z_{5j+1}, z_{5j+3}\}$
	z_{5j+4}	$\{z_{5j+3}\}$	w_0	$\{x_0, x_{5m+2}\}$
	w_{5m+1}	$\{x_{5m}, x_{5m+1}\}$	w_{5m+2}	$\{x_{5m+1}, x_{5m+2}\}$
	y_{5m}	$\{x_{5m}, z_{5m}\}$	y_{5m+1}	$\{x_{5m+1}\}$
	y_{5m+2}	$\{x_{5m+2}, z_{5m+2}\}$	z_{5m+1}	$\{z_{5m}, z_{5m+2}\}$
$5m+4$	w_{5j+1}	$\{x_{5j}, x_{5j+1}\}$	w_{5j+2}	$\{x_{5j+1}, x_{5j+2}\}$
	w_{5j+3}	$\{x_{5j+2}, x_{5j+3}\}$	w_{5j+4}	$\{x_{5j+3}, x_{5j+4}\}$
	$w_{5(j+1)}$	$\{x_{5j+4}, x_{5(j+1)}\}$	y_{5j}	$\{x_{5j}\}$
	y_{5j+1}	$\{x_{5j+1}, z_{5j+1}\}$	y_{5j+2}	$\{x_{5j+2}\}$
	y_{5j+3}	$\{x_{5j+3}, z_{5j+3}\}$	y_{5j+4}	$\{x_{5j+4}\}$
	z_{5j}	$\{z_{5j+1}\}$	z_{5j+2}	$\{z_{5j+1}, z_{5j+3}\}$
	z_{5j+4}	$\{z_{5j+3}\}$	w_0	$\{x_0, x_{5m+3}\}$
	w_{5m+1}	$\{x_{5m}, x_{5m+1}\}$	w_{5m+2}	$\{x_{5m+1}, x_{5m+2}\}$
	w_{5m+3}	$\{x_{5m+2}, x_{5m+3}\}$	y_{5m}	$\{x_{5m}\}$
	y_{5m+1}	$\{x_{5m+1}, z_{5m+1}\}$	y_{5m+2}	$\{x_{5m+2}\}$
	y_{5m+3}	$\{x_{5m+3}, z_{5m+3}\}$	z_{5m}	$\{z_{5m+1}\}$
	z_{5m+2}	$\{z_{5m+1}, z_{5m+3}\}$		

4.2. The Graph of Convex Polytope B_n

The graph of convex polytope B_n comprises $2n$ 4-gonal faces, n trigonal faces, n pentagonal faces and a pair of n-gonal faces (see Figure 10). It can also be obtained by the combination of graph of convex polytope Q_n [16] and a graph of prism D_n [15]. Alternatively, it has the vertex set

$$V(B_n) = \{v_j, w_j, x_j, y_j, z_j \mid j = 0, \ldots, n-1\},$$

and the edge set

$$E(B_n) = \{v_j v_{j+1}, w_j w_{j+1}, y_j y_{j+1}, z_j z_{j+1} \mid j = 0, \ldots, n-1\} \cup$$
$$\{v_j w_j, w_j x_j, w_{j+1} x_j, x_j y_j, y_j z_j \mid j = 0, \ldots, n-1\}.$$

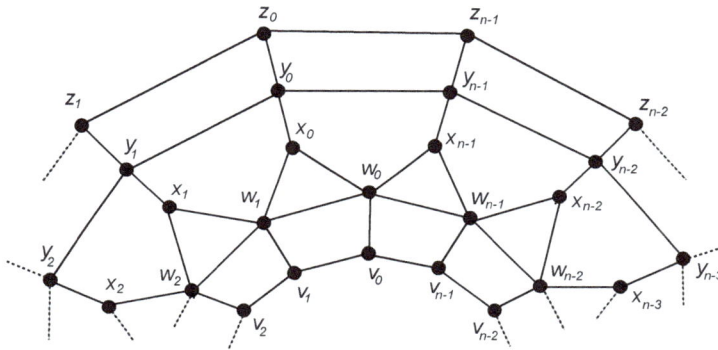

Figure 10. The graph of convex polytope B_n.

Imran et al. [18] showed that the metric dimension of the convex polytope B_n is three. Next, we prove a tight upper bound on the binary locating-dominating number of B_n.

Theorem 6. *The binary locating-dominating number of B_n is bounded above by $2n$, i.e.,*

$$\gamma_{l-d}(B_n) \leq 2n,$$

and this upper bound is tight.

Proof. Let $S \subset V(B_n)$ such that $S = \{w_j, y_j \mid j = 0, ..., n-1\}$. Next, we show that S is a binary locating-dominating number of B_n. It can be seen that

$$S \cap N[v_j] = [w_j], \ S \cap N[x_j] = [w_j, w_{j+1}, y_j], \text{ and } S \cap N[z_j] = [y_j].$$

Note that all these intersections have at least one element and they are distinct as well. This shows that S is a binary locating-dominating set of B_n. Therefore, we obtain that $\gamma_{l-d}(G) \leq 2n$.

Using the CPLEX solver on the integer linear programming formulation with constraints (1), (2), (4) and (6), we obtain the optimal solutions: $\gamma_{l-d}(B_7) = 14$, $\gamma_{l-d}(B_8) = 16$, $\gamma_{l-d}(B_9) = 18, \ldots, \gamma_{l-d}(S_{15}) = 30$. This shows that the upper bound is tight. $\square$

4.3. The Graph of Convex Polytope T_n

The graph of convex polytope T_n consists of $4n$ trigonal faces, n 4-gonal faces and a pair of n-sided faces (see Figure 11). Mathematically, we have

$$V(T_n) = \{w_j, x_j, y_j, z_j \mid j = 0, \ldots, n - 1\}$$

and

$$E(T_n) = \{w_j w_{j+1}, x_j x_{j+1}, y_j y_{j+1}, z_j z_{j+1} \mid j = 0, \ldots, n - 1\} \cup$$
$$\{w_{j+1} x_j, w_j x_j, x_j y_j, y_i z_j, y_{j+1} z_j \mid j = 0, \ldots, n - 1\}.$$

It can also be obtained by the combination of the graph of convex polytope R_n [15,19] and the graph of an antiprism.

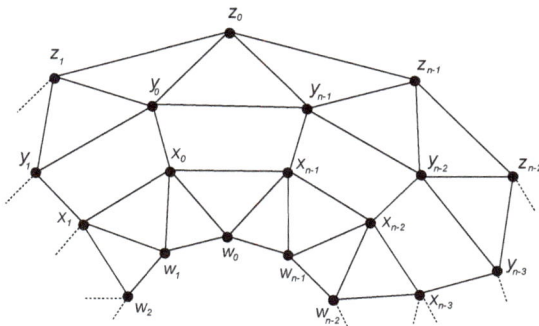

Figure 11. The graph of convex polytope T_n.

Theorem 7. *For the graph of convex polytope T_n, we have*

$$\gamma_{l-d}(T_n) \leq \left\lceil \frac{7n}{5} \right\rceil,$$

and this upper bound is tight.

Proof. Let S be a proper subset of the vertex set of T_n, such that

$$S = \begin{cases} \{x_{5j}, x_{5j+1}, x_{5j+2}, x_{5j+3}, x_{5j+4}, z_{5j+1}, z_{5j+3} \mid j = 0, \ldots, m - 1\}, & n = 5m; \\ \{x_{5j}, x_{5j+1}, x_{5j+2}, x_{5j+3}, x_{5j+4}, z_{5j+1}, z_{5j+3}\} \cup \\ \{x_{5m}, z_{5m}\} \mid j = 0, \ldots, m - 1\}, & n = 5m + 1; \\ \{x_{5j}, x_{5j+1}, x_{5j+2}, x_{5j+3}, x_{5j+4}, z_{5j+1}, z_{5j+3}\} \cup \\ \{x_{5m}, x_{5m+1}, z_{5m+1}\} \mid j = 0, \ldots, m - 1\}, & n = 5m + 2; \\ \{x_{5j}, x_{5j+1}, x_{5j+2}, x_{5j+3}, x_{5j+4}, z_{5j+1}, z_{5j+3}\} \cup \\ \{x_{5m}, x_{5m+1}, x_{5m+2}, z_{5m}, z_{5m+2}\}, & n = 5m + 3; \\ \{x_{5m}, x_{5m+1}, x_{5m+2}, x_{5m+3}, z_{5m+1}, z_{5m+3}\} \cup \\ \{x_{5j}, x_{5j+1}, x_{5j+2}, x_{5j+3}, x_{5j+4}, z_{5j+1}, z_{5j+3} \mid j = 0, \ldots, m - 1\}, & n = 5m + 4. \end{cases}$$

We show that S is a binary locating-dominating set of T_n. We need to discuss the following two possible cases:

Case 1: When $n = 5m$.

In order to show S to be a binary locating-dominating set, we need to show that the neighborhoods of all vertices in $V \backslash S$ are non-empty and distinct. Table 3 shows these neighborhoods and their intersections. Although some formulas for some intersections can be somewhat similar, but they are distinct.

Case 2: When $n = 5m + 1$.

As in the previous case, the the neighborhoods of all vertices in $V \backslash S$ are non-empty and distinct shown in Table 3. Thus, from the above discussion, we can say that Case 3, Case 4 and Case 5 are analogous to the above-mentioned cases.

Note that $|S| = \lceil \frac{7n}{5} \rceil$. This implies that $\gamma_{l-d}(T_n) \leq \lceil \frac{7n}{5} \rceil$.

Next, we use the CPLEX solver for the ILP formulation with constraints (1), (2), (4) and (6) and obtain the following optimal solutions: $\gamma_{l-d}(T_6) = 9$, $\gamma_{l-d}(T_7) = 10$, $\gamma_{l-d}(T_8) = 12$, $\gamma_{l-d}(T_9) = 13$, $\ldots$, $\gamma_{l-d}(T_{21}) = 30$, $\ldots$, $\gamma_{l-d}(T_{29}) = 41$. This shows the upper bound in Theorem 7 is tight. □

Table 3. Binary locating-dominating vertices in T_n.

n	$v \in V \backslash S$	$S \cap N[v]$	$v \in V \backslash S$	$S \cap N[v]$
$5m$	w_{5j}	$\{x_{5j}, x_{5(j-1)+4}\}$	w_{5j+1}	$\{x_{5j}, x_{5j+1}\}$
	w_{5j+2}	$\{x_{5j+1}, x_{5j+2}\}$	w_{5j+3}	$\{x_{5j+2}, x_{5j+3}\}$
	w_{5j+4}	$\{x_{5j+3}, x_{5j+4}\}$	y_{5j}	$\{x_{5j}, z_{5j+1}\}$
	y_{5j+1}	$\{x_{5j+1}, z_{5j+1}\}$	y_{5j+2}	$\{x_{5j+2}, z_{5j+3}\}$
	y_{5j+3}	$\{x_{5j+3}, z_{5j+3}\}$	y_{5j+4}	$\{x_{5j+4}\}$
	z_{5j}	$\{z_{5j+1}\}$	z_{5j+2}	$\{z_{5j+1}, z_{5j+3}\}$
	z_{5j+4}	$\{z_{5j+3}\}$		
$5m+1$	w_{5j+1}	$\{x_{5j}, x_{5j+1}\}$	w_{5j+2}	$\{x_{5j+1}, x_{5j+2}\}$
	w_{5j+3}	$\{x_{5j+2}, x_{5j+3}\}$	w_{5j+4}	$\{x_{5j+3}, x_{5j+4}\}$
	$w_{5(j+1)}$	$\{x_{5j+3}, x_{5(j+1)}\}$	y_{5j}	$\{x_{5j}, z_{5j+1}\}$
	y_{5j+1}	$\{x_{5j+1}, z_{5j+1}\}$	y_{5j+2}	$\{x_{5j+2}, z_{5j+3}\}$
	y_{5j+3}	$\{x_{5j+3}, z_{5j+3}\}$	y_{5j+4}	$\{x_{5j+4}\}$
	z_{5j}	$\{z_{5j+1}\}$	z_{5j+2}	$\{z_{5j+1}, z_{5j+3}\}$
	z_{5j+4}	$\{z_{5j+3}\}$	w_0	$\{x_0, x_{5m}\}$
	y_{5m}	$\{x_{5m}, z_{5m}\}$		
$5m+2$	w_{5j+1}	$\{x_{5j}, x_{5j+1}\}$	w_{5j+2}	$\{x_{5j+1}, x_{5j+2}\}$
	w_{5j+3}	$\{x_{5j+2}, x_{5j+3}\}$	w_{5j+4}	$\{x_{5j+3}, x_{5j+4}\}$
	$w_{5(j+1)}$	$\{x_{5j+3}, x_{5(j+1)}\}$	y_{5j}	$\{x_{5j}, z_{5j+1}\}$
	y_{5j+1}	$\{x_{5j+1}, z_{5j+1}\}$	y_{5j+2}	$\{x_{5j+2}, z_{5j+3}\}$
	y_{5j+3}	$\{x_{5j+3}, z_{5j+3}\}$	y_{5j+4}	$\{x_{5j+4}\}$
	z_{5j}	$\{z_{5j+1}\}$	z_{5j+2}	$\{z_{5j+1}, z_{5j+3}\}$
	z_{5j+4}	$\{z_{5j+3}\}$	w_0	$\{x_0, x_{5m+1}\}$
	w_{5m+1}	$\{x_{5m}, x_{5m+1}\}$	y_{5m}	$\{x_{5m}, z_{5m}\}$
	y_{5m+1}	$\{x_{5m+1}, z_{5m}\}$	z_{5m}	$\{z_{5m+1}\}$
$5m+3$	w_{5j+1}	$\{x_{5j}, x_{5j+1}\}$	w_{5j+2}	$\{x_{5j+1}, x_{5j+2}\}$
	w_{5j+3}	$\{x_{5j+2}, x_{5j+3}\}$	w_{5j+4}	$\{x_{5j+3}, x_{5j+4}\}$
	$w_{5(j+1)}$	$\{x_{5j+3}, x_{5(j+1)}\}$	y_{5j}	$\{x_{5j}, z_{5j+1}\}$
	y_{5j+1}	$\{x_{5j+1}, z_{5j+1}\}$	y_{5j+2}	$\{x_{5j+2}, z_{5j+3}\}$
	y_{5j+3}	$\{x_{5j+3}, z_{5j+3}\}$	y_{5j+4}	$\{x_{5j+4}\}$
	z_{5j}	$\{z_{5j+1}\}$	z_{5j+2}	$\{z_{5j+1}, z_{5j+3}\}$
	z_{5j+4}	$\{z_{5j+3}\}$	w_0	$\{x_0, x_{5m+2}\}$
	w_{5m+1}	$\{x_{5m}, x_{5m+1}\}$	w_{5m+2}	$\{x_{5m+1}, x_{5m+2}\}$
	y_{5m}	$\{x_{5m}, z_{5m}\}$	y_{5m+1}	$\{x_{5m+1}, z_{5m+2}\}$
	y_{5m+2}	$\{x_{5m+2}, z_{5m+2}\}$	z_{5m+1}	$\{z_{5m}, z_{5m+2}\}$
$5m+4$	w_{5j+1}	$\{x_{5j}, x_{5j+1}\}$	w_{5j+2}	$\{x_{5j+1}, x_{5j+2}\}$
	w_{5j+3}	$\{x_{5j+2}, x_{5j+3}\}$	w_{5j+4}	$\{x_{5j+3}, x_{5j+4}\}$
	$w_{5(j+1)}$	$\{x_{5j+3}, x_{5(j+1)}\}$	y_{5j}	$\{x_{5j}, z_{5j+1}\}$
	y_{5j+1}	$\{x_{5j+1}, z_{5j+1}\}$	y_{5j+2}	$\{x_{5j+2}, z_{5j+3}\}$
	y_{5j+3}	$\{x_{5j+3}, z_{5j+3}\}$	y_{5j+4}	$\{x_{5j+4}\}$
	z_{5j}	$\{z_{5j+1}\}$	z_{5j+2}	$\{z_{5j+1}, z_{5j+3}\}$
	z_{5j+4}	$\{z_{5j+3}\}$	w_0	$\{x_0, x_{5m+2}\}$
	w_{5m+1}	$\{x_{5m}, x_{5m+1}\}$	w_{5m+2}	$\{x_{5m+1}, x_{5m+2}\}$
	w_{5m+3}	$\{x_{5m+2}, x_{5m+3}\}$	y_{5m}	$\{x_{5m}, z_{5m}\}$
	y_{5m+1}	$\{x_{5m+1}, z_{5m+1}\}$	y_{5m+2}	$\{x_{5m+2}, z_{5m+3}\}$
	y_{5m+3}	$\{x_{5m+3}, z_{5m+3}\}$	z_{5m}	$\{z_{5m+1}\}$
	z_{5m+2}	$\{z_{5m+1}, z_{5m+3}\}$		

5. Conclusions

In this paper, we focus on a class of geometric graphs which naturally arise from the structures of convex polytopes. Besides finding exact values for the binary locating-dominating number of two infinite families of graphs of convex polytopes, we also find tight upper bounds on other three infinite families of convex polytopes. An integer linear programming model for binary locating-locating number is used to find tightness in the obtained upper bounds.

Generalized Petersen graphs and certain families of strongly regular graphs can be considered for further research on this problem.

Author Contributions: H.R., S.H. and X.-F.P. contributed equally to this paper.

Funding: This research was funded by the Startup Research Grant Program of Higher Education Commission (HEC) Pakistan under Project # 2285 and grant No. 21-2285/SRGP/R&D/HEC/2018 received by Sakander Hayat. APC was covered by Hassan Raza who is funded by a Chinese Government Scholarship.

Acknowledgments: The authors are grateful to the anonymous reviewers for a careful reading of this paper and for all their comments, which lead to a number of improvements of the paper.

Conflicts of Interest: The authors declare no conflict of interest.

References

1. Haynes, T.W.; Hedetniemi, S.; Slater, P. *Fundamentals of Domination in Graphs*; CRC Press: Boca Raton, FL, USA, 1998.
2. Haynes, T.W.; Henning, M.A.; Howard, J. Locating and total dominating sets in trees. *Discret. Appl. Math.* **2006**, *154*, 1293–1300. [CrossRef]
3. Charon, I.; Hudry, O.; Lobstein, A. Extremal cardinalities for identifying and locating-dominating codes in graphs. *Discret. Math.* **2007**, *307*, 356–366. [CrossRef]
4. Honkala, I.; Hudry, O.; Lobstein, A. On the ensemble of optimal dominating and locating-dominating codes in a graph. *Inf. Process. Lett.* **2015**, *115*, 699–702. [CrossRef]
5. Honkala, I.; Laihonen, T. On locating-dominating sets in infinite grids. *Eur. J. Comb.* **2006**, *27*, 218–227. [CrossRef]
6. Seo, S.J.; Slater, P.J. Open neighborhood locating-dominating sets. *Australas. J. Comb.* **2010**, *46*, 109–119.
7. Seo, S.J.; Slater, P.J. Open neighborhood locating-dominating in trees. *Discret. Appl. Math.* **2011**, *159*, 484–489. [CrossRef]
8. Slater, P.J. Fault-tolerant locating-dominating sets. *Discret. Math.* **2002**, *249*, 179–189. [CrossRef]
9. Hernando, C.; Mora, M.; Pelayo, I.M. LD-graphs and global location-domination in bipartite graphs. *Electron. Notes Discret. Math.* **2014**, *46*, 225–232. [CrossRef]
10. Slater, P.J. Domination and location in acyclic graphs. *Networks* **1987**, *17*, 5–64. [CrossRef]
11. Slater, P.J. Locating dominating sets and locating-dominating sets. In *Graph Theory, Combinatorics, and Algorithms, Proceedings of the Seventh Quadrennial International Conference on the Theory and Applications of Graphs, Kalamazoo, MI, USA, June 1–5, 1992*; Alavi, Y., Schwenk, A., Eds.; John Wiley & Sons: New York, NY, USA, 1995; Volume 2, pp. 1073–1079.
12. Charon, I.; Hudry, O.; Lobstein, A. Identifying and locating-dominating codes: NP-completeness results for directed graphs. *IEEE Trans. Inf. Theory* **2002**, *48*, 2192–2200. [CrossRef]
13. Charon, I.; Hudry, O.; Lobstein, A. Minimizing the size of an identifying or locating-dominating code in a graph is NP-hard. *Theor. Comput. Sci.* **2003**, *290*, 2109–2120. [CrossRef]
14. Lobstein, A. Watching Systems, Identifying, Locating-Dominating and Discriminating Codes in Graphs. Available online: http://perso.telecom-paristech.fr/~lobstein/debutBIBidetlocdom.pdf (accessed on 16 November 2018).
15. Bača, M. Labelings of two classes of convex polytopes. *Util. Math.* **1988**, *34*, 24–31.
16. Bača, M. On magic labellings of convex polytopes. *Ann. Discret. Math.* **1992**, *51*, 13–16.
17. Imran, M.; Baig, A.Q.; Ahmad, A. Families of plane graphs with constant metric dimension. *Util. Math.* **2012**, *88*, 43–57.
18. Imran, M.; Bokhary, S.A.U.H.; Baig, A.Q. On the metric dimension of rotationally-symmetric convex polytopes. *J. Algebra Comb. Discret. Appl.* **2015**, *3*, 45–59. [CrossRef]

19. Imran, M.; Bokhary, S.A.U.H.; Baig, A.Q. On families of convex polytopes with constant metric dimension. *Comput. Math. Appl.* **2010**, *60*, 2629–2638. [CrossRef]
20. Malik, M.A.; Sarwar, M. On the metric dimension of two families of convex polytopes. *Afr. Math.* **2016**, *27*, 229–238. [CrossRef]
21. Kratica, J.; Kovačević-Vujčić, V.; Čangalović, M.; Stojanović, M. Minimal doubly resolving sets and the strong metric dimension of some convex polytopes. *Appl. Math. Comput.* **2012**, *218*, 9790–9801. [CrossRef]
22. Salman, M.; Javaid, I.; Chaudhry, M.A. Minimum fault-tolerant, local and strong metric dimension of graphs. *arXiv* **2014**, arXiv:1409.2695.
23. Simić, A.; Bogdanović, M.; Milošević, J. The binary locating-dominating number of some convex polytopes. *ARS Math. Cont.* **2017**, *13*, 367–377. [CrossRef]
24. Hayat, S. Computing distance-based topological descriptors of complex chemical networks: New theoretical techniques. *Chem. Phys. Lett.* **2017**, *688*, 51–58. [CrossRef]
25. Hayat, S.; Imran, M. Computation of topological indices of certain networks. *Appl. Math. Comput.* **2014**, *240*, 213–228. [CrossRef]
26. Hayat, S.; Malik, M.A.; Imran, M. Computing topological indices of honeycomb derived networks. *Rom. J. Inf. Sci. Tech.* **2015**, *18*, 144–165.
27. Hayat, S.; Wang, S.; Liu, J.-B. Valency-based topological descriptors of chemical networks and their applications. *Appl. Math. Model.* **2018**, *60*, 164–178. [CrossRef]
28. Bange, D.W.; Barkauskas, A.E.; Host, L.H.; Slater, P.J. Generalized domination and efficient domination in graphs. *Discret. Math.* **1996**, *159*, 1–11. [CrossRef]
29. Sweigart, D.B.; Presnell, J.; Kincaid, R. An integer program for open locating dominating sets and its results on the hexagon-triangle infinite grid and other graphs. In Proceedings of the 2014 Systems and Information Engineering Design Symposium (SIEDS), Charlottesville, VA, USA, 25 April 2014; pp. 29–32.
30. Hanafi, S.; Lazić, J.; Mladenović, N.; Wilbaut, I.; Crévits, C. New variable neighbourhood search based 0-1 MIP heuristics. *Yugoslav J. Oper. Res.* **2015**, *25*, 343–360. [CrossRef]
31. Bača, M. Face anti-magic labelings of convex polytopes. *Util. Math.* **1999**, *55*, 221–226.
32. Imran, M.; Siddiqui, H.M.A. Computing the metric dimension of convex polytopes generated by wheel related graphs. *Acta Math. Hung.* **2016**, *149*, 10–30. [CrossRef]
33. Miller, M.; Bača, M.; MacDougall, J.A. Vertex-magic total labeling of generalized Petersen graphs and convex polytopes. *J. Comb. Math. Comb. Comput.* **2006**, *59*, 89–99.
34. Imran, M.; Bokhary, S.A.U.H.; Ahmad, A.; Semaničová-Feňovčíková, A. On classes of regular graphs with constant metric dimension. *Acta Math. Sci.* **2013**, *33B*, 187–206. [CrossRef]
35. Raza, H.; Hayat, S.; Pan, X.-F. On the fault-tolerant metric dimension of convex polytopes. *Appl. Math. Comput.* **2018**, *339*, 172–185. [CrossRef]
36. Erickson, J.; Scott, K. Arbitrarily large neighborly families of congruent symmetric convex polytopes. *arXiv* **2001**, arXiv:math/0106095v1.

MDPI

St. Alban-Anlage 66

4052 Basel

Switzerland

Tel. +41 61 683 77 34

Fax +41 61 302 89 18

www.mdpi.com

Symmetry Editorial Office

E-mail: symmetry@mdpi.com

www.mdpi.com/journal/symmetry